AF518377

Trees of
WESTERN GHATS
INDIA

Authors

Dr. S. Karuppusamy obtained his M.Sc. (1996) degree from the Madurai Kamaraj University, Madurai and Ph.D. (2003) from the Gandhigram University, Dindigul, Tamil Nadu. He was a Post-Doctoral Fellow of Sri Krishnadevaraya University (2004-2008) Anantapur. At present he is working as Associate Professor in Botany, The Madura College, Madurai, Tamil Nadu State. He has published more than 150 research papers in nationally and internationally reputed journals. He is the author of Caralluma (sensu lato) – Anti-obesity plants and Orchids of Tamil Nadu. He is also the co-author of Flora of Andhra Pradesh, Flora of Eastern Ghats (India), Orchids of Eastern Ghats, Monograph on Brachystelma and Ceropegia in India (CRS Press), Camptothecin and camptothecin producing plants (Academic Press, Elsevier). He is the recipient of Young Scientist Award from the Indian Botanical Society (2006) and Rolla S Rao Award from the Indian Association of Angiosperm Taxonomy (2012). He also receives the Prof. Y.S. Murthy gold medal award from Indian Botanical Society for contribution of Angiosperm Taxonomy in India. He is the member of Species Survival Commission (IUCN) and Member of Western Ghats Species Specialist Group (IUCN).

Mr. P. Bharath Simha Yadav is a PhD scholar at the Botanical Research Centre, The Madura College, Madurai, Tamil Nadu. With extensive expertise in field of plant taxonomy, particularly in the identification of plants, birds and snakes, he has made significant contributions to the field of plant ecology and quantification studies, GIS mapping, and photography. His remarkable achievements include the discovery of two new plant species and the publication of 22 research articles covering diverse research fields of avian studies, herpetology, and plant taxonomy. He has been actively involved in consultancy work for biodiversity assessment and wildlife mitigation across five major projects and has collaborated with government institutes on four more significant biodiversity and carbon studies. With four and a half years of research experience across various projects in Andhra Pradesh, Telangana, Tamil Nadu, and Kerala, his work has primarily focused on plant biodiversity, bird ecology, and plant studies. His academic a Master's Degree in Botany from Osmania University, Telangana and a Bachelor's degree in BZC (Botany, Zoology, and Chemistry) from Palamuru University, Telangana.

Trees of Western Ghats
India

Vol. 1

(Achariaceae – Lythraceae)

S. KARUPPUSAMY
&
P. BHARATH SIMHA YADAV

Department of Botany
The Madura College (Autonomous)
Madurai, Tamil Nadu – 625 011.

2025

Regency Publications
A Division of
Astral International Pvt. Ltd.
New Delhi – 110 002

ISBN: 978-93-5919-767-8

Publisher's Note:

Every possible effort has been made to ensure that the information contained in this book is accurate at the time of going to press, and the publisher and author cannot accept responsibility for any errors or omissions, however caused. No responsibility for loss or damage occasioned to any person acting, or refraining from action, as a result of the material in this publication can be accepted by the editor, the publisher or the author. The Publisher is not associated with any product or vendor mentioned in the book. The contents of this work are intended to further general scientific research, understanding and discussion only. Readers should consult with a specialist where appropriate.

Every effort has been made to trace the owners of copyright material used in this book, if any. The author and the publisher will be grateful for any omission brought to their notice for acknowledgement in the future editions of the book.

Published by : **Regency Publications**
A Division of
Astral International Pvt. Ltd.
– ISO 9001:2015 Certified Company –
4736/23, Ansari Road, Darya Ganj
New Delhi-110 002
Ph. 011-43549197, 23278134
e-mail: info@astralint.com
Website: www.astralint.com

Printed at : **Replika Press Pvt. Ltd.**

PREFACE

Trees are lungs of the earth and Western Ghat mountains are heart of India. Forests of the Western Ghats making ever fertile ecosystem to sustain thousands of flora and fauna including human beings. This book described the wealth of trees in the Western Ghats with their precise diversity, identification, nomenclature, uses and threat status. It is much needed to describe the updated account of trees in Western Ghats biogeographic zone for understanding the knowledge on various aspects used by botanists, foresters, ecologists, conservationists, ethnobotanists, natural lovers and other interested stack holders of floristic resources.

Trees are of immense ecological importance to define and form the major structural components of forest ecosystems that cover approximately 31% of the world's land surface. Forests play a major role in the Earth's biogeochemical processes, influencing soil production, hydrological, nutrient and carbon cycles, as well as the global climate. Forests contain about 50% of the world's terrestrial carbon stocks and over 75% of the world's accessible freshwater is obtained from forested catchments. Forests provide habitat for a wide range of other species supporting at least half of the Earth's known terrestrial plant and animal species. Reports frequently state that up to 80% of terrestrial plants and animals occur in forests but this cannot yet be verified accurately. Trees are not only found in forests, but also occurs in woodlands, savannahs, shrublands, grasslands, deserts, wetlands, coastal and rocky ecosystems as well as artificial and urban environments.

Most of the tree species at greatest risk from climate change are not currently conservation priorities, finds an IUCN assessment that introduces a pioneering method to assess the vulnerability of tree species to climate change. The IUCN Red List continues to provide the best possible assessment of extinction risk. The focus is on providing Red Listed Trees of Western Ghats with recent IUCN assessment criteria. This data will be useful for botanists, forest department and conservationist to evaluate accurate policies for making conservation strategies. The book is also provided the list of introduced and exotic tree species in the Western Ghats. It is much useful for assessing the impact of exotic and alien tree flora.

With keeping all the above concepts in mind, the arduous task of writing these two volumes of book is to made lighter to some extent by the selfless and unsolicited help rendered by a few botanists and forest officials. Authors extent the immense thanks to Prof. T. Pullaiah (Rtd), Sri Krishnadevaraya University, Anantapur, Andhra Pradesh for his encouragement and inspiration for preparing this book. Thanks are due to Dr. V. Ravichandran, Botanical Survey of India, southern circle, Coimbatore, all my colleagues in the department of Botany, The Madura College, Madurai for their cooperation. Special thanks are also due to all the forest officials in the Tamil Nadu, Kerala, Karnataka and Maharashtra Forest divisions at various times for their help and cooperation.

Dr. S. Karuppusamy

Mr. Bharath Simha Yadav

CONTENTS

TREES OF WESTERN GHATS, INDIA

Introduction

Trees are of vital importance ecologically, culturally and economically which is closely associated natural ecosystem dynamics. Until now there have been surprising gaps in knowledge of the diversity, distribution, utilization and conservation status of trees in local, regional, national and global scale. The tree diversity on the earth planet is an important resource for food, shelter, cloth and agriculture to sustain the basic necessities of humankind. Trees define forest composition, structure, landscape, distribution and thus provide habitat for about 50% of the world's known terrestrial plants and animal species. Tree diversity is a key driver of taxon-level and overall forest associated functional components, determines forest composition, climate and soil which is significant component of carbon storage in many ecosystems such as woodlands, grasslands, artificial arboretum and urban environment. The current scenario of interlinked biodiversity crisis and challenges of global climate change cannot be addressed without knowing the diversity, distribution and conservation management of tree species (BGCI, 2021).

Tree species that have evolved over millions of years, adapting to changing climates, can no longer survive the onslaught of human threats continues. How short-sighted are we to allow the loss of tree species on which global society is ecologically and economically dependent. If we could only proper approach to respect trees, undoubtedly many environmental challenges would greatly be achieved benefits. Preserving the tree diversity of forests assures their productivity and potentiality increases the accumulation of carbon and nitrogen in the soil, which helps to sustain soil fertility and mitigate global climate change. Recent studies evidenced that increased tree diversity enhanced soil carbon storage by 30% to 32% and enhanced nitrogen storage by 42% to 50 % on a decadal timescale. Trees are visible and measurable indicators of the health of an ecosystem. Trees alter and change the environment they are in, making them key ecosystem engineers. In addition, trees provide many ecosystem services through water purification, erosion prevention, flood sustenance, carbon sequestration, air temperature control, and regulation of air quality. There is also considerable evidence and increasing acceptance that trees can play a substantial part in tackling the climate crisis.

Trees provide a wide range of products of huge value to local livelihoods, national economies and global trade. Timber, fuelwood, wood pulp, medicinal and aromatic products, fruits and nuts are amongst the most valuable products derived from trees, many of which are predominantly sourced from the wild, poorly recorded and are under-valued. There are 58,497 tree species globally (BGCI, 2021, GlobalTreeSearch - https://tools.bgci.org/ global_tree_search.php), provide a complete list of the world's tree species and their country level distribution, as a tool for assessing, monitoring, and managing tree species diversity and forests on a global, regional, and national level. The information, derived from a range of plant databases, scientific references and tree experts is kept up-to-date to reflect updates in taxonomy and nomenclature as well as known distributions of the species. Plant taxonomists and botanists continue to catalogue tree species diversity and understand the relationships between species groups. Some areas of the world remain poorly known botanically, and many taxa still remain undescribed.

The Western Ghats

The Western Ghats, otherwise called the Sahyadri hill ranges represented with extraordinarily unique in biodiversity of flora and fauna. The continuous hill ranges are running parallel to west coast of Peninsular India about 1,600 km starting from the Tapthi river in Gujarat downwards to Tamil Nadu through Maharashtra, Goa, Karnataka and Kerala. The Western Ghats is one among the 36 biodiversity worlds hotspots recognized by Norman Myers (Marchese, 2015). It has been identified rich number endemic species diversity and a major endemic center in the Indian subcontinent (Nayar *et al.*, 2014). The hill ranges geologically categorized into two segments that is northern basaltic and southern gneissic, where the high rainfall belts of montane forests, soils are varied from rich humus peat to lateritic in the lower elevation. Generally, soils are acidic due to high humus contents. Climatic conditions of the Western Ghats are highly influenced with the altitudinal gradients and physical proximity of the Arabian Sea. The Western Ghats experienced a tropical climate and being a warm and humid during most of the months in a year with an average temperature ranging from 20°C in the southern to 24°C in the northern Ghats, and in the higher elevations experienced from subtropical climates to temperate and rare occasions of the frost. Further, it has been recorded that the coldest months in the southern Western Ghats coincide with the wettest climate. These enriched amidst weather condition prevails the vast assemblage of floristic and faunastic communities of Western Ghats. The average elevation of Ghat peaks reached about 1,200 m ranging from 300 to 2,695 m with some important ranges at Kalsubai (1,646 m) and Salher (1,567 m) in northern Western Ghats, Dodabetta (2,637 m), Mukurthi (2,554 m) and Anamudi (2,695 m) in southern Western Ghats. The Western Ghats is one of a UNESCO World Heritage Site and is one among the eight "hottest hotspots" of world biological diversity (Myers *et al.*, 2000; UNESCO, 2012). The range runs parallel to north-south along the western edge of the Deccan Plateau known as Great Escarpment of India and separates the Deccan plateau from a narrow coastal plain, called Konkan, along the Arabian Sea coast.

The varied climatic and topographic heterogeneity from 2695 m of highest peak to sea level and a distinct monsoon pattern with rainfall gradient of annual precipitation <50 cm in sheltered high altitude valleys in the east to >700 cm over the west-facing hill slopes, the combined effect of climate gave rise to a tremendous diversity of varied life forms and vegetation types. The forest types including tropical wet evergreen forest, montane stunted evergreen hill peak forest and grasslands, lateritic plateaus, moist deciduous and dry deciduous forests, dry thorn forests, and low-level savannah grassland. All these forests types are considered as critical habitats for many plants and animal species. The typical sholas and grasslands of the southern Western Ghats are unique as well as highly vulnerable to future climate change. Forests of the Western Ghats have been home to host of extraordinary tree diversity for millions of years. The riparian forests along the rivers and streams of the Ghats sheltered high levels of plant and animal diversity in addition to acting as relict ecotone forests, where confined the lowland *Dipterocarp* forests and *Myristica* swamps on the western slopes are highly threatened. Around 4000 and add species of flowering plants with about 27% of total plant species of the country are known to occur in the Western Ghats. A total of 955 species of evergreen trees, about 56% are endemic to the Ghats. The diversity of lower plant groups especially bryophytes are highly impressive in the Ghats with around 1000 species, of which 682 species are mosses with 28% of endemism and 280 species are liverworts with 43% endemism.

Fomration of Western Ghats

Geological age of the Western Ghats is estimated older than the Himalayas mountains. The oldest part of the Western Ghats region named as Gondwanaland. The Gondwanaland lost its haze of a distant past by the subsequent geological events. The *Gonds* a tribal community inhabited in Madhya Pradesh, Odisha and northern Andhra Pradesh of present day are probable ancestors and those who have custodians of the mother land of Indian Peninsula and a part of a supercontinent later it called as Gondwanaland. The oldest fossil plant *Glossopteris* was found in Gondwanaland firstly and subsequently it was recorded all over India, Southern Africa, South America, Australia and Antarctica. Continental drift theory explains the initial landmass braking apart due to various changes on the mantle of the earth crust during the formation of earth. The Pangaea ('all lands') primarily made up of Gondwana in the south and Laurasia in the north. Laurasia consists of North America, Europe and Asia. The remaining part of the earth crust broke away about 200 to 150 million years ago (m.y.a.) and formed Gondwanaland including continents of Australia, India, South America, Africa and Antarctica. When the Carboniferous and the Permian period (300-260 m.y.a) most part of the Gondwanaland was covered by ice. For those times a very little evidences of animals and plants actually living on Gondwana landmass. The ice was melted in the late Permian (240 m.y.a.) and the Triassic (225 m.y.a.), slowly animals were able to colonize and continues warm until 40 m.y.a. The part of Antarctica now froze over the period again. During the drift of Gondwanaland began at the time that Dinosaurs were the dominant over on land (Cretaceous - 120 m.y.a.), though it was a spontaneous process but they were unlikely to be affected by it.

The moving of continents is very slow and a large chunk from Gondwanaland broke up and from Antarctica in the south. This chunk fissured to form S. America and Africa to its west and Australia to its east. The primary Gondwanaland is a triangular stable "shield" area known as the Deccan Plateau that covers the Peninsula of India (and which includes Sri Lanka). Most geologists supported this view as having formed Indian Peninsula as a part of Gondwanaland. The Indian Plate consisted a deep seated volcanic hot-spot of united Gondwana islands. The underwent heat beneath evolved basaltic magma which gives rise to lithosphere of the crust. This tilt rose the altitude of western portion and the major rivers of the peninsular India flow eastward. Around 45 m.y.a. the Peninsular shield moved towards the Asian landmass after having undergone many uplifts and modifications. This uplift causes the Asian Plate to form the elevated Tibetan plateau. The Gondwana landmass and the Asian Plate was obliterated and the Tethys sea caught up between the two moving landmasses subsequently by its sediments raised up to form the Himalayas.

Deccan Volcanic Episode

The events of turbulence, upheavals and extinctions occurred from late Cretaceous to the early Tertiary period known as Deccan Volcanic Episode. The huge fissures erupted from the enormous lava basalts that was flooded over the Deccan region between 60-65 m.y.a. Palaeoecologists considered the recent palaeo-magnetic and isotopic data revealed the possibility of magma outbursts in short duration of about 2 million years around 65 m.y.a. It is evidenced from the absence of a typical cone-like peaks in southern Peninsula and it is formed by the vast spread of basaltic rock points to a series of outbursts confined to magma spread out in layer after layer. It was cited by Birbal Sahni, the renowned palaeo-botanist as "this terrible drama of fire and thunder". About 90% of the fauna and flora extinct in the Cretaceous - Tertiary period due to this remarkable episode. The huge range of extinction includes large reptiles (Dinosaurs) and ammonites at the end of Cretaceous. It is believed that the magnitude and suddenness of volcanic activity and abrupt changes in environment, is causing the mega extinction of plant and animal species. The Deccan basalts lava may have played a crucial role in the extinction of the Dinosaurs by releasing eruption fumes and it may lead to great change the global climate and the demise of the large mammals including Dinosaurs. The original basalt covered by the lava flows as high as 1.5 million sq. km. The total volume of the basalt is estimated approximately 5,12,000 cubic km. The term "Deccan Trap" was first introduced by Col. W. H. Sykes for explaining lithological rock basalts. An entire Western Ghat landscape characterized under the Deccan Volcanic Province of Peninsular India. The upland region of Deccan Trap was not preserved any sedimentological records during stratigraphic period from post-Eocene to pre-Pleistocene which modifies lateritic covers extensively. Hence, it has been reconstructed the Tertiary history of Deccan Plateau landscape due to the absence of any sedimentological records (Elizen *et al.*, 2013).

The Peninsular India separated by the Deccan volcanic episode from the Gondwanaland about 150 million years ago. In this event the landmass was started moving towards northern direction. The continuous northward drift was ended up to 100 million years and finally colliding the Peninsula with the Asian mainland around 45 million years ago. It was major geological transformations as the Peninsula drifted over the Asian plate and act as localised volcanic center where the earth's lithosphere 200-300 km across and it has been remained active for several million years. Further, it has been generated heat beneath basaltic magma which raised the lithosphere and by crustal arching around 120-130 million years ago. By the event resulted in the upliftment of the Western Ghats in easterly direction. Subsequently, there was a series of volcanic eruptions around 65 m.y.a. and transformed the extensive Deccan Traps. These volcanic episodes added multitude of magma layers largely to the northern Western Ghats (Subramanya, 1987).

The Western Ghats are formed as a result of a domal uplift and the underlying rocks are aged around 2000 million years old. The oldest rocks of Gondwana crust are found in the Nilgiris and the High altitudes of the southern Western Ghats ranges. Indian Peninsula broke along its axis of weakness and the western range drifted westward towards the Arabian sea and this region rise to the present-day hill ranges of the Western Ghats and the western sea coast. The exposed face of the eastern unsubmerged plate was lifted during the Eocene period (in between 45 and 65 m.y.a.) form the scarp of the Western Ghats. At this time Peninsular India ended its northern drift and conjoined with Asian mainland and it is the well-marked eastward tilt permanently changed. The western faulting processes led to 'river capture' and easterly diversion of drainage to the west in many instances. The rivers of Sharavathy and Kali in Karnataka are some classical examples of westerly drainage due to uplift and faulting. The Western Ghats metamorphosed with tectonically active center, fast rates of uplift, high summit altitudes, steep slopes, deep gorges and potential for erosion with correspondingly high sedimentation yields.

The uplift of the western part of the Indian Peninsula leads to several effects. There has been taken place the orientation and direction of the big rivers towards the east. This event started a chain of regressive erosion on the steep and abruptly in western slopes. By this orientation of the Western Ghats into chiseled into steps following the sedimentation of basaltic beds and carved by deep valleys. The Sharavathy river is cleaved by vertical cliffs towards Agumbe (Karnataka). The Western Ghats (WG) escarpment was separated from the Seychelles and it is closely joined with the late Cretaceous by Deccan flood basalt eruption. It has been oriented the low-lying seaward coastal plains in the west and elevated the Deccan plateau to the east (Fig. 1). The Western Ghat escarpment is one of the larges volcanic continental margin with over 1500 km stretch of paralleling to sea coast and elevation immediate start of sea level ranging over 1 km. Therefore, Western Ghats is a classic example to understand the relationship between volcanic episodes, uplift and extension of plateaus and high ranges (Radhrakrishna *et al.*, 2019).

The northern Western Ghats from Tapti valley down to Goa formed horizontal beds of massive Deccan Trap. The western face of Western Ghats has appeared in a steeply-cut valley towards the Arabian Sea. The Deccan scarp of ranges in this region presents a vertical profile over 1000 m elevation of successive sedimentation layers. From Goa to Mahabaleshwar known as 'Arthur's Seat', the horizontal traps give place to steeply dipping gneisses and schists beds. In this section the Ghats lose their abrupt steep and an average height falls lesser than 2000 ft. The western flowing rivers effected by cutting deep gorges and canyons on sedimented rocks. The waters flows separating the easterly and westerly Gangavali drainage by nearly a 100 km inland as near Hubli. The scarp again lying back to the Arabian coast of south Honnavar forming the significant peaks of Kodachadri (1,343 m) and Kudremukh (1,892 m).

Topography

The Western Ghats are metamorphosed basaltic rock mountains and formed with faulted edge of an elevated plateaus. The Volcano-Sedimentary Archaean rock shield is mainly composed of crystalline, mostly gneisses, granites and charnockites alternating with Schist-belts. The Schist-belts are mainly sedimented in 'Dharwar - Schists' (Bourgeon, 1987). The Indian Peninsula formed a great east-west dissymmetry and composed of a fragment of the Archaean Gondwanaland. The Western Ghats may have raised begun during Jurassic period and further episode of tectonic drifts combined with segregating by river flows and have constituted present day topography. On the Western side, the Ghats appear abruptly rising from the Arabian Sea whereas they merge through a raised hill peaks with the Deccan Plateau in the east. They show escarpment formed by coalesced amphitheatre shaped valley heads. The Ghat is composed of steep valleys, narrow gorges and high water falls. The long stretch of hill the Western Ghats ranges have a major separation in Palghat gap separating the Nilgiris range from the Anamalai hills. In the southern section, the range lies close to the coast till it joins the Nilgiri mountains. Northern Nilgiri plateau of eastern flank merged with somewhat raised into the elevated plateau of Karnataka but the western slopes suddenly elevated from the lower coast. The Coorg (Kodagu) district of Karnataka state is more concerned the Western Ghats that occupies a major position in the humid tropical belt of the Ghats and is situated to the South-West in Karnataka state. The Southern section of the Ghats, the Brahmagiri (1,360 m) forms the southern boundary with the Wayanad plateau of Kerala and there is another Brahmagiri peak near Bhagamandala which is the source of Cauvery river. The highest peak of Brahmagiri district is Tadiandamol (1,734 m). Madikeri plateau (Mercara) at an average elevation of about 1,050 m extends northwards as far as Somwarpet, a distance of 30 km but on the east slopes down to the Cauvery.

The Nilgiris is the meeting ground of three mountain systems of Peninsular India *viz*, the Western Ghats, the Southern Ghats and the Eastern Ghats. The abrupt rise of the Nilgiri from the surrounding area is very striking. The South-Western part is more hilly and is traversed by bold ranges and intersected by deep valleys. The rest of the Nilgiri is extremely undulating. The highest peak in this range is Dodabetta (2,637m). The Palghat Gap, 24 km wide is the most prominent break in the continuity of the range. South of Palghat gap, the Ghats are very steep and rugged on both the slopes. The Ghats raised prominent hill peaks in the Anamalais - Palanis block which is formed the hill ranges in the Nellaiampathi plateau to the west, the Anamalais plateau of the north, and the Palani hills plateau in the north east. The higher ranges of the Anamalais consist of a series of plateaus of high elevations. These are covered with rolling downs and dark evergreen forests and are cut off by deep valleys. All these hill ranges branch off from the southern highest peak Anaimudi about 2,695 m. To the south of Anamlais plateau West-East oriented hills ranges, the Ghats reappear further changes. The elevated hill blocks are slanting towards the west of the Periyar plateau, formed by the flow of Periyar river. The Cardamom range adjoined eastern part of the Periyar, the plateau is known for cardamom plantations. This middle hill ranges attains a peak at Devar Malai (1,922 m) and ended the hill range in the east by a steady cliff around 1,000 m high. The Southwest-Northeast orientation of Varushanadu massif is detached from Periyar-Anamalais plateaus and continued by the Andipatti hills, which together with the Palani hills embrace the Cumbam valley.

The usual orientation of the Ghats is northwest-southeast directions and there are three more or less arcuate projections extending eastward into the South Madras plains. The Nilgiri range oriented north-east between the Coimbatore plains through Gudalur Ghat to the Mysore plateau, and continued northward into Sathyamangalam – Kollegal hills and Biligirirangan hills of Karnataka, the Palani range (Kodaikanal) running also north-east and continued east of southeast, after a narrow pass at Dindigul, in the east of Sirumalai –Ammayanayakkanur-Ayyalur hills, and southwest of the low Varshanad – Antipatti range running northeast towards the Vaigai valley. Lying between arcuate projections of the Nilgiri range and the Anamalai- Palani range is the Palghat Gap, with an elevation of 250-1,000 ft., the highest point in the gap being a little over 1000 ft at Pollachi (Jacob and Narayanaswami, 1953).

In Kerala, the range adjoins the laterite terraces which gradually slope down towards the coast. This coastal zone (30 - 50 m) constituting Kerala is made up of converge shaped hills with rounded summits and extends up to Kanniyakumari. Here the tabular reliefs are not found. Kerala Ghats is generally isolated from the Deccan plateau by rocky mountainous belts. The mountainous belts consist of a series of hilly ranges clothed with dense forest vegetation. The eastern border of Kerala mountains bounded by an almost unbroken stretch of mountainous wall. The continuity of the Ghats is broken by a prominent depression, the Palghat Saddle. On the lower south -western side of the Nilgiri plateau lies the Verdant Silent Valley. The Ghats area receives both south-west and north-east monsoons. A major proportion of the annual precipitation is received from the south-west monsoon and north-east monsoon brings light showers.

Lithology

The Western Ghats can be conveniently divided into three distinct geographic regions based on lithology and structure.

Northern Konkan Region - mainly composed of Deccan Basalt/trap rocks belong to Tertiary age: The northern part of Western Ghats hill ranges covering Maharashtra and Goa is better known as *"Konkan"*. It consists of horizontal to nearly horizontal sheets of basalts, cut into a steep scarp on the western side and eastward gentle slope. The highest peak of the northern range is Mahabaleshwar with 4,540 feet above mean sea level. It is a flat topped and slightly elevated hilly terrain compared to Deccan plateau. The steep western margin is usually terraced and resembling the *ghat* or landing – a *stair* from which consider the name derived from the region.

Central Sahyadri or Malnad Region - represented by Metamorphosed Volcano-sedimentary sequence of Archaean and Proterozoics of Dharwar Super Group (DSG): This hill range forms the central part of the Western Ghats covering Karnataka state starting from the South of Konkan (south of about 16° N latitude) up to Kasaragodu in Kerala. The highest peak of middle Western Ghats region is Mullayanagiri with an elevation of 1,892 m above mean sea level. The major rivers like Cauveri, Sharavathi, Chandranadi, Varahi, Tunga, Bhadra, Kalinadi, Gangavalli, Swamanadi, Sita Nadi and Netravathi originate from this Sahyadri hill range.

Southern Malabar or Nilgiri Region - comprising high grade granulite terrain of Archaean age. The southernmost part of the Western Ghats clothed over the Kerala and Tamil Nadu states, it extends from south of Kasaragodu down to Kanyakumari (Cape Comorin) which are known as Malabar hill ranges. These hill ranges are essentially made up of high-grade granulite rocks of Archaean age. The general foliation of these formations coincides with the Western Ghats and its ranges varies 40° to sub vertical towards northeast direction. The highest peak of Anaimalai is a major portion of this region at 2,695 m above mean sea level. Periyar and Idamalayar rivers form the main drainage system in this region. A part of central Kerala in India Peninsula is a basaltic product of prolonged late Mesozoic and Cenozoic formation of overprinted on Archaeo-Proterozoic basement. The Gondwana supercontinent is a formation of post-late Mesozoic breakup episodes (Radhakrishna *et al.*, 1999), subsequently by the migration of Gondwana plate northwards and finally colloid with Eurasian plates and resulted the raising of the Himalayan highlands (Müller *et al.*, 2012; Chatterjee *et al.*, 2013; Yatheesh *et al.*, 2013; Ramkumar *et al.*, 2016).

Rock Composition

The Western Ghats are constituted of the following litho units i) Gneissic ii) Chlorite schist iii) Phyllites:

Gneissic: A northern and eastern part of the hill range is comprised of gneisses/migmatites of PGC represented by gray gneisses of granodiorite to trandhjemitic composition. Gneissic generally occur as boulder outcrops in low-lying mounds. The failed slopes near Anegundi, Hariharpura, Thirthahalli, and Megaravalli are represented by this lithounit. The gneisses occasionally contain enclaves of hornblende schist and traversed by pegmatite veins. The gneisses are highly weathered and fractured particularly in the region of failed slope.

Chlorite schists: It is the second major lithouint of the Western Ghats. The rock is light green in colour and shows well-developed schistocity with slaty cleavages, often shows pinching and swelling structure. The general trend of the schistocity plane is NW-SE, with varying dip of 35° to 50°E. It is essentially comprised of quartz, chlorite and biotite mica. The rock is highly friable in nature susceptible for weathering and denudation. The shear planes frequently contain clayey gauge material often associated with heavy seepages. The presence of stretched lineation, pinch and swell structures are indicating that rock has undergone higher degree of deformation.

Phyllites: Phyllite forms the major litho unit in Agumbe, Hosur and Begar slide areas. The rock is highly foliated brownish yellow in colour, soft and traversed by three to four sets of joints. In the failed slopes region, the rock is highly sheared and show pinch and swell structures, often filled with clay, indicating that it is intensely deformed and weathered.

Crystalline Rocks: The Western Ghats are recognized three major Archean rock types such as Archaean sediment (Oldest rocks), Tertiary bed rocks (Upper Miocene to Pliocene) and Quaternary sedimented rocks (recent rock deposits). All they are belonging to crystalline types and they have aligned in north-south orientation.

Precambrian Rock System

Basic dykes: The oldest rocks are made up of mylonitised crusts. They mainly found in southern Western Ghats which composed of dolerite. The basic dykes generally formed from fine to medium grained, free form of olivine in Cochin area. The coarse-grained crystalline forms are made by gabbro's sediments. This kind of exposed rocks also found in Cochin area. This characteristic old rocks crusts as dyke formed during the Deccan episodic breakup of Gondwanaland plate at 90 and 65 m.y.a under various exomantle thermal regimes. The cooling interval between pre and post dyke formation in later Cenozoic, possibly under the influence of thermal pressure that led to the upliftment of the Western Ghats mantle. This event may extend down to the crust-mantle interface sediment. It leads to form younger geophysical lineaments by reorientation along the weak planes of the charnockite rock beds of the Precambrian. Later formed the cold crust with devoid of any dykes. It may be confined only in the upper-middle crust of Cambrian period. They may be formed by various reasons of past episodes such as the back-thrust Gondwanaland and Eurasian plate collision; the effect of epeirogeny forces and isostatic uplift of the Western Ghats and the flexural isostatic uplifts by surface loads of late Mesozoic basaltic lavas and Cenozoic sedimentation crusts (Ajaykumar *et al.*, 2016).

Residual laterites: The eastern boundary of the Kerala state constitutes a linear zone of lateritic crystalline rocks to the west direction. These are Varkalai formation exposed on the surface of laterite, which exhibits characteristics variations from the typical lateritic. These lateritic formations preserve the ancient features of the parent rock with less compact. Beneath the laterite layer, the kaolin layer is sedimented in the depth of undecomposed rock gradation (Pascoe, 1962; Krishnan, 1982). The lateritic features of Maharashtra have been pointed by Patil (1992) where the 'laterites' are nothing but the 'red beds', and it has been considered as the secondary laterites (Pascoe, 1962; Sahasrabudhe and Deshmukh, 1981; Umarjikar, 1983; Tandale, 1987; Sahasrabudhe and Rajguru, 1990).

The Varkali formation: It is represented by the most conspicuous characteristic sedimentary bed rock composition found in Varkala. The cliffs near the Varkala seashore consist of clayey sandstone, white and variegated clay with thin layer of carbonaceous lignite.

Recent deposits: After the post Deccan episode mainly developed sand and silt layers in northern parts of Quilon in southern Western Ghats. It extends the coast of Alleppey and all along the seashore of Kerala which deposits the lacustrine mud banks. It contains valuable mineral sands for the economic point of view.

Rivers and Water Bodies

The Western Ghats is the source of origin of the several important perennial rivers of India namely the Godavari, Bhima, Koyna and Krishna originate. Northen Ghats of Maharashtra has originated more than 11 west flowing rivers namely Damanganga, Surya, Vaitarna, Ulhas, Savitri, Kundalika, Patalganga, Vashisti, Shastri, Karli, and Terekhol (Fig. 2). There are several smaller water-flowing courses joining the creeks in the Ghats (WGEEP, 2011). The Western Ghats are regarded the main water source of south India and it supplies more than 60% of surface water to Karnataka (Malhotra and Prasad, 1984), 70% to Kerala, 60% to Tamil Nadu, 50% to Maharashtra, 25% to Goa and 10% to Gujarat (Putty and Madhusoodhanan, 2014).

Important other rivers originate from the Western Ghats are the Godavari, Tungabhadra, Krishna, Thamiraparani and Kaveri and they flow into the Bay of Bengal. The west flowing rivers include Periyar, Bharathappuzha,

Netravati, Sharavathi, Mandovi and Zuari are drain into the Arabian Sea and the Laccadive Sea. A number of these rivers feed the backwaters to Kerala and Maharashtra. The Kaveri and Krishna are comparatively slow moving due to larger rivers. Several larger tributaries originate from the Ghats and merge with some other rivers include the Tunga River, Bhadra river, Bhima River, Malaprabha River, Ghataprabha River, Hemavathi river, Kabini River. There is some smaller riverian course flow from the Ghats namely Chittar River, Gadananathi River, Manimuthar River, Kallayi River, Kundali River and the Pachaiyar River (Jacob and Narayanaswami, 1953).

The several of streams originating from the Western Ghats that are joining to the rivers Krishna and Cauveri in monsoon months only. Across the many rivers have been constructed the dams for production of electricity and irrigation purposes. The major reservoirs constructed in Western Ghats region include Lonavala and Walwahn in Maharashtra; V.V. Sagar, K.R. Sagar, Bhadra, Lingamakki in Karnataka; Mettur, Upper Bhavani, Lower Bhavani, Mukurthi, Parson's Valley, Porthumund, Avalanche, Emerald, Pykara, Sandynulla, Karaiyar, Servalar, Kodaiyar, Gadananathi River, Manimuthar and Glenmorgan in Tamil Nadu; and Kundallay and Maddupatty in Kerala. Of which Lonavla, Walwahn, Upper Bhavani, Mukurthi, Parson's Valley, Porthumund, Avalanche, Emerald, Pykara, Sandynulla, Glenmorgan, Kundally and Madupatty reservoirs used for fisheries and irrigation purposes.

The several manmade reservoirs in the Western Ghats created for tourism, fisheries and small-scale hydroelectric power stations. The known reservoirs are in Ooty (34.0 ha) in Nilgiris, and in Kodaikanal (26 ha) and Berijam in Palani Hills. The Pookode lake of Wayanad in Kerala is a beautiful scenic with boating and garden establishment for tourism. Two smaller lakes in Kerala are Devikulam (6.0 ha) and Letchmi (2.0 ha) in the Munnar.

Soil Types

With a range of parent rock materials occurring in the Peninsula and a wide range of climatic factors, a wide variety of soils is developed. Development of soils is greatly contributing by climate comprising of precipitation, humidity, evaporation, temperature, wind and biological agencies comprising vegetation with other biological components. The works of Bourgeon (1989) is extensively analyzed the edaphic features of Western Ghats. The Ghats appear to be homogeneous but can be distinguished into three major subdivisions for the understanding the major compartments: (a) the escarpment (b) the coastal hinter land and (c) the residual reliefs in the coastal area which is the retreat of the escarpment. The upper horizon of the soil is reddish brown, humiferous clayey and about 20 cm thick. This horizon is largely porous with plenty of plant root system and abundant of microbial activity. The deeper horizons are of variable thickness (50 cm to several meters). It is red in colour with clayey texture. The empty space is generally filled with micro-aggregates. The humiferous zone is slightly acidic, highly saturated with water and rich in exchangeable mineral elements. The deeper layers are with coarse elements, acidic and with medium saturation. The foot of the escarpment shows two types of soils *viz*, soils on debris and soils on regolith. Soils on debris contains high proportion of alluvial with coarse elemental compositions. It shows two horizons, upper dark brown humiferous layer and lower yellow brown horizon. These soils are acidic and contains about 5% of organic matter with sufficient amount of exchangeable bases.

The soils on regolith are more weathered and show a humiferous horizon of about 10 cm thick which is dark brown with sandy loam. It is followed by a yellowish-brown alluvial horizon of about 20 cm thick with sandy loam. Deeper horizons show red and yellow mottles in the form of vertical streaks. These soils are acidic to very acidic. With a wide range of climatic factors, a wide range of soil types are seen.

Raychaudury *et al.* (1963) categorized the south Indian hill soils into four main types *viz*, red soils, black soils, lateritic soils and alluvial soils. Red soils vary considerably in texture, some are sandy soils, a great many are barns of different kinds and a few are clay soils. The red barns and the red sandy soils are two prominent classes in the region. Laterite is generally reddish in colour and often has a vermicular structure. These soil types are formed due to high rainfall with alternating wet and dry seasons. The subsoil contains hydrated oxides of aluminum and iron. These soils classes are well organized on the historic summits of Ghats. The black soils are named after its black colour and are more fertile than red soil. They contain a considerable quantity of lime, magnesia, potash, iron and alumina. Alluvial soils are formed by transportation by streams and rivers and deposited over plains, valleys or along coastal belts. These soils are light in texture and well drained.

Climate

Climate of the Western Ghats is highly influenced by altitudinal gradation and latitudinal distance from the equator. Usually, the climate is tropical humid in the lower and reaches temperate in high altitudes above 2000 m mean sea level. An altitude of 1,500 m (4,921 ft) and above in the northern Ghats and 2,000 m (6,562 ft) and above in the southern Ghats have attained temperate climate. Average annual temperature noted around 15°C but in high altitudes of Ghat area touch the freezing point in winter months. Mean annual temperature ranges from 20°C in the south to 24°C in the north. There has also been recorded the coldest periods in the southern Western Ghats with wettest in the same period. The works of Pascal (1988) and Pascal *et al.* (2004) have been extensively used to explain the climate in the region. The broad picture of the climate of Western Ghats is determined by its geographical situation. The southern Peninsular hill ranges is a meteorological unit in itself and has a great diversity of climates with contrasting ombrothermic patterns. The Western Ghats has a tropical-warm and humid climate. The weather systems of the Ghats have vast variation due to heterogeneity of the region largely depending on geographical positions and physiographical features. The distinctive monsoon pattern with alternation of seasons is well known factor for maintenance of climatic cycle of the Western Ghats.

The Western Ghats situated in the wind ward direction, receives a heavy rainfall of more than 500 cm per year. The increase of temperature in May month over the mainland of Asia plate, the air pressure decreases in that area and the air circulation in the Indian area and the neighboring seas becomes more and more intense. Almost the whole of India is quickly over-run by cool and moisture laden air known as the 'South-West monsoon'. The south-west monsoon from the Arabian sea and the north-east monsoon from the Bay of Bengal are two sets in different seasons prevails the more precipitation of the Western Ghats. The south-west monsoon strikes the west coast receiving heavy rain to the coastal districts and mountains of the Western Ghats. It is during the south - west monsoon that approximately 80 to 85% of total rainfall of Western Ghats occurs. In June, rainfall exceeds 760 mm along the west coast of the peninsula. In July also, most of the west coast gets about 760 mm of rain fall. It fell down to the eastern side of the Ghats, being less than 125 mm. The distribution of rain fall in August follows the same general pattern as in July but the amounts are generally less. The distribution of September rainfall over the area is similar to that of August but amounts are still less. Subsequently the monsoon withdraws and retreats by October, by this time starts the winter season of northern part of the Ghats and north-east monsoon begins where it is prevailing dry climate in southern Ghats. The dry winter season is followed by the premonsoon transition period of a hot summer extending generally from March to the beginning of June. Next to rainfall, temperature is perhaps the most important climatic factor.

Rainfall

The heterogeneity in the physical features of Western Ghats has a profound influence on the climate of the region (Nair, 1991). The Western Ghats are lying in the tropical monsoon tract characterized with well distinct wet summers and dry winters (Putty and Madhusoodhanan, 2014). The normal annual rainfall pattern recorded in the Western Ghats that exists a systematic variation in rainfall across the region. The western side of the Western Ghats receives threshold of southwest annual monsoon about 203-254 cm, and the eastern side is rainshadow area have lesser annual rainfall (Subramanyam and Nayar, 1974). The Ghat area below 15^0 N is influenced both south-west and north-east monsoons, north-east monsoon being more effective deeper south, where tropical cyclones originating in the Bay of Bengal bring substantial rains. The influence of south-west monsoon increases gradually towards north, parts of Kerala and the rest of Ghats getting more than 90% of their rains during this season. The pre-monsoon rainfall that occurs in the months of April and May due to local convection is never widespread. Throughout the length of Ghats and coast, daily intensities are very high. In areas which Net Annual Rainfall (NARF) above 5000 mm, daily rainfall exceeding 250 mm is quite common feature and even where NARF is less than 2000 mm, falls exceeding 100 mm contribute about 25% of the annual rainfall. On very heavy rainy days (150 mm) in the region, it rains 20 – 23 hours and the contribution of low intensity rainfall is always much greater (Thipperudrapa, 2009). Even in very heavy rainfall area (NARF 5000 mm) falls less than 6 mm/15 minutes (24 mm/ hour) last 95% of the time and contribute 75% of the total rain (Putty and Madhusoodhanan, 2014).

Temperature

The temperatures in the Western Ghats are highly variable depending on the seasons and other physiographic factors. In winter season, the temperature decreases but in summer months increases. Temperature, humidity and rainfall are largely determining the features of extent vegetation. Naturally March, April and May months are hot months. But these months mellowed by the effects of temperature and humidity which formed cloud and rain of the south-west monsoon. The cold weather months from November to February prevails dry continental winds over the Ghats. December and January are the coldest months in high altitudes of the Ghats and the mean temperature ranges from 5 - 15°C. The mean temperature is somewhat highest in the months of March to May in the eastern side of the Ghats due to dry winds prevailing to coastal side.

Moisture

The atmospheric moisture is an important factor for vegetation concerned. The characteristic wind flow from the sea is the main factor which determines the hydro vapour content of air at any place. Generally, humidity increases towards the Arabian sea in the west border of the Western Ghats. In the hot summer season, dried air is blown at the foot hills, especially in the border of the hills. The commencement of oceanic winds on the coast begun to increase the humidity. In cold weather seasons, temperature as well as humidity pressure decreases along the coastal region towards the interior. In summer months, during the hot afternoons usually relative humidity is very low. Humidity is varying in monsoon seasons from sea level to mountain summits. Hence humidity variability is of general characteristic features of the Ghats in all over months. During the rainy season, as usual weather is saturated but alternatively in summer hot winds blow humidity is low.

Biogeography of Western Ghats

The Western Ghats have been served as refugia for wet ever-green elements since the periods of volcanism, the post reformation of volcanic period might have led the evergreen species dispersed from southern Western Ghats to central and northern Western Ghats. The detailed phytogeographic analysis of Western Ghats has been recorded by Subramanyam and Nayar (1974). Mani (1974) has characterized the Western Ghats as one of the oldest regions of evolutionary center of flora and fauna thereby, highlighted the significance of the Western Ghats formation in oriental biogeography. Biogeography explicit the origin, evolution and distribution pattern of biological species through the time course (Parthasarathy, 1988). This concept can be examined in a historical and biogeographical framework that could be compared in modern molecular phylogenetics with conjunction molecular prospecting and event-based biogeographical explorations. The modern molecular approach on the phylogeny of Western Ghats endemics, nested within a well-separated clades in the central Western Ghats and northern Western Ghats and also differentiating from southern Western Ghats species. Additionally, the experiments proved the ancestral areas of the deeper center of origin divergent in the phylogeny that could be assessed in southern Western Ghats and dispersals from southern to northern in the post-volcanic period. The Peninsular Indian plate probably departed from ancient Gondwana elements and taken place to Asian plate after breaking of the Gondwana supercontinent. It has been clearly addressed in the "Biotic ferry" model or "Out-of-India" hypothesis (Joshi and Karanth, 2013). The biodiversity of Western Ghats received many relatively young invaded intrusive elements from other adjoining parts such as South China, Malayan archipelago, Ethiopia, Mediterranean and Himalayan Pleistocene relicts. Therefore, it is interesting to recognize the biodiversity components of the Western Ghats that has been close relationship with Gondwana affinities. A few phylogenetic studies have been explored the complex evolutionary origins of selected Western Ghats biota (Joshi and Karanth, 2011a).

Recent molecular phylogenetic studies have been proved the ancient lineages of the amphibians and centipedes in the southern Western Ghats (Biju and Bossuyt 2003; Bossuyt *et al.* 2004; Vijayakumar *et al.*, 2016). Possibly, physiographical or ecological barriers might have played a crucial role in the present distribution of biotic groups. Distribution pattern of plants, birds, dragonflies, arboreal frogs (*Philautus*), and fishes of the Western Ghats suggested that Palghat gap might have a potential geographical barrier for dispersal (Fraser 1936; Subramanyam and Nayar 1974; Ali and Ripley 1987; Dahanukar *et al.* 2004; Biju *et al.* 2005). A few molecular-based biogeographic approaches have confirmed the role of the Palghat gap in orientation of current distribution of flora and fauna. It is strongly supported by microsatellite data on the distribution of Asian Elephant population across the Palghat gap (Vidya *et al.*, 2005).

Plant populations of *Eurya nitida* and *Gaultheria fragrantissima* are structuring their distribution patterns across Palghat gap and this gap act as an important geographical barrier for population distribution even within the species (Bahulikar *et al.*, 2004; Apte *et al.*, 2006). Hill peaks and deep valleys are appearing to be a significant barrier for high-elevation endemic bird White-bellied Shortwing (*Brachypteryx major*) (Robin *et al.*, 2010). Thus, these studies have been concluded that Palghat gap is serving as a spatial geographical barrier to gene flow of plants and animal species distributed across the gap. All these studies have narrowly focused on intraspecific variations in the species and elucidated the impact of Palghat gap in distribution of interspecific populations. More interestingly, a study on a group of caecilians showed that the gap did not function as a spatial barrier for the species and may have traditionally considered as limited dispersal ability (Gower *et al.*, 2007). Nevertheless, experimental conclusion from a molecular phylogenetic approach which included multiple taxonomic groups of Indian and Sri Lankan species have been pointed toward the impact of Palghat gap in shaping species assemblages (Bossuyt *et al.*, 2004).

The four other biogeographic subdivisions of the Western Ghats have also been restricted species distribution pattern and they are serving as ecological barriers in addition to the Palghat gap. These subdivisions are spatially separated and differing by their climatic conditions and vegetal composition (Subramanyam and Nayar, 1974). Species in every subdivision might also be closely related to each other than the species from other subdivisions. This phenomenon would be reflected in the molecular phylogenetic clades by species from each subdivision forming distinct divergent clades. Up to date, there is no published molecular work that addressed the ecological barrier for species distribution in the Western Ghats.

Earlier Explorations of Vegetations in Western Ghats

Western Ghats, especially the Malabar Coast is perhaps the sole center of spice trade, notably pepper, ginger and cardamom. Greek, Arab and later European traders, lured by the spices, found their way to the Western Coast of India and it has been created well known world history of commerce as an important episode of colonial survival. The Portuguese settlement at Goa and the Dutch possessions of Malabar, interested in the exploration of vegetation of this region, they contributed for the first time to the scientific documentation of the plant resources of the region. In 1565, Garcia de Orta enumerated a list of medicinal plants of India. Later Heinrich von Rheede has published *Hortus Malabaricus,* a monumental work on the plants of Malabar between 1678-1703. Robert Wight, in the middle of the nineteenth century, published a series of books with illustrations on South Indian plants. Between 1872 and 1894, Hooker, assisted by several botanists, published 6 volumes of *Flora of the British India* and gave the first comprehensive treatise on the Flora of Indian subcontinent. This flora accelerated the publication of provincial floras like the

Flora of the Madras Presidency by Gamble and subsequently completed by C. E. C. Fischer (1915-1936), and the *Flora of the Presidency of Bombay* by Cooke (1901-1908). The remarkable contribution for the biodiversity survey of southern India was made by Richard Henry Beddome. Rama Rao (1914) enumerated the flowering plants of Travancore. Other workers like Blatter, Bourdillon, Cleghorn, Dalzell, Fischer, Fyson, Gibson, Lawson, Santapau and Talbot have made greatest contributions for understanding the flora of the Western Ghats. The geological antiquity, evolutionary history and biogeographic patterns, with special emphasis on the endemism of the flora and fauna of the Western Ghats, have been accounted by several authors in different times (Govindarajulu and Swamy, 1958; Blasco, 1970; Subramanyam and Nayar, 1974; Nayar, 1977; Ahmedullah and Nayar, 1987; Rao, 1978; Pascal, 1988; Nair and Daniel, 1986; Nair, 1991).

The Western Ghats is recognized one among the 'hottest hotspots' of world biodiversity with major tropical evergreen-forested area of Indian subcontinent (Rodgers and Panwar, 1988). The floristic richness and plant diversity of the region has been accounted out by Gamble and Fischer (1915-1936), Fyson (1932), Nair and Daniel (1986), Nayar (1996). Consequently, several other State and District floras (Cooke, 1901-1908; Fyson, 1932; Ahuja and Singh, 1963; Manilal, 1988; Matthew, 1981-1984 and 1999; Mohanan and Henry, 1994; Nayar, 1996; Ramachandran and Nair, 1988; Rao, 1985-86; Sasidharan and Sivarajan, 1996; Rao and Razi, 1981; Saldanha and Nicolson, 1976; Saldanha, 1984, 1953; Keshava Murthy and Yoganarashiman, 1990; Yoganarashiman *et al.*, 1981; Matthew, 1999-2003; Rao *et al.*, 2004; Sasidharan, 2006; Sasidharan and Sivarajan, 1996; Nayar *et al.*, 2008; Ganesan *et al.*, 2019) highlighted the diversity and richness of the flowering plants in the region. Approximately 1200 species of lower plant groups are estimated to represent here in this zone. Out of 5800 species of flowering plants, 2100 are endemic in this megaendemic area (Yoganarasimhan, 2000; Nair and Henry, 1983; Singh *et al.*, 2015). It has been covered approximately 27 % of the total Indian flora. The endemic centers of the Ghats are Agasthyamalai (200 km^2) with 2000 species; The Nilgiris *ca* 2611 species, while Silent Valley (90 Km^2) supports 1300 species. Most of the District floras published in recent past reveal that most of them have more than 1200 species (Santapau, 1953, 1957; Rao and Razi, 1981; Keshava Murthy and Yoganarasimhan, 1990; Saldanha, 1984; Saldanha and Nicolson, 1976; Yoganarasimhan *et al.*, 1981; Manilal, 1988; Ramachandran and Nair, 1988; Chandrabose *et al.*,1988; Mohanan and Henry, 1994; Mohanan and Sivadasan, 2002; Ramaswamy *et al.*, 2001; Sasidharan and Sivarajan, 1996; Sasidharan, 2004).

Southern Western Ghats

The first authentic botanical account of southern Western Ghats can be found in the *Hortus Indicus Malabaricus* of Hendrik Andrian van Rheede published between 1678 and 1703. C. A. Barber, R. H. Beddome, T. F. Bourdillon, K. C. Jacob, M. A. Lawson, V. Narayanaswamy, M. Rama Rao, K. Venkoba Rao and R. Wight were the pioneer plant explorers of southern most Western Ghats. Wight's *Icones Plantarum Indiae Orientalis* (1840-1853) and *Illustrations to Indian Botany* (1840-1850) included accounts of many plants from this range. Bourdillon (1893) gave a detailed report on the Forests of Travancore, which formed the basis of his later monumental volume on the *Forest Trees of Travancore*. This work stimulated the taxonomic research and inspired many of the taxonomists who have worked out provincial floras like Dalzell and Gibson (1861), Cooke (1901-1908), Talbot (1909-11), Gamble and Fischer (1915-36), Parkinson (1923), and Fyson (1932). Rao (1914) enumerated the flowering plants of Travancore which contains a list of 3,535 plant species. Fyson (1915) has contributed a major and important work on the hill top flora of Southern India and also his important contributions have been compiled an illustrated book on *The Flora of the Nilgiri and Pulney Hill-Tops* (1915, 1921). He has published the revised volumes of the same in 1932. He has described the wild and introduced flowering plants around the hill stations of Kotagiri, Kodaikanal and Ootacamund.

Botanical explorations carried out by different workers in the recent past in various states including different districts of southern parts of India, mainly included the works of Saldanha (1984, 1996) *'Flora of Karnataka'* containing over 3400 species of flowering plants; Manilal (1988) *'Flora of Silent Valley'*; Sasidharan and Sivarajan (1996) *'Flowering Plants of Trissur Forest'*. M. Mohanan published the Flora of Thiruvananthapuram, which formed an authentic report on the floristic diversity of southern part of Western Ghats. Mohanan and Sivadasan (2002) published an exclusive floristic study of Agasthyamalai, one of the three endemic hotspots recognized in Kerala. A study of Shenduruny flora by Sasidharan (1997) with emphasis on endemic species resulted in recognition of 951 taxa, of which 309 are Western Ghats endemics.

Northern Western Ghats

North of Kali river is the Deccan trap having relatively fragile rocks with hills not rising beyond 1,650 m and having flat hill tops showing rocky outcrops of herbaceous vegetation. Monsoon herbs are abundant and show higher endemism as compared to the woody endemics of southern Western Ghats. Several workers have contributed the floristic account of northern Western Ghats include Cooke (1908), Blatter and McCann (1935), Caius (1936), Desai and Murthy (1950), Santapau (1953), Saldanha (1984, 1953, 1957), Saldanha and Santapau (1956), Auti *et al.* (2020), Bhushan *et al.* (2020), Potdar *et al.* (2012). Karthikeyan *et al.* (1989) enumerated the grass flora of northern Western Ghats that reported about 1254 species belonging to 260 genera. Some important floristic explorer of northern Western Ghats included Saldanha (1984 and 1996) *'Flora of Karnataka'* containing over 3400 species of flowering plants; Kulkarni (1988) *'Flora of Sindhudurg'*; Sahni (1990) *'Gymnosperms of India and Adjacent countries'*; Almeida (1996, 1998, 2001, 2003, 2009) *'Flora of Maharashtra'*; Naik (1998) *'Flora of Marathwada of Maharashtra'*. Many researchers have reported about the floral diversity of plateaus in northern Western Ghats (Bor, 1960; Bharucha and Ansari, 1963; Patel, 1968; Sharma *et al.*, 1984; Mahabale, 1987; Karthikeyan *et al.*, 1981; Lakshminarasimhan and Sharma, 1991; Karthikeyan and Anandkumar, 1993; Kothari and Moorthy, 1993: Karthikeyan, 1996; Diwakar and Sharma, 1999; Singh and Karthikeyan, 2000, 2001; Yadav and Sardesai, 2002; Patil, 2003; Singh *et al.*, 2001; Watve, 2013; Vinayaka and Krishnamurthy, 2016). Report on two basaltic plateaus of northern Western Ghats has been provided by Rahangdale and Rahangdale (2014, 2018). Their documentation chiefly focuses on flowering plant diversity from Durgawadi Plateau (600 taxa) and Naneghat Plateau (249 taxa). Specialised vegetation (Lekhak and Yadav, 2012), a new species (Malpure and Yadav 2009) and endemic flora (Joshi and Janarthanam 2004) were also reported from lateritic plateaus of northern Western Ghats.

Botanical explorations as carried out by different workers in the recent past in various parts of Western Ghats. In fact, these publications have made worth-while contributions to the existing account of plant wealth of the Western Ghats. At present the Botanical Survey of India (BSI), some State and Central Universities, Forest and Environment Departments, Conservational organizations are actively engaged in floristic revisions, vegetation structure, ecological interactions, rate of deforestation and afforestation, and conservation programmes of Western Ghats.

Vegetation

The vegetation component of the Western Ghats is one of the most complex and diverse types found in India. A part of megadiversity region, the flora and vegetation of this tract was under constant surveillance of botanists since the 19th century, more active among earlier ones being Hooker and Thomson (1855), Beddome (1869-1874), Bourdillon (1908), Rama Rao (1914), Gamble and Fischer (1915-1936) and others. Of the later contributions of Champion (1936), and Champion and Seth (1968), seem to be the most comprehensive and outstanding. They classified the forests of India into six major types with several subdivisions and edaphic or seral variations. The forests of Western Ghats belonging to 4 major types with more than 40 subtypes in various proportions of the forested areas from the north Maharashtra to south of Kanyakumari. The entire Western Ghats areas records about 7,402 flowering plant species in 1,480 genera under 210 families. Form which 5,588 species are indigenous, 376 species are exotics naturalized and 1,438 species are cultivated or planted. Of the 5,588 indigenous species of plants recorded in India, 2,253 plant species are endemic and 1,273 plant species are endemic to the Western Ghats. Apart from the above, there are 593 taxa with subspecies and variety status recorded for the Western Ghats (Nayar *et al.*, 2014). The original forest vegetation of Western Ghats and natural biodiversity can be seen in pockets of legally protected areas or within locally semiprotected regions like scared groves. Rest of the Western Ghats landscapes are degraded in various degrees of stages depending on the grading of disturbance and it is resulted in six major vegetation groups namely rocky outcrops, open grasslands with scrub, dense shrubbery, dwarf canopy forests, riparian forests and tall woodland forests. Composition of biological community pertaining to each of these vegetation classes are more or less definite with some overlaps.

The Western Ghats holds rich repository of biodiversity with indigenous vegetation. There is no comprehensive account on endemic plant diversity of the Western Ghats but sporadic publications are available for literature. Chatterjee (1940) enumerated only 34 endemic genera of dicotyledons from Indian Peninsula. Rao (1972) liste about 64 plant genera are confined to Indian floristic region, from which nearly 60 plant genera are narrowly distributed to Peninsular India and Sri Lanka. Subramanyam and Nayar (1974) accounted endemic plants of the Western Ghats. Nayar (1977) enumerated that about 2,100 endemic flowering plants occur in Peninsular India including the Western Ghats, which contributed about 32% of its flora. However, Ahmedullah and

Nayar (1986) gave the first exhaustive list on the endemic Plants of Peninsular India. They have accounted 1,940 endemic plant species including infraspecific taxa of Peninsular India. Three volumes of *Red Data Books of Indian Plants* (Nayar and Sastry, 1987, 1990), some 90 endemic plant taxa were included from Northern Western Ghats alone. Nayar (1996) has recorded 2,150 endemic plants in Peninsular India. Tetali *et al.* (2000) have recorded 439 endemic taxa of plants for Maharashtra state alone. Mishra and Singh (2001) have given elaborate account of 215 endemic and threatened taxa from Maharashtra. Irwin and Narasimhan (2011) have reviewed endemic genera of India with about 49 endemic genera, of which 40 are confined to Peninsular India. A total of 159 plant species (including infraspecific taxa) are confined to northern Western Ghats of India. Most of them are confined their distribution in small geographical area and they are facing high risk of extinction (Gaikwad *et al.*, 2014).

The Western Ghats, being one of the 36 global biodiversity hotspots (Marchese, 2015), is very rich in plant diversity and endemism (Nayar, 1996). Most of the arboreal taxa (63%) of the Western Ghats are narrowly confined to the region and characteristically the evergreen forests have distributed a very high percentage of endemic plant species (Ramesh, 2001). The mega centers of Western Ghats, has been further classified in to eight microcenters of endemism (Nayar, 1996). The Nilgiris forms a part of one of the micro-endemic centers along with its contiguous regions such as Silent valley-Wayanad and Kodagu. This whole region ranks second in terms of total number of endemic species next to Agasthyamalai (Narasimhan and Irwin, 2009).

Forest types of Western Ghats

Champion and Seth (1968) have classified the Indian forests into 5 major forest groups and 16 minor forest groups based on temperature and rainfall (climatic types) and further categorized more than 200 subforest types. Champion and Seth's forest classification was widely accepted in India but large number of forest subtypes is not feasible for managing forest classification at a time. Forest Survey of India FSI (2012) has proposed the Indian forest atlas with digitized forest classification of India for the first time. The official Indian forest maps constructed according to the Champion and Seth's classification including States and UnionTerritories have been prepared on 1:50,000 scale. In order to prepared further upscale of forest distribution map done by FSI, and have been attempted to reclassification forest types in India and to rework the forest types based on ground survey. According to Champion and Seth (1968), the forest types of the Western Ghats have listed out in Table 1.

Tropical forests of Western Ghats have been divided into four series of climatic forest types including dense and multilayered moist tropical forests due to high temperature and rainfall. These areas are comprising characteristically with wet evergreen forests composed of dense growth of tall trees, rich in climbers, lianas, epiphytes and shrub communities but poor in ground vegetation and grasses. The semievergreen forests are made by dense growth of intermixed tall deciduous and evergreen trees with rich layer of ground herbs, grasses and ferns. The moist deciduous forests constituted by the dominance of deciduous trees with a lower storied evergreen trees and shrubs. The littoral and swamp forests near the sea coast characterized by the dominance of halophytic evergreen trees, shrubs and herbaceous vegetation (Singh and Chaturvedi, 2017).

Dry tropical vegetation comprised with dry-deciduous forest, where the abundance of shruby vegetation and open canopy of small trees, the climate experienced around 6 months of dry period in the annual cycle. The thorn forests vegetation in lower elevational gradient experienced more than 6 months of dry period every year, which are characterized by very sparse distribution of small thorny trees with abundance of shrubs. Ground floor vegetation appears during only in rainy season when trees and shrubs also flushed leaves. The dry-evergreen forests prevailed high temperature and low rainfall experienced during the summer season. Bamboos are usually absent from these dry forests but grasses and small thorny trees are abundant. The dry deciduous forests located on the leeward side of the Western Ghats Ranges within an elevational gradient of 300-900 meters where mean annual rainfall around 900-2,000 millimeters. They were distributed across the southern Ghats of Karnataka and Tamil Nadu. The high altitude Western Ghat ranges increased the moisture content from the southwest monsoon, hence the eastern slopes of the Deccan Plateau receive relatively little rainfall approximately 900 to 1,500 millimeters. The undulating lower range valleys have very shallow soils. Thorny vegetation become more common in buffer zone of foot hills where grazing pressure is usually high.

The moist deciduous forests occupied the largest areas of the Western Ghats with an elevational range of 500-900 meters, where experienced mean annual rainfall of 2,500-3,500 millimeters. The narrow strip of moist deciduous forests is on windward side of the Ghat range, where the southwest monsoon promotes to develop dense wet evergreen forests. On the leeward side of the Ghats has more dry weather with less steep, which caused the rain shadow area developed in a broader and uneven strip of moist deciduous forests. This forest further extends into the Deccan Plateau of northern side. Rainfall pattern of the leeward side of Ghats are influenced by complex climatic conditions, with some areas experiencing less than one-fifth of the 3,000 millimeters or more of average annual precipitation usually higher elevation of the mountains.

Montane sub-tropical forests are occupied between the altitudes of 1000 m and 2000 m where experienced cooler climate than tropical but warmer than temperate regions. Semi-xerophytic evergreen vegetations are predominantly colonized and the forests are classified into broad-leaved hill forests, which prevailed the abundance of climbers and epiphytic ferns, orchids and dense growth of evergreen broad-leaved trees. The pine forests comprising with the open formations of pine trees which are usually cultivated and dry-evergreen forests are mainly characterized by small leaved evergreen trees and thorny xerophytes.

Montane temperate forests are occupied in areas of low temperature with comparatively high humidity in high altitudes of the Ghats. The temperate forests comprised with dense tall coniferous trees or evergreen angiosperm trees and their branches clothed with epiphytic mosses, lichens and ferns. These forests are further classified into three groups: montane wet temperate forests which are characterized in a northern and a southern vegetation subtype (Singh and Chaturvedi, 2017). The habitat types of the southern Western Ghats typical tropical evergreen forests include the wet montane evergreen forests and *shola*-grassland complexes in the higher altitudes (1,900-2,200 meters). In the montane evergreen forests, vegetations are highly diverse, multistoried with rich epiphytes, at a stunted low canopy at 15 to 20 meters (Ganesh *et al.*, 1996). The montane evergreen forests holding rich diversity of endemism usually more than half of the tree species are confined only in this forest type, especially major plant families of Dipterocarpaceae and Ebenaceae. The majority of monotypic species are found in this region. The distribution and richness of endemic plants is not uniform within this forest type, where some areas having higher accumulation of endemic plants. For the convenient of descriptions of the forest types of the Western Ghats, the regions divided in to two ranges are such as northern Western Ghats and southern Western Ghats.

Northern Western Ghats covered the states of Gujarat, Maharashtra and Goa, whereas in southern Western Ghats occupied in the states of Karnataka, Kerala and Tamil Nadu (Plate 1).

Table 1. Forest types of Western Ghats, India (Champion and Seth, 1968).

<table>
<tr><th>Forest categories</th><th>Major forest Group</th><th>Subgroup</th><th>Subtypes</th></tr>
<tr><td rowspan="14">I. Moist tropical forests</td><td rowspan="5">Tropical wet evergreen forests</td><td rowspan="2">Southern tropical wet evergreen forests</td><td>Southern hill top evergreen tropical forests (1A/C3)</td></tr>
<tr><td>West coast tropical evergreen forests (1A/C4)</td></tr>
<tr><td rowspan="3">Edaphic and seral wet evergreen forest types</td><td>Pioneer Euphorbiaceous scrub (1A/2S1)</td></tr>
<tr><td>Cane bakes (1A/E1)</td></tr>
<tr><td>Wet bamboo bakes (1A/E2)</td></tr>
<tr><td rowspan="6">Tropical semievergreen forests</td><td rowspan="3">Southern tropical semievergreen forests</td><td>West coast semievergreen forests (2A/C2)</td></tr>
<tr><td>Tirunelveli semievergreen forests (2A/C3)</td></tr>
<tr><td>West coast semievergreen Diptercarp forests (2A/2S1)</td></tr>
<tr><td rowspan="5">Edaphic and seral semievergreen forests types</td><td>Cane brakes (2B/E1)</td></tr>
<tr><td>Wet bamboo brakes (2B/E2)</td></tr>
<tr><td>Moist bamboo brakes (2B/E3)</td></tr>
<tr><td></td><td>Leteritic semievergreen forests (2B/E4)</td></tr>
<tr><td></td><td>Secondary moist bamboo brakes (2B/2S1)</td></tr>
<tr><td rowspan="3">Tropical moist deciduous forests</td><td rowspan="3">South Indian moist deciduous forests</td><td>Slightly moist and moist teak forests (3B/C1c)</td></tr>
<tr><td></td><td>Southern moist mixed deciduous forests (3B/C2)</td></tr>
<tr><td></td><td>Southern secondary moist mixed deciduous forests (3B/2S1)</td></tr>
</table>

	Littoral and swamp forests	Littoral forests	Littoral forests (4A/L1)
		Tidal swamp forests	Mangrove scrub and mangrove forests (4B/TS1)
		Tropical fresh water swamp forests	Submontane hill valley swamp forests (4C/FS2)
		Tropical riparian fringing forests	Riparian fringing forests (4E/RS1)
II. Dry tropical forests	Tropical dry deciduous forests	Southern tropical dry deciduous forests	Dry teak forests (5A/C1b)
			Very dry teak forests (5A/C1a)
			Southern dry mixed deciduous forests (5A/C3)
		Degradation stage of tropical dry deciduous forests	Dry deciduous scrub (5/DS1)
			Dry savannah forests (5/DS2)
			Euphorbia scrub (5/DS3)
			Dry grassland (5/DS4)
		General edaphic types of deciduous forests	*Boswellia* forests (5/E2)
			Hardwickia forests (5/E4)
			Butea forest (5/E5)
			Dry bamboo brakes (5/E9)
		General seral types of deciduous forests	Dry tropical riverain forests (5/1S1)
			Secondary dry deciduous forests (5/2S1)
	Tropical thorn forests	Southern tropical thorn forests	Southern thorn forests (6A/C1)
			Karnatak umbrella thorn forests (6A/C2)
			Southern thorn scrub (6A/DS1)
			Southern *Euphorbia* scrub (6A/DS2)
		General edaphic, degraded and seral types of thorn forests	*Cassia auriculata* scrub (6B/DS1)

	Tropical dry evergreen forests	Dry evergreen forests	Tropical dry evergreen forests (7/C1)
			Tropical dry evergreen scrub (7/DS1)
III. Montane subtropical forests	Subtropical broad-leaved hill forests	Southern tropical broad-leaved hill forests	Nilgiri subtropical hill forests (8A/C1)
			Ochlandra reed brakes (8A/E1)
			South India subtropical hill savannah (8A/DS1)
IV. Montane temperate forests	Montane wet temperate forests	Southern montane wet temperate forests	Southern montane wet temperate forests (11A/C1)
			Southern montane wet temperate scrub (11A/DS1)
			Southern montane wet grass land (11A/DS2)

Northern Western Ghats

***Tropical wet evergreen forests*:** This forest type consists of two storied canopies with ground vegetation. The top canopy is composed the tree species mainly of *Beilschmiedia dalzellii, Carallia brachiata, Diospyros malabarica, D. montana, D. sylvatica, Mangifera indica, Myristica malabarica, Tetrapilus dioicus, Syzygium cumini,* and *Symplocos macrophylla* subsp. *rosea.* The lower canopy is represented by *Actinodaphne angustifolia, Tritaxis glabella, Ixora brachiata, I. nigricans, Litsea stocksii, Mallotus resinosus* and *Memecylon umbellatum.* The number of climbing species among the trees are *Piper hookeri, P. trichostachyon, Stephania japonica, Ancistrocladus heyneanus* and *Premna obtusifolia* var. *pubescens.* The ground floor of the forest is clothed only in the openings of the canopy, along the sides of streams, which is grown by *Achyranthes coynei, Asystasia dalzelliana, Canscora diffusa* var. *diffusa, Curcuma pseudomontana, Chlorophytum tuberosum, Blumea stricta, Desmodiastrum racemosum* var. *rotundifolium, Ecbolium ligustrinum, Habenaria foliosa, Leucas deodikarii, Impatiens dalzellii, I. pulcherrima, Rhinacanthus nasutus, Paracaryopsis coelestina, Senecio bombayensis, Rubia cordifolia, Smithia bigemina, S. setulosa, S. purpurea, Vigna dalzelliana, Sida rhombifolia, Peristylus lawii, Zingiber neesanum* and *Arisaema murrayi.*

***Tropical semi-evergreen forests*:** This forest type occurs mostly on upper hills slope from 450 meters to 1050 meters above the MSL in the Western Ghats.

The major tree species are *Actinodaphne lanceolata, Mangifera indica, Memecylon umbellatum, Syzigium cumini, Terminalia paniculata, T. chebula* and *Tetrapilus dioicus*, etc. Foot hills of Western Ghats towards Konkan are covered by semi-evergreen forest. The common trees that form the first storey of this forest are *Albizia lebbeck, Atalantia racemosa, Bombax ceiba, Calophylum inophyllum, Carallia brachiata, Ficus racemosa, Mangifera indica, Syzygium cumini, Terminalia bellirica*, and *T. chebula*. The dominant endemics found in this evergreen class were *Knema attenuata, Hopea ponga, Reinwardtiodendron anamallayanam, Holigarna arnottiana, Holigarna grahamii, Diospyros candolleana, Ixora brachiata* and *Flacourtia montana*. The climax evergreen forests of the Western Ghats composed of the characteristic tree species such as *Palaquium ellipticum, Vateria indica, Dipterocarpus indicus, Poeciloneuron indicum* and *Dysoxylum malabaricum* were found only in this very high evergreen class. Major endemics observed in the forest types are *Holigarna grahamii, Holigarna beddomei*, and *Monoon fragrans*. Other dominant endemics recorded were *Garcinia indica, Blachia denudata, Tritaxis glabella, Sageraea laurina, Polyspora obtusa, Hydnocarpus pentandrus, Drypetes venusta*, and *Diospyros* saldanhae. These forest vegetations developed in the high rain fall areas of Ghats usually seen in northern Sahyadri Plateaus like Mahabaleshwar, Matheran, Bhimashankar, Amboli Ghat, etc.

Tropical Semi-evergreen forests are primarily distributed in the states of Maharashtra, Goa, and Karnataka in the Western Ghats, within range of 300-900 meters elevational gradients. In this forest types includes the secondary evergreen dipterocarp forests, lateritic semievergreen forests, bamboo brakes, and riparian forests (Champion and Seth, 1968). These forests established with varied composition and structure from north to south and especially from eastern to western slopes of the Ghats. The predominant tree species includes: *Aporosa cardiosperma, Careya arborea, Celtis timorensis, Elaeocarpus tuberculatus, Hopea parviflora, Holigarna arnottiana, Hydnocarpus pentandrus, Mesua ferrea, Lagerstroemia microcarpa, Memecylon umbellatum, Tetrapilus dioicus, Terminalia paniculata, Syzygium* spp., and *Vateria indica*. These forest types are also have noted high levels of tree diversity and endemism.

Bhimashankar wildlife sanctuary of Maharashtra is constituted the top canopy of trees such as *Caryota urens, Ficus racemosa, Firmiana colorata, Garcinia indica, Mangifera indica, Mallotus philippensis, Memecylon umbellatum, Tetrapilus dioicus, Sterculia guttata, Syzygium cumini* and *Xantolis tomentosa*. The lower canopy is composed by *Atalantia racemosa, Callicarpa tomentosa, Tritaxis glabella,*

Litsea deccanensis, L. ghatica, L. stocksii, Hymenodictyon obovatum, Ixora brachiata, and *Mallotus resinosus*. Common climbers are included *Cyclea peltata, Diploclisia glaucescens, Gnetum ula, Oxyceros rugulosus, Piper hookeri, P. trichostachyon, Gymnema cuspidatum, Stephania japonica, Salacia macrosperma* and *Tinospora sinensis*. The ground floor is occupied by *Curcuma angustifolia, C. decipiens, C. pseudomontana, Crepidium versicolor, Nervilia concolor, N. plicata, Peristylus lawii, Rubia cordifolia, Zingiber neesanum* and *Z. nimmonii*. This forest vegetation support to grow more diversity of epiphytes when compared to other forest types. The common epiphytic flora seen in *Aerides crispa, A. maculosa, Bulbophyllum fimbriatum, Porpax braccata, Porpax microchilos, Dendrobium aqueum, D. barbatulum, Hoya wightii, Remusatia vivipara, Oberonia brachystachys, Smithsonia viridiflora* and *Thunia alba* var. *bracteata* as well as other ferns and mosses (Savita and Sanjaykumar, 2017).

A decrease in the altitude in the south of Maharashtra as well as disturbance resulted in the appearance of deciduous elements and the evergreens vegetation makes way for the semievergreen type of vegetation. These forests resemble the evergreen forests in their structure except for the profuse under growth. The tree species, which dominate in these forests are *Schleichera* sp., *Memecylon* sp., *Madhuca* sp., *Magnolia* sp., *Semecarpus* sp., *Syzygium* sp., *Persa* sp., *Hopea* sp., *Hydnocarpus* sp., *Ceiba* sp., *Mappia* sp., *Polyalthia* sp., *Stereospermum* sp., *Pterocarpus* sp. and *Terminalia* sp. The shrubs like *Ixora* spp., *Allophylus cobbe. Calamus* spp., *Ardisia solanacea* and *Alangium salvifolium* dominate the ground flora. The opening of the canopy has resulted in the establishment of a large number of herbs, including bryophytes, pteridophytes and orchids.

***Tropical Moist Deciduous Forest*:** This type of forest is found all along the eastern side of Sahyadri Ghats where the average annual rainfall ranges from 130 to 180 cm and average annual temperature from 24^{0}C to 27^{0}C. The forests of Allapalli subdivision in Gadchiroli districts also fall under this vegetation. Usually, the top of northern Sahyadri ranges have constituted moist deciduous forests, which become drier towards east. The dominant tree species throughout the Ghats occupies upper canopy such as *Tectona grandis,* and other deciduous species like *Albizia lebbeck, Bombax ceiba, Dalbergia lanceolaria, D. latifolia, Mangifera indica, Senegalia chundra, Senegalia ferruginea, Tamarindus indica, Terminalia bellirica* and *T. chebula*. These tree species compositions are found in mid elevation of Bhimashankar Wildlife Sanctuary, especially along the hill slopes surrounding rocky areas and on the western side. The part of the sanctuary in the southern side of Konkan region has constituted the dry deciduous nature of vegetation. The top canopy trees are composed by small trees of *Bridelia retusa, Firmiana simplex, Erythrina stricta, Ficus racemosa, Lasiosiphon glaucus, Heterophragma quadriloculare, Mallotus philippensis, Terminalia anogeissiana, T. chebula, T. bellirica, T. cuneata, Tectona grandis* and *Xantolis tomentosa*. This vegetation is intermixed with *Butea monosperma, Boswellia serrata* and *Lannea coromandelica*. The understoried shrybby vegetation comprises of *Atalantia racemosa, Breynia retusa, Carissa spinarum, Helicteres isora, Holarrhena pubcescens, Bergera koenigii, Pavetta crassicaulis and Strobilanthes callosa*. The lower elevation of dry zones species like, Balanites aegyptiaca were also notices at a few places in the Konkan region. The tree vegetation supports diverse species of climbers and ground cover. The common climbers are *Asparagus racemosus, Dioscorea pentaphylla* var. *pentaphylla, Hemidesmus indicus* var. *indicus, Gymnema sylvestre* and *Jasminum malabaricum*. The ground vegetation covered *Curculigo orchioides, Drimia indica, Euphorbia fusiformis, Hypoxis aurea, Pancratium parvum, Zingiber neesanum* and grasses (Savita and Sanjaykumar, 2017).

These vegetation types are found in Chandrapur, Gadchiroli, Bhandara, Gondia and Amaravati districts in Nagpur and also on the slopes of northern part of the Western Ghats spreading from Nandurbar, Dhule, Nashik, Thane to Kolhapur districts. There are two sub-types represented in this group.

Moist Teak bearing Forests: An important and valuable forests in commercial view point of the Maharashtra State, restricted to Project Tiger area in Melghat region with major tree of *Tectona grandis*, and associate with *Adina cordifolia, Dalbergia latifolia, Mitragyna parvifolia, Madhuca indica, Terminalia tomentosa, Pterocarpus marsupium, Bombax ceiba* and *Dendrocalamus strictus*.

***Moist Mixed deciduous Forests*:** Teak is present occasionally in the characteristic evergreen component intermixed with major component of semi-evergreen elements. The major tree species are *Pterocarpus marsupium, Bombax ceiba, Terminalaia bellarica, Dalbergia latifolia, Syzigium cumini, Terminalia tomentosa*, and *Lagerstroemia parviflora*.

***Tropical Dry Deciduous Forests*:** These forests are situated on eastern slopes of the Western Ghats where the mean annual rainfall ranges from 50-150 cm and mean annual temperature from 26^{0}C to 27.3^{0}C. The forests of the districts of Kolhapur, Sangli, Satara, Pune, Ahmednagar, Nasik, Dhule, Jalgoan, Aurangabad, Jalna, Parbhani, Nanded, Yavatmal, Buldhana, Akola, Amravati, Wardha, Nagpur, Chandrapur, Bhandara and Gadchiroli are fully and partially covered by this type of vegetation. The common trees are *Albizia* *lebbeck, Azadirachta indica, Dalbergia latifolia, Euphorbia neriifolia, Ficus amplissima, Mangifera indica, Senegalia chundra, S. ferruginea, Tectona grandis,* etc. These forest types occupied about 62% of total forested area of Maharashtra state. The Ghat region of Nagpur, Amravati and Uttara Maharashtra and their eastern hill slopes represented major portion of the tropical dry deciduous forests. The dominant trees are Teak *(Tectona grandis),* Tiwas *(Ougeinia oojeinensis),* Khair *(Senegalia catechu),* Shivan *(Gmelina arborea),* Dhavda *(Terminalia anogeissiana)* in this type of forests. In this region recognized two types of following subforests vegetations.

a. ***Dry Teak Bearing Forests*:** Principal predominant species is *Tectona grandis* and their associate trees are *Senegalia catechu* (Khair), *Terminalia anogeissiana* (Dhawada), *Gmelania arborea* (Shivan), *Ougeinia oojeinensis* (Tiwas), etc.

b. ***Dry Mixed Forests*:** This forest types occurs mostly on elevational hill slope and valleys from 450 meters to 1050 meters m.s.l. in the Western Ghats. The main tree species are *Actinodaphne lanceolata* (Pisa), *Mangifera indica* (mango), *Memecylon umbellatum* (Anjani), *Tetrapilous dioicus* (Parjamun), *Syzigium cumini* (Jambul), *Terminalia paniculata* (Kinjal), *Terminalia chebula* (Hirda), etc.

***Tropical Thorn Forests*:** The forest types occupied in drier parts of Desh, Khandesh, Vidarbha and Marathwada districts, where the average annual rainfall ranges from 35-80 cm and mean temperature from 26°C to 27°C. Major portion of these forests are heavily degraded by low soil fertility associated with lower annual rainfall. The thorn forests are characteristically blank, scattered, restricted to shallow soils and buffered by cultivable lands on all sides. These are also subjected to heavy grazing, lopping and illicit felling. The trees usually have short boles and the usual height is 6-9 m. There is an ill-defined lower storey of smaller trees and large shrubs, mostly spiny and often with other xerophytic adaptations. During rainy seasons various herbaceous plants are also found. *Azadirachta indica, Balanites roxburghii, Bauhinia racemosa, Butea monosperma, Cassia auriculata, Dolichandrone falcata, Euphorbia neriifolia, Ixora arborea, Lantana camara. Mimosa hamata, Vachellia nilotica, Vachellia leucophloea,* and *Ziziphus mauritiana*. These portions of the forests are predominantly occupied *Euphorbia* and *Cassia* scrub. A few places where the scrub vegetation is dominated by grasslands mostly on gradual hill slopes near the foot hills of northern Ghats. Intermittent areas are seen in coppiced trees, as an indicator of remnants or earlier canopied forests over these areas. The zones are most exploited and highly disturbed hardy species by heavy degradation. The ground cover clothed with dominant herbs and grasses include *Apluda mutica, Heteropogon contortus, Themeda cymbaria* along with shrubs of *Carrisa spinarum, Gnidia glauca, Meyna laxiflora*, and also sparse trees of *Mangifera indica, Syzygium cumini, Terminalia elliptica*, etc (Ghate, 2014).

Montane subtropical broad leaves hill forests**:** These forests are found in the higher ghats of Sahyadri areas, restricted to patches or narrow strips usually over 900 m altitude. The mean annual rainfall of this region is above 300 cm and the mean annual temperature is around 20°C. The soil is usually rich in humic contents and the climate is humid. Actually, there is no sharp distinction between the moist mixed deciduous type and montane broad leaved hill type, the latter probably represent the post climax of the former type. Most of the forests of this type have fundamentally been altered by shifting cultivation or lopping, but where properly developed it has dense evergreen elements mixed with the deciduous elements. The trees are usually of low to medium average height (5 to 15 m) with comparatively smaller girth (up to 2 m) and spreading canopies. Top storey consists of trees like *Senegalia ferruginea, Albizia procera, Bombax ceiba. Celtis timorensis. Cinnamomum macrocarpum, Dalbergia lanceolaria, D. latifolia, Terminalia bellerica* and *T. chebula*. The members of second storey are *Mallotus philippensis, Mangifera indica, Memecylon umbellatum* and *Bambusa bambos*.

Littoral and Swamp Forests: This forests type occurs along the creeks and littoral areas of Sindhudurg and Thane districts. Coastal areas have mostly sandy rocky beaches do not possess muddy flats and evergreen mangrove forests. There are few patches of mangrove forests along coastal areas of Sindhudurg and Goa. The mangrove forest of soft tidal mud flats in lower elevation near the sea level supports to develop evergreen shrubs or small trees. The mangrove species are typically occupied in this region include *Avicennia* spp. and *Rhizophora* spp. The vegetation incorporates closed evergreen forest of moderate height. The representative's species of this vegetation are *Acanthus ilicifolius, Aegiceras corniculatus, Bruguiera parviflora, Casuarina equisetifolia, Calophyllum inophyllum, Spinifex littoreus, Vitex trifolia* and *Xylocarpus granatum*. Other than the above dominant species, *Saraca asoca, Pandanus spinifructus, Calamus pseudotenuis* and *Lagenandra* sp. were found. The northernmost distribution of *Myristica* swamps has been reported in the Western Ghats areas of 'Nirankarachi Rai', in Goa. The predominant tree species were reported here *Myristica fatua, Holigarna arnottiana, Stereospermum colais*, etc. (Sreedharan and Indulkar, 2018).

Southern Western Ghats

The forests of southern Western Ghats, according to Champion and Seth (1968) fall under four major categories viz., (1) Moist tropical forests, (2) Dry tropical forests, (3) Moist subtropical forests and (4) Montane temperate forests, with about 30 sub divisions. Chandrasekaran (1962, 1973) classified little modification from Champion and Seth, they classified the forests of southern Western Ghats in to five categories: (1) Tropical wet evergreen forests with three climax forest types viz., (a) Low-level evergreen forests, (b) High-level evergreen forests and (c) Low tropical ghat forests; seven secondary types viz., (a) semievergreen forests, (b) Secondary evergreen forests, (c) Moist deciduous forests, (d) Open deciduous forests, (e) Wet bamboo (reed) brakes, (f) Moist bamboo brakes and (g) low-level grasslands, and 6 edaphic forest types such as: (a1) Myristica swamps, (b1) Tropical valley freshwater swamps, (c1) Tropical riverine forests, (d1) Cane brakes, (e1) Xylia mixed forests and (f1) Laterite scrubs; (2) Tropical deciduous forests of their two climax types as (a) Wayanad Plateau deciduous forests and (b) Moist mixed deciduous forest; two secondary forms as: (a1) Moist savannah and (b1) Secondary bamboo forests and single edaphic type namely Post Climax evergreen forests; (3) Tropical dry deciduous forests having only climax and secondary forest types. The climax types are (a)Tropical dry deciduous forests and (b) Mixed dry deciduous forests with sandal; (4) Montane subtropical forests having only one climax formation viz., Subtropical wet hill forest and (5) Montane temperate forests with one climax type namely Wet temperate forests and one secondary forest as high level (montane) grasslands.

***Moist tropical forests*:** They occur in warm humid areas characterized by high rainfall, temperature, humidity and uniform photoperiods throughout the year. Winter in such forests will be of a short period. Occurrence of frost or fog is uncommon. These forests are mostly confined to altitude between 200 and 1500 m in areas with several broken hills, long spurs and extensive ravines. Trees are generally tall and broad-leaves.

This forests type is confined to southern the Western Ghats, which are occupied at an altitude of 1500 m or above. The vegetation is characterized by three-layered structure i.e. forest floor, with rich herbaceous ground cover; under storey layer with small or medium-sized trees, dense epiphytic growth of orchids and pteridophytes; upper tall trees canopy layer continuous layer to attain a height more than 45 m. Some tree species also emerge out of the canopy. Species of *Artocarpus, Durio, Dysoxylum, Elaeocarpus* and *Lophopetalum* are common in these forests.

Eighteen tree species are indicators of wet evergreen forests in the southern Western Ghats such as *Bhesa indica, Calophyllum austro-indicum, Cinnamomum keralense, Cullenia exarillata, Diospyros atrata, D. bourdillonii, Dipterocarpus bourdilloni, D. indicus, Ficus beddomei, Gluta travancorica, Hopea parviflora, Prioria pinnata, Lophopetalum wightianum, Mangifera indica, Nageia wallichiana, Palaquium ellipticum, Strombosia ceylanica* and *Vateria indica*. Some the tree species are considered as better indicators of the ecological conditions among the above associates, and their ecological amplitude are very narrow such species of *Bhesa indica, Cullenia exarillata, Dipterocarpus bourdillonii, D. indicus, Gluta travancorica*, and *Nageia wallichiana*.

The two types of tropical wet evergreen forests met with are (a) West coast tropical evergreen forests and (b) Southern hilltop tropical evergreen forests.

West coast tropical evergreen forests: This type is generally found in valley of high altitudes between 200 and 1500 m. *Cullenia – Mesua – Palaquium* series (Gadgil and Meher-Homji, 1986b) is the major association here. This association is considered biologically the richest diversity of plants. The top canopy is formed of trees like *Acrocarpus fraxinifolius, Aglaia elaeagnoidea, Ailanthus triphysa, Antiaris toxicaria, Artocarpus heterophyllus, A. hirsutus, Bischofia javanica, Calophyllum calaba, Canarium strictum, Cullenia exarillata, Dillenia bracteata, Diospyros bourdillonii, D. candolleana, Dimocarpus longan, Dipterocarpus indicus, Donella lanceolate, Drypetes venusta, D. longifolia, Dysoxylum malabaricum, Elaeocarpus tuberculatus, Epiprinus mallotiformis, Garcinia gummi-gutta, Holigarna arnottiana, H. beddomei, Hopea parviflora, H. racophloea, Litsea oleoides, Lophopetalum wightianum, Mangifera indica, Mastixia arborea, Mesua ferrea, Myristica dactyloides, Ostodes zeylanicus, Palaquium ellipticum, Paracroton pendulus* subsp. *zeylanicus, Poeciloneron indicum, Reinwardtiodendron anamalaiense, Syzygium chavaran, S. gardneri, S. parameswaranii, Terminalia travancorensis, Tetramelos nudiflora* and *Vateria indica*.

The second storey has medium-sized trees adapted to partial shady conditions such as *Actinodaphne bourdillonii, A. tadulingamii, Aglaia elaeagnoidea, A. lawii, A. malabarica, Aphanamixis polystachya, Apodytes dimidiata, Aporosa cardiosperma, Beilschmiedia bourdillonii, Psydrax dicoccos, Carallia brachiata,*

Cryptocarya beddomei, Dendrocnide sinuata, Diospyros paniculata, D. pruriens, Elaeocarpus serratus, Flacourtia montana, Glochidion ellipticum, Polyspora obtusa, Goniothalamus cardiopetalus, Gymnacranthera farquhariana, Glyptopetalum zeylanicum, Hydnocarpus pentandrus, Knema attenuata, Litsea coriacea, L. floribunda, Macranga peltata, Mallotus philippensis, Melicope lunu-ankenda, Mimusops elengi, Myristica malabarica, Nageia wallichiana, Mappia nimmoniana, Monoon fragrans, Symplocos cochinchnensis, Turpinia malabarica, Xanthophyllum flavescens, etc.

The third storey possesses small trees and large shrubs like *Acronychia pedunculata, Agrostistachys indica, Antidesma montanum, Atalantia wightii, Callicarpa tomentosa, Canthium angustifolium, C. travancoricum, Celtis timorensis, Chassalia curviflora, Clausena indica, Chionanthus mala-elengi* ssp. *mala-elengi, Elaeocarpus munroi, Euonymus indicus, Flacourtia montana, Grewia flavescens, Humboldtia unijuga* var. *trijuga, Ixora elongata, I. malabarica, I. nigrescens, Isonandra lanceolata, Lasianthus jackianus, Leea indica, Leptonychia caudata, Nothopegia beddomei, Pandanus thwaitesii, Pavetta hispidula, Psychotria nudiflora, Sarcococca coriacea, Tabernaemontana gamblei, Tournefortia heyneana,* etc.

The fourth storey is formed by shrubs, undershrubs and herbs like *Acranthera grandiflora, Alpinia galanga, A. malaccensis, Amomum muricatum, Argostemma courtallense, Asplenium indicum, Begonia malabarica, Boesenbergia pulcherrima, Barleria courtallica, Blachia denudata, Hellenia speciosa, Curcuma aurantiaca, C. neilgherrensis, Calamus gamblei, Elatostema acuminatum, E. lineolatum, Elettaria cardamomum, Murdannia trichocarpa, Munronia pinnata, Mycetia acuminata, Neanotis monosperma, Neurocalyx calycinus, Ophiorrhiza brunonis, O. eriantha, O. hirsutula, O. mungos, Orophea uniflora, O. zeylanica, Pavetta zeylanica, Psychotria congesta, Saprosma corymbosa, Sarcandra glabra* ssp. *brachystachys, Sonerila rheedei, Indianthus virgatus, Thottea siliquosa, Tarenna monosperma, Zingiber neesanum, Z. roseum,* etc.

Southern hilltop tropical wet evergreen forests: This type of forests generally seen above 1200 m in valleys surrounding ridges. The trees are usually shorter compared to the previous type and are with fuller and rounded crowns. This forest area is considered a transitional stage from tropical to subtropical climax. The top storey is characterized by trees like *Bischofia javanica, Calophyllum austroindicum, Cullenia exarillata, Elaeocarpus serratus, E. tuberculatus, Ficus arnottiana, Garcinia cambogioides, Glochidion ellipticum, Gluta travancorica, Polyspora obtusa, Holigarna arnottiana, Mesua ferrea, Machilus glaucescens, Prunus zeylanica, Semecarpus anacardium, Syzygium zeylanicum,* etc.

The second storey consists of medium-sized trees of younger forms of the upper storey along with *Alstonia venenata, Alseodaphne semecarpifolia* var. *parvifolia, Baccauera courtallensis, Beilschmiedia bourdillonii, Cinnamomum filipedicellatum, Deberegeasia ceylanica, Elaeocarpus glandulosus, Harpullia arborea, Humboldtia brunonis, Litsea bourdillonii, L. floribunda, L. oleoides, Memecylon angustifolium, Meliosma pinnata, Mallotus tetracoccus, Otonephelium stipulaceum, Toona ciliata, Heynea trijuga, Tupinia malabarica,* etc.

The third storey vegetation composed of large shrubs and small trees merge with the lower ground layer formed of herbaceous species. The herbs, shrubs, and smaller trees of this storey are *Aglaia tomentosa, Aporosa acuminata, Ardisia pauciflora, Casearia tomentosa, Clausena austroindica, Gomphandra tetrandra, Lepisanthes erecta, Meiogyne pannosa, Phlogacanthus grandis, Strobilanthes foliosus, S. heyneanus, S. ciliata, Oreocnide integrifolia, Orophea uniflora, Pavetta oblanceolata, Isodon wightii, Pogostemon paniculatus, P. travancoricus, Psychotria anamallayana, P. flavida, Rauvolfia verticillata, Rostellularia quinqueangularis, Sarcandra glabra* ssp. *brachystachys, Tarenna alpestris, T. canarica, Thottea siliquosa* and *Wendlandia bicuspidata*. The common climbers are *Aganosma cymosa, Calamus gamblei, Elaeagnus conferta, Embelia ribes, Jasminum cordifolium, Millettia rubiginosa, Moullava spicata, Mucuna pruriens* var. *hirsuta, Salacia beddomei, Shuteria vestita* and *Vigna dalzelliana*. The trunks of trees are more carpeted with mosses and lichens compared to the earlier type.

Pascal *et al.* (2004) categorized the wet evergreen forests in to three main groups such as A, B and C, based on elevational ranges, which are low, medium and high elevational level wet evergreen forest types respectively.

***Low elevation forests (to 800 m)*:** In the Western Ghats, the low elevation forests are characteristically composed by the presence of *Dipterocarpus indicus* and *Strombosia ceylanica*, occurring above 800 m. Further north of the Ghats, *Dipterocarpus indicus* is a characteristic canopy tree species in the other low elevation evergreen forests. However, it has been found more sporadic patches because of over exploitation of the tree species. *Strombosia ceylanica* has distributed widely throughout the Western Ghats, but it is a common canopy tree in the low elevation types. Hence, these two tree species were selected for naming of the low elevation types. Some other tree species were contributing to the low elevation forests such as *Diospyros bourdillonii, Dipterocarpus bourdilloni, Ficus beddomei, Hopea parviflora, Prioria pinnata* and *Lophopetalum wightianum*. Among these trees, *Dipterocarpus bourdillonii* has markedly separated the locations to the north of the Ariankavu Pass and to the south which are distributed more prominent presence among the *Prioria pinnata*. This tree species shows a peculiar distribution pattern in the north of the Ariankavu Pass, but it has become prominent again in the dry evergreen forest in the east and south. Further it has been classified into two low elevation forest types based on the composition of tree species (*sensu* Oldeman, 1974) with a canopy height reaching about 35-45 m.

Dipterocarpus indicus – Prioria pinnata – Strombosia ceylanica type: This forest type occurs to south of the Ariankavu Pass (8°20′N to 9°00′N), where the period of the dry climates varies from 2 to 3 months. This forest is less disturbed now mainly represented above 8°40′N. at present it is mostly in the fragmented form or in a highly degraded where in *Dipterocarpus indicus* that has become very rare due to overexploitation. In the eastern side of the southern Western Ghats, the forest type is restricted to moist valleys in Tambraparni river (east of Agasthyamalai), at an elevation 450 to 750 m. This region is more or less as a transitional stage to medium elevation type. The characteristic endemic species are represented in this region namely *Palaquium bourdillonii, Diospyros barberi* and *Memecylon subramanii* in the lower strata of the forest. These forest community have been identified under *Dipterocarpus-Prioria-Strombosia* type are both occurring in humid areas. One group is characteristically by the abundance of local endemics (e.g. *Hopea racophloea*), and the other forest corresponds to a specific ecosystem adapted usually to water logged areas.

a. *Hopea racophloea–Humboldtia decurrens type:* This forest community is composed by *Hopea racophloea* and *Humboldtia decurrens*. Other common trees are found in intermixed canopy: *Artocarpus gomezianus, Cynometra* sp., *Ficus beddomei, Holigarna nigra, Otonephelium stipulaceum* and *Vateria indica. Poeciloneuron indicum* has been occupied rarely in patches. This forest community is confined only in humid valleys of Kallar and Shendurni rivers, located west of Agasthyamalai zone.

b. *Myristica swamps*: The swamps are water logged muddy flat terrains throughout the year. The soil types are generally sandy loam in this region. The unique feature of this ecosystem found the abundance of Myristicaceae, especially with two species such as *Gymnacranthera canarica* and *Myristica fatua* var. *magnifica*. They produce numerous stilt roots and pneumatophores as in mangroves. Only a few such swamps in the entire Western Ghats near Kulathupuzha in Travancore, rare patches of relatively well preserved *Myristica* swamps can be observed. Along with Myristicaceae members, other riparian woody species represented such as *Carallia brachiata, Elaeocarpus tuberculatus, Humboldita* sp., *Lophopetalum wightianum, Lagerstroemia speciosa, Syzygium montana,* and *Tetrameles nudiflora.*

c. *Dipterocarpus indicus-Diptercarpus bourdillonii–Strombosia ceylanica type*: *Dipterocarpus indicus* and *Stromosia ceylanica* are characteristically found in this forest. Here *Prioria pinnata* has been replaced by the dipterocarp *Dipterocarpus bourdillonii.* The dipterocarps are also again sporadic patchy due to overexploitation. They generally distributed at lower elevations (<600 m) and it is the most threatened species in southern Western Ghats. Usually, this forest community is now encountered only along streams sides and in some inaccessible valleys.

These forests were encountered for many species of hardwoods and softwoods for several decades. Several of their important species are *Vateria indica, Palaquium ellipticum, Calophyllum polyanthum, Otonephelium stipulaceum, Donella lanceolata, Dipterocarpus indicus, Semecarpus auriculatus* and *Poeciloneuron indicum.* However, several other species in the forest are having wider ecological amplitude *viz. Artocarpus gomezianus, Antiaris toxicaria, Bombax ceiba, Monoon fragrans* and *Pterygota thwaitesii.* This forest covers wide area between 9°N and 11°N in the Western Ghats. This forest community is found in north of Ariankavu Pass, and it is highly discontinuous. The continuous stretch of *Dipterocarpus indicus-Dipterocarpus bourdillonii–Strombosia ceylanica* forest between 9°10′N and 9°35′N, and other smaller fragments of secondary moist deciduous forest and in degraded stages enclaved by either thickets or savannas, where intermixed with several other deciduous tree species.

Medium elevation types (800 - 1450 m): The medium elevation forests may be comprised between 800 to 1450 m, or lower elevation (650 m) depending on local climatic variations and exposure. These forests are structurally similar to low elevation ones but varies with canopy reaching 35-45 m with four structural layers. In the upper elevation limit, these forests have become stunted with two to three layers canopy (<18 m). They varied floristically from the low elevation types by distribution of certain indicator tree species or the variation in relative abundance of species. The floristic composition of the zone showed that *Cullenia exarillata* is a good indicator of the elevation and its range usually from 700 to 1400 m, very rarely below to 550 m in occurrence of moist valleys. *Mesua ferrea* and *Palaquium ellipticum* are well representing this forest and also occur at lower elevation but with a fewer frequency. In the south of the Ariankavu Pass, the original floristic composition is replaced by locally colonized tree species: *Gluta travancorica* and *Calophyllum austroindicum*. In other areas of southern Ghats has been identified specifically with the presence of *Nageia wallichiana* and *Diospyros atrata* in medium elevation.

a. *Cullenia exarillata – Mesua ferrea – Palaquium ellipticum – Gluta travancorica* type: This unique forest type is located only in Agasthyamalai and Mahendragiri regions and south of the Ariankavu Pass between 8° 20′N and 8° 50'N. Some tree species were rarely found in the medium elevation type of further north of the region. The major tree species are *Atuna travancorica, Calophyllum austroindicum, Diospyros barberi, Garcinia rubroechinata, Garcinia imbertii* and *Garcinia travancorica* at top canopy and subcanopy and lower canopy with *Phlogacanthus grandis, Goniothalamus rhynchantherus, Memecylon subramanii, Memecylon gracile, Octotropis travancorica, Popowia beddomeana* and *Monosis travancorica*.

b. *Nageia wallichiana* facies: This composition is found by the abundance of some indigenous Gymnosperm tree species in South India, specifically *Nageia wallichiana*. This species distribution is confined only in eastern side of the Western Ghats between 8° 20′N and 9° 30′N and above 1000 m. Other Angiosperm tree species were characteristically found in this type: *Actinodaphne campanulata, Aglaia bourdillonii, Diospyros atrata, Elaeocarpus venustus, Eugenia floccosa,* and *Syzygium microphyllum*. An endemic palm *Bentinckia condapanna* is usually found along the margins of the valleys near the cliffs.

a. *Cullenia exarillata – Mesua ferrea – Palaquium ellipticum* type: This forest type is found along the southern Western Ghats between the Ariankavu Pass and the Brahmagiri ghat, and north of the Palghat Gap (Pascal, 1982). The lower limit of the forest type locally varies from 550 to 700 m elevation and in the eastern side established only above 800 m. This tree species composition has been found along the southern part of the Periyar plateau and Elamalai region. Most part of the forest areas have been converted into cardamom plantations. The surrounding of Idukki dam of the northern part of Periyar plateau, these forests are highly fragmented and established secondary forests like thickets and savannas. Other tree species are common as canopy trees or emergent that are *Aglaia lawii, Cinnamomum keralaense, Dimocarpus longan, Diospyros sylvatica, Drypetes venusta, Litsea oleoides* and *Syzygium gardneri.*

There are species with second and third layer with wide distribution such species like *Agrostistachys borneensis, Diospyros paniculata, Homalium travancoricum, Myristica dactyloides,* and *Discospermum apiocarpum.* Other common tree species found in the elevation are *Bhesa indica, Diospyros nilagirica, Drypetes venusta, Litsea bourdillonii, Litsea keralana,* and *Semecarpus travancoricus.* This forest composition may extend up to the Palghat gap. Beyond the Palghat Gap, the vegetation changed somewhat with other canopy trees and understoried shrubs. The understorey vegetation constitutes *Ardisia pauciflora, Goniothalamus wightii, Tabernaemontana gamblei, Psychotria anamalayana,* and *Lasianthus jackianus. Nageia wallichiana* is very rarely distributed between the Kottaimalai and Devarmalai.

High elevation type (1400-1800 m): *Bhesa indica – Gomphandra coriacea – Litsea* spp. type: In this forest type is characterizable by the major canopy of *Bhesa indica* above 1350 m especially on the exposed terrain. The abundant lower storied tree species is *Gomphandra coriacea* because it was used to name the forest type, some part of the forest dominates the lower storey by different other species of *Litsea.*

This type is established along the high elevation eastern ridges of the Ghats in between Kottaimalai and Devarmalai and further extends to Megamalai and Cardamom hills. Several species in the lower elevation forests disappear for example *Cullenia exarillata, Palaquium ellipticum, Diospyros* spp. and *Agrostistachys borneensis.* The dominant family is Annonaceae at lower strata in both low and medium elevations, but they are completely absent at high elevations. Some other tree species like *Gomphandra coriacea, Hydnocarpus alpinus* are less important at lower elevations, but they are much significant at higher elevations. Several other tree species are *Heptapleurum capitatum, Mastixia arborea, Archidendron clypearia, Cocculus laurifolius, Acronychia pedunculata, Isonandra lanceolata, Meliosma simplicifolia* and *Symplocos* spp. very commonly established in this forest type. Structurally, these forests constituted two storied stunted canopy layers which seldom exceeds 15 m.

Champion and Seth (1968) revised forest classification of India concerned on the whole part of the Western Ghats which belongs the Southern tropical wet evergreen forests (sub-group 1A) and they subdivided into two groups depending on the exposure to wind, edaphic and climatic conditions. The West coast tropical evergreen forests (1A/C4) and the Southern hilltop tropical evergreen forests (1A/C3). These forest group 1A/C4 extends from the extreme southern part of India up to in Maharashtra. The lower plains and foot hills at the sea level established the two medium elevational forest types. All these forests are growing under the more unfavourable corresponds rather than the group 1A/C3. In higher altitude, there were divided into two major categories around 1500 m: Montane subtropical forests (sub-group 8A Southern subtropical broad-leaved hill forests) and Montane temperate forests (sub-group 11A Southern montane wet temperate forests). Below the elevation of 2000 m, the sub-group 11A is found therefore limited and it has not been individualized in this classification due to specific indicator tree species replaced by other local population.

Chandrasekharan (1962 b, c, d) has been proposed a detailed classification based on plant communities of the forest types in Kerala. He separated the low-level evergreen forests from the high-level forests at about 450 m msl. It will extend from 450 m to 1050 m. and above 900 m the low tropical ghats evergreen forests starts to appear on the upper slopes to top of ghat ridges. In between 1050-1200 m and 1500-1650 m established the montane sub-tropical forests, and above 1500 m elevation occurs the montane temperate forests. He has identified that there was no marked zonal variation between the forest types, as they were overlapping directly with one another.

***Tropical Semievergreen forests*:** These forests form a bridge between wet evergreen and moist-deciduous forests; usually occur as a narrow strip between 450 and 1200 m elevations in southern Western Ghats and greatly altered due to biotic influences. This forest is composed of evergreen species as lower storey and deciduous components as dominants. Trees such as *Haldina cordifolia, Albizia odoratissima, Artocarpus hirsutus, Emblica officinalis, Hopea parviflora, Holoptelea integrifolia, Hydnocarpus alpinus*; shrubs and stragglers or climbers including *Clerodendrum, Glycosmis, Strobilanthes, Entada, Strychnos* and *Dioscorea* are commonly occurring in these forests.

Of the three main semievergreen forests recognized in the southern Western Ghats such as West coast semievergreen forests which are found along the banks and rivers and streams adjoining evergreen forests in Kerala. These are confined to the area rich in alluvial soils. This type has a mixture of both evergreen and deciduous trees as it is an intermediate between these two types. The stratification in this forest type is not as marked as in the evergreen types. However, two storey can be easily discernible. Champion and Seth (1968) considered it to be a climatic climax while Chandrasekaran (1962 a,b,c) and others considered it to be a transitional stage with much biotic interference.

The dominant elements of the first storey are *Ailanthus triphysa, Alstonia scholaris, Antiaris toxicaria, Artocarpus heterophyllus, A. hirsutus, Baccaurea courtallensis, Bombax ceiba, Buchanania lanceolata, C. inophyllum, Carallia brachiata, Chukrasia tabularis, Dimocarpus longan, Diospyros buxifolia, Elaeocarpus tuberculatus, Grewia tiliifolia, Holigarna arnottiana, Hopea parviflora, H. ponga, Hydnocarpus alpinus, H. pentandra, Humboldtia vahliana, Ixora brachiata, I. nigricans, Knema attenuata, Lagerstroemia microcarpa, L. speciosa, Madhuca nerifolia, Mangifera indica, Mastixia arborea, Mallotus nudiflorus, Nothopegia travancorica, Otonephelium stipulaceum, Pterocarpus marsupium, Schleichera oleosa, Sterculia guttata, Syzygium gardneri, Tamilnadia uliginosa, Terminalia bellirica, T. paniculata, Vateria indica, Vitex altissima, V. pinnata* and *Xylia xylocarpa.*

The lower storey is formed by smaller trees, shrubs and undershrubs like *Aglaia edulis, A. lawii, Agrostistachys borneensis, Arenga wightii, Bischofia javanica, Cinnamomum verum, Clerodendrum infortunatum, Diospyros montana, Elaeocarpus serratus, Flacourtia montana, Flemingia semialata, F. strobilifera, Ficus nervosa, Homonoia riparia, Mallotus philippensis, Saraca asoca, Xanthophyllum flavescens,* etc. Common climbers are *Abrus melanospermus, Adenia hondala, Senegalia pennata, Senegalia rugata, Anamirta cocculus, Anodendron parviflorum, Aristolochia indica, Spatholobus parviflorus, Guilandina bonduc, Getonia floribunda, Combretum latifolium, Elaeagnus indica, Entada pursaetha, Gouania microcarpa, Mezoneuron cucullatum, Cissus repanda, Olax imbricata, Salacia oblonga, Strychnos colubrina, S. minor, Tiliacora acuminata, Tinospora cordifolia, Uvaria narum,* etc. along with some of those found in the evergreen type. More common herbs found in the ground cover are *Centotheca latifolia, Costus speciosus, Geophila repens, Gymnostachyum febrifugum, Mussaenda frondosa, Ophiorrhiza mungos* and *Spodiopogon rhizophorus.*

***Tropical Moist Deciduous forests*:** These forests established between semi-evergreen and dry deciduous forests at an elevation of 500 m altitude. Usually these are located in the dampy valleys or in moist habitats with shallow or porous soil. Deciduous trees along with some evergreen ground vegetation including climbers and canes are common characteristic feature yet the proportion of teak is very limited in these forests. *Hardwickia binata, Pterocarpus marsupium, Schleichera oleosa, Terminalia anogeissiana* and *Terminalia bellerica* are some of the common species occurring in these forests. Commonly found in Papanasam hills, Kalakad, Periyakulam, Gundar valley and lower elevations of Kodaikkanal.

The moist deciduous forests of Western Ghats particularly bearing teak are variously classified as moist, very moist and slightly moist teak forests. The other two types are the southern moist mixed deciduous forests and southern secondary moist mixed deciduous forests. These two forests are considered together as there is little difference in the floristic composition. The main differences are degree of degradation.

Southern moist mixed deciduous forests and Southern secondary moist mixed deciduous forests: These forests occur at lower elevation (500-900 m) receiving less rainfall (250-350 cm). They have a mixed composition with a few evergreen trees as well. The characteristic composition found is *Tectona–Dillenia–Lagerstroemia–Terminalia* series (Gadgil and Meher-Homji, 1986b). The canopy is rather open and defoliation beings during December. These forests have the following dominant trees: *Achronychia pedunculata, Albizia amara, A. lebbeck, A. odoratissima, A. procera, Alstonia scholaris, Bombax ceiba, B. insigne, Bridelia retusa, Buchanania lanzan, Careya arborea, Chukrasia tabularis, Dalbergia latifolia, D. sissoides, Dillenia pentagyna, Adina cordifolia, Hymenodictyon obovatum, H. orixense, Lagerstroemia microcarpa, L. speciosa, Lannea coromandelica, Macaranga peltata, Mastixia arborea, Melia dubia, Mitragyna parvifolia, M. tubulosa, Olea dioica, Pongamia pinnata, Pterocarpus marsupium, Pterosepermum reticulatum, Radermachera xylocarpa, Scleropyrum pentandrum, Stereospermum colais, Tectona grandis, Terminalia anogeissiana, Terminalia auriculata, T. bellerica, Tetrameles nudiflora, Vitex altissima, Xylia xylocarpa*, etc. Scattered amidst dense grasses with *Phyllanthus emblica* are found.

The middle layer of the forests constitutes of *Piliostigma malabaricum, B. racemosa, Bridelia retusa, Buchanania lanzan, Butea monosperma, Careya arborea, Cipadessa baccifera, Cassia fistula, Clausena austroindica, Catunaregum spinosa, Cleistanthus collinus, Cycas circinalis, Dalbergia sissoo, Eriolaena quinquelocularis, Ficus callosa, F. exasperata, F. hispida, F. racemosa, Garuga pinnata, Gmelina arborea, Grewia tiliifolia, Macaranga peltata, Miliusa tomentosa, Kydia calycina, Mappia nimmoniana, Schleichera oleosa, Spondias pinnata, Sterculia foetida, S. guttata, S. villosa, Trema orientalis, Mallotus nudiflorus, Wrightia arborea, Zanthoxylum rhetsa, Ziziphus xylopyrus*, etc.

The lower layer forming the underground has *Abelmoschus moschatus, Abutilon indicum, Allophylus cobbe, Baliospermum solanifolium, Canthium rheedei, Clerodendrum serratum, C. infortunatum, Cynoglossum zeylanicum, Ecbolium viride, Embelia tsjerium-cottam, Flemingia strobilifera, Gomphostemma heyneanum,*

Gardenia gummifera, Glycosmis pentaphylla, Helicteres isora, and several members of Acanthaceae, Orchidaceae and ferns. The frequent climbers are *Asparagus gonoclados, Spatholobus parviflorus, Guilandina bonduc, Getonia floribunda, Cynanchum callialatum, Capparis rheedei, Combretum albidum, Connarus wightii, Dalbergia horrida, D. volubilis, Dioscorea pentaphylla, D. tomentosa, D. wallichii, Diploclisia glaucescens, Erythropalum scandens, Jasminum azoricum, Naravelia zeylanica, Senegalia pennata, Senegalia torta, Tiliacora acuminata,* etc. The ground floor is covered with *Apluda mutica, Bothriochloa pseudoischaemum, Brachiaria distachyos, Centotheca lappacea, Ottochloa nodosa, Panicum trypheron, Rottboellia exaltata,* etc. The understorey is of evergreen type in nature giving the forest as a whole a more or less evergreen appearance in most part of the year. Species like *Tectona grandis, Terminalia* spp., *Garcinia* spp., *Mallotus* spp., *Macaranga peltata, Macaranga indica, Artocarpus heterophyllus, Hopea parviflora, Ixora* spp., *Pavetta indica, Xylia xylocarpa* and *Tabernaemontana heyneana* dominate the stand. These forests are rich in ground flora. The stand is dominated by members of Combretaceae such as *Terminalia tomentosa* and *T. paniculata* especially in Ghats of Karnataka.

Moist teak forests: These are found in lower elevations in the Nilambur and Attapadi valleys of Kerala, where the rainfall is 2000-3000 mm. Besides teak, these forests may also have *Albizia odoratissima, Bombax ceiba, Butea monosperma, Careya arborea, Cassia fistula, Catunaregam spinosa, Cleistanthus collinus, Dalbergia latifolia, D. sissodes, Grewia tiliifolia, Holarrhena pubescens, Hydnocarpus pentandra, Kydia calycina, Lagerstroemia parviflora, Mallotus philippensis, M. tetracoccus, Mitragyna parvifolia, Pterocarpus marsupium, Radermachera xylocarpa, Terminalia anogeissiana,* etc. The floor is covered by gregarious growth of grasses, the dominant ones being species of *Cymbopogon* and *Themeda*. Amidst grasses may be present some herbs and subshrubs like *Alpinia malaccensis, Cipadessa baccifera, Clerodendrum infortunatum, Curcuma angustifolia, Helicteres isora, Leea indica, Solanum erianthum,* etc. Large portions of these forests have been occupied by exotic weeds like *Lantana camara* and *Chromolaena odorata*.

Xylia xylocarpa dominant lateritic deciduous forests: This type of forests is seen in lower ranges and lateritic soil in sufficient depth and is an edaphic climax type dominated by *Xylia xylocarpa*. Other species frequently seen also include *Canthium parviflorum, Cassia fistula, Catunaregum spinosa, Strychnos nux-vomica* and *Terminalia paniculata*.

Low-lying marshy deciduous forests: These types of vegetation seen in northern parts of Kerala (Sivarajan and Mathew, 1997), such areas usually dry up during summer leaving the floor with the top soil contains large proportion of clay. These areas are unsuitable for farming, hence where dominated by grasses and weeds forming grazing grounds for cattle. A few trees of *Cordia wallichii, Adina cordifolia, Tamilnadia uliginosa* and *Terminalia alata* may be seen along with hydrophilous species in the undergrowth. These are *Ardisia solanacea, Artanema longifolia, Crotalaria quinquefolia, Limnophila repens, Ludwigia perennis, Rotala indica*, species of *Cyperus* and *Fimbristylis*, grasses like *Brachiaria repens, Cyrtococcum trigonum, Digitaria longiflora, Echinochloa crus-galli, Eragrostis atrovirens, Isachne globosa, I. miliacea, Ischaemum rugosum, Panicum paludosum, P. psilopodium, Paspalidium flavidum, Rottboellia exaltata, Sacciolepia indica*, etc.

Swamp forests: The characteristic forest types occur in low level-hill valleys of the southern Western Ghats as *Myristica* swamps. The *Myristica* swamps was first reported by Krishnamoorthy (1960) from the Travancore of the southern Western Ghats. This forest type is restricted in the valleys of Shendurney, Kulathupuzha and Anchal forest ranges in Kerala. Champion and Seth (1968) categorized the forest vegetation as tropical fresh water swamp forests (4C/FS1). Pascal (1988) and Rodgers and Panwar (1992) have described the special vegetation as most critically needed for conservation. In the northern Ghats, swamps have also been reported from Uttara Kannada of Karnataka (Chandran *et al.*, 1999) and Satari region in Goa (Santhakumaran *et al.*, 1995).

Myristica swamp forests: The characteristic forest with chief composition of *Myristica magnifica, Gymnacranthera farquhariana* and other *Myristica* species occasionally found that are *Myristica malabarica* and *Knema attenuata*. They formed characteristically pneumatophores or breathing roots, which are specialized survival features of trees in waterlogged conditions. They also produced superficial lateral roots emerged out the air and loop back into the muddy soil which are called breathing roots. The intermixed undergrowth of the forests are spiny plants such as *Pandanus* sp. and *Calamus* sp. and along the herbs like *Alpinia malaccensis, Legenandra ovata, Phrynium* sp., *Codariocalyx motorius, Selaginella* sp., *Ochlandra* sp., *Gnetum* sp., *Helicteres* sp., *Bauhinia* sp., and rare occurrence of creepers as *Chilocarpus* sp., *Kunstelaria* sp. and *Piper* sp. (Senthilkumar *et al.*, 2014).

Myristica swamps are highly restricted, vulnerable and fragmented ecosystems in the Western Ghats. These swamps cover up less than 0.01% of the total forest area of the Western Ghats. The biodiversity of the unique ecosystem is solely depending on the existence of the forest type. The plant communities varied in degrees of environmental factors and species composition differed in specific niches. Therefore, this forest type is very important for conservation because the vegetation composition of the special ecosystem is composed of Myristicaceae trees along with narrow endemic tree species like *Lophopetalum wightianum, Vateria indica, Holigarna beddommi, Semecarpus auriculata, Syzygium stocksii,* etc. On the margin of the forests are seen *Humboldtia vahliana, Hopea parviflora, Machilus glaucescens, Vateria indica,* etc. with sometimes gregarious growth of *Strobilanthes* spp.

Tropical riverine forests: Several rivers and their tributaries together form a specialized habitat for a plant community remarkably similar in appearance but a totally dissimilar taxonomic affinity. Trees common to the habitat are *Agrostistachys borneensis, Calophyllum apetalum, Cleidion javanicum, Ficus heterophylla, Garcinia wightii, Helicia nilagirica, Holigarna arnottiana, Homonoia riparia, Hydnocarpus alpinus, H. pentandra, Lagerstroemia microcarpa, L. speciosa, Lophopetalum wightianum, Madhuca neriifolia, Neolamarckia cadamba, Ochreinauclea missionis, Ochlandra* spp., *Rapanea wightiana, Syzygium hemisphericum, S. heynanum, S. occidentale, Tetrameles nudiflora, Vateria indica, Vitex alata* and *V. leucoxylon.*

Wet bamboo brakes: This type occurs in places wherever there is a brake in the canopy such as slides of streams, higher slopes or gaps created by felling. Being very aggressive, reeds colonize very rapidly and form large populations especially in gaps that are partially shaded. Common such reeds are *Ochlandra ebracteata, O. scriptoria, O. travancorica* and *O. wightii.*

Low level grasslands: The grasslands found below 1000 m are usually classified as low-level grasslands. These include secondary grasslands of the lower elevations also. They are scattered amidst moist deciduous and dry deciduous forests based on which they are divided as (a) Dry and (b) Wet grasslands.

Dry grasslands: They are limited in occurrence. Northern part particularly Kasargod, Kozhikode, parts of Palakaad and Wayanad Districts have dry rocky areas which are covered by grasses intermixed with scrub jungles and dry deciduous forests. Grasses frequent here are *Arundinella cannanorica, A. mesophylla, Bhidea burnsiana, Danthonidium gammiei, Dimeria bialata, Heteropogon contortus, Ischaemum indicum* and *Pseudanthistiria heteroclita.*

Vast areas in the dry and moist deciduous forests are covered by grasses which are formed consequent to biotic interference like fire, grazing and indiscriminate felling of trees. In such areas are found *Arundinella ciliata, A. mesophylla, Chrysopogon orientalis, C. zeylanicus, Cymbopogon flexuosus, Digitaria ciliaris, Eragrostis tenuifolia, Eulalia trispicata, Heteropogon contortus, Ischaemum timorense,* etc. Gregarious amidst these grasses occur the fire-resistant *Phoenix loureiri.* More common shrubs associated with these grasses are *Hedyotis purpurascens, Uraria rufescens* and herbaceous species of *Biophytum, Crotalaria, Cyanotis, Fimbristylis, Leucas, Phyllanthus, Rhynchosia, Striga* and *Vernonia.*

Wet grasslands: These are characteristics of low-lying wetlands of the coastal regions particularly in Alappuzha district of Kerala. The common grasses adapted to this situation are *Brachiaria mutica, Imperata cylindrica, Eriochloa procera, Paspalum conjugatum, Paspalidium geminatum, Panicum repens,* etc. Grasses found in areas with tidal action adapted to saline conditions are *Sporobolus virginicus, Paspalum scrobiculatum, Zoysia matrella,* etc.

Tropical dry deciduous forests: Dry deciduous forests are restricted to well-drained hill slopes or undulating terrain between 600 and 1000m elevations. Some of the characteristic species include *Tectona grandis, Bridelia retusa, Pterocarpus marsupium, Chloroxylon swietenia, Erythroxylon monogynum, Gardenia gummifera, Buchanania lanzan, Grewia* spp., *Albizia amara, A. lebbeck, Wrightia tinctoria, Morinda coreia, Phoenix sylvestris,* etc. These forests are distributed widely in Tamil Nadu especially lower elevations of Kodaikkanal and Kerala of Chinnar, Palakkad and Kannur.

Southern dry mixed deciduous forests: They occur in drier areas of southern Western Ghats and it is characterized by the light understorey canopies, these forests occur at low elevations between 300 and 900 m and receive less rainfall, 100 – 200 cm. the undergrowth is dense as light penetrates the canopy sufficiently but epiphytes and ferns are rare. The common association here is *Terminalia –Terminalia anogeissiana–Tectona grandis* series. Dominant trees are *Albizia amara, Terminalia anogeissiana, Bauhinia* spp., *Buchanania lanzan, Butea monosperma, Commiphora caudata, Dalbergia latifolia, Dillenia pentagyna, Ehretia laevis, Garuga pinnata, Gmelina arborea, Adina cordifolia, Phyllanthus emblica, Pterocarpus marsupium, Semecarpus anacardium, Terminalia chebula,* etc.

The smaller trees and shrubs are *Benkara malabarica, Carissa spinarum, Chassalia curviflora, Desmodium gangeticum, D. laxiflorum, D. triangulare, Erythroxylum monogynum, Holarrhena pubescens, Leea indica, Phyllanthus polyphyllus, Flueggea leucopyrus, Senegalia pennata, Solanum americanum, S.*

torvum, Writghtia tinctoria and *Zingiber* spp. The herbaceous flora after the monsoon may have *Corchorus aestuans, Crotalaria evolvuloides, C. heyneana, C. nana, Desmodium heterocarpum, D. triflorum, Indigofera trifoliata, Mimosa pudica, Sida acuta, S. alnifolia, S. rhombifolia, Rhychosia densiflora* and grasses like *Apluda mutica, Eragrostis japonica, E. tenella, Oplismenus compositus* and *Themeda triandra.* Gregarious growth of the bamboo *Dendrocalamus strictus* is a distinguishing feature in the forest type.

The dominant species in the Ghat area of Karnataka shows teak bearing forests comprising *Tectona grandis, Terminalia anogeissina* and *Terminalia* spp., and in the non-teak forests, *T. anogeissiana* and *Terminalia* spp., *Diospyros* spp. and *Sterculia* spp. are dominant. Important timber species like *T. grandis, Lagerstroemia microcarpa, Terminalia* spp., *Dalbergia latifolia* and *Pterocarpus marsupium* are abundant in these forests. Bamboo *(Dendrocalamus strictus* and *Bambusa arundinacea)* generally present along the streams and watercourses are more towards the southern region and less in northern region of the Ghats. These dry deciduous forests are abundant with shrubs like *Gardenia* spp., *Catunaregam dumetorum, Microcos paniculata, Flacourtia indica, Ziziphus* spp., *Capparis* spp., *Canthium* spp. and *Carissa* spp. The rainy season results in profuse ground flora. Grasses mainly constitute these annuals. Climbers like *Hemidesmus indicus, Ichnocarpus frutescens, Tinospora cordifolia, Vincetoxicum indicum, Cocculus hirsutus, Aristolochia* spp., *Decalepis hamiltonii* and *Cryptolepis buchananii* are common in these forests. Sandal wood trees are found in the dry deciduous and scrub forests of Mysore, Hassan and Chikmagaluru districts.

Tropical thorn forests: These forests are also known as scrub jungles occurring in the dry plains and low elevations or foot hills of Western Ghats, particularly in Dindigul, Madurai and Tirunelveli. The annual rainfall ranges from 500 to 700 mm. Tropical thorn forests in Tamil Nadu are the most exposed forests to human disturbances. Some of the dominant species of these forests include *Albizia amara, Azadirachta indica, Butea monosperma, Cassia fistula, Cordia dichotoma, Diopyros candolleana, Holarrhena pubcescens, Limonia acidissima, Senegalia chundra, Strychnoa nux-vomica, Terminalia anogeissiana, Tectona grandis, Terminalia bellerica, Zizyphus mauritiana, Z. xylopyrus. Z. oenoplia,* etc. The common stragglers include *Capparis zeylanica, Carissa spinarum, Flacourtia indica, Maytenus emarginata* and *Toddalia asiatica*. More common climbers are *Cardiospermum halicacabum, Ceropegia candelabrum, C. juncea, Cissus quadrangularis, C. repanda, Cyclea peltata, Hemidesmus indicus, Ichnocarpus* *frutescens, Jasminum angustifolium, Vicetoxicum indicum,* etc. The area being dry for most part of the year, the ground is almost barren.

After the monsoon it is covered by herbs like *Achyranthes aspera, Allmania nodiflora, Apluda mutica, Aristida setacea, Blumea mollis* and species of *Barleria, Blepharis, Cleome, Cymbopogon, Glinus, Indigofera, Leucas, Mollugo* and *Hedyotis* and a few shrubs like *Canthium coromandelicum, Catunaregum spinosa, Pavetta indica, Ixora coccinea, Pseudaidia speciosa,* etc. The other dominant species of these scrub lands in Karnataka are species of *Albizia, Pterolobium, Canthium, Capparis, Cadaba* and *Ziziphus.* Shrubs like *Scutia myrtina, Maytenus emarginata, M. heyneana, Dodonaea viscosa, Cipadessa baccifera, Terenna asiatica* and *Gmelina asiatica* are also common.

Tropical dry evergreen forests: Tropical dry evergreen forests are usually confined to coastal regions of Tamil Nadu and patches forests in Kerala coasts. These forests occur along the sandy beaches, coastal plains with isolated hillocks of nearby coastal area usually contains red lateritic soil. The forests are dense towards the west and these thickets thin out as scrub land mixed with open grasslands towards the coast. Notable trees include *Ficus benghalensis, Syzygium cumini* and *Pongamia pinnata*. However, the typical dry evergreen forests are categorized under sub-type 7/C1, which are constituted by common trees like *Manilkara hexandra, Memecylon umbellatum, Diospyros ferrea, Chloroxylon swietenia, Albizia amara* and other evergreen species. Further, the sub-type 7/DS1 were described under tropical dry evergreen scrub, where dominated trees are *Memecylon edule, Ziziphus xylopyrus, Dichrostachys cinerea, Psydrax dicoccos, Carissa spinarum, Albizia amara, Buchanania axillaris, Dodonaea viscosa* and other associate species. The phyto-sociological composition of the tropical dry evergreen forest vegetation were also mentioned by several other authors as *Manilkara hexandra* series, generally *Manilkara hexandra – Drypetes sepiaria – Chloroxylon swietenia – Memecylon umbellatum* series, the *Manilkara hexandra – Memecylon umbellatum – Drypetes sepiaria –Pterospermum suberifolium – Carmona microphylla* facies of the *Albizia amara* community, the *Manilkara hexandra – Chloroxylon swietenia* vegetation type within the *Albizia amara* zone and the *Memecylon edule – Atalantia monophylla* series. This forest is floristically varied by a good representation of unique and preferential species, exclusively confined to the vegetation type. The preferential woody species are *Atalantia monophylla, Lannea coromandelica, Lepisanthes tetraphylla, Manilkara hexandra, Memecylon umbellatum, Psydrax dicoccos, Pamburus missionis, Sapindus* *emarginatus, Putranjiva roxburghii, Dolichandrone falcata, Buchanania axillaris,* etc. The common climbers and woody liana are *Combretum albidum, Ventilago maderaspatana, Grewia orientalis* and *Hugonia mystax.*

Southernmost tip of Western Ghats with some remnants patches of tropical dry evergreen forest occurs with core species of *Albizia amara, A. lebbeck, Atalantia monophylla, Azadirachta indica, Cassia fistula, Chionanthus mala-elengi, Crateva magna, Dalbergia lanceolaria, Diospyros ebenum, D. ferrea, Drypetes sepiaria, Lannea coromandelica, Lepisanthes tetraphylla, Manilkara hexandra, Psydrax dicoccos, Pterospermum suberifolium, Sapindus emarginata* and *Syzygium cumini.* The common shrubs are *Benkara malabarica, Cadaba fruticosa, Canthium parviflorum, Capparis brevispina, Carissa spinarum, Carmona retusa, Ecbolium viride, Flacourtia indica, Glycosmis mauritiana, Gmelina asiatica, Ixora pavetta, Memecylon umbellatum, Opuntia dillenii, Phyllanthus reticulatus, Catunaregam spinosa, Flueggea leucopyrus* and *Tarenna asiatica.* Climbers and stragglers are *Adenia wightiana, Allophylus cobbe, Asparagus racemosus, Capparis zeylanica, Cassytha filiformis, Cayratia pedata, Cissus vitiginea, Cissus quadrangularis, Cocculus hirsutus, Combretum albidum, Dioscorea oppositifolia, Grewia flavescens, Gymnema sylvestre, Hugonia mystax, Jasminum angustifolium, Passiflora foetida, Reissantia indica, Rivea hypocrateriformis, Cynanchum sarcomedium, Scutia myrtina, Senegalia intsia, Solanum trilobatum, Solena amplexicaulis, Strychnos minor, Symphorema involucratum, Vincetoxicum indicum, Ventilago maderaspatana, Zanthoxylum asiaticum,* and *Ziziphus oenoplia.* In between found the twiners like *Abrus precatorius, Aristolochia indica, Canavalia cathartica, Ichnocarpus frutescens* and *Ipomoea sagittifolia.* Occasionally found the *Borassus flabellifer* in the sandy vegetation. Some common perennial succulent and tuberous plants are *Caralluma adscendens, Curculigo orchioides, Gloriosa superba, Sansevieria roxburghiana* and *Theriophonum fischeri.*

Subtropical broad leaved hill forests: These forests are also referred as "stunted rain forests" found between 1100 and 1800 m elevations of Nilgiris and Kanyakumari. Though the attributes resemble tropical rain forest the components are not as dense and the trees are smaller with less shapely trunks, often with epiphytes. Species such as *Syzygium arnottianum, S. montanum, Microtropis ovalifolia, Dillenia* sp., *Antidesma* sp. etc. commonly occur in these forests.

***Subtropical montane forests*:** These are the southern subtropical hill forests. They occur mostly on plateaus in the upper reaches of slopes in higher altitudes (1000 – 1800 m). A transitional stage between tropical wet evergreen and temperate shola forests, they are miniature sholas with a floristic composition of both these types. Though they may not be as luxuriant as the wet evergreen forests, where the trees are taller above 20-30 m when compare to wet temperate forests. The stunted trees are attributed to the high velocity of winds and less favourable conditions of soil. The stratification is not clearly marked. The dominant elements in the top storey are *Actinodaphne bourdillonii, A. bourneae, A. lanata, A. salicina, Bischofia javanica, Callicarpa tomentosa, Chionanthus courtallensis, C. mala-elengi, Celtis tetrandra, Cinnamomum macrocarpum, C. sulphuratum, C. verum, C. wightii, Cryptocarya lawsonii, C. neilgherrensis, Daphniphyllum neilgherrense, Dimocarpus longan, Elaeocarpus serratus, E. tuberculatus, Eurya nitida, Fagraea ceilanica, Ficus arnottiana, Glochidion neilgherrense, Gordonia obtusa, Ligustrum robustum, Maesa indica, Neolitsea cassia, Machilus glaucescens, Pittosporum neelgherrense, Syzygium cuminii, S. densiflorum, Toona ciliata* and *Turpinia malabarica.*

The second storey may have *Actinodaphne campanulata, Brysophyllum tetrandrum, Cinnamomum filipedicellatum, C. sulphuratum, C. travancoricum, Gomphandra tetrandra, Heptapleurum wallichianum, Hydnocarpus pentandra, Memecylon angustifolium, Pavetta brevifolia, Pittosporum napaulense, Psydrax dicoccos, Rapanea wightiana, Strobocalyx arborea, Symplocos cochinchinensis* subsp. *laurina, Ternstroemia gymnanthera, Wendlandia thyrsoidea,* etc. The third storey is formed of younger forms of the upper two storeys along with shrubs and undershrubs like *Canthium neilgherrense, Hedyotis purpurascens, H. ramarowii, H. swertioides, Knoxia heyneana, Pavetta hispidula, Psychotria nigra, Saprosma fragrans,* etc.

The ground is often covered by *Arundinella tuberculata, Elatostemma lineolatum, Strobilanthes heyneanus, S. wightianus, Pogostemon purpurascens, Polygala arillata, Viola pilosa* in addition to pteridophyte species of *Asplenium, Osmunda, Selaginella,* etc. The common epiphytes are orchids like *Aerides crispa, A. ringens, Bulbophyllum fischeri, B. tremulum, Coelogyne nervosa, Dendrobium herbaceum, Porpax nana, Pinalia polystachya, Porpax reticosa, Liparis elliptica, Oberonia santapaui, Trichoglottis tenera,* etc. *Impatiens auriculata* and certain *Peperomia* spp. are also common. More common ground orchids are *Calanthe masuca, Habenaria longicorniculata, H. plantaginea, Peristylus aristatus, P. densus, Pecteilis gigantea, Phaius luridus,* etc. *Lagerstroemia lanceolata* was the most dominant tree in southern region of Karnataka and the other dominant trees are

Elaeocarpus tuberculatus, Dimocarpus longan, Canarium strictum, Hopea glabra. It is interesting to note that a deciduous tree is the most dominant one indicating that the forests have been subjected disturbances resulting in modification of forest structure and composition (Sathish *et al.*, 2013).

The Western Ghats of Karnataka with heavy rainfall supported tropical evergreen forests. The characteristic climatic conditions provided the prolonged wet periods, and established storied canopy with poor under growth. Thick layer of decaying biomass on the ground is another feature of these forests. Trees attained heights greater than 30 m and girths exceeding 300 cm. The canopy is extremely dense and the differentiation of definite canopy layers probably does not exist. These forests are confined to the higher altitudes of the slopes and in the regions of low disturbance of the Ghats. The windward sides of the Ghats harbor more of this type of vegetation as they received rains in most part of the year. The characteristic vegetation composition made by the tree species of *Myristica, Cinnamomum, Bischofia, Dipterocarpus, Vateria, Holigarna, Hopea, Litsea, Machilus, Artocarpus, Prioria, Poeciloneuron, Dysoxylum* and *Aglaia* as the top canopy trees. The middle canopy is formed by species of *Diospyros, Tetrapilus, Humboldtia* and *Knema,* while the lower canopy is formed by species of *Garcinia, Mappia, Memecylon, Canthium* and *Euodia.* The species like *Poeciloneuron indicum, Prioria pinnata, Knema attenuata., Hopea canarensis, Dipterocarpus indicus, Vateria indica* and *Holigarna* species are endemic to the Western Ghats and are found distributed in these primary forests.

***Montane temperate forests*:** These forests are also known as "Hill shola forests" occur above 1500 m altitude. The height of the forest is relatively low. Trees are short and their branches are densely covered with lichens, mosses, ferns and epiphytes. Montane forests are usually found in patches in more sheltered areas like valleys interspersed with grasslands. These forests occur in Anamalais, Nilgiris, Palanis and Tirunelveli hills. Some of the notable species includes *Byrsophyllum tetrandrum, Hedyotis leschenaultiana, Rhododendron arboreum* subsp. *nilagiricum, Eugenia floccosa, Euphorbia santapaui, Tetrapilus polygamus, Cryptocarya stocksii* etc.

These are two types (a) Southern montane wet temperate (sholas) and (b) southern montane wet grasslands (High altitude grasslands). Both are closely associates with each other.

Southern montane wet temperate forests (sholas): These types of forests confined to depressions and moist sheltered wet grasslands, the sholas are unique type of vegetation distinct from other types by various features. They are small, isolated, compact and sharply defined evergreen woodlands scattered amidst vast stretches of grasslands at higher altitudes above 1500 m. The sholas in their best vegetation composition can be observed in the Silent vaeely adjacent to the Nilgiris and in the high Ranges. A fast-receding community (Vishnu Mittre and Gupta, 1968), the sholas are characterized by stunted trees with short boles reaching less than 15 m high. The crowns of these trees are usually dense and rounded with entire and coriaceous leaves usually reddish when young. The trunks are not buttressed and branches are crooked and densely clothed with lichens, mosses, ferns and epiphytic orchids. The stunted growth and slow spread of this vegetation is generally attributed to the high velocity of winds and unsuitable soil nearby (Adriel, 1964). A mixture of both temperate and tropical elements are found here that is another remarkable feature. The occurrence of sino-Himalayan *Rhododendron* is such forests in the High Ranges. Unlike tropical evergreen forests, this type has only two well marked storeys. The dominant trees in the top canopy are *Actinodaphne bourneae, Cinnamomum wightii, C. perrottetii, Callicarpa tomentosa, Daphniphyllum neilgherrense, Elaeocarpus tuberculatus, Euonymus indicus, Glochidion neilgherrense, Polyspora obtusa, Litsea wightiana, Melicope lunu-ankenda, Meliosma simplicifolia, Microtropis ramiflora, Neolitsea cassia, Pheobe wightii, Syzygium calophyllifolium, S. densiflorum,* etc. Along the fringes are found light-loving *Elaeocarpus munroi, E. recurvatus, Eurya nitida, Prunus zeylanica, Rhododendron arboreum* var. *nilagiricum, Rhodomyrtus tomentosa, Symplocos cochinchinensis* subsp. *laurina* and *Ternstroemia gymnanthera.*

The second storey is with smaller trees like *Actinodaphne salicina, Apollonias arnottii, Cryptocarya stocksii, Gomphandra coriacea, Hydnocarpus alpina, Ixora notoniana, Ligustrum lucidum, Memecylon malabaricum, Pittosporum tetraspermum, Rapanea wightiana, Heptapleurum capitatum, Symplocos racemosa, Symplocos foliosa, Symplocos pendula, Syzygium gradneri, Turpinia cochinensis, Vaccinium leschenaultii, V. neilgherrense, Viburnum erubescens, Strobocalyx arborea* and *Wendlandia thyrsoidea.*

The ground layer may have shrubs, subshrubs and herbs such as *Anotis monosperma, Canthium neilgherrense, Elatostemma lineolatum, Gnidia glauca, Hedyotis stylosa, Hypericum mysorense, Hypochaeris glabra, Laportea bulbifera, Lasianthus cinereus, L. jackianus, Linocera ligustrina, Maesa indica, Memecylon avescens, Micromeria biflora, Polygala arillata, Psychotria congesta, Sarcocca saligna,*

Senecio lavandulifolius, Rotala rotundifolia, Rungia latior, Viola pilosa, etc. Lianas are not common. The common climbers are *Clematis wightiana, Elaeagnus indica, E. kologa, Linocera leschenaultia, Piper* spp., *Pentapanax leschenaultia, Rosa leschenaultiana, Rubia cordifolia, Rubus racemosus, Heptapleurum venulosum, Heptapleurum wallichii, Senecio candicans, S. corymbosa, S. intermedius, Vincetoxicum pauciflora*, etc. Parasites are rare but epiphytes are abundant. More common epiphytes are *Aerides ringens, Bulbophyllum tremulum, Bulbophyllum sterile, Chiloschista pusillum, Coelogyne mossiae, C. odoratissima, Dendrobium barbatulum, D. herbaceum, D. heterocarpum, D. heyneanum, Porpax dalzellii, P. polystachya, Liparis viridiflora, Oberonia proudlockii, O. santapaui, Sirhookera latifolia, Bulbophyllum bonaccordensis, Bulbophyllum stocksii* and *Vanda spathulata. Hoya ovalifolia, Peperomia tetraphylla* and ferns are also common. Mosses and lichens completely cover the branches and trunks of trees. On the fringes of water courses *Ochlandra beddomei* may be found.

The status of shola forests has been a subject of dispute. While Champion and Seth (1968) and Chandrasekharan (1962 a,b,c) considered them as temperate vegetation, but Meher-Homji (1965) treated them as subtropical montane type.

Southern montane wet grasslands *(High altitude grasslands)*: These forests associated with and dominating the shola woodlands are vast stretches of grasslands, a characteristic feature of the high mountains above 1500m. The Eravikulam National Park, Munnar and Devikulam in the High Ranges of Idukki, Silent Valley of Kerala, Avalanch of Nilgiris, Thirukurungudi range of Tirunelveli hills and Muthukuzhivayal of Kalakadu-Mundanthurai Tiger Reserve in Tamil Nadu are such extensive grasslands in southern Western Ghats. These are mixed with herbs and shrubs in varying proportions and hence called shrub savannahs. These montane grasslands have superficial resemblance to the 'patenas' of Sri Lanka (Sreekumar and Nair, 1991). The chief communities of grasses are (i) *Sehima – Dichanthium* types (Gupta *et al.*, 1967), wherein *Heteropogon contortus, Sehima nervosum* and *Dichanthium* species are the dominant elements. (ii) *Cymbopogon – Themeda* type, the dominant species are *Cymbopogon flexuosus* and *Themeda cymbaria*, (iii) *Saccharum – Imperata – Phragmites* association consisting of grasses like *Phragmites karka, Saccharum spontaneum, Imperata cylindrica, Paspalum conjugatum*, etc. A number of temperate grasses such as *Agrostis schmidii, A. zenkeri, Brachypodium sylvaticum, Bromus diandrus, Poa gamblei* along with temperate species of *Exacum, Gentiana,*

Bupleurum, Dipsacus, Ranunculus are also found. Some other common grass species in the areas are *Agrostis peninsularis, A. setosa, Andropogon polyptychus, Arthraxon castratus, Bromus ramosus, Chrysopogon zeylanicus, Cymbopogon flexuosus, Eragrostis nigra, E. tenuifolia, Indochloa oligantha, Isachne bourneorum* and *Tripogon bromoides*. The sedges found amidst these grasses are *Kyllinga cylindrica, Mariscus cyperinus, Pycreus angulatus*, etc. *Anaphalis neelgerryana, Erigeron bonariensis, Crotalaria barbata, C. formosa, C. ovalifolia, Hedyotis leschenaultiana, Helichrysum* spp., *Impatiens* spp., *Memecylon lawsonii, Osbeckia* spp., *Ranunculus peninsularis* and *Senecio* spp. are common. The ecological status of the shola-grasslands has been worked extensively by various workers (Bor, 1938; De Rosaryo, 1945; Gupta, 1960; Legris, 1960; Chandrasekharan, 1962 a,b,c; Noble, 1967).

The manifestation of these grassy vegetation cover is primarily constituted by climatic, edaphic and chiefly by latitudinal influence. Whereas in *Sehima-Dichanthium* composition is tropical, and *Dichanthium-Cenchrus-Lasiurus, Phragmites-Saccharum-Imperata* and *Themeda-Arundinella* compositions are sub-tropical. Usually, the temperate alpine grass vegetation is varied from the above types. In the altitudinal gradients separates *Themeda-Arundinella* grass cover which is confined to the northern hill ranges from *Phragmites-Saccharum-Imperata* and from *Dichanthium-Cenchrus-Lasiurus* covers, generally occurs in plains. In low altitudes, moisture level separates *Phragmites-Saccharum-Imperata* type from the *Dichanthium-Cenchrus-Lasiurus* type (relatively dry habitats).

Earlier investigations described the vegetation of these grassland types by the most ubiquitous plant species composition (Ranganathan, 1938; Meher-Homji, 1965). Generally, Botanical exploreres have categorized the shola grasslands based on dominance of certain grass communities (Gupta *et al.*, 1967; Blasco, 1970). An investigation on the ecology of Nilgiri tahr described, the tahr habitat dominated by two grass species such as *Eulalia phaeothrix* and *Dicanthium polyptychum* (Rice, 1988). There is no clear and exhaustive inventory of the plant species composition of the montane grasslands of the Western Ghats (Shetty and Vivekanadan, 1971). However, the grasslands with high rates of endemism associated with the island-like phenomena of mountain tops (Karunakaran *et al.*, 1998). The grassland communities are considered undoubtedly a high conservation priority. These grasslands are considered to be frost-related climactic climax (Meher-Homji, 1965, 1967), or considerable amount of disturbance controlled by sub-climax (Bor, 1938;

Noble 1967), or existing dual climax (Ranganathan, 1938). Some other workers considered these grassland forests are to be edaphic climax vegetation caused by differential soil moisture (Ranganathan, 1938), and there exists distinct edaphic microclimatic variations between the forests and grasslands (Jose *et al.*, 1994).

Meher-Homji (1965, 1967) was found that exhaustively on the phytogeographical study of the Western Ghats shola species, which showed distinctly two principal floristic elements, that is tropical and extra-tropical (or temperate). These typical shola species are found mostly from tropical regions of the Western Ghats and Sri Lanka, or Indo-Malayan region, while woody species established along the fringes of the shola forests and in adjacent open high-altitude grasslands supported subtropical to temperate origins (Meher-Homji, 1967). The high-altitude grasslands or montane grasslands may evoke to explain the frost-induced distributional pattern. The grassland community in order to adapt for climate specific vegetation which controlled by environmental niche and temperature (Meher-Homji, 1967). Biogeographic similarities showed that high altitude montane grasslands of the Western Ghats are more related to the Western Himalayan vegetation than to the tropical montane grasslands of Sri Lanka (Karunakaran *et al.* 1998).

Global Biodiversity Hotspot

Endemism

The nature of endemism in the Western Ghats have been discussed by many botanists in different times since the past (Chatterjee, 1940, 1962; Nayar, 1980, 1982; Rao, 1972; Subramanyam and Nayar, 1974; Ahmedulla and Nayar, 1987; Nair and Daniel, 1986; Nayar and Ahmedulla, 1984; Bhushan *et al.*, 2020). The Western Ghats are categorized only next to Himalayan ranges in having high number of endemic species. Although the Western Ghats are a part of the Gondwanaland, they protected by Arabian sea coast along the western side, Vindhya and Satpura mountain ranges on the northern side, semi-arid Deccan plateau on the eastern side and Indian Ocean on the southern end which act as barrier for migration of plants. Hence, these kind of oceanic island in supporting with large number of endemic species. The rich summits of the Western Ghats with their characteristic climates depend upon islands as regards the distribution of endemic plant species (Subramanyam and Nayar, 1974; Blasco, 1970, 1971). There are about 56 (now 60) endemic genera with around 2100 (38 %) narrow endemic species in the Peninsular India (Nayar, 1982). Among these

49 genera are monotypic. Unlike the Himalayas, most of the endemic species in the Western Ghats are palaeoendemics. In the southern Western Ghats, especially Agasthyamalai range are having richest in endemism followed by Wayanad and Anamalai hill ranges. Of the 8,080 higher plants species in the Western Ghats 1720 species of flowering plants are narrowly endemic. Nearly every third plant species are rare, threatened, and several are believed to be in the verge extinction for example *Dalbergia travancorica* and *Blepharistemma serratum*. Of these 44 genera are 'monotypic', they have represented only one species for example *Agasthiyamalaia pauciflora, Erinocarpus nimmonii, Sanjappa cynometroides, Solenocarpus indicus,* etc., In the family Orchidaceae reported approximately 100 endemic species in the Western Ghats (Singh *et al.*, 2019). Acanthaceae with 24 genera (*Andrographis, Gymnostachyum, Phlogacanthus* and *Strobilanthes,*) and Apocynaceae with 16 genera (*Decalepis, Kamettia, Seshagiria*) and 40 other families are considered as regards large number of endemic plants concerned. From the 22 aroborescent taxa having more than 5 endemic species in the Western Ghats. The evergreen tree species about 370 species (60 % of the total evergreen species) are enumerated to be endemic to the Western Ghats (Pascal, 1991).

The genus *Crotalaria is a largest genus* with about 30 % of the species which are endemic to the region. Some other genera having high degree of endemic species in the Western Ghast are *Eriocaulon* 50 species; *Ceropegia* 33 species; *Iscahemum 34 species; Sonerila* 31 species; *Syzygium* 28 species: *Memecylon* 21 species; *Habenaria* 20 species; *Dimeria* 21 species; *Bulbophyllum* 19 species; *Oberonia* 16 species; *Isachne* 15 species; *Dichanthium* 11 species. Some of the woody endemic genera are *Blepharistemma, Erinocarpus, Meteoromyrtus, Otonephelium, Poeciloneuron,* and *Symplocos*. One among the unique feature the Western Ghats is the representation of rich number of arborescent endemic genera and species.

Northern Western Ghats accounted about 181 local endemics including 4 monospecific genera (Datar and Watve, 2018). They have found that a majority of the endemic species are therophytes, which complete their life cycle in short period during monsoon. Western Ghats of Maharashtra revealed that the distribution of 159 higher plant species with 81 genera and 31 families (Mishra and Singh, 2001; Yadav and Kamble, 2008). The genus *Ceropegia* has the high number of species (17) with all other endemic genera such as *Glypochloa, Dipacdi* and *Eriocaulon* (Gaikwad *et al.*, 2014). Most of these endemic taxa restricted to small range of geographical area and are rare in occurrence. Most of the narrow endemic taxa have been recorded only by their original type collection, which could not be recollected even after repeated botanical investigation by several botanists in their habitat of occurrence in last decades. Hence, they are assessed in the category of Data Deficient (DD) by IUCN.

According to Chatterjee (1940), only 34 endemic dicotyledonous genera have been listed from Peninsular India. Rao (1972) enumerated about 164 endemic genera to the Indian floristic region, from which nearly 60 genera are distributed to Peninsular India and Sri Lanka. Nayar (1977) accounted that a total of 2,100 endemic flowering plant species represented in Peninsular India, they contributed about 32% of its flora. However, Ahmedullah and Nayar (1986) have reported the exhaustive list on the endemic Plants of Peninsular India. They have enlisted 1,940 endemic species including infraspecific taxa from Peninsular India. From the three volumes of Red Data Book of Indian Plants (Nayar and Sastry 1987, 1988, 1990) have listed 90 endemic taxa which included from the northern Western Ghats. Nayar (1996) has revised the endemism in India and reported about 2,150 endemic plants of Peninsular India. Tetali *et al.* (2000) have been recorded 439 endemic taxa of the northern Western Ghats.

Western Ghats harbor a number of endemic plants but highest number of endemic species were noted in Agasthyamalai Biosphere Reserve of southern most Western Ghats. Sholas of Western Ghats are reservoirs of endemic species which possessed unique nature of climate and ecosystems. Several pockets of shola are isolated with patches of grasslands were concentrated number of trees and herbaceous species in small patches of shola vegetation. The best example is *Elaeocarpus blascoi* at the time of exploration of the Flora of Palani hills noted only a single tree in the habitat (Matthew, 1997). This evergreen vegetation patches restore the number of endemic plants.

Monotypic genera are represented by only one species with its genus having no other closely allied species anywhere else in the world and hence they have high priority for conservation significance. There has been reported about 236 monotypic genera in India, of which 44 genera are distributed in the Western Ghats (Singh *et al.*, 2015) (Table 9.1). Rana and Ranade (2009) have enumerated a detailed account of monotypic genera in India. According to them, Poaceae with 32 monotypic genera is the largest endemic family in India followed by Fabaceae (15 monotypic genera) and Asteraceae (with 12 monotypic genera). The list of arborescent monotypic tree species given in table 2.

Table 2. Arborescent monotypic tree species of Western Ghats.

Agasthiyamalaia pauciflora (Bedd.) S. Rajkumar & Janarth.	Calophyllaceae
Bentinckia condapanna Berry	Arecaceae
Blepharistemma serratum (Dunnst.) Suresh	Rhizophoraceae
Deccania pubescens (Roth.) Triveng.	Rubiaceae
Erinocarpus nimmonii J. Graham	Malvaceae
Hardwickia binata Roxb.	Fabaceae
Prioria pinnata (Roxb. ex DC.) Breteler	Fabaceae
Meteoromyrtus wynaadensis (Bedd.) Gamble	Myrtaceae
Octotropis travancorica Bedd.	Rubiaceae
Otonephelium stipulaceum (Bedd.) Radlk.	Sapindaceae
Ougeinia oojeinensis (Roxb.) Hochr.	Fabaceae
Poeciloneuron indicum Bedd.	Calophyllaceae
Sanjappa cynometroides (Bedd.) E.R.Souza & Kiranraj	Fabaceae
Solenocarpus indicus Wight & Arn.	Anacardiaceae

Rare, endangered and threatened plants

Unique features of Western Ghats are exceptionally high degree of endemism with rich biological diversity. A large number of plants are rare, endangered and threatened due to heavy biological pressure and human interference. At least 350 globally threatened plant species occur in the Western Ghats (IUCN Red Data List, 2024). Out of 1500 narrow endemic plants of Western Ghats, 150 are classified as Vulnerable (VU), 173 are as Endangered (EN) and 36 as critically endangered (CR). About 15 plant species are possibly extinct since last decades.

Among the 5,500 seeded plants accounted from the southern Western Ghats, 2,015 species are listed as endemics (Kiran Raj *et al.*, 2003). In the Southern Western Ghats constituting southern parts of Karnataka, Kerala and southern Tamil Nadu are having richest biological diversity. There has been estimated over 4,500 flowering plants, of which 1,500 are the Western Ghats endemics which are distributed narrowly to this region. However, a precise account on the Rare, Endangered and Threatened (RET) flowering plants of the southern Western Ghats are being inadequate. A database launched by KFRI of Kerala estimated that 760 RET species distributed over 332 genera in 109 families of the southern Western Ghats (http://redplants.kfri.res.in).

The largest endemic families of Western Ghats are Acanthaceae (24 genera with 127 species), Apocynaceae (16 genera with 70 species), Asteraceae (22 genera with 71 species), Euphorbiaceae (20 genera with 54 species), Fabaceae (36 genera with 97 species), Rubiaceae (30 genera with 150 species), Poaceae (52 genera with 193 species), Orchidaceae (27 genera with 94 species) and other 113 families contributed to more than 1000 species that showed the richest endemic nature of Western Ghats.

Narrow endemic plants of Agasthyamalai are *Buchanania barberi, Dialium travancoricum, Humboldtia unijuga* var. *trijuga, Litsea venulosa, Memecylon agastyamalaianum, Memecylon sivadasanii, Pavetta bourdillonii, Symplocos macrophylla* ssp. *namboodirianus, Syzygium bourdillonii, Sanjappa cynometroides, Sageraea thwaitesii* with characteristic woody habit substantiate the relic nature of the flora. The recent explorations are also continuously discovering many new species from that area such as *Memecylon sivadasanii, M. agastyamlaianum, M. manickamii, Ixora agasthyamalayana, Litsea beei* and *Polyalthia shendurunii.* There has been recorded 309 endemic species among them 100 are endangered (Warrier *et al.*, 2020). Some notable endangered species are *Eugenia floccosa* (EN), *E. discifera* (EN), *Hopea utilis* (EN), *Bentinckia condapanna* (VU), *Vateria indica* (CR) and *Syzygium travancoricum* (CR). Genera with more than 15 endemic species are listed.

Nilgiris is an important geographic location known for floristic endemism in southern part of Peninsular India. There are three zones where concentrated number of endemic species such as Nilgiris, Nilgiri-Wayanad and Sigur plateau. Wayanad microclimatic center is represented by narrow endemic plant species; *Aglaia perviridis, Cynometra beddomei, Goniothalamus wynaadensis, Eugenia wynadensis* and *Phyllanthus megacarpus*. Some of the important endemic species concentrated in Nilgiri are *Dalbergia congesta, Embelia gardneriana* and *Ilex gardneriana*.

Silent valley vegetation complex is rich with valuable gene pool of endemic species, where represented some notable endemic tree species are *Dalbergia beddomei, Cullenia exarillata*, etc. Palani hill complex of Kodaikkanal Wildlife Sanctuary covers patches of evergreen forests in high altitudes were composed of 49 endemic species solely endemic to this region. The significant endemic species of this hill ranges are *Actinodaphne bourneae* and *Elaeocarpus blascoi*. Kanuvai hills of Coimbatore district revealed the RET species that has been reported 51 endemic species with 39 genera, 28 families and two subfamilies. Among them *Berberis nilghiriensis* one of the Critically Endangered species and others including the area are *Elaeocarpus recurvatus, Litsea wightiana* var. *tomentosa* and *Dalbergia congesta* (Prabhu Kumar *et al.*, 2012; Warrier *et al.*, 2020).

Anamalai is the highest peak of southern Western Ghats covers all ranges of forests in different altitude where endemic taxa are represented. Some important endemic species are *Cryptocarya anamalayana* and *Phyllanthus anamalayanus*. About 352 species of endemic tree species have been reported from the Western Ghats, 50 species (14%) belonging to 19 families were listed as threatened categories and also included in the Red Data Book (Nayar and Sastry, 1987-1990). Five species with indeterminate status due to unavailability of information since the type collection were recorded prior to 1997 (Ramesh and Pascal 1997). These species are: *Memecylon sisparense, Miliusa nilagirica, Palaquium bourdillonii, Phaeanthus malabaricus* and *Poeciloneuron pauciflorum* (Puyravaud *et al.*, 2003).

Many endemic herbaceous genera are confined to the northern Western Ghats of Maharashtra and these include *Calacanthus, Glypochloa, Haplanthodes, Hubbardia, Pinda, Seshagiria*, etc. *Ceropegia, Chlorophytum, Dipcadi, Crinum* are showing higher representation of endemism in this region. Exclusively endemic species of the region are *Abutilon ranadei, Amorphophallus konkanensis. Ceropegia sahyadrica, C. rollae, C. vincaefolia, Chlorophytum indicum, Crinum brachynema, C. woodrowii, Curcuma caulina, Dicliptera nasikensis, Dipcadi minor* and *Euphorbia katrajensis*. Southern zone of the northern Western Ghats comes into Karnataka, the vegetation dominated by thorny scrub species with stunted growth. The genera such as *Bhidea, Karnataka, Polyzygos* are endemic to the region. Some species are restricted to this range are *Calamus nagabettai, Callipedium magdalenii, Dyschoriste vagans, Hopea canarensis, Impatiens barberi, Isachne meeboldii, Ischameum tumidum, Justicia neesii*, and *Nervilia hispida*.

Conservation

Trees are one of the most familiar forms of ecosystem component and represent the largest part of earth biomass. They are growing in most regions of the terrestrial world. About 60,000 species of trees in existence in the world with most diverse habitats on the planet. A huge number of other associated species like epiphytic plants, fungi, birds, mammals, invertebrates, amphibians and reptiles, depend on their existence. Protection of trees leads to enormous benefits to humans and wildlife alike. It is therefore very surprising to learn that so little was known about their conservation status and shocking to know that deforestation rates remain so high. Many tree species are on the brink of extinction, some represented by one last living individual.

The Western Ghats consists of more than 30% of all plants, fish, herpetofauna, bird, and mammal species with less than 6% of total land area of India (Daniels, 2002). The Ghat region has a spectacular assemblage of wild flora and fauna, and the area is serving for several globally significant wildlife sanctuaries, tiger reserves, and national parks. A total of 58 protected areas all along the Ghats consisting of 3 Biosphere Reserves, 14 National Parks and 44 Wildlife Sanctuaries within its boundaries. The legally protected total area cover is 13,595 sq km contributing 9.06% of the Western Ghats (Rodgers *et al.*, 2002). The Western Ghats is one of the highest levels of protected cover area in India and another one is being the Andaman and Nicobar Islands (Rodgers and Panwar, 1988).

Table 3. Protected areas of Western Ghats

Protected areas	District(s) covered	Year declared	Area (Km^2)
Biosphere Reserves Agasthyamalai Biosphere Reserve	Tirunelveli and Kanyakumari district of Tamil Nadu and Thriuvananthapuram and Kollam districts of Kerala.	2001	3500 Km^2
Nilgiri Biosphere Reserve	Kodagu district of Karnataka, Palakkad and Wayanad districts of Kerala, Erode, Nilgiri and Coimbatore districts of Tamil Nadu.	1986	5520 Km^2
National Parks Indira Gandhi National Park	Coimbatore and Tirupur districts of Tamil Nadu.	1989	958 Km^2
Anshi National Park	Uttara Kannada district of Karnataka	1987	475.018 Km^2
Anaimudi Shola National Park	Idukki district of Kerala	2003	7.5 Km^2
Chandoli National Park	Sangli district of Maharashtra	2004	741.22 Km^2

Protected areas	District(s) covered	Year declared	Area (Km^2)
Grass hill National Park	Idukki district of Kerala and Coimbatore district of Tamil Nadu	1989	650 Km^2
Eravikulam National Park	Idukki and Ernakulam districts of Kerala.	1978	97 Km^2
Karian Shola National Park	Idukki district of Kerala and Coimbatore and Tirupur districts of Tamil Nadu.	1989	12.82 Km^2
Pambadum Shola National Park	Idukki district of Kerala and Dindigul district of Tamil Nadu.	2003	13.24 Km^2
Mathikettan Shola National Park	Idukki district of Kerala	2003	12.82 Km^2
Mudumalai Wildlife National Park	Nilgiri district of Tamil Nadu	1940	321 Km^2
Mukuruthi National Park	Nilgiri district of Tamil Nadu	2001	78.46 Km^2
Bandipur National Park	Chamrajnagar district of Karnataka	1973	868.63 Km^2
Silent Valley National Park	Palakkad, Malapuram districts of Kerala and Nilgiri district of Tamil Nadu.		89.52 Km^2
Tiger Reserves Anamalai Tiger Reserve	Coimbatore and Tirupur districts of Tamil Nadu.	2005	1479.87 Km^2
Bandipur Tiger Reserve	Chamarajnagar district of Karnataka	1974	874 Km^2
Bhadra Tiger Reserve	Chikkamagaluru and Shivamogga districts of Karnataka.	1998	572 Km^2
Dandi-Anshi (Kali) Tiger Reserve	Uttara Kannada district of Karnataka	1987	875 Km^2

Protected areas	District(s) covered	Year declared	Area (Km^2)
Kalakad Mundanthurai Tiger Reserve	Tirunelveli, Kanyakumari	1988	817 Km^2
Kali Tiger Reserve (Anshi-Dandeli Tiger Reserve, 2007)	Uttara Kannada district, Karnataka.	2015	400 Km^2
Mudumalai Tiger Reserve	Nilgiri district of Tamil Nadu.	2007	322 Km^2
Nagerhole Tiger Reserve	Kodagu and Mysore district of Karnataka.	1999	643 Km^2
Sathyamangalam Tiger Reserve	Erode district of Tamil Nadu.	2010	1411.6 Km^2
Srivilliputhur Megamalai Tiger Reserve	Virudhunagar, Madurai and Theni districts of Tamil Nadu.	2021	1016.57 Km^2
Sahyadri Tiger Reserve	Satara, Sangli and Kolhapur districts of Maharashtra	2008	741.22 Km^2
Periyar Tiger Reserve	Idukki and Pathanamthitta districts of Kerala.	1978	952 Km^2
Elephant Reserves Anamalai Elephant Reserve	Coimbatore and Tirupur district of Tamil Nadu	2003	1457 Km^2
Anaimudi Elephant Reserve	Idukki district of Kerala	2002	3728 Km^2
Agasthyamalai Elephant Reserve	Kanyakumari, Tirunelveli districts of Tamil Nadu and Thiruvananthapuram and Kollam districts of Kerala	1976	1,197.48 Km^2
Coimbatore Elephant Reserve	Coimbatore district of Tamil Nadu	2003	566 Km^2
Nilgiri Elephant Reserve	Kodagu district of Karnataka, Palakkad and Wayanad districts of Kerala, Nilgiri and Erode districts of Tamil Nadu	2003	5520 Km^2

Protected areas	District(s) covered	Year declared	Area (Km^2)
Nilambur Elephant Reserve	Palakkad, Malappuram and Kozhikode district of Kerala	2002	1419 Km^2
Periyar Elephantt Reserve	Idukki district of Kerala	2002	3742 Km^2
Srivilliputhur Elephant Reserve	Virudhunagar district of Tamil Nadu	2002	1249 Km^2
Wildlife Sanctuaries Gangaikondan Spotted Deer Wildlife Sanctuary	Tirunelveli district of Tamil Nadu	2013	28.8 Km^2
Indira Gandhi Wildlife Sanctuary	Coimbatore and Tirupur districts of Tamil Nadu.	1976	841.49 Km^2
Chimmini Wildlife Sanctuary	Thrissur district of Kerala	1984	85.067 Km^2
Peechi-Vazhani Wildlife Sanctuary	Thrissur district of Kerala	1958	14.79 Km^2
Idukki Wildlife Sanctuary	Idukki district of Kerala	1976	105.364 Km^2
Palani hills Wildlife Sanctuary and National Park (Kodaikanal Wildlife Sanctuary)	Dindigul and Theni district of Tamil Nadu.	2008	736.87 Km^2
Wayanad Wildlife Sanctuary	Wayanad district of Kerala	1973	344.44 Km^2
Kalakad Wildlife Sanctuary	Tirunelveli district of Tamil Nadu.	1988	223.58 Km^2
Kanyakumari Wildlife Sanctuary	Kanyakumari district of Tamil Nadu	2002	402.4 Km^2
Kodaikanal Wildlife Sanctuary	Dindigul and Theni districts of Tamil Nadu	2013	608.95 Km^2
Megamalai Wildlife Sanctuary	Theni and Madurai districts of Tamil Nadu.	2009	26910.8
Nagarahole National Park	Kodagu and Mysore districts of Karnataka	1999	6,000 Km^2

Protected areas	District(s) covered	Year declared	Area (Km^2)
Chinnar Wildlife Sanctuary	Idukki district of Kerala	1984	90.44 Km^2
Eravikulam-Rajamala Wildlife Sanctuary	Idukki and Ernakulam districts of Kerala	1975	97 Km^2
Mudumalai Wildlife Sanctuary	Nilgiri district of Tamil Nadu.	1940	217.76 Km^2
Parambikulam Wildlife Sanctuary	Palakkad and Thrissur district of Kerala	1973	643.66 Km^2
Radhanaagari Wildlife Sanctuary	Kolhapur district of Maharashtra	2020	351 Km^2
Shendurney Wildlife Sanctuary	Kollam district of Kerala	1984	172.403sq. Km^2
Mundanthurai Wildlife Sanctuary	Tirunelveli district of Tamil Nadu.	1977	582207.5
Nellai Wildlife Sanctuary	Tirunelveli district of Tamil Nadu	2015	356.7 Km^2
Neyyar Wildlife Sanctuary	Thiruvananthapuram district of Kerala.	1958	128 Km^2
Kurinjimala Sanctuary	Idukki district of Kerala	2006	32 Km^2
Peppara Wildlife Sanctuary	Thiruvananthapuram district, Kerala	1983	75 Km^2
Aralam Wildlife Sanctuary	Kannur district of Kerala	1984	55 Km^2
Pushpagiri Wildlife Sanctuary	Kodagu district of Karnataka	1987	102.92 Km^2
Kudremukha Wildlife Sanctuary	Chikmagaluru district of Karnataka	1987	600.32 Km^2
Someshwara Wildlife Sanctuary	Udupi and Shivamogga districts of Karnataka	1974	314.25 Km^2
Koyna Wildlife Sanctuary	Satara district of Maharashtra	1985	423.55 Km^2
Mookambika Wildlife Sanctuary	Udupi district of Karnataka	1974	370.37 Km^2
Thalakkaveri Wildlife Sanctuary	Kodagu district of Karnataka	1987	105 Km^2

Protected areas	District(s) covered	Year declared	Area (Km²)
Brahmagiri Wildlife Sanctuary	Kodagu district of Karnataka	1974	181.29 Km²
Sathyamangalam Wildlife Sanctuary	Erode district of Tamil Nadu.	2013	1411.16 Km²
Srivilliputhur Grizzled Squirrel Wildlife Sanctuary	Virudhunagar district of Tamil Nadu.	1989	485.2 Km²
Vallanad Black Buck Sanctuary	Tirunelveli and Thoothukudi districts of Tamil Nadu.	2007	164.1 Km²
Mundigekere Bird Sanctuary	Uttara Kannada district of Karnataka	2000	21 Km²
Thattekad Bird Sanctuary	Ernakulam and Idukki districts of Kerala	1983	25 Km²
Karimpuzha Wildlife Sanctuary	Malappuram district of Kerala	2020	227.97 Km²

Documenting the conservation status of a group of nearly 60,000 species is the largest initiative in the history of the IUCN Red List of Threatened Species. It represents a huge effort by hundreds of individuals under the coordination of Botanic Gardens Conservation International and the IUCN Species Survival Commission who should be thanked for this key contribution to nature conservation. This compilation is an important step towards the providing baseline data for all tree species of Western Ghats, which will constitute a ready reckoner of the areas or regions will be able to report on progress and assessment towards the protection of the world's biodiversity. An entire community composed of botanists, foresters and conservationists stands ready to assist with this massive task and is already at work. They need to be associated to concrete efforts which should not be exclusively led by foresters and companies for which the number of trees planted is their sole driver.

At present seven national parks in the Western Ghats region with a total area of about 2,073 sq. km (1.3% of the region) and 39 wildlife sanctuaries covered an area of about 13,862 sq. km (8.1%). While *in situ* conservation of several of indigenous species is partly protected by the establishment of legally conserved areas like the Nilgiris Biosphere Reserve, Agasthyamalai Biosphere Reserve, Kalakad-Mundandurai Tiger Reserve, Indira Gandhi National Park,

Silent Valley National Park, Bandipur National Park, Kudremukh National Park, Nagarahole National Park, Giant Squirrel Wildlife Sanctuary, Idukki Wildlife Sanctuary, Eravikulam Wildlife Sanctuary, Kodaikkanal Wildlife Sanctuary, Kanyakumari Wildlife Sanctuary, Wayanad Wildlife Sanctuary, Periyar Wildlife Sanctuary, Parimbikulam Wildlife Sanctuary, Shendurney Wildlife Sanctuary, Peppara Wildlife Sanctuary, etc. Though many number of endemic species are being outside the reserve, where do not find any *in situ* programmes for conservation. Even within the protected Biosphere Reserves, there is a severe threatening pressure by the invasive weeds. The Agasthyamalai peak (1868 m) in southernmost highest range of the Western Ghats recognized with one of the most popular hotspots, the significance of this area is contributed about 12% of the known plants are narrowly endemic to the region. Ponmudi and Chemunji are another highest ridge in the Agasthyamalai region which contributes a number of endemic biodiversity. In the same region included two more other Wildlife Sanctuaries, viz. Peppara Wildlife Sanctuary and Neyyar Wildlife Sanctuary which are also main part of the Agasthyamalai Biosphere landscape area and thus the major component of conservational vegetation complex in the southern Western Ghats with riches gene pool of endemic floras. As of now the total protected area cover is 19,244 Km^2 forests under in situ conservation, roughly 11.7% of the Western Ghats and 35% of their original forest cover (Table 3).

Methodology

The two volumes have given the revisionary updated enumeration/ monograph on trees of Western Ghats. Gathering field data, herbarium consultation and published literature on the trees of Western Ghats accounted systematically for precise identification. Reworked on the nomenclature, systematics, distribution, utilization and conservation status for each tree taxa by searching specific collections deposited in different herbaria (SGH, TBGT, CALI, RHT, KFRI, MH, CAL, XHT) for confirmation of the identification and distribution of tree species in Western Ghats. Consulting International Plant Name Index (IPNI), Plants of the World Online (POWO), World Flora Online (WFO), Germplasm Resource Information Network (GRIN), International Union for Conservation of Nature (IUCN) for scrutinizing the nomenclature, diversity status and utilization of each species. Cultivated, planted and naturalized trees are also enumerated and given the list of species at the end of each respective families.

Primary checklist of the tree species recorded from the Western Ghats has been complied by field studies and relevant literature. Woody plants attaining a height of more than 4 m with a clear bole are considered as trees. Being a manual on the trees, every effort has been taken to make it useful to botanists, biologists, foresters and other users. Most of the forest trees being lofty, collection of flowering and fruiting specimens is rather difficult. Flowering and fruiting specimens collected during the filed observation of senior author last 15 years of experience in the Western Ghats as well as the published literature were made use of for the preparation of the extensive check list of tree species. Authentic specimens and type specimens of tree species available at BSI archive (https://ivh.bsi.gov.in/phanerogams). Also checked the scanned images of types of the tree species from Royal Botanic Garden Herbarium, Kew, herbarium catalogue (https://apps.kew.org/herbcat/navigator.do) were also studied for confirming the identity of rare trees. The herbarium specimens prepared during the study are deposited in the Sri Ganesan Herbarium (SGH), Department of Botany, The Madura College (Autonomous), Madurai.

How to follow this book?

Tree taxa have been arranged systematically by alphabetical order of their respective families and each family enumerated the key to the genera, then key to the species if genera with multiple species. For each tree species given accepted (legitimate) binomial name according to recent changes in Plants of World Online (POWO) and APG-IV classification. Valid binomial name given with full citation followed the synonyms and basionym if available. Vernacular names in Kanada, Marathi, Malayalam and Tamil have given at the end of citation. Precise description of each taxa with characteristic features, flowering and fruiting phenology, distribution in Western Ghats, India and other parts of the world, IUCN threat status according to recent changes in IUCN reassessment (IUCN, 2024), uses of the taxa and explanatory nomenclatural or taxonomic note were given. If necessary taxonomic and nomenclatural note were given for some important taxa at the end of the species description. It has been given the diversity status and distribution scale from the recent global Species Information System (SIS, IUCN, 2024). The threat categories extinct (EX), extinct in the wild (EW), regionally extinct (RE), critically endangered (CR), endangered (EN), vulnerable (VU), near threatened (NT), least concern (LC), data deficient (DD), not evaluated (NE) and not applicable (NA) based on recent IUCN red-list assessment (IUCN, 2024).

Identification method

- Dichotomous parallel keys to recorded tree families, genera, species and infraspecific taxa based on reproductive and vegetative characters are provided.
- The families are arranged alphabetic order with delimitation following their current concepts by updated classification of APG-IV. Genera under families and species under genera are arranged in alphabetical order.
- Latest revisions/monographs have been followed with respect to the circumscription and delimitation of genera and species.
- The nomenclature is updated with their basionym and synonym(s), if any, are also provided.
- As local and vernacular names are preferred in forestry and timber trade, every effort has been made to include the local names in Kannada, Marathi, Malayalam and Tamil.
- Colour photographs of rare tree species of Western Ghats are provided for maximum number of trees species.

Abbreviations used

APG-IV: Angiosperms Phylogenic Group, IV-classification (2016).

BSI: Botanical Survey of India.

CAL: Central National Herbarium, Howrah, Kolkatta.

CALI: Calicut University Herbarium, Department of Botany, Calicut University, Kerala.

CR: Critically endangered

DD: Data deficient

EN: Endangered

EW: Extinct in the wild

EX: Extinct

FSI: Forest Survey of India.

GRIN: Germplasm Resource Information Network.

IUCN: International Union for Conservation of Nature

Kan.: Kannada

KFRI: Kerala Forest Research Institute, Thrissur, Kerala.

LC: Least concern

Mal.: Malayalam

Mar.: Marathi

MH: Madras Herbarium, Southern Circle, Coimbatore.

NA: Not applicable

NE: Not evaluated

NT: Near threatened

POWO: Plant of World Online.

RE: Regionally extinct

RET: Rare, Endanger and Threatened

RHT: The Rapinat Herbarium, St Joseph College, Tiruchirappalli, Tamil Nadu.

SGH: Sri Ganesan Herbarium, Department of Botany, The Madura College (Autonomous), Madurai.

SIS: Species Information System

Tam.: Tamil

TBGT: Tropical Botanical Garden and Research Institute, Palode, Thiruvananthapuram, Kerala.

Vern.: Vernacular names

VU: Vulnerable

WFO: World Flora Online

XHT: St. Xavier College Herbarium, St. Xavier College, Palayamkottai, Tirunelveli.

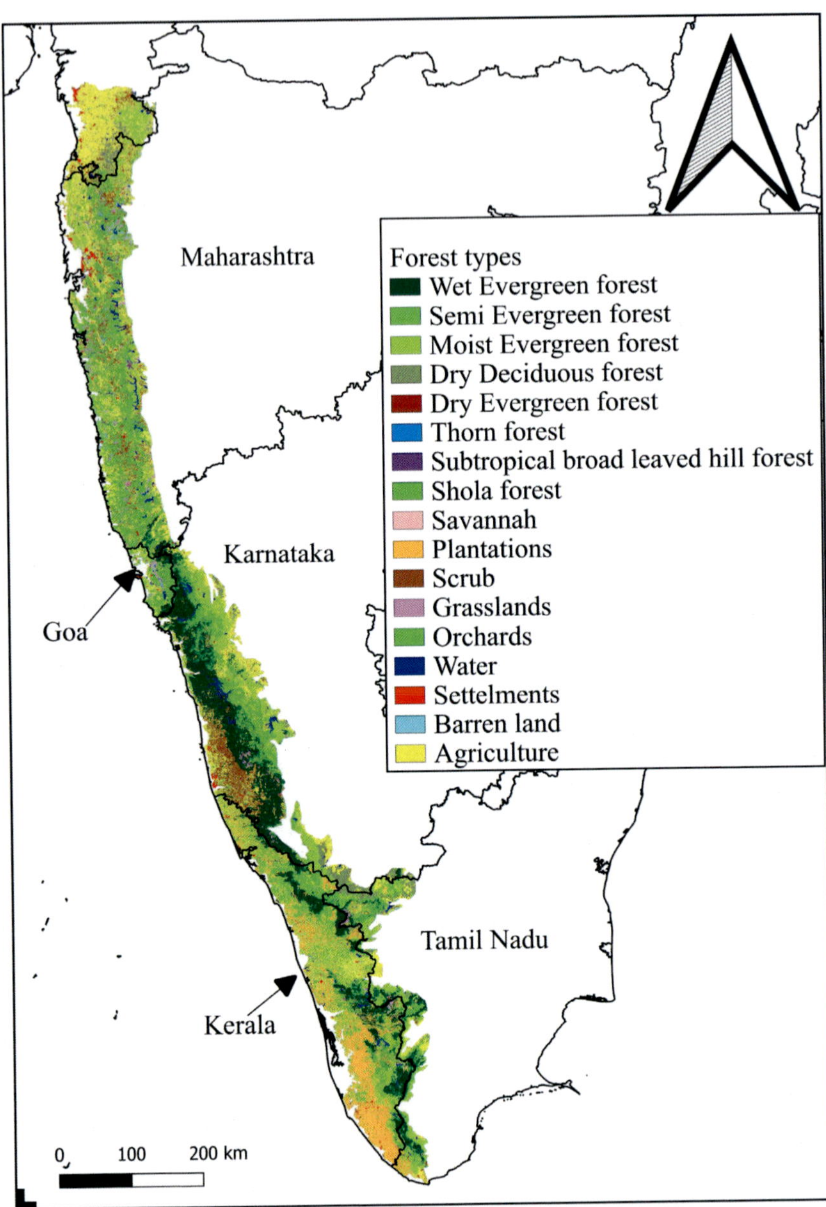

Fig 1. Forest types of Western Ghats.

Semi Evergreen - Maharasthra

Wet Evergreen Forest - Kerala

Savannah woodland - Tamil Nadu

Riparian Forest

Grasslands - Karnataka

High altitude grasslands - Tamil Nadu

Shola forest - Tamil Nadu

Waterfalls

Threats

Excessive tourism

Cultivation & Plantation

Domestic animals grazing

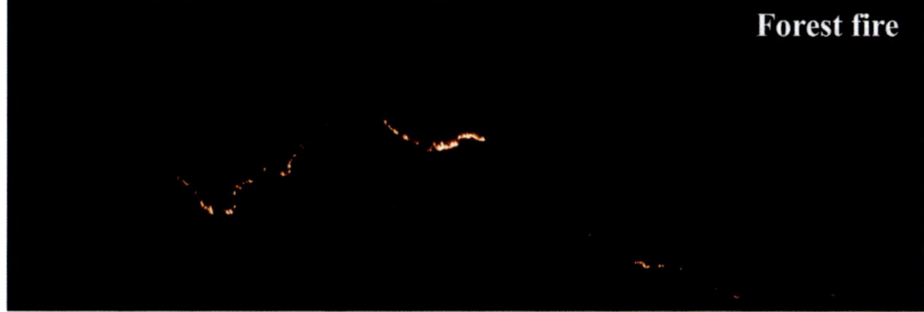
Forest fire

Key to the Families

IA. Unbranched palm trees; inflorescence usually long branched rachis bearing on spadix.. **Arecaceae**

IB. Branched trees with definite basal trunk; inflorescence bearing on leaf axile or terminal of the branchlets, other than spadix 1

1a. Flowers with two or more seriate perianths; usually outer sepaloid and inner petaloid 2

1b. Flowers with uniseriate or perianth absent; if petals present either sepaloid or petaloid...82

2a. Inner perianth lobes usually free, petals divided up to the base; stamens attached on the thalamus 3

2b. Inner perianth lobes united, petals connate in various levels; stamens attached on the petals 65

3a. Ovary hypogynous... 4

3b. Ovary epigynous or perigynous...57

4a. Carpels free, ovary 1-celled 5

4b. Carpels united, ovary 2-many celled...21

5a. Ovule single, placentation basal or marginal 6

5b. Ovules many, placentation parietal or axial 11

6a. Leaves 3-nerved; flowers unisexual; seeds horse-shoe shaped ....**Menispermaceae**

6b. Leaves penni-nerved; flowers bisexual; seeds not horse-shoe shaped....7

7a. Petals and sepals fleshy, trimerous, usually petaloid; fruit a follicle .... **Magnoliaceae**

7b. Petals and sepals not fleshy, 4-5-merous, sepaloid; fruit other than follicle 8

8a. Fruits winged (samaroid), compressed, papery .... **Simaroubaceae**

8b. Fruits not winged, usually fleshy...9

9a. Stamens produced connective appendage or hooded anther; fruitlets aggregate or free; endosperm ruminate .... **Annonaceae**

9b. Stamens connective not produced appendages; fruits not aggregate; endosperm not ruminate 10

10a. Leaves opposite, rarely alternate; stamens more than 20 ....**Calophyllaceae**

10b. Leaves usually alternate; stamens less than 10 .... **Anacardiaceae**

11a. Ovary with distinct gynophore and gynandrophore present; petals clawed at base .... **Capparidaceae**

11b. Ovary usually sessile; petals not clawed at base 12

12a. Carpels 3 or more 13

12b. Carpel single 17

13a. Carpels deeply 3-10-lobed; ovary embedded in a disc; styles connate .... **Ochnaceae**

13b. Carpels not lobed; disc absent; styles free 14

14a. Stamens quite free from the petals 15

14b. Stamens attached to the base and deciduous with the petals .... **Pentaphylacaceae**

15a. Stipule leafy-fleshy; petals white, narrowly ribbon-shaped .... **Hamamelidaceae**

15b. Stipules membranous, not fleshy; petals absent or very broadly ovate 16

16a. Leaves palmately lobate; perianth absent; stem with resinous exudate.... .... **Altingiaceae**

16b. Leaves not lobed, margin strongly serrate; perianth well developed, broad yellow or white petals .... **Dilleniaceae**

17a. Fruit dehiscent 18

17b. Fruit indehiscent 23

18a. Stamens monadelphous or diadelphous; fruits legume .... **Fabaceae**

18b. Stamens free; fruits capsule 19

19a. Seeds enclosed with pulp 21

19b. Seeds not enclosed with pulp .... **Culsiaceae**

20a. Inflorescence long spike; petals absent; capsule 4-valved .... **Tetramelaceae**

20b. Inflorescence panicles or cymes; petals present; fruits usually pulpy .... **Flacourtiaceae**

21a. Ovary 1-2-celled; sepals usually persistent 31

21b. Ovary 2-more celled; sepals not persistent .... **Zygophyllaceae**

22a. Stem not flattened or articulated; leaves well developed 23

22b. Stem fleshy succulent, articulate; leaves reduced into spines... **Cactaceae**

23a. Petals with scales at base; fruits indehiscent berry .... **Achariaceae**

23b. Petals without basal scales; fruits otherwise 24

24a. Leaves well developed 25

24b. Leaves scaly or minute .... **Tamaricaceae**
25a. Hypanthium prominent, formed by calyx 26
25b. Hypanthium not prominent, calyx simple 29
26a. Stamens more than 15 27
26b. Stamens 10 or less than 29
27a. Flowers irregular; stamens in one unilateral bundles ....**Chrysobalanaceae**
27b. Flowers regular; stamens arranged around the ovary 28
28a. Stamens 5-10; stipules absent; style terminal .... **Connaraceae**
28b. Stamens many; stipules prominent; style lateral .... **Rosaceae**
29a. Stamens opposite to the petals 30
29b. Stamens alternate to the petals 32
30a. Leaves exstipulate; fruits usually 1-celled .... **Olacaceae**
30b. Leaves stipulate; fruits usually 2-celled .... **Vitaceae**
31a. Leaves estipulate; ovary incompletely 2-3-celled; sap pungent .... **Pittosporaceae**
31b. Leaves stipulate; ovary 1-celled; sap not pungent .... **Violaceae**
32a. Leaves simple; stamens in single row; ovule pendulous 33
32b. Leaves unipennate; stamens in 2-rows; ovule erect .... **Berberidaceae**
33a. Petals 2-lobed, imbricate; raphe of seed ventral .... **Dichapetalaceae**
33b. Petals entire, valvate; raphe of the seed dorsal 34
34a. Flowers to 7 mm diam; stamens alternate to the petals .... **Icacinaceae**
34b. Flowers less than 4 mm diam; stamens opposite to the petals 35
35a. Sepals and petals fleshy, valvate; fruit drupaceous .... **Nyssaceae**
35b. Sepals and petals not fleshy, valvate or imbricate; fruit other than drupe 36
36b. Locules of ovary 1 or 2 ovules 37
36a. Locules of ovary 3 or more ovules 38
37a. Petal free; filaments flattened, covered with club-shaped hairs; fruit flattened .... **Stemonuraceae**
37b. Petals connate at base; filaments cylindrical, not covered with hairs; fruit indehiscent, 5-ribbed .... **Metteniusaceae**
38a. Seeds bright coloured or embedded in a cottony wool .... **Bixaceae**
38b. Seeds not bright coloured or not embedded in a cottony wool 39
39a. Leaves trifoliolate or unipinnate 40

39b. Leaves simple 41
40a. Styles 3; leaves opposite ....**Staphyleaceae**
40b. Styles 5; leaves alternate ....**Averrhoaceae**
41a. Leaves opposite; seeds prominently arillate ....**Celastraceae**
41b. Leaves alternate; seeds not arillate 42
42a. Flowers small less than 1 cm diam; capsule deeply bilobed ....**Centroplacaceae**
42b. Flowers more than 2 cm diam; capsule not lobed 43
43a. Stamens more than 15 44
43b. Stamens 10 or less than 45
44a. Stamens on or at the apex of staminal tube; anther one-celled.................. ....**Malvaceae**
44b. Stamens shortly connate at the base; anthers two-celled 47
45a. Flora disc absent...46
45b. Floral disc present...51
46a. Leaves entire; petals entire or minutely lobed; fruits nut or capsule ....**Dipterocarpaceae**
46b. Leaves serrate; petals laciniate or fimbriate; fruits drupe or woody tubercled pyrenes ....**Elaeocarpaceae**
47a. Flowers irregular; sepals and petals unequal, petals usually 3; fruits capsule ....**Polygalaceae**
47b. Flowers regular; sepals equal not petaloid; petals not adnate to the staminal column...48
48a. Stamens as many as and opposite to the petals 49
48b. Stamens twice or more than the petals, usually alternate to the petals ....**Theaceae**
49a. Ovule single in each locule; placentation basal......................**Rhamnaceae**
49b. Ovule two in each locule; placentation axile 50
50a. Leaves unipennate; ovary not stipitate ....**Sabiaceae**
50b. Leaves 2-3-pinnate; ovary stipitate ....**Moringaceae**
51a. Petals with an erect ligule at base ....**Erythroxylaceae**
51b. Petals without ligule at base 52
52a. Leaves with aromatic glands; anthers gland tipped ....**Rutaceae**
52b. Leaves and anthers without glands 53

53a. Leaves simple; filaments adnate to the petals .... **Aquifoliaceae**
53b. Leaves usually pinnate, 3-more foliolate 54
54a. Filaments united to form a distinct staminal tube .... **Meliaceae**
54b. Filaments usually free or rarely united at base 55
55a. Petals scaly at base .... **Sapindaceae**
55b. Petals without scales at base 56
56a. Bark exudes resinous; fruits fleshy capsule .... **Burseraceae**
56b. Bark not exudes any sap; fruits indehiscent drupe or berry ....
.... **Staphyleaceae**
57a. Stamens more than 15 58
57b. Stamens less than 10 62
58a. Leaves alternate 59
58b. Leaves opposite or subopposite 60
59a. Stamens 3-12; anthers usually opening by pores; petals contorted in bud .... **Melastomataceae**
59b. Stamens many; anthers opening by slits; petals imbricate in bud ....
.... **Lecythidaceae**
60a. Leaves with intramarginal nerves, often glandular punctate; bark peeling off in strips .... **Myrtaceae**
60b. Leaves without intramarginal nerves, not glandular punctate; bark not as above 61
61a. Fruits typical capsule; seeds winged .... **Lythraceae**
61b. Fruits fleshy berry; seeds not winged .... **Sonneratiaceae**
62a. Ovary one-celled 63
62b. Ovary 2 or more celled 64
63a. Ovules solitary in locule; stamens 4-5 or over 15 .... **Cornaceae**
63a. Ovules 2-3 in a locule; stamens 8 or 10 .... **Combretaceae**
64a. Leaves simple; petals lacerate at apex .... **Rhizophoraceae**
64b. Leaves digitate; petals entire .... **Araliaceae**
65a. Ovary epigynous 66
65b. Ovary hypogynous or perigynous 70
66a. Inflorescence capitula; calyx into pappus; fruits achene .... **Asteraceae**
66b. Flowers solitary or inflorescence other than capitula; fruits other than achene 67

67a. Plants with punctate hairs; inflorescence terminal umbellate corymbs .... **Viburnaceae**

67b. Plants without punctate hairs; inflorescence otherwise 68

68a. Leaves opposite, stipules prominent, inter or intra-petiolar ...**Rubiaceae**

68b. Leaves alternate, stipules absent or inconspicuous 69

69a. Ovary 1-celled; stamens 5 .... **Primulaceae**

69b. Ovary 2-many celled; stamens usually 10 or more ....**Symplocaceae**

70a. Stamens more than the number of corolla lobes......................................71

70b. Stamens as many as the number of corolla lobes 74

71a. Flowers unisexual ....**Ebenaceae**

71b. Flowers bisexual 72

72a. Ovary 5-celled; ovules many in locule; fruit a capsule 73

72b. Ovary 2-8-celled; ovules solitary in locule; fruits indehiscent berry.......... .... **Sapotaceae**

73a. Young inflorescence formed strobili like leafy obovate bracts; flowers deep red to pink coloured; corolla campanulate ....**Ericaceae**

73b. Young inflorescence with woolly tomentose, not formed strobili like bracts; flowers pinkish purple; corolla urceolate ....**Actinidiaceae**

74a. Corolla actinomorphic 75

74b. Corolla zygomorphic 81

75a. Leaves opposite or whorled 76

75b. Leaves alternate .... **Boraginaceae**

76a. Stamens 2-4; corolla regular 77

76b. Stamens 4-6; corolla regular or zygomorphic...78

77a. Petals 4-9, lobes imbricate in bud; stamens 2, alternate with carpels .... **Oleaceae**

77b. Petals 4, lobes valvate in bud; stamens 4, alternate with petals ....**Salvadoraceae**

78a. Calyx persistent, enclosed the fruits partly or fully ....**Lamiaceae**

78b. Calyx not persistent, not enclosed the fruit 79

79a. Trees with milky latex; corolla often coronal appendaged at mouth ....**Apocynaceae**

79b. Trees without milky latex; corolla not with appendaged 80

80a. Corolla tubular, leaves prominently 3-ribbed .... **Loganiaceae**

80b. Corolla widely campanulate, wide mouthed; leaves penni-nerved .... **Gentiaceae**

81a. Fruit dehiscent; seeds winged .... **Bignoniaceae**

81b. Fruits indehiscent; seeds not winged .... **Verbenaceae**

82a. Trees with flowers bisexual 83

82a. Trees with flowers unisexual 86

83a. Ovary epigynous; stamens attached to the perianth lobes ... **Santalaceae**

83b. Ovary hypogynous; stamens free from the perianth lobes 84

84a. Perianth lobes valvate, recurved or revolute; stamens 4, uniseriate .... **Proteaceae**

84b. Perianth lobes imbricate, not recurved; stamens 2-4-seriate 85

85a. Perianth short tubular, united at base only; inner series of the stamens prominently glandular .... **Lauraceae**

85b. Perianth long tubular; stamens not glandular ....**Thymeleaceae**

86a. Inflorescence catkin like spike; flowers with perianth absent 87

86b. Inflorescence not as above; flowers with perianth present 89

87a. Leaves unipennate; catkin with broad leafy bracts; endosperm ruminate. .... **Juglandaceae**

87b. Leaves simple; spike with small bracts; endosperm not ruminate 88

88a. Ovary 1-celled; locules with 4 or more ovules; seeds with silky hairs .... **Salicaceae**

88b. Ovary 2-6-celled; locule with single ovule; seeds without silky hairs....... .... **Betulaceae**

89a. Ovary with one celled; ovule solitary 90

89b. Ovary with 2-3-celled; ovules 1-2 in each locule 94

90a. Locule with 1-ovule; latex often present; seeds carunculate or not ....**Euphorbiaceae**

90b. Locule with 2-ovules; latex usually absent; seeds not carunculate 91

91a. Leaves crowded on the top of branchlets; male flowers without subtending bracts and sepals .... **Daphniphyllaceae**

91b. Leaves distributed on the branchlets; male flowers with bracts and sepals 92

92a. Fruit indehiscent drupe; trees dioecious ....**Putranjivaceae**

92b. Fruit capsule; trees monoecious 93

93a. Leaves alternate; sepals 3-6 in male flowers; stamens 6-many .... **Phyllanthaceae**

93b. Leaves opposite; sepals 2-5 in male flowers; stamens 5-6 .... **Picrodendraceae**

94a. Stamens connate in column; seeds arillate ....**Myristicaceae**

94b. Stamens free; seeds without aril 95

95a. Stamens alternating with staminodes; fruits prominently 2-winged .... **Hernandiaceae**

95b. Staminodes absent; fruits not winged 96

96a. Trees with milky or watery exudates; flowers arranged on receptacle; stamens single 97

96b. Trees without any exudates; flowers in cymes or fascicles; stamens 4-8... 98

97a. Plants with watery sap; sometimes with stinging hairs, fruits urticle....... .... **Urticaceae**

97a. Plants with milky sap; not with stinging hairs; fruit syconia ... **Moraceae**

98a. Filaments inflexed in bud; fruits dry winged or drupaceous ... **Ulmaceae**

98b. Filaments not inflexed in bud; fruits usually ovoid drupes .... **Cannabaceae**

SYSTEMATIC ENUMERATION OF TREES

ACHARIACEAE Harms

Hydnocarpus Gaertn.

1a. Sepals 4; petals 8 or more, stamens 15-many 2

1b. Sepals 5; petals and stamens 5 (rarely 15) 3

2a. Leaves shiny, tertiary veins irregularly branched; petals 8 .... **H. kurzii**

2b. Leaves not shiny, tertiary veins seldom branched; petals 12 .... **H. macrocarpus**

3a. Branchlets stellately puberulent; leaves entire, glabrous 4

3b. Branchlets simply puberulent; leaves serrate to remotely crenate-serrate, minutely appressed pubescent .... **H. pentandrus**

4a. Berry globose, not angular, apex rounded; peduncle 5–8 mm long; pedicel up to3.5 cm long 5

4b. Berry ovoid-oblong, 5-angular, apex beaked; peduncle 50–65 mm long; pedicel 35–50 mm long .... **H. longipedunculatus**

5a. Stamens 15; pedicel of female flowers longer than the male .... **H. pendulus**

5b. Stamens 5; pedicel of male and female flowers sub-equal ..... **H. alpinus**

Hydnocarpus alpinus Wight, Icon. Pl. Ind. Orient. 3: t. 942. 1845. Vern.: *Attuchankala, Kattumaroti, Malamaroti, Maravetti, Pinervety, Vilamaram* (Mal.); *Aatuchenakalai, Koranguthalai, Neer vetti, Maravetti* (Tam.); *Maratuti* (Kan.).

Trees to 15 m tall; bark smooth, greyish-brown, slightly rough; branchlets puberulus; young leaves copper red. Leaves simple, alternate, drooping, ovate, elliptic-oblong or ovate-lanceolate, 8-25 x 5-10 cm, base oblique, round or acute, apex acute or acuminate, margin entire, glabrous, glossy, coriaceous; lateral nerves 7-10 pairs, intercostae reticulate, prominent; stipules lateral, deciduous; petiole 5-10 mm, stout, grooved above and glabrous. Flowers unisexual 22-25 mm across, yellowish-white, solitary or in stout axillary fascicles; pedicel 1.5-2 cm long, deflexed, pubescent. Sepals 5, 8 mm long, oblong, pubescent, imbricate. Petals 5, 1 cm long, narrow, glabrous, with a linear scale at the base, as long as petals, sparsely hairy. Stamens 5-15; filaments glabrous; connectives broad. Ovary 1-celled, tomentose, stigmas 5, free, radiating, recurved. Berry globose, 5-7 cm across, densely tomentose, dark brown. Fl. & Fr.: February – April.

Western Ghats: Common in evergreen forests of Maharashtra, Goa, Karnataka, Kerala, Tamil Nadu. World: Sri Lanka.

IUCN: VU-D1. (IUCN, 2024)

Hydnocarpus kurzii (King) Warb., H.G.A. Engler & K.A.E. Prantl, Nat. Pflanzenfam. 3(6a): 21. 1893. Synonyms: *Taraktogenos kurzii* King., *Hydnocarpus kurzii* f. *latifolius* N.Mukh., *Hydnocarpus kurzii* ssp. *australis* Sleumer Vern.: *Maravetti, Neeradimuthu, Niratimuthu* (Tam.).

Trees, 10-20 m tall; branchlets tawny tomentose, soon glabrescent. Leaves alternate, bifarious, variable, lanceolate, oblong-lanceolate, elliptic-oblong, obovate-oblong or even elliptic, 90-25 x 3-7 cm, usually cuneate, rarely rounded and somewhat inequilateral at base, obtusely short or abruptly long acuminate, or even obtuse at apex, entire, membranous and minutely appressed pubescent when young, soon glabrescent; lateral nerves 7-9 pairs; petioles swollen and slightly geniculate at upper end, 1-3 cm long. Male flowers 8-12 mm across, pale yellow, 3-5 each in two compact fascicles on a common peduncle 7-15 mm long; peduncles 2-3 mm long, pedicels 7-10 mm long, all strigosely fulvous tomentose. Sepals ovate-orbicular, cuneate at base, ciliate along margins, 4 mm long; epipetalous scales obovate, densely white pilose at apex. Stamens 20-30; filaments ca 4 mm long, densely pilose. Female flowers 2-5 in compact fascicle on ca 5 mm long peduncle, rarely solitary; pedicels to 12 mm long, strigosely fulvous tomentose. Sepals and petals often fewer than in male flowers. Staminodes 10-16. Ovary ovoid, obscurely 6-7-ribbed, densely fulvous tomentose, with 4 parietal placentae; stigmas 4. Berry globose, 8-10 cm across; pericarp reddish brown, velvety tomentose, afterwards scurfy. Seeds 12-18, angular-ovoid. Fl. & Fr.: February – November.

Western Ghats: Kerala, Maharashtra. World: India, Bangladesh, Myanmar.

IUCN: DD (IUCN, 2024).

Hydnocarpus longipedunculatus Robi, Sasidh & Jose, Webbia 69: 243. 2014.

Trees to 12 m tall; branchlets angular, dense rusty-stellate tomentum when young. Leaves alternate, 9–25 x 3–8 cm, oblong, elliptic-lanceolate, glabrous above, minutely stellate-pubescent on the midrib and nerves beneath, lateral nerves 4–6; petiole 5 mm long, base swollen, slightly channelled above and rounded below; stipules lanceolate, 5 mm, acuminate, rusty-stellate tomentose, caducous. Inflorescence pendulous, in axillary pseudo-cymes. Male flowers greenish-white, fragrant; peduncle 5–6.5 cm long, sparsely stellate–pubescent; pedicel slender, 3.5–5 cm long, straight. Sepals 5, elliptic, 12 x 5 mm, slightly incurved, green, three-nerved. Petals 5, elliptic-lanceolate, 15 x 5 mm, white, strongly incurved; scales 14 x 2 mm, linear, pale brown. Stamens 5, filaments 5 mm long, Female flowers creamy-white, fragrant; peduncle slender, flattened towards apex, 5–6.5 cm long; pedicel 3.5–5 cm long, densely stellate-pubescent. Sepals, petals and scales same as in male flowers. Ovary 7 mm long, densely tomentose; ovules many; style 3 mm long, stigma shortly five-lobed. Berry ovoid-oblong, 5–10 x 2.5–6 cm, densely covered by stellate hairs. Seeds 5–9, 3–4 x 2 cm, angular-ovoid to rounded, pale brown, glossy. Fl. & Fr.: November – July.

Western Ghats: Kerala (Endemic).

IUCN: DD (IUCN, 2024).

Hydnocarpus macrocarpus (Bedd.) Warb., H.G.A. Engler & K.A.E. Prantl, Nat. Pflanzenfam. 3(6a): 21. 1893. Synonyms: *Asteriastigma macrocarpa* Bedd., *Taraktogenos macrocarpa* (Bedd.) N.P. Balakr., *Hydnocarpus macrocarpus* ssp. *malabaricus* Sleumer Vern.: *Malamaroti, Malamkummati* (Mal.); *Kurangu thalai maram, Vellanaku* (Tam.).

Evergreen trees to 20 m tall; bark smooth, whitish brown; young branchlets terete, minutely pubescent. Leaves simple, alternate, estipulate; petiole 10-25 cm, slender, grooved above, pubescent when young; lamina 13.5-30 x 4.5-10 cm, broadly oblong, base acute, obtuse or subtruncate, unequal side, apex obtusely acuminate, margin entire, glabrous, coriaceous; lateral nerves 6-7 pairs, prominent, intercostae scalariform. Flowers polygamous, white in axillary fascicles or on old branches. Sepals 4, imbricate. Petals 12-16, greenish-yellow with 3-lobed scales at the base inside. Stamens many. Ovary sessile,

1-celled; stigmas 6, bilobed. Berry globose, 12-15 cm across, dark brown, woody; seeds many, angular. Fl. & Fr.: June – October.

Western Ghats: Kerala, Tamil Nadu. World: India, Myanmar, Thailand, Vietnam

IUCN: NE (IUCN, 2024).

Uses: Fruit is used in traditional medicine.

Hydnocarpus pendulus Manilal, T.Sabu & Sivar., Trop. Pl. Sci. Res. 1: 355. 1983.

Trees up to 10 m tall. Leaves oblong-lanceolate, 15 x 5 cm, acute at apex, glabrescent on both surfaces. Flowers in axillary pendulous rusty tomentose clusters, white; pedicels of male flowers very slender, to 3 cm long, those bisexual one stout to 3.5 cm long. Sepals 5, oblong, 11-13 mm long, subequal, reflexed, slightly pubescent. Petals ovate-lanceolate, 13 mm long, folded inwardly at margin, enclosing linear or oblong-linearscale with ciliate-hairy margin and as long as or slightly longer than petals. Stamens 5, erect, to mm long; filaments broad below, narrow above; pistillode absent in male flowers. Ovary ovoid, 7-8 mm long, faintly ribbed, rusty tomentose. Berry globose, to 8 cm across, hard, woody, rusty-tomentose, 25 seeded. Fl. & Fr.: April – May.

Western Ghats: Evergreen forests of Kerala (Endemic)

IUCN: NE (IUCN, 2024).

Hydnocarpus pentandrus (Buch.-Ham.) Oken., Allg. Naturgesch. 3(2): 1381. 1841. Synonyms: *Chilmoria pentandra* Buch.-Ham., *Hydnocarpus laurifolius* Sleumer, *Hydnocarpus wightianus* Blume Vern.: *Kodi kotti Maravatti, Maravetti, Niralam, Nirvatta, Nirvetti, Thamana* (Mal.); *Malai neer vetti, Maravetti* (Tam.).

Trees to 25 m tall; bark surface pale brown mottled with white patches, smooth, brittle; branchlets pubescent. Leaves simple, alternate, elliptic, elliptic-lanceolate or elliptic-oblong, 8–17 x 3–8 cm, apex acuminate, base acute, obtuse or round, margin crenate, serrate or entire, glabrous, coriaceous; lateral nerves 6–8 pairs, pinnate, prominent, intercostae scalariform, prominent; petiole 8–10 mm, stout, pubescent, grooved above; stipules small, lateral, cauducous. Flowers unisexual, 5–10 mm across, greenish-yellow, solitary or in small fascicle. Sepals 5, broadly ovate, pubescent. Petals 5, broadly ovate, greenish-yellow, fringed with soft white hairs, with a scale at the base; scales half as long as petals, densely hairy. Stamens 5–15; filaments subulate, hairy at the base; anthers of female flowers possess no pollen; ovary globose, 1-celled, tomentose; stigma sessile and two lobbed. Berry globose, 5–7 cm across, rind thick, brown, rough with uneven surface. Seeds numerous, yellowish, obtusely angular. Fl. & Fr.: March – August.

Western Ghats: Maharashtra, Goa, Karnataka, Kerala, Tamil Nadu (Endemic).

IUCN: VU-D1. (IUCN, 2024).

ACTINIDIACEAE Engl. & Gilg

Saurauia Willd.

1a. Leaves with glabrous or sparsely tomentose beneath; sepals subequal.... .... **S. fasciculata**

1b. Leaves with ferrugineous tomentose beneath; sepals outer 3 smaller, inner two larger .... **S. napaulensis**

Saurauia fasciculata Wall., Pl. Asiat. Rar. 2: 40, t. 148.1831. Synonym: *Ternstroemia fasciculata* Wall, *nom.nud*.

Small trees to 10 m tall; bark brown; branchlets with tubercular dots and subulate scales. Leaves simple, alternate, lanceolate-ovate to oblanceolate-elliptic, base obtuse to cuneate, margins serrate or subentire, apex obtuse to acuminate, membranous, coriaceous or chartaceous, glabrous or sparsely tomentose beneath, midrib impressed above and prominent, lateral veins parallel, veinlets reticulate, petiole glabrous, tomentose or scaly, exstipulate. Inflorescence axillary, cymes or panicles. Flowers bisexual, pedicellate, bracteate. Sepals 5, subequal, free, ovate-suborbicular, persistent. Petals 5, imbricate, obovate-orbicular, base connate, white, red, yellow or pink. Stamens numerous, adnate to the base of the petals, filaments slender, anthers versatile, 2 loculed, dehiscing by apical pores. Ovary globose, 3-5 locular, ovules numerous in each locule, axile placentation; style 3-5, free or connate till middle; stigma simple or discoid. Berry globose-ovoid, glabrous or hairy, usually ribbed, rarely dry and subdehiscent. Seeds numerous, small, brown. Fl. & Fr.: May – June.

Western Ghats: Nilgiri. India: Arunachal Pradesh, Assam, Sikkim, Tamil Nadu, West Bengal. World: Nepal, India, China.

IUCN: NE (IUCN, 2024).

Saurauia napaulensis DC., Mem. Soc. Phys. Geneve 1: 421. 1822. Synonyms: *Saurauia paniculata* Wall. ex G.Don, *Saurauia sterculifolia* Griff., *Tonshia polypetala* Buch.-Ham. ex D.Don, *Ternstroemia paniculata* Wall., *Ternstroemia racemosa* D.Don, *Zanthoxylum serra* Turcz.

Trees to 20 m tall; branchlets brown pubescent to glabrescent, with setose hairs intermixed with subulate scales. Leaves narrowly elliptic to oblong-obovate, 13-36 x 7-15 cm, leathery, abaxially sparsely ferruginous tomentose, adaxially glabrescent, sometimes with scales at base of midvein, lateral veins 28-40 pairs, base obtuse to subrounded to cuneate, margin finely serrate, apex acuminate to acute; petiole 2.5-5 cm, pubescent to glabrescent. Inflorescences 12-33 cm long, axillary, sparsely scaly and puberulent; peduncles with 1 or 2 bracts at base of each branch; pedicels 1.7-2.5 cm, with 2 nearly opposite bracteoles below middle; bracteoles caducous, lanceolate, 2-4 mm. Flowers pink to purplish, 0.8-1.5 cm across. Sepals unequal, outer 3 smaller, elliptic to broadly elliptic; inner 2 larger, broadly elliptic to orbicular, 5-7 mm. Petals oblong, 8 x 3 mm, connate at base, recurved at apex. Stamens 50-90. Ovary globose to depressed-globose; styles 4 or 5, connate below middle. Berry green to yellowish, globose to depressed-globose, ribbed or slightly ribbed. Fl. & Fr.: April - August

Western Ghats: Kodaikanal, Nilgiris. India: Uttar Pradesh, West Bengal. World: Bhutan, China, Laos, Malaysia, Myanmar, Nepal, Thailand, Vietnam.

IUCN: LC (IUCN, 2024).

ALTINGIACEAE Horan.

Liquidambar L.

Liquidambar styraciflua L., Sp. Pl. 2: 999. 1753. Vern.: *Sempisin maram* (Tam.).

Trees to 41 m tall; bark scaly, smooth. Leaves palmately lobed, main lobes sometimes again dentate-lobed, 7-19 x 4.4-16 cm; surfaces glabrous, except young leaves hairy on veins and main vein-axils at base with persistent reddish-brown simple hairs; stipules linear-lanceolate, 3-4 mm, early deciduous, leaving 2 stipular scars adaxially near base of petiole; petioles 60-100 mm long. Staminate flowers in pedunculate clusters, 3-6 cm. Perianth absent. Stamens 4-8 per flower, 150-176 per cluster, falling after anthesis. Pistillate flowers without perianth; hypanthium disclike, with 5-8 staminodes around cycle of disc lobes. Ovary 1-2-locular; styles 2; stigmas introrsely curved. Capsular heads brown at maturity, globose, 2.5-4 cm diam. Seeds apically winged, 8-10 mm, marked with resin ducts; aborted seeds brownish, 1-2 mm, unwinged, irregular, resembling sawdust. Fl. & Fr.: February – March

Western Ghats: Kodaikanal (Tamil Nadu), Naturalized. World: Trpical America, Cultivated elsewhere.

IUCN: LC (IUCN, 2024).

Uses: It is commonly called sweet gum tree, which has many commercial, medicinal, edible and residential uses.

ANACARDIACEAE R. Br

Key to the genera

1a. Leaves simple 2

1b. Leaves uni-pnnate 8

2a. Ovary superior 3

2b. Ovary inferior .... **Holigarna**

3a. Stamens as many as spetals, only one fertile .... **Mangifera**

3b. Stamens double the number of petals, all fertile 4

4a. Flowers bisexual; style single; fruit stalk not enlarged 7

4b. Flowers dioecious or polygamous; style 1-3; fruits stalk enlarged or not. 5

5a. Fruit stalk stount, not enlarge or not seated on fleshy cup ... **Nothopegia**

5b. Fruit stalk enlarged form the fleshy hypocarp .... 6.

6a. Flowers dioecious; hypocarp thick cup-shaped..................... **Semicarpus**

6b. Flowers polygamous; hypocarp fleshy form the pseudofruit .... **Anacardium**

7a. Stamens 4-6; drupe one-celled, oblique .... **Gluta**

7b. Stamens 8-10; drupe 2-valved, globose .... **Buchanania**

8a. Ovary 1-celled 9

8b. Ovary 2-6-celled .... **Spondias**

9a. Flowers bisexual; petals valvate; style single .... **Solenocarpus**

9b. Flowers dioecious; petals imbricate; styles 3 10

10a. Leaflets 7-9, margin entire .... **Lannea**
10b. Leaflets 3-5, margin undulate or lobed ....**Searsia**

Anacardium L.

Anacardium occidentale L., Sp. Pl.: 383 (1753). Synonyms: *Acajuba occidentalis* (L.) Gaertn., *Anacardium microcarpum* Ducke, *Anacardium occidentale* var. *americanum* Jacq., *Anacardium occidentale* var. *gardneri* Engl. Vern.: *Geru* (Kan.); *Kasumavu, Kappamavu, Parangimavu* (Mal.); *Andiparuppu maram, Kolla maram, Munthiri* (Tam.).

Trees to 4-10 m tall; branchlets glabrous to subglabrous. Leaves obovate, 8-11 x 6-8.5 cm, leathery, glabrous on both sides, base broadly cuneate, margin entire, apex rounded, truncate to retuse, lateral veins 12 pairs, reticulate venation pattern prominent on both sides; petiole 1-1.5 cm long. Inflorescence paniculate, 10-20 cm, glabrous to densely grayish sericeous; floral subtending bracts ovate-lanceolate, 5-10 mm, keeled, grayish sericeous abaxially, glabrous adaxially. Flower sessile to shortly pedicellate. Sepals 5, grayish sericeous abaxially, 4 x 1.5 mm. Petals 5, greenish yellow to red, linear-lanceolate, 7-9 x 1.2 mm, minutely pubescent to subglabrous adaxially. Stamens 7-10, larger one 8-9 mm in male flowers, 5-6 mm in bisexual flowers, sterile stamens 3-4 mm. Ovary to 2 mm diam, glabrous; style 4-5 mm. Fleshy hypocarp 3-7 x 4-5 cm, purplish red at maturity; drupe reniform, 2-2.5 x 1.5 cm, shiny. Fl. & Fr.: September–December

Western Ghats: Throughout. India: Almost all the states. World: Native to south America.

IUCN: LC (IUCN, 2024).

Uses: Cultivated this tree in commercial scale for its edible nut (Cashew nut).

Buchanania Spreng.

1a. Leaves scarcely coriaceous; inflorescence glabrous ....**B. axillaris**
1b. Leaves strongly coriaceous; inflorescence pubescent 2
2a. Leaves glabrous abaxially, apex acuminate ....**B. lanceolata**
2b. Leves hairy abaxially; apex obtuse or rounded 3
3a. Bark rough; petals hairy; leaves usually 5-10 cm wide ....**B. lanzan**
3b. Bark sooth; petals glabrous; leaves usually 3-5 cm wide 4

4a Ovary disc glabrous; leaves densely tomentose throughout abaxillay .... **B. abrahamiana**

4b. Ovary disc villous; leaves glabrous except midrib abaxially .... **B. barberi**

Buchanania abrahamiana E.S.S.Kumar & Shareef, Ann. Bot. Fennici 59: 23-27. 2022.

Trees to 17 m tall; bark smooth, slightly fissured, greyish; branchlets pendulous, pubescent when young. Leaves simple, alternate, broadly elliptic-oblong, 10-20 x 4.5-8.5 cm, cuneate or obliquely cuneate at base, obtuse or emarginate at apex, slightly revolute at margin, galbrous above, densely tomentose beneath, coriaceous; lateral nerves 15-20 pairs, subhorizontal, raised abaxially; petiole 1.3-2.5 cm long, subterte, densely tomentose. Inflorescence terminal panicles, 9-22 cm long, densely pubescent. Flowers scented, 6 mm in diam; bract ovate-deltoid, rufus abaxially, glabrous adaxially. Clyx lobes 5, pale green, ovate, 1 x 0.8 mm, villous. Petals 5, creamy white, ovate-oblong, 3 x 1.5 mm, glabrous. Stamens 10, biseriate; filaments 1-1.5 mm long, glabrous; anthers ovate, 0.5-0.8 mm long, ontuse at apex. Carples 4-5, immersed in disc, ellipsoid, only carpel developed, others suppressed. Ovary pilose, uni-ovuled, pendulous; style lateral, glabrous, short; stigma truncate. Drupe globose or subglobose, 1.8 x 1.6 cm, smooth, pinkish brown when ripe. Seed 1, stony, lenticular, smooth. Fl. & Fr.: November – April.

Western Ghats: Evergreen forests of Thiruvananthapuram district in Kerala (Endemic).

IUCN: NE (IUCN, 2024).

Buchanania axillaris (Desr.) Ramam., C. J. Saldanha & Nicolson, Fl. Hassan Dist. 374. 1976. Synonyms: *Mangifera axillaris* Desr., *Buchanania angustifolia* Roxb. Vern.: *Yemme murukalu gida, Maradi mara* (Kan.); *Kattuma, Kal mangai, Kolamaavu, Mudama, Saraparuppu* (Tam.).

Tree to 10 m tall; branches stout; bark rough, deeply fissured. Leaves linear-oblong, elliptic or elliptic-lanceolate, rounded or acute at base, entire at margin, obtuse, rounded or emarginate at apex, 7.5-15 x 3-5 cm, glabrous on both surfaces, reticulate; lateral nerves 12-15 pairs; petioles 1.75 -3 cm long, slender. Flowers 3-6 mm across, in axillary and terminal glabrous branched panicles, white. Sepals 5, lobes semiorbicular, 1 mm long. Petals 5, oblong or ovate, 2.5 mm long. Stamens 10; anthers oblong. Ovary pilose. Drupes obliquely globose, 1.2 x 0.4 cm, slightly compressed. Fl. & Fr.: May – September.

Western Ghats: Deciduous forests of Maharashtra, Karnataka, Kerala and Tamil Nadu. India: Andhra Pradesh, Telangana and Madhya Pradesh. World: Sri Lanka.

IUCN: NE (IUCN, 2024).

Uses: Seed kernal (Saraiparuppu) is used as edible and as tonic.

Buchanania barberi Gamble, Bull. Misc. Inform. Kew 1916: 135. 1916. Vern.: *Malamavu* (Tam.).

Trees to 20 m tall; bark dark brown with vertical deep fissures and horizontal cracks. Leaves simple, alternate, estipulate, 8-13 x 3-5 cm, elliptic, boradly ovate, oblong, obovate or elliptic-oblong, base attenuate or cuneate, apex obtuse or acute, margin entire, rusty villous beneath when young, coriaceous; lateral nerves 16-20 pairs, parallel, prominent; petiole 8-10 mm, slender. Flowers bisexual, small, white in terminal and axillary rusty villous panicles of 10 cm long; bract and bracteoles ovate-acute; pedicel 2-3 mm long. Sepals short; lobes acute, villous, persistence, imbricate. Petals 4-5, 2.5 cm long, oblong, acute, glabrous. Stamens 8-10, inserted at the base of the disc, 1 mm long; filaments slender; disc thick, crenulate. Carpels 5-6, free, seated in the cavity of disc; only 1 fertile, 1 mm long, villous at base, ovule 1, pendulous. Drupe single seeded. Fl. & Fr.: March – January.

Western Ghats: Semievergreen forest of Kerala and Tamil Nadu (Endemic)

IUCN: CR-B2ab(iii,v); C2a(i,ii); D. (IUCN, 2024)

Buchanania lanceolata Wight, Pl. Ind. Orient. 1: t. 237. 1839. Synonym: *Buchanania lonchophylla* Voigt Vern.: *Kulamavu, Malama* (Mal.); *Kaatuma* (Tam.).

Trees to 20 m tall; bark dark grey, smooth; exudation resinous. Leaves simple, alternate, elliptic-lanceolate, elliptic-oblanceolate, elliptic-obovate, linear-lanceolate, oblanceolate or oblong-lanceolate, 10-23.5 x 2.5-4.5 cm, base acute or cuneate, apex acute or acuminate, margin entire, glabrous, coriaceous; lateral nerves 13-20 pairs, pinnate, ascending, prominent, secondary laterals prominent, intercostae reticulate, prominent; petiole 20-25 mm, slender, grooved above. Flowers bisexual, white, in axillary and terminal rusty tomentose panicles. Sepals 4-5, shortly lobed, lobes 3-5, fid, ovate, pubescent, persistent, imbricate. Petals 4-5, ovate-obtuse, reflexed, glabrous. Stamens 8-10, 1.2 mm long, free, inserted at the base of disc; filaments subulent; anthers sagittate; disc thick, 8-10 crenate. Carpels 5-6, free, superior, seated in the cavity of the disc, only one fertile, hairy at base, ovule pendulous from a basal funicle; style short; stigma truncate. Drupe laterally compressed, 18 mm long, 2 valved, red. Fl. & Fr.: October – January.

Western Ghats.: Kerala, Karnataka. World: India, Myanmar.

IUCN: VU-D1. (IUCN, 2024).

Buchanania lanzan Spreng., Schrader, J. Bot. 2: 234. 1800. Synonyms: *Buchanania latofolia* Roxb. *Buchanania cochinchinensis* (Lour.) M.R. Almeida, *Toluifera cochinchinensis* Lour. *Buchanania lanzan* var. *palodensis* E.S.S.Kumar, Shareef & Raj Vikr. Vern.: *Charoli, Kole maavu, Murkali* (Kan.); *Kaattumangai, Sareiparuppu, Moraetha* (Tam.).

Trees to 18 m tall; bark thick, surface black or dark brown, rough, tessellate the cracks being deep and narrow, somewhat resembling crocodile hide. Leaves simple, alternate, estipulate, 10-23.5 x 5-12 cm, broadly oblong, base round or acute, apex obtuse or emarginate, margin entire, glabrous above and densely tomentose beneath, coriaceous; lateral nerves 10-20 pairs, pinnate, prominent, pubescent, intercostae reticulate, prominent; petiole 12-22 mm, stout, glabrous; Flowers bisexual, greenish-white, sessile, in pubescent terminal and axillary panicles. Sepals 5, persistent. Petals 5, oblong, reflexed, pilose. Stamens 10, inserted at the base of fleshy disc; filaments free, glabrous; disc cupular, 5 lobed, pilose. Carpels 5-6, free, superior, tomentose, 4 reduced to thread like processes, one fertile, ovule one, pendulous; style lateral; stigma truncate. Drupe oblong, 10-13 x 8 mm, laterally compressed, black; stone hard, 2 valved. Seed one. Fl. & Fr.: December – July.

Western Ghats: Deciduous forest to semi-evergreen forests of Gujarat, Maharashtra, Goa, Karnataka, Kerala, Tamil Nadu. India: Eastern Ghats and drier parts of the country. World - India, Myanmar, Indonesia, Malaysia.

IUCN: NE (IUCN, 2024).

Uses: Seeds are used as expectorant and tonic. The oil extracted from the seed kernal is applied for removing scar and pimples. The bark is furnished the natural varnish and tannin sources.

Gluta L.

Gluta travancorica Bedd., Fl. Sylv. S. India: t. 60. 1870. Vern.: *Thodappei* (Mal.); *Shenkurinji, Chenkurinji* (Tam.).

Evergreen trees to 35 m tall; bark thick, greyish-brown, smooth; exudation black, acrid. Leaves simple, alternate, crowded at the tips of branchlets, estipulate, 8-18 x 3.5-7 cm, elliptic-ovate, obovate-spathulate or spathulate, base cuneate or attenuate, apex obtuse, sometimes emarginate; margin entire, glabrous, coriaceous; lateral nerves 13-18 pairs, pinnate, prominent; intercostae reticulate, prominent; petiole 10-20 mm, stout, glabrous, winged. Flowers bisexual, cream coloured, 8 mm across, in axillary and terminal panicles. Sepals spathaceous, splitting irregularly, deciduous. Petals 4-6, inserted on the disc and adnate with it below, imbricate, spreading in flower; disc elongate, cylindric. Stamens 4-6, inserted on the disc; filaments filiform; anthers dorsifixed. Ovary obliquely ovoid, superior, apocarpous, sessile or stipitate, pubescent, 1-celled, 1-ovuled, pendulous; style lateral, filiform; stigma simple. Drupe globose, 2.5-3.5 cm across, brown; pericarp thick-scurfy. Seed adherent to the pericarp. Fl. & Fr.: June – August.

Western Ghats: Evergreen forests above 1000 m of Kerala and Tamil Nadu (Endemic).

IUCN: NT (Lower Risk/near threatened). (IUCN, 2024).

Uses: Timber is used for making furniture.

Holigarna Buch. -Ham. ex Roxb.

1a. Leaves quite glabrous beneath; drupes fully enclosed by hypocarp 2
1b. Leaves pubescent or tomentose beneath; drupes half-embended with hypocarp 4
2a. Drupes glabrous.......... **H. arnottiana**
2b. Drupes tomentose 3
3a. Leaves obovate, apex obtuse; stigma simple .... **H. ferruginea**
3b. Leaves spathulate, apex notched; stigma flagged .... **H. nigra**
4a. Petiolar spur early deciduous, leaf apex acuminate.......... **H. beddomei**
4b. Petiolar spur persistant, leaf apex acute.......... **H. grahamii**

Holigarna arnottiana Hook.f., Fl. Brit. India 2: 36. 1876. Synonyms: *Katou-tsjeroe arnottianum* (Hook.f.) Kuntze, *Holigarna longifolia* Wight & Arn., *Holigarna racemosa* Hohen. ex Engl., *Mangifera racemosa* Lam. Vern.: *Holigar, Huligeri, Kudugeru* (Kan.); *Cheru* (Mal.); *Bibu, Holgeri* (Mar.); *Karuncherai, Kaatucharai* (Tam.).

Evergreen trees to 20 m tall; bole buttressed; bark thick, pale brown, rough; exudation milky, black on drying. Leaves simple, alternate, crowded at the tips of branchlets, estipulate, 8-25 x 2.5-7.5 cm, oblanceolate, base cuneate or attenuate, apex obtusely acute, margin entire, glabrous, coriaceous; lateral nerves 10-20 pairs, parallel, intercostae reticulate, prominent; petiole 7-15 mm, stout, tomentose. Flowers polygamous, yellowish-brown, 2-3 mm across, in axillary panicles, tomentum of inflorescence pale golden brown. Sepals 5, cupular, 5-lobed. Petals 5, cohering at the base and with the edge of the disc, densely villous inside, valvate; disc lining the calyx tube, obscure in bisexual flowers. Stamens 5, inserted outside the disc; filaments subulate; anthers red, versatile. Ovary inferior, 1-celled, ovule pendulous; styles 3, divergent; stigmas capitate. Drupe obliquely-ovoid, 2.5 cm long, glabrous, almost entirely included within hypocarp, resinous, acrid. Seed parietal, single. Fl. & Fr.: December – august.

Western Ghats: Evergreen forests above 700 m in Maharashtra, Goa, Karnataka, Kerala, Tamil Nadu (Endemic).

IUCN: NE (IUCN, 2024).

Uses: Oleoresin extracted from the wood and fruit is used as varnish to wooden furnitures. The wood is used for making household implements. Fruits and bakrs are used in local medicine to treat some ailments.

Holigarna beddomei Hook.f., Fl. Brit. India 2: 36. 1876. Synonym: *Katou-tsjeroe beddomei* (Hook.f.) Kuntze Vern.: *Paal vidinjan* (Tam.).

Large trees to 24 m tall; bark thick, smooth, grey; exudation black on dryness. Leaves simple, alternate, estipulate, 25-38 x 14-17 cm, oblanceolate, broadened above the middle, base cuneate, apex acuminate, margin entire, ciliate, pubescent beneath, coriaceous; lateral nerves 20-30 pairs, parallel, midrib on upper surface ciliate, intercostae reticulate, prominent; petiole 10-25 mm, villous, stout, villous-pubescent. Flowers small, polygamous. Sepals 5, cupular, 5-fid, rusty tomentose. Petals 5, oblong, cohering at the base, villous inside; disc lining the calyx tube, with tuft of hairs at the centre. Stamens 5, inserted outside the disc; filaments subulate; anthers oblong-cordiform, versatile. Ovary inferior, 1-celled, ovule pendulous from near the top of the cell; styles usually 3, divergent; stigmas capitate. Drupe obliquely ovoid, hypocarp enclosing half way. Seeds parietal, single. Fl. & Fr.: January – March.

Western Ghats: Evergreen forest above 1000 m in Karnataka, Kerala, Tamil Nadu (Endemic).

IUCN: NE (IUCN, 2024), Endangered (Narasimhan and Irwin, 2021).

Holigarna ferruginea March., Rév. Anacardiac.: 171. 1869. Synonym: *Katou-tsjeroe ferrugineum* (Marchand) Kuntze Vern.: *Haisikutti, Karunchaeri* (Tam.).

Evergreen trees to 25 m tall; bark greyish-white, smooth; exudation acrid. Leaves simple, alternate, crowded at the tip of branchlets, estipulate, 7-17.5 x 4-10 cm, obovate, base attenuate, apex obtuse or emarginate, margin entire, glabrous, coriaceous; lateral nerves 7-12 pairs, parallel, somewhat ascending, prominent, midrib slightly pubescent when young; intercostae reticulate, prominent; petiole 10-20 mm, stout, glabrous. Flowers small, polygamous, in axillary and terminal panicled racemes, tomentum of inflorescence dark purple. Sepals 5, cupular, shortly 5-toothed. Petals 5, cohering at the base and with the edge of the disc, densely villous inside, valvate, ciliate along the margins; disc lining the calyx tube, obscure in bisexual flowers. Stamens 5, inserted out side the disc; filaments subulate; anthers versatile. Ovary inferior, 1-celled, ovule pendulous; style 3, divergent; stigmas capitate. Drupe obovoid, included within hypocarp, rusty tomentose. Fl. & Fr.: January – September.

Western Ghats: Evergreen forests of Goa, Karnataka, Kerala, Tamil Nadu (Endemic including Andaman & Nicobar).

IUCN: NE (IUCN, 2024), vulnerable (Narasimhan and Irwin, 2021).

Uses: Oleoresin extracted from the stem is used for painting varhish.

Holigarna grahamii (Wight) Kurz., J. Asiat. Soc. Bengal, Pt. 2, Nat. Hist. 41: 305 (1872) Synonyms: *Semecarpus grahamii* Wight., *Holigarna wightii* N.P. Balak., *Holigarna grahamii* Hook.f., *Katou-tsjeroe grahamii* (Hook.f.) Kuntze. Vern.: *Bipti, Balwuli* (Mal.); *Sarai* (Tam.).

Evergreen trees to 30 m tall; bark thick, surface yellowish-grey, mottled with green and white, smooth, lenticellate; exudation scanty; branchlets brown tomentose. Leaves simple, alternate, clustered at the tips of branchlets, oblanceolate, 15-70 x 7-25 cm, suddenly broad above the middle, base cuneate or acute, apex acuminate, margin entire, glabrous above, densely golden brown tomentose below; lateral nerves many, parallel, intercostae reticulate, prominent; petiole 15-50 mm long, stout, broadened at base, brown tomentose. Flowers small, polygamous, dull white, in terminal rusty tomentose panicles. Sepals 5, cupular, shortly 5-toothed. Petals 5, cohering at the base and with the edge of the disc, villous within; disc lining the calyx tube, obscure in bisexual flowers. Stamens 5, inserted outside the disc; filaments subulate; anthers versatile. Ovary inferior, 1-celled, ovule pendulous from near the top of the cell; styles usually 3, divergent; stigmas capitate. Drupe ovoid, included in hypocarp, vertically shallowly grooved, reddish-purple. Fl. & Fr.: December – June.

Western Ghats: Evergreen forest of Maharashtra, Goa, Karnataka, Kerala, Tamil Nadu. World: India, Myanmar.

IUCN: NE (IUCN, 2024), vulnerable (Narasimhan and Irwin, 2021).

Holigarna nigra Bourd., Indian Forester 30: 95. 1904. Vern.: *Cherei* (Mal.); *Karunjarai* (Tam.).

Trees to 30 m tall; bark thick, grey, smooth; exudation black. Leaves simple, alternate, estipulate, 7.5-15 x 3.7-5 cm, spathulate, base cuneate, apex obtuse or emarginate, margin entire, dark green, glabrous, coriaceous; lateral nerves 6-9 pairs, parallel, intercostae reticulate, prominent; petiole 20-25 mm long, stout, glabrous. Flowers polygamous, purple-white, in terminal and axillary dark purple brown tomentose racemes and panicle of 30.5 cm long in males and 7.5 cm long in female flowers. Sepals 5, villous, cupular, lobes 5. Petals 5, white within, clothed with white hairs. Stamens 5, filaments white, anthers black; disc 5 lobed. Ovary inferior, obconic, tomentose, 1-celled; styles 3, recurved; stigma capitate. Drupe, obovoid, 4 x 1.5 cm, glabrous. Fl. & Fr.: January – July.

Western Ghats: Evergreen forests above 1500 m in Karnataka, Kerala, Tamil Nadu (Endemic).

IUCN: NE (IUCN, 2024), vulnerable (Narasimhan and Irwin, 2021).

LANNEA A. Rich.

Lannea coromandelica (Houtt.) Merr., J. Arnold Arbor. 19: 353 (1938). Synonyms: *Dialium coromandelicum* houtt. *Odina wodier* Roxb., *Lannea grandis* (Dennst.) Engl., *Lannea wodier* (Roxb.) Adelb. Vern.: *Godda, Gugul* (Kan.); *Kalasan* (Mal.); *Othiyan, Odiya, Wodier* (Tam.).

Deciduous trees to 10-12 m tall; branchlets densely covered with ferruginous stellate hairs. Leaves imparipinnately compound, with 7-9-leaflets, ovate or oblong-ovate, 5.5-9 x 2.5-4 cm, adaxially glabrous, membranous or papery, abaxially mixed pubescent and ferruginous stellate-haired, base oblique, subrounded, margin entire, apex long acuminate or caudate-acuminate, lateral veins 6-10 pairs, slightly impressed adaxially, prominent abaxially; leaflets petiolule 1-3 mm, with ferruginous stellate hairs; main rachis terete, sparsely covered with ferruginous stellate hairs. Inflorescence paniculate or racemose, appearing before leaves, gathered at branch apices or on short shoots in leaf-scar axils, male inflorescence 15-30 cm, female inflorescence smaller; subtending bracts 1-2 mm with ciliate margins. Flowers small, yellow or purplish. Sepals 5, ovate to broadly ovate, 1 mm long, glabrous with ciliate margins. Petals ovate-oblong, 2.7 x 1.5 mm, recurved at anthesis. Stamens 8, subequal to petals in male flower, reduced and sterile in female flower. Disc annular. Ovary glabrous, ovoid, 4-locular, usually only 1 ovule fertile. Drupe ovoid to slightly reniform, purplish red at maturity, 6-10 × 0.5-1 mm. Fl. & Fr.: January – July.

Western Ghats: Common in foot hills to dry deciduous forests of Gujarat, Maharashtra, Goa, Karnataka, Kerala, Tamil Nadu. India: Most of the states. World: Sri Lanka, Nepal, Bangladesh, Pakistan, Myanmar, China, Bhutan, Laos, Cambodia, Vietnam, Thailand.

IUCN: LC (IUCN, 2024).

Uses: Gum exudate extracted from the bark is used for calico-printing, confectioneries in match stick industries. Wood is useful in pulp industry.

Mangifera L.

1a. Panicle glabrous; flowers yellow 2
1b. Panicle tomentose; flowers creamy .... **M. indica**
2a. Lateral veins 10-13 pairs; petiole to 3 cm long .... **M. austroindica**
2b. Lateral veins 16-20 pairs; petiole 3-7 cm long.......................... **M. sylvatica**

Mangifera austroindica Kosterm., Mangoes 127. 1993. Vern.: *Maa maram* (Tam.).

A shola tree to 20 m tall; bark rough, longitudinally fissured; branchletes green, cylindric, glabrous. Leaves elliptic-lanceolate, 7-15 x 3-6 cm, glabrous on both sides, margin entire, uneven sometimes; lateral nerves 10-13 pairs, Flowers yellow on branched glabrous panicles. Fruits unknown (Description adopted from the Isotype, *K.M. Matthew*, 26941[K]). Fl. & Fr.: December – May

Western Ghats: Lesserknown forest tree in Karnataka and Tamil Nadu (Endemic).

IUCN: NE (IUCN, 2024).

Uses: Fruits are said to be edible.

Mangifera indica L., Sp. Pl.: 200 (1753). Synonyms: *Mangifera amba* Forssk., *Mangifera anisodora* Blanco, *Mangifera viridis* Bojer Vern.: *Maa maram* (Tam.).

Trees, 10-25 m tall; branchlets brown, glabrous. Leaves oblong to oblong-lanceolate, 12-30 x 3.5-6.5 cm, leathery, deep green adaxially, light green abaxially, glabrous on both sides, base cuneate to obtuse, margin entire, undulate, apex acute to long acuminate, lateral veins 20-25 pairs, midrib prominent on both sides, reticulate venation obscure; petiole 2-6 cm, grooved apically, inflated basally. Inflorescence paniculate, terminal, 20-35 cm, glabrous to tomentose-pilose; bracts 1.5 mm long, lanceolate pubescent. Sepals ovate-lanceolate, 2.5-3 x 1.5 mm, glabrous to pubescent, acuminate. Petals light yellow with prominent red tree-shaped pattern adaxially, oblong or oblong-lanceolate, 3.5-4 x 1.5 mm, glabrous, recurved at anthesis. Fertile stamen 1, 2.5 mm long, with ovate anther; staminodes 4, 0.7-1 mm. Disc inflated, fleshy, 5-lobed. Ovary oblique, ovate, 1.5 mm in diam. at anthesis; style to 2.5 mm, eccentric. Drupe oblong to subreniform, greenish yellow to red, 5-10 x 3-4.5 cm; fleshy mesocarp bright yellow; endocarp more or less compressed. Fl. & Fr.: December – June.

Westrn Ghats: Cultivated for its fruits in Gujarat, Maharashtra, Goa, Karnataka, Kerala, Tamil Nadu. World: India, Sri Lanka, Myanmar, Laos, Cambodia, Vietnam, Thailand, Malesia.

IUCN: DD (IUCN, 2024).

Mangifera sylvatica Roxb., Fl. Ind. 2: 438 (1824)

Trees to 6-20 m tall; branchlets dark brown, glabrous. Leaves lanceolate to oblong-lanceolate, 15-24 x 3-5.5 cm, papery or thinly leathery, glabrous on both sides, base cuneate, margin entire, apex acuminate, lateral veins 16-20 pairs, prominent on both sides; petiole 3-7 cm, inflated basally. Inflorescence terminal paniculate, 15-33 cm long, glabrous, with slender primary branches; floral subtending bracts ovate-lanceolate, 1 mm long. Pedicels slender, 3-8 mm, glabrous, articulate. Sepals ovate to lanceolate, 3.5 x 1.5 mm, concave, glabrous. Petals white with 3-5 dark brown veins adaxially, lanceolate to linear-lanceolate, 7 x 1.5 mm, recurved and twisted apically at anthesis. Fertile stamen 1, 4 mm long; staminodes 1 or 2, shortly subulate to dentate. Ovary globose, 1.5 mm in diam; style subterminal. Drupe oblique, ovate, 6-8 x 4-5 cm, apex elongated into a curved beak; mesocarp thin; endocarp large, rounded to elliptic. Fl. & Fr.: February – May.

Western Ghats: Semi-evergreen forests areas fo Maharashtra. India: Most of the states. World: Nepal, Bangladesh, Myanmar and Thailand.

IUCN: LC (IUCN, 2024).

Nothopegia Blume

1a. Branchlets and leaves rusty villous; lateral nerves 25-30 pairs 2

1b. Branchlets and leaves glabrous; lateran nerves 10-13 pairs 4

2a. Leaves alternate, apex abruptly acuminate, petiole densely villous........3

2b. Leaves opposite, apex caudate acuminate, petiole quite glabrous............ .... **N. aureo-fulva**

3a. Racemes supra-axillary; drupes ovoid; ripened fruit pale green .... **N. travancorica**

3b. Racemes axillary; drupes globose; ripened fruit pinkish-red .... **N. monadelpha**

4a. Leaf apex acuminate 5

4b. Leaf apex acute or obtuse 7

5a. Flowers pentamerous .... **N. sivagiriana**

5b. Flwoers tetramerous 6

6a. Leaves thinly coriaceous, lanceolate; petals ovate................ **N. beddomei**

6b. Leaves thick coriaceous, obovate; petasl oblong 7

7a. Flowers small, 3 mm diam; sepals pubscent ....**N. racemosa**

7b. Flowers large, 7 mm diam; sepals glabrous .... **N. vajraveluei**

8a. Petals and stamens glabrous; drupes ovovid, pericarp thick..................... ....**N. colebrookiana**

8b. Petals and stamens pubescent; drupes compressed pole-wise; pericarp thin .... **N. heyneana**

Nothopegia aureo-fulva Bedd. ex Hook.f., Fl. Brit. India 2: 40. 1876

Evergreen trees to 10 m tall; bark rough, greyish, lenticellate. Leaves opposite to subopposite, 12-20 x 2-4.5 cm, linear oblong to lanceolate, acuminate at apex, around at base, glossy above; nerves 20-30 pairs, close and parallel, vein reticulations raised below, midrib hairy at base. Flowers in short axillary racemes, 1-3 cm long, polygamous. Sepals 4. lobes 0.05-0.1 cm long, hairy outside. Petals 4, oblong, 0.15-0.2 cm, brownish. Disc 4-lobed. Stamens 4, attached between lobes of the disc; filaments hairy at base. pitillode present in male flowers. Ovary 0.2-0.3 cm across, ovoid, 1-celled; ovules 1 in each cell; stigma capitate. Drupe globose, 0.3-0.5 cm diam.; stylar apex persistent in fruits. Fl. & Fr.: June – May.

Western Ghats: Evergreen forest above 1200 m in Kerala and Tamil Nadu (Endemic).

IUCN: CR- B1+2c (IUCN, 2024).

Nothopegia beddomei Gamble, Fl. Madras: 265. 1918. Vern.: *Ambata* (Kan.); *Aera, Cheru* (Mal.); *Pecherru maram, Sigatai* (Tam.).

Trees to 15 m tall; bark thick, grey-brown, mottled with white and green, smooth, brittle. Leaves simple, alternate, estipulate, 4-15 x 1.2-4.5 cm, elliptic or elliptic-oblong, acute at both ends, margin entire, glabrous, coriaceous; lateral nerves 10-25 pairs, parallel, prominent, intercostae subscalariform, prominent; petiole 3-10 cm, slender, glabrous, grooved above. Flowers small in axillary racemes, polygamous, sometimes subterminal. Sepals small, 4-lobed, margins ciliate with many dark longitudinal swellings on ventral surface, persistent, imbricate. Petals 4, oblong, glabrous, spreading, imbricate; disc annular, 4-lobed. Stamens 4, inserted under the margin of the disc; filaments subulate, anthers oblong, introrse. Ovary ovoid, superior, 1-celled, style short; stigma capitate; ovule pendulous from near the top of the cell from a flattened funicle. Drupe fleshy, ovoid, acute at apex, blue; pericarp thick filled with black resin cells. Seed pendulous. Fl. & Fr.: March – September.

Western Ghats: Evergreen forests of Goa, Karnataka, Kerala and Tamil Nadu. World: Sri Lanka.

IUCN: NE (IUCN, 2024).

Key to the varieties

1a. Leaf apex caudate-acuminate; sepals and filaments villous ....var. **beddomei**

1b. Leaf apex shortly acuminate; sepals and filaments glabrous ....var. **wynaadica**

Nothopegia beddomei Gamble var. **wynaadica** J.L. Ellis & V. Chandras, Bull. Bot. Surv. India 12: 257. 1972.

Fl. & Fr.: March – July.

Western Ghats: Evergreen forests of Kerala (Endemic).

IUCN: NE (IUCN, 2024).

Nothopegia colebrookeana (Wight) Blume, Mus. Bot. Lugd. Bot. 1: 203. 1850. Synonyms: *Nothopegia monadelpha* var. *colebrookiana* (Wight) K.M.Matthew, *Pegia colebrookeana* Wight, *Tapirira colebrookeana* (Wight) Hook.f. Vern.: *Ulagera* (Kan.); *Sonemous* (Mal.); *Amberi* (Mar.); *Peyicheri* (Tam.).

Trees to 4-9 m tall; branchlets glabrous; scars at lower portion of twigs, rusty pubescent. Leaves obovate to elliptic-lanceolate, 5-20 x 2.5 - 3.75 cm, acute to short acuminate, glabrous above, puberulous beneath, coriaceous; petioles 1-1.5 cm long, glabrous, longitudinally channelled. Inflorescence axillary racemes, 3-4 cm long, arising from axes of leaf. Male flower buds 1 mm in diam. Sepals4, lobes 0.5 x 0.7 mm, densely pubescent above. Petals oblong, 4, 0.8 x 0.5 mm, incurved, ciliate. Stamens 4, 0.5 mm long, glabrous. Disc 0.5 mm in diam. Carpels absent. Female flowers not seen. Drupes ovoid, 1.25 cm in diam. Fl. & Fr.: January – May.

Western Ghats: Deciduous and evergreen forests of Kerala and Tamil Nadu. World: Sri Lanka

IUCN: NE (IUCN, 2024).

Nothopegia heyneana (Hook.f.) Gamble, Fl. Madras: 265. 1918 var. **heyneana.** Synonyms: *Nothopegia colebrookiana* var. *heyneana* Hook.f. Vern. *Peyicheri* (Tam.).

Trees, 4-5 m tall; branches glabrous. Leaves elliptic to elliptic-oblong, 4-14 x 0.7-3.5 cm, obtuse to obtusely acute, puberulous beneath, coriaceous; petioles 0.4-0.7 cm long, glabrous. Racemes axillary, 1.5-3 cm long, rusty, puberulous. Flowers unisexual. Male flowers 2 mm across, sessile. Sepals 4, lobes 0.8 x 0.6 mm, margin ciliate with many dark longitudinal swellings on ventral surface. Petals 4, 1.5 x 0.8 mm, incurved, pubescent beneath. Stamens 4, 1.2 mm long, pubescent. Disc 1 mm in diam., 4-lobed. Female flowers 3 mm across; pedicels 1.5 mm long, pubescent. Ovary 3 mm in diam., sessile, glabrous; style short, thick; stigma globose. Drupes ovoid, 1-1.5 cm in diam., apiculate, compressed. Fl. & Fr.: April – September.

Western Ghats: Moist deciduous to evergreen forests of Kerala and Tamil Nadu (Endemic). India: Andhra Pradesh and Odisha.

IUCN: NT (Low risk/near threatened) (IUCN, 2024).

Key to the varieties

1a. Leaves elliptic or elliptic-oblong; petioles ca 0.7 cm long; racemes 3 cm or more in length ....var. **heyneana**

1b. Leaves linear-lanceolate; petioles ca 0.4 cm long; racemes less than 2.5 cm ....var. **linearifolia**

Nothopegia heyneana (Hook.f.) Gamble var. **linearifolia** Chandra & Ghosh, in Bangladesh J. Bot. 9(2): 173- 175. 1980. Vern.: *Kutrala Peyicheri* (Tam.).

Fl. & Fr.: July

Western Ghats: Evergreen forests above 900 m in Tamil Nadu (Endemic).

IUCN: NE (IUCN, 2024).

Nothopegia monadelpha (Roxb.) Forman, Kew Bull. 53: 754. 1998. Synonyms: *Blackburnia monadelpha* Roxb., *Blackburnia oppositifolia* Planch. Vern.: *Chorapala, Era, Snemous* (Mal.); *Peyicheri* (Tam.).

Evergreen trees to 12 m tall; bark greyish-brown, smooth; exudation white, acrid, branchlets dark, lenticellate, ultimate branchlets pubescent. Leaves simple, alternate, 5-15 x 2.5-5 cm, elliptic, elliptic-obovate, elliptic-oblong or obovate, apex acuminate, base acute, narrowed or attenuate, margin entire, glabrous, coriaceous, lateral nerves 15-21 pairs, parallel, prominent, intercostae scalariform, prominent; petiole 10-17 mm, stout, glabrous, grooved above. Flowers small, polygamous, greenish-white, 5-6 mm across, in axillary rufous racemes. Sepals small, 4-lobed; lobes orbicular, tomentose, persistent, imbricate. Petals 4, oblong, puberulus inside, densely tomentose outside, suberect with recurving tips; disc annular, 4-lobed. Stamens 4, inserted under the margin of the disc; filaments subulate, glabrous. Ovary sessile, ovoid, 1-celled, ovule pendulous from near the top of the cell; style short; stigma capitate. Drupe globose, 1-2 cm across, fleshy, pinkish-red, filled with black resin-cells. Fl. & Fr.: March – May.

Western Ghats: Evergreen forests of Karnataka, Kerala and Tamil Nadu. World: Sri Lanka.

IUCN: NE (IUCN, 2024).

Key to the varieties

1a. Leaves lanceolate-ellipotic, apex obtuse to acuminate; fruit pedicel up to 1 cm long .... var. **monadelpha**

1b. Leaves obovate-elliptic, apex round shortly mucronate; fruit pedicel up to 1.5 cm long .... .var. **macrocarpa**

Nothopegia monadelpha (Roxb.) Forman var. **macrocarpa** (Hook.f.) Karthik. & Moorthy. Fl. Pl. India: 81. 2009. Synonym: *Nothopegia colebrookeana* var. macrocarpa Hook.f.

Western Ghats: Rare in evergreen forests of Tamil Nadu (Endemic).

IUCN: NE (IUCN, 2024).

Nothopegia racemosa (Dalz.) Ramamurthy, Saldanha & Nicolson, Fl. Hassan Dist. 377. 1976. Synonyms: *Nothopegia castaneifolia* (Roth) Ding Hou., *Glycycarpus* *racemosus* Dalz., *Nothopegia dalzellii* Gamble, *Nothopegia colebrookeana* sensu Hook. f.

Trees up to 8 m tall; bark greyish, finely fissured; blaze deep red, latex black and scanty; branchlets glabrous. Leaves simple, alternate, 12-20 x 3.5-6.5 cm, elliptic-oblong or oblanceolate, glaucousbeneath, base acute, apex acuminate, margin entire, coriaceous; lateral veins 16-20, prominen; petiole 10-15mm long, slightly twisted, stout, glabrous. Flowers very small, polygamous, creamish-white in colour, borne in axillary, pubescent racemes. Sepals 4, broadly ovate, pubescent, persistent, imbricate. Petals 4, oblong, glabrous except ciliate hairs on the mid vein, inwardly curved; disc annular, fleshy, 4-lobed. Stamens 4; filaments puberulous. Ovary sessile, ovoid, single-celled; style short; stigma obscurely 3-lobed. Drupe globose, fleshy, light-green when immature, purple-violet when ripe, transversely oblong, striate, depressed atapex. Seed pendulous. Fl. & Fr.: May-August

Western Ghats: Moist deciduous to evergreen forests of Maharashtra, Goa, Karnataka, Kerala and Tamil Nadu. World: Sri Lanka.

IUCN: NE (IUCN, 2024).

Uses: The wood is used as timber. Fruit is edible.

Nothopegia sivagiriana Manickam & Murugan, J. Econ. Taxon. Bot. 32: 522. 2008. Vern.: *Sivagiri Peyicheri* (Tam.).

Trees, up to 8 m high; branchlets angular. Leaves simple, 5-15 x 3-5 cm, alternate, base acute, margin entire, apex acuminate, oblong - lanceolate, petioles 5-12 mm long. Flowers bisexual, pentamerous, 3 -4 mm, across; pedicels 2-3 mm long; bracts lanceolate, 1-1.5 mm long, base broad, apex acute. Calyx tube, 1-1.5 mm long, lobes 5, imbricate, ovate, 1-2 mm long, margin ciliate, apex acute. Corolla lobes 5, oblong, 6 x 2.5 mm, acute base. Stamens 5, filaments1-2 mm long, anthers oblong, 1.5 mm. Ovary globose, 1.5 mm long. Drupe globose, 1-1.5 cm across. Seeds 1. Fl. & Fr.: July – December.

Western Ghats: Rare in evergeen forests of Tamil Nadu (Endemic).

IUCN: NE (IUCN, 2024).

Nothopegia travancorica Bedd. ex Hook.f., Fl. Brit. India 2: 40. 1876. Vern.: *Thiruvidangur Peycheri* (Tam.).

Small trees to 7 m tall; branches rusty tomentose. Leaves elliptic-oblong or linear oblong, 9-16 x 2.5-4 cm, caudate-acuminate, midrib pubescent at base, coriaceous, alteral nerves 30-35 pairs; petioles 1-1.2 cm long, rusty pubescent. Racemes supra-axillary, 1 cm long. Male flowers 2 mm across. Sepals 4, 4-lobed; lobes broadly ovate-lanceolate, 2 x 0.5 mm, glabrous with 5-6 thick, dark swellings on ventral side. Petals 4, oblong, 2.5 x 0.8 mm, glabrous above, villous at base. Stamens 4, 2 mm long; filaments villous; anthers sagittate, sparsely villous. Disc 4-lobed, 0.8 mm in diam., glabrous. Female flowers 5 mm across. Sepals and petals as in male flowers. Stamens much shorter, 0.8 mm long. Ovary 1.5 mm in diam., sessile; style indistinct; stigma slightly lobed. Drupes ovoid, 1 x 1.8 cm, compressed. Fl. & Fr.: September – August.

Western Ghats: Evergreen forests at 1000 m in Karnataka, Kerala and Tamil Nadu (Endemic).

IUCN: NE (IUCN, 2024).

Nothopegia vajravelui K.Ravik. & V. Lakshm., Rheedea 9: 56. 1999. Vern.: *Paal vellayil, Mullamaram, Megamalai peyicheri* (Tam.).

Trees up to 10 m tall; branches glabrous, terete. Leaves alternate, obovate or elliptic-oblong to oblanceolate, 6-15 x 1.5-5 cm, obtusely acuminate at apex, acute-obtuse at base, entire, wavy, thickly coriaceous, glabrous; petioles up to 1.3 cm long, twisted, canaliculate. Racemes axillary up to 2 cm long, 7-12-flowered; penduncles up to 2 cm long, 4-5-angled, rusty hairy. Flowers unisexual, then male, or bisexual, 1 mm across. Sepals 4-5, lobes ovate-triangulate, 1 mm long, glabrous, greenish yellow. Petals oblong, 5.5 x 2 mm, glabrous. Stamens 4-5, alternating with petals; filaments equal or subequal, 2 mm long, pilose. Disc annular, 2 mm across, 4-5-lobed. Bisexual flowers: Sepals, petals, stamens and disc as in male or unisexual flowers. Ovary globose, up to 2.5 mm in diam.; styles cylindric, stout; stigma capitate. Drupes globose or ovoid, up to 2.2 x 1.2 cm, acute at both ends, glabrous, vertically striate, 1-seeded. Seeds ovoidglobose, pale yellow. Fl. & Fr.: April – August.

Western Ghats: Evergreen forsts Megamalai Wildlife Sanctuary of Tamil Nadu (Endemic).

IUCN: NE (IUCN, 2024).

Searsia F.A. Barkley

1a. Leaves trifoliolate; panicle or heads in terminal and axillary 2

1b. Leaves 3-5-foliolate; panicle branched terminal ....**S. paniculata**

2a. Leaflets pubescent beneath; drupe globose glabrous ....**S. mysorensis**

2b. Leaflets velvety canescent beneath; drupe ovoid, grey velvety hairy .... **S. tomenetosa**

Searsia mysorensis (G.Don) Moffett., Bothalia 37: 170. 2007. Synonyms: *Rhus mysorensis* G.Don., *Rhus sinuata*Thunb., *Toxicodendron mysurense* (G.Don) Kuntze., *Rhus indica* Wight & Arn. Vern.: *Hulmari, Sabale, Sappi* (Kan.); *Amboni* (Mar.); *Chippamaram* (Mal.); *Mulluvedanthi, Neikizhuvai, Chippamaram* (Tam.).

Small tree up to 2-5 m tall; branchlets woody spinescent. Leaves trifoliolate, leaflets 8-34 x 6-23 mm, the terminal the largest, obovate-cuneate, pubescent, the upper half irregularly lobed, lower entire; upper surface of a darker colour; petiole 8-18 mm long, pubescent. Panicle slender, 30-65 mm long, terminal and axillary. Flowers unisexual, yellowish. Bracts to 1 mm long, lanceolate, pubescent. Pedicel 2-4 mm long, pubescent. Petals 5, to 1 mm long, oblong-obtuse, glabrous, sometimes apex acutish. Disc faintly 5-lobed. Stamens 1 mm long, included, abortive in female flowers; filaments slender; anthers suborbicular. Ovary ovoid, less than 1 mm long, glabrous styles 3, spreading. Drupe subglobose, compressed, 4-5 mm broad, yellowish-brown, epicarp crustaceous, shiny. Fl. & Fr.: August – December.

Western Ghats: Scrub forests to deciduous forests of Maharashtra, Karnataka, Kerala and Tamil Nadu. India: Andhra Pradesh, Delhi, Punajb, Rajasthan and Uttar Pradesh. World: Myanmar, Pakistan and Africa.

IUCN: NE (IUCN, 2024).

Uses: It is dye yielding plant, brown dye is obtained from the wood and is used for cotton colouring.

Searsia paniculata (Wall. ex G. Don) Moffett., Bothalia 37: 171. 2007. Synonyms: *Rhus paniculata* Wall. ex G. Don., *Terminthia paniculata* (Wall. ex G. Don) C.Y. Wu & T.L. Ming., *Toxicodendron paniculatum* (Wall. ex D. Don) Kuntze Vern.: *Mulluvedanthi* (Tam.).

Small trees to 2-6 m tall; branchlets glabrous, lenticellate. Leaves palmately 3-foliolate or rarely 5-foliolate; leaflets sessile, oblong or oblanceolate, lateral leaflets 3-7 x 1.5-3 cm, terminal leaflet 6-11 x 2-4 cm, papery, base broadly cuneate, entire or shallowly undulate, apex obtuse, lateral veins prominent on both sides; petioles 2.5-4 cm, glabrous, channeled above. Inflorescence paniculate, terminal, 12-20 cm, with many slender branches, yellow hirsute. Flowers light yellow, 4- or 5-merous. Sepals glabrous, lobes obtuse, ovate, 0.5 mm. Petals elliptic, 1.5 x 0.8 mm, with featherlike venation pattern. Stamen filaments 0.5 mm; anthers ovoid, 0.3 mm. Disc 10-lobed. Ovary globose, 0.5-0.8 mm in diam., glabrous; styles 3, free; stigma capitate. Drupe subglobose, slightly compressed, 4 mm in diam; exocarp orange-red at maturity, finally separating; mesocarp glutinous. Fl. & Fr.: June – August.

Western Ghats: Common in foot hills up to 800 m in Tamil Nadu. India: Andhra Pradesh and West Bengal. World: Bhutan, Myanmar and China.

IUCN: NE (IUCN, 2024).

Searsia tomentosa (L.) F.A. Barkley., C.L. Lundell, Fl. Texas 3: 104. 1943. Synonyms: *Rhus tomentosa* L., *Toxicodendron tomentosum* (L.) Kuntze., *Rhus bicolor* Licht. ex Schult. Vern.: *Aprica Nei Kiluvai* (Tam.).

Small tree to 5 m tall; branchlets lenticellate. Leaves 3-foliolate, markedly discolorous; leaflets more or less broadly lanceolate, elliptic or obovate, dark grey-green above, silvery-white and densely velvety with midrib and lateral veins prominent below; margin entire or with a few large teeth in the apical half. Flowers in dense terminal and axillary heads, greenish, furry, unpleasantly scented. Drupe ovoid, 4 x 6 mm, in large clusters, covered in grey velvety hairs. Fl. & Fr.: October – December.

Western Ghats: Deciduous forests of Tamil Nadu (Naturalised). World: Native to Tropical Africa.

IUCN: LC (IUCN, 2024).

Uses: Cultivated as an ornamental tree.

Semecarpus L.f.

1a. Leaf-base auricled .... **S. auriculatus**

1b. Leaf-base entire 2

2a. Leaf entire, apex obtuse, round or emarginate 3

2b. Leaf entire or wavy, apex obtusely and abruptly acuminate .... **S. kathalekanensis**

3a. Leaf puberulent and glaucous beneath, and panicle tomentose .... **S. anacardium**

3b. Leaf glabrous beneath, and panicle almost glabrous ... **S. travancoricum**

Semecarpus anacardium L., Suppl. Pl.: 182. 1782 var. **anacardium**. Synonyms: *Anacardium latifolium* Lam., *Anacardium longifolium* Lam., *Semecarpus latifolius* Pers. Vern.: *Alakkucheru, Cherkuru, Cheru, Cherumaram, Perkuru, Sambiri, Temprakku, Thenkotta* (Mal.); *Munthiri, Saerankottai, Saenkottai* (Tam.).

Trees to 12 m tall; bark thick, grey to greyish-brown with tessellate scales, furrows longitudinal and horizontal; exudation acrid. Leaves simple, alternate, crowded at end of branchlets, obovate-oblong, elliptic-oblong or obovate, 15-50 x 10-15 cm, base acute, obtuse, round or truncate, apex obtuse, round or emarginate, margin entire, glabrous above, puberulent and glaucous beneath, coriaceous; lateral nerves 10-20 pairs, parallel; intercostae reticulate, prominent; petiole 20-45 mm long, stout, pubescent. Flowers polygamous, greenish-yellow, 6-8 mm across, in terminal pubescent panicles. Sepals 5, lobes ovate, margin ciliate, imbricate, deciduous. Petals 5, imbricate. Disc broad, annular, villous. Stamens 5 inserted below the disc; filaments filiform, glabrous, in male flowers as long as or longer than petals, in female flowers shorter than petals; anthers oblong. Ovary superior, densely villous, 1-celled, ovule pendulous from a basal funicle; styles 3, divergent. Drupe fleshy, 2-5 cm long, purplish-black, seated on an orange hypocarp as long as the drupe; seed pendulous. Fl. & Fr.: July – December.

Western Ghats: Evergreen forests of Gujarat, Maharastra, Goa, Karnataka, Kerala and Tamil Nadu. India: Andhra Pradesh, Arunachal Pradesh, Assam, Bihar, Gujarat, Madhya Pradesh, Odisha, Sikkim, Uttar Pradesh and West Bengal. World: Nepal, Malaysia and Australia.

IUCN: NE (IUCN, 2024).

Uses: Dye obtained from the fruits used for permanemt marking on cloth. Gum is used in traditional medicine.

Key to the varieties

1a. Leaves acute at base, puberulent or galacous beneath ...var. **anacardium**

1b. Leaves cuneate at base, often tomentose beneath ....var. **cuneifolia**

Semecarpus anacardium L. var. **cuneifolia**. DC. Prodr. 2: 62. 1825.

Fl. & Fr.: December – May.

Western Ghats: Evergreen forests of Maharashtra and Tamil Nadu (Native of Himalaya). World: Myanmar

Semecarpus auriculatus Bedd., Fl. Sylv. S. India: t. 232. 1872. Synonyms: *Cassuvium auriculatum* (Bedd.) Kuntze Vern.: *Charae* (Mal.); *Vellei cheri* (Tam.).

Trees to 25 m tall; bark thick, greyish-brown, smooth; blaze red; exudation watery, turns black on drying. Leaves simple, alternate, clustered at apex, estipulate, 10-28 x 1.7-6.7 cm, oblanceolate, base auriculate, apex acuminate, margin entire, glabrous, coriaceous; lateral nerves 13-17 pairs, parallel, ascending, intercostae reticulate, prominent; petiole 2-7 mm, stout, glabrous. Flowers small, polygamous; male flowers yellow and female flowers green, inconspicuous, in axillary glabrous panicles. Sepals 5, lobes 4-5, deciduous. Petals 5, sometimes 4 in male flowers, 5 in bisexual flowers, ovate, glabrous. Disc annular, broad. Stamens 5, inserted outside the disc; filaments filiform, glabrous, in bisexual flowers shorter than petals, in male flowers, as long as or longer than petals; anthers oblong. Ovary superior, glabrous, 1-celled, ovule pendulous from a basal funicle; styles 3, divergent. Drupe obliquely ovoid, 18 x 12 mm, fleshy, seated on a fleshy cup like receptacle, black, pericarp thick, loaded with acrid resin. Seeds pendulous. Fl. & Fr.: November – March.

Western Ghats: Evergreen forests of Karnataka, Kerala and Tamil Nadu (Endemic).

IUCN: NE (IUCN, 2024).

Uses: Wood is used for making match boxes.

Semecarpus kathalekanensis Dasappa & Swam., Phytotaxa 247: 299. 2016. Vern.: *Dodda ele holegare, Hedagalu* (Kan.).

Trees, up to 20 m tall; bark smooth, greyish brown, mottled with numerous prominent lenticels, blaze light brown; latex acrid, watery, turning black after exposure. Leaves simple, alternate, spiral, 45-100 x 15-22 cm, largely oblong-lanceolate or obovate, apex obtusely and abruptly acuminate, base cuneate and oblique, margin entire and wavy; secondary nerves 20 pairs, tertiary nerves reticulate; petiole 5-10 cm long. Inflorescence axillary panicles with unisexual greenish flowers; male inflorescence 30-50 cm long; female inflorescence 5-15 cm long, minutely rusty pubescent. Drupe obliquely reniform, compressed, 2 cm diam, 1-seeded. Fl. & Fr.: November – May.

Western Ghats: Rare in evergreen forests of Karnataka (Enedemic).

IUCN: CR- C2a(i) (IUCN, 2024).

Semecarpus travancoricus Bedd., Fl. Sylv. S. India: t. 232. 1872. Synonyms: *Cassuvium travancoricum* (Bedd.) Kuntze Vern.: *Kadu-koppi* (Kan.); *Avukaram* (Mal.); *Kaatu Shenkottai, Thiruvidangur saenkottai* (Tam.).

Trees to 30 m tall; bark surface greyish-brown blotched with black, smooth, glabrous; exudation black on drying. Leaves simple, alternate, clustered at the tips of branchlets, estipulate, 20-60 x 10-15 cm, obovate or obovate-oblong, base acute, apex obtuse, round or emarginate, margin entire, revolute, glabrous, coriaceous; lateral nerves 12-21 pairs, parallel, ascending, intercostae reticulate, prominent; petiole 20-110 mm, stout, glabrous. Flowers polygamous, greenish-yellow in axillary and terminal glabrous panicles. Sepals 5, imbricate, deciduous. Petals 5, ovate, glabrous, imbricate. Disc broad annular. Stamens 5, inserted below the disc; filaments filiform, as long as or longer than the petals in male flowers, shorter than petals in bisexual flowers; anthers oblong. Ovary superior, glabrous, 1-celled; ovules pendulous from a basal funicle; styles three, divergent. Drupe obliquely oblong, 3-3.7 cm long, fleshy, black, striate, seated on a short broad furrowed hypocarp. Seeds pendulous. Fl. & Fr.: September – July.

Western Ghats: Evergreen forests above 900 m in Kerala and Tamil Nadu (Endemic).

IUCN: NE (IUCN, 2024).

Solenocarpus Wight & Arn.

Solenocarpus indicus Wight & Arn., Prodr. Fl. Ind. Orient. 1: 171. 1834. Synonyms: *Spondias indica* (Wight & Arn.) Airy Shaw & Forman Vern.: *Kaatu Maamaram* (Tam.).

Small trees up to 25 m tall; bark brown, rough, branchlets slender. Leaves pinnately compound; leaflets to 6 pairs, opposite or subopposite, 6-7 x 2-3 cm, ovate-lanceolate, apex acuminate, sessile except terminal one, many-nerved, base oblique; petiolule of terminal leaflet 12 mm long. Panicle to 20 cm long, axillary or terminal, spreading. Flowers 3 mm across, white, many, bisexual. Sepals5, lobes small, imbricate, cauducouus. Petals 5, ovate to oblong-lanceolate, recurved. Stamens 10, inserted outside the disk. Ovary sessile, unilocular; ovule solitary, pendulous. Drupe obovoid, 8 x 5 mm, smooth. Fl. & Fr.: January – July.

Western Ghats: Evergreen forests of Maharashtra, Karnataka, Kerala and Tamil Nadu (Endemic).

IUCN: NE (IUCN, 2024).

Spondias L.

Spondias pinnata (L.f.) Kurz., Prelim. Rep. Forest Pegu, App. A: 44. 1875. Synonyms: *Mangifear pinnata* L.f., *Spondias mangifera* Willd., *Spondias acuminata* Roxb. Vern: *Ambaada* (Ori); *ambate, Amate, Poondi, Tindeeka* (Kan.); *Ambalam,*

Manpuli, Kattampazham (Mal.); *Amada, Ambada, Dholamba* (Mar.); *Kaattu Mamaram, Kinjam, Pulima, Mampulichai, Puli ilai, Puli maa* (Tam.).

Deciduous trees to 10-15 m tall; branchlets yellowish brown, glabrous. Leaves 30-40 cm long, imparipinnately compound with 5-11 opposite leaflets; leaflet ovate-oblong to elliptic-oblong, 7-12 x 4-5 cm, papery, glabrous on both sides, base cuneate to rounded, often oblique, margin serrate or entire, apex acuminate, lateral veins 12-25 pairs; petiole 10-15 cm, petiole and rachis glabrous; leaflet petiolule 3-5 mm. Inflorescence paniculate, terminal, 25-35 cm, glabrous, basal first order branches 10-15 cm. Flower sessile or subsessile, white, glabrous. Sepals 5, lobes triangular, 0.5 mm. Petals ovate-oblong, 2.5 x 1.5 mm, apically acute. Stamens 1.5 mm long. Ovary subglobose, 1 mm across; styles 4 or 5, free, 0.5 mm. Drupe ellipsoid to elliptic-ovoid, yellowish orange at maturity, 3.5-5 x 2.5-3.5 cm; inner part of endocarp woody and grooved, outer part fibrous; mature fruit usually with 2 or 3 seeds. Fl. & Fr.: December – July.

Western Ghats: Deciduous foresrts of Gujarat, Maharashtra, Goa, Karnataka, Kerala and Tamil Nadu. India: Andaman & Nicobar Islands, Assam, Bihar, Meghalaya, Tripura, Uttar Pradesh ad West Bengal. World: India, Sri Lanka, China and Tropical Asia.

IUCN: NE (IUCN, 2024).

Uses: Fruit is edible. Wood is used for making packing boxes. The bark is used in local medicine to treat diarrhoea and dysentery. The bark is also a good source of tannin.

Planted:

Pistacia terebinthus M. Bieb.; Exotic: Native to Northern Africa, West Asia, South Europe.

Rhus simarubifolia A. Gray; Exotic: Native to Fiji Islands.

Schinus mole L.; Exotic: Native to Southern America

Schinus terebinthifolia Raddi; Exotic: Native to Brazil

Spondias cytherea Sonn. Vern.: *Kasi-ambate* (Kan.), *Vilayati ambada* (Mar.). Exotic: Native to Pasific Islands

ANNONACEAE Juss.

Key to genera

1a. Flowers solitary or in pairs; carpel solitary ....**Cyathocalyx**
1b. Flowers usually more than 2, in cymes or fascicles 2
2a. Inner petals longer as long as or shorter than outer ones 3
2b. Inner petals longer than outer ones 4
3a. Carpels more than 15 ....**Goniothalamus**
3b. Carpels less than 15 6
4a. Stamens 6-12; carples 3-15 ....**Orophea**
4b. Stamens more than 12; carpels more than 15 5
5a. Stamens usually 24; monocarps globose, orange coloured ....**Popowia**
5b. Stamnes many; monocarps ellipsoid, orange red .... **Meiogyne**
6a. Receptacle hollow .... **Xylopia**
6b. Receptacle not hollow 7
7a. Aggregate fruit single compact; ovule single in each ovary .... **Annona**
7b. Aggregate fruit separated monocarps; ovules more than 2 in each ovary 8
8a. Outer petals similar amd almost as long as sepals .... .11
8b. Outer petals neither similar nor as long as sepals 9
9a. Outer petals ovate or strap-shaped, finely tomentose or pubescent.......... .... **Monoon**
9b. Outer petals ovate-rhomboid, glabrous 10
10a. Petals whorls unequal, tip rounded or broad; seed raphe deeply grooved..**Huberantha**
10b. Petals whorls equal, tip acute; seed raphe minutely grooved **Polyalthia**
11a. Monocarps woody ....**Alphonsea**
11b. Monocarps not woody 12
12a. Petals imbricate in bud; stamens with connective produced .... **Sageraea**
12b. Petals valvate in bud; stamens with anthers concealed by connective..13
13a. Inner petals saccate at base .... **Miliusa**
13b. Inner petals not saccate ....**Mitrephora**

Alphonsea Hook.f. & Thoms.

1a. Flower dense in fascicles; ripe carpels broadly globose or ovoid, to 1.5 cm long .... **A. lutea**

1b. Flowers few in fascicles; ripe carpels broadly subglobose or ovoid, to 2.5 cm long .... **A. zeylanica**

Alphonsea lutea (Roxb.) Hook.f. & Thoms., Fl. Ind. 1: 153. 1855. Synonyms: *Uvaria lutea* Roxb., *Uvaria russelii* Wall., *Alphonsea madraspatana* Bedd., *Alphonsea sclerocarpa* Thw. Vern.: *Karukarumbai maram* (Tam.).

Trees to 10-20 m tall; bark dark brown, corky, fissured and fibrous, branchlets minutely rufous-tomentose. Leaves simple, alternate, ovate-oblong to oblong-elliptic, 8-14 x 3-6.5 cm across, base obtuse or rounded, margin entire, apex shallow acuminate or acute, coriaceous, dark green, glabrous above, sparsely minutely pubescent beneath; lateral veins 7-11 on either side of the midrib, almost parallel, impressed above and beneath; petiole pubescent when young, 5-8 mm long. Inflorescence very short raceme. Flowers in dense fascicles, leaf opposed, subcapitate, 1.2-1.7 cm across, pedicels 5-7 mm long; bracts very small. Sepals 3, connate into 3 angled shallow cup, tomentose outside and glabrous inside, about 2-3 mm long. Petals 6, subequal, ovate, apex acute, thick, tomentose outside, glabrous inside, 10 x 5 mm, reflexed at the tips. Stamens many, 1.5 mm long, connectives broad and blunt. Carpels 1-5, oblong, strigose, 2.5 mm long; stigma sessile, capitates. Riped carpels, broadly globose or ovoid, obtuse at both ends, 1.5 cm long, puberulous, stalk stout. Seeds 3-6 in 2 rows, oval-oblong, pointed. Fl. & Fr.: March – September.

Western Ghats: Dry deciduous forests up to 800 m in Kerala and Tamil Nadu. India: Andhra Pradesh, Arunachal Pradesh, Assam and Odisha. World: Sri Lanka and Myanmar.

IUCN: LC (IUCN, 2024).

Uses: Wood is used for timber and firewood.

Alphonsea zeylanica Hook.f. & Thoms., Fl. Ind. 1: 153. 1855. Synonym: *Guatteria acutifolia* Wight ex Wall. Vern.: *Ceylon Karukarumbai maram* (Tam.).

Trees to 8 m tall. bark dark brown, corky, fissured and fibrous, branchlets puberulous, droopy when young. Leaves simple, alternate, lanceolate-elliptic to narrow ovate, 5-9 x 1.5-3 cm across, base acute, margin entire, apex acuminate, coriaceous, dark green, glabrous shining above; lateral veins 6-10 on either side of the midrib, impressed above and beneath, reticulate veinlets; petiole pubescent when young, 4-5 mm long. Inflorescence very short raceme. Flowers bisexual, few in extra axillary fascicles or leaf opposed fasciles, subcapitate, 1-1.5 cm across; pedicels pubescent, about 5-7 mm long; both basal and submedian bracts small. Sepals 3, very small, base connate, tomentose outside and glabrous inside. Petals 6, subequal, ovate, apex acute, yellowish green, glabrous inside, reflexed at the tips. Stamens many, 1.5 mm long, connectives broad, apiculate. Carpels 5-6, oblong, 1 mm long; stigma sessile, capitate. Riped carpels subglobose or ovoid, obtuse at both ends, 2.5-3.5 cm long, talk stout. Seeds 4-6 in 2 rows. Fl. & Fr.: January –April.

Westren Ghats: Deciduous forests of Karnataka, Kerala and Tamil Nadu. World: Sri Lanka.

IUCN: NE (IUCN, 2024).

Note: *Alphonsea zeylanica* Hook.f. & Thoms. is excluded from the Western Ghats in the Flora of Kerala (Daniel, 2005) and mentioned that it is narrowly endemic to Sri Lanka, erroneously reported from southern India due to confusion with *A. sclerocarpa* Thwaites.

Annona L.

1a. Leaves glabrous; petals 6 2
1b. Leaves hairy or puberulous beneath; petals 3 3
2a. Leaves ovate; petiole to 7 mm long; fruit spiny .... **A. muricata**
2b. Leaves elliptic or ovate-lanceolate; fruit smooth .... **A. glabra**
3a. Leaves to 9 cm wide; tomentose beneath .... **A. cherimola**
3b. Leaves to 5 cm wide; sparsely hairy only on nerves .... **A. squamosa**

Annona cherimola Mill., Gard. Dict., ed. 8; Annona No.5. 1768. Synonyms: *Annona pubescens* Salisb. in Prodr. Stirp. Chap., *Annona tripetala* Aiton in Hort. Vern.: *Mulluseetha* (Tam.).

Trees to 8 m tall; branchlets rusty tomentose. Leaves simple, alternate, elliptic, ovate or ovate-lanceolate, 5-14 x 4-9 cm, base round or obtuse, shortly decurrent on petiole, apex acute, acuminate or obtuse, margin entire, pubescent above and velvety tomentose beneath, sub-coriaceous; lateral nerves 11-14 pairs, pinnate, intercostae reticulate; petiole 6-12 mm long, slender, glabrous. Flowers green, solitary or upto 5 in extra-axillary or leaf opposed clusters; pedicels 1-1.3 cm long, tomentose. Sepals 3, 2-4 mm long, connate, tomentose on outer side, glabrous within. Petals 3+3, 1.5-3 cm long, greenish with out, puberulous, basally purple-spotted with in, fleshy. Stamens many; connectives concealing the anthers. Pistils many, linear-oblong, stigma spathulate. Fruit an aggregate of berry, 12 x 8 cm, globose or subglobose, turning yellow or reddish, smooth; areoles rather flat, separated by a reticulation of often raised ridges; pulp yellowish. Seeds black-brown. Fl. & Fr.: December-June

Western Ghats: Cultivated and naturalized. World: Native to south America

IUCN: LC (IUCN, 2024).

Annona glabra L., Sp. Pl.: 537. 1753. Synonyms: *Cassia ketschta* Hasselq., *Asimina arborea* Raf., *Annona australis* A.St.-Hil. Vern.: *Kattatha* (Mal.); *Periya Paanchipazham* (Tam.).

Trees to 12 m tall; branches glabrous, lenticellate. Leaves ovate, elliptic-ovate, elliptic or oblong, 6-20 x 3-8 cm, glabrous at maturity, secondary veins 7-12, reticulate, base obtuse to rounded and slightly decurrent onto petiole, apex acute to obtuse; petiole 0.8-2.5 cm. Inflorescences internodal or terminal on a short branchlet, 1-flowered. Flowers greenish, pedicel 1.5-2 cm, glabrous. Sepals broadly ovate, 3-4 x 3-4 mm, free, glabrous. Petals greenish yellow to pale yellow and inside basally with a red spot, broadly ovate, 1.5-3 x 1.3-2.5 cm, outside glabrous, inside minutely puberulent, apex obtuse; inner petals outside yellow, inside carmine, 1.2-2.5 x 0.7-1.5 cm, minutely puberulent. Stamens oblong, 3-4 mm; connectives apically slightly convex. Carpels connate at anthesis, glabrous. Syncarp yellow to orange, ovoid, 5-12 x 5-8 cm, smooth, apex rounded. Seeds pale reddish brown, 1.3-1.5 cm. Fl. & Fr.: January – April

Western Ghats: Cultivated and naturalized. World: Native to Tropica America, West Tropical Africa and cultivated elsewhere.

IUCN: LC (IUCN, 2024).

Uses: Cultivated for its edible fruits.

Annona muricata L., Sp. Pl.: 536. 1753. Vern.: *Mullaathi, Mullenchaka* (Mal.); *Mullu seetha, Mullu Paanchipazham* (Tam.).

Trees to 10 m tall; bark pale brown; young twigs glabrescent. Leaves simple, alternate, estipulate, 7-14.5 x 3-5.5 cm, elliptic, oblong, obovate, oblong-obovate, elliptic-oblong or elliptic-obovate, base acute, apex acute to acuminate, coriaceous, margin entire; lateral veins 8-10 pairs, slender, pinnate, prominent, intercostae reticulate, domatia present; petiole 4-8 mm long, slender, glabrous, grooved above. Flowers yellowish-green, solitary, axillary or from mature branches. Sepals 3, triangular, persistent. Petals 3+3 in two rows, ovate-acute, yellow, thick, glabrous, outer ones 2.5-3.5 x 2-2.5 cm, base cordate, apex acuminate; inner petals 1.5 x 1 cm, shortly stipitate. Stamens many, 4-5 mm long, linear, filaments broad at base, with capitate top of the connective. Ovary superior, 4 mm long, linear, slightly curved, strigose; style broad at base; stigma entire. Fruit an aggregate of berry, ovoid to obovoid, 15-25 x 10-15 cm, green, covered with curved spines, stalks 2-3 cm long, stout. Seeds many, reddish-brown, 1.5 cm long. Fl. & Fr.: April-October

Western Ghats: Cultivated and naturalized. World: Colombia, Costa Rica, Mexico, Panama and cultivated elsewhere.

IUCN: LC (IUCN, 2024).

Annona squamosa L., Sp. Pl.: 537. 1753. Synonyms: *Annona asiatica* L., *Annona glabra* Forssk., *Xylopia glabra* L. Vern.: *Aathi, Amirthunka, Athamaram, Seethapazham* (Mal.); *Paanchipazham, Seetha* (Tam.).

Trees to 6 m tall; bark brownish or brownish-black; branchlets seriaceous, glabrous when mature. Leaves simple, alternate, estipulate, 5-17 x 2-7 cm, ovate, ovate-lanceolate or elliptic-oblong, base acute, obtuse, cuneate, or round, apex acute or obtuse, margin entire, glabrous, glaucous beneath, coriaceous; lateral nerves 8-12 pairs, pinnate, intercostae reticulate; petiole 6-20 mm long, stout, glabrous, grooved above. Flowers axillary, solitary or a few together, leaf-opposed; pedicels 1-3 cm long, glabrous. Sepals 3, 2-3 x 3-4 mm, broadly ovate, shortly acuminate, pubescent outside. Petals 3+3 in two rows, outer 3 petals, 1.5-3 x 3-0.5 cm, linear-oblong, keeled inside, thick, greenish-white or yellow, reddish at base inside, pubescent outside, inner three petals usually missing or rudimentary, ovate, 1 mm long. Stamens many, 1 mm long, anther thecae narrow, with ovoid top of connectives. Carpels many, superior, 1 mm long, subconnate, ovoid, pubescent, 1-ovuled; style oblong; stigma entire. Fruit an aggregate of berry, 8-10 cm across, ovoid, greenish, glabrous, tuberculate with rounded tips, glaucous, pulp white. Seeds many, black, shiny. Fl. & Fr.: June-October

Western Ghats: Cutlivated and naturalized. World: Colombia, Costa Rica, Mexico, Panama and cultivated elsewhere.

IUCN: LC (IUCN, 2024).

Note: *Annona* trees are frequently planted in estates and orchards for its delicious fruits. They escaped from the cultivation and naturalized growing in the edges of forests and roadside. Hence it has been included in the enumeration. **Annona reticulata** L. is less cultivated in forests and it is mostly seen in plain areas.

Cyathocalyx Champ. ex Hook.f. & Thoms.

Cyathocalyx zeylanicus Champ. ex Hook.f. & Thoms., Fl. Ind. 1: 127. 1855. Synonyms: *Soala litoralis* Blanco Vern.: *Ceylon Manoranjitham* (Tam.).

Evergreen trees to 10-20 m tall; bark greyish-brown, smooth; branches horizontal, young branches drooping. Leaves simple, alternate, elliptic, oblong, elliptic-oblong or oblong-lanceolate, 13-27 x 4-9 cm, base obliquely obtuse, cuneate or acute, apex acuminate or caudate-acuminate, margin entire, coriaceous, glabrous, shiny above; lateral nerves 8-13 pairs, intercostae reticulate, midrib and nerves pubescent beneath when young; petiole 7-10 mm, slender, canaliculate above. Flowers solitary, or in pairs, extra axillary, pale green; pedicels stout, 1-5 cm long, brown pubescent when young. Sepals combined in to a 3 lobed cup, rusty tomentellous, glabrous inside. Petals 3+3 in two rows, constricted above base, thinly fleshly, clawed, ovate-lanceolate, subacute, golden pubescent outside, puberulous inside; outer petals 2.5-4 x 7-0.8 cm; inner petals as long as or slightly longer and narrower than outer. Stamens numerous, conective broad, concealing the anthers, cuneate at the base, truncate at the top, 2 mm long. Carpels solitary on a concave torus, cylindric, 4 mm long. Ovary pubescent, ovules many, style short, stigma peltate, dark red. Berry ovoid, 7 x 5 cm, slightly constricted in the middle, fleshy, yellowish, turning black, glabrous. Seeds 8-10 in 2 rows, compressed, testa reddish-brown, bony, smooth. Fl. & Fr.: January – June.

Western Ghats: Evergreen forests of Karnataka, Kerala and Tamil Nadu. World: Sri Lanka.

IUCN: NE (IUCN, 2024).

Goniothalamus Hook.f. &Thoms.

1a. Connective of stamens apiculate at apex 2

1b. Connective of stamens flat or convex at apex 3

2a. Stigma entire .... **G. rhynchantherus**

2b. Stigma bifid .... **G. wynaadensis**

3a. Outer petals never exceeding 2 cm in length 4

3b. Outer petals always exceeding 2 cm in length 7

4a. Leaves oblong lanceolate, monocarps shortly stalked .... **G. cardiopetalus**

4b. Leaves lanceolate, narrowly lanceolate to ovate or elliptical lanceolate, rarely obovate-lanceolate. 5

5a. Sepals persistent in fruits .... **G. wightii**

5b. Sepals deciduous in fruits 6

6a. Stigma funnel shaped .... **G. keralensis**

6b. Stigma bifid .... **G. salicinus**

7a. Trees, 6-20 m tall. Sepals 8-13 mm long .... **G. thwaitesii**

7b. Small tree, upto 8 m tall. Sepals 5–6 mm long .... **G. sericeus**

Goniothalamus cardiopetalus (Dalz.) Hook.f. & Thoms., Fl. Ind. 1: 107. 1855. Synonyms: *Polyalthia cardiopetala* Dalz., *Polyalthia wynaadensis* Bedd. ex Talbot., *Atrategia thomsonii* Bedd. ex Hook.f.

Small trees to 5 m tall; young branches puberulous. Leaves alternate, oblong-lanceolate or oblanceolate, cuneate at base, abruptly acuminate at apex, slightly undulate along margins, 10-23 x 2.5-7.5 cm, coriaceous, glabrous, brownish beneath; lateral nerves 8-15 pairs, prominently raised beneath, looping near margins; petioles 5-10 mm long, glabrous. Flowers solitary or a few in superposed clusters, axillary; pedicels 5-10 mm long, pubescent when young; bracts minute at the base of pedicels. Sepals 3, almost orbicular, connate upto middle, free at tips, 6 mm across, brownish pubescent. Petals 3+3 in two rows, outer 3 cucullate, oblong-ovate, clawed at base, acute at apex, 10-15 x 4-6 mm, brown tomentose; inner ones triangular, united into a mitriform column, 6-9 x 5 mm. Stamens many, 2 mm long; connectives capitate at apex. Carpels many, linear-oblong, 3 mm long, brown-pubescent; 1-ovuled; styles 2-3 mm long, bifid, elongate, curved; stigma bilobed. Monocarps ellipsoid, 1.5 x 0.7 cm; seed 1. Fl. & Fr.: November – March.

Westen Ghats: Evergreen forests of Karnataka, Kerala and Tamil Nadu (Endemic).

IUCN: VU (Low risk/near threatened) (IUCN, 2024).

Uses: Essential oil can be extracted from the bark.

Goniothalamus keralensis E.S.S. Kumar, Shaju, Royet Raj Kumar, Taiwania, 58(3): 171. 2013.

Small trees up to 4 m tall; branchlets terete, black, minutely hairy when young andglabrous on maturity. Leaves linear-lanceolate, 12-28 x 2-4 cm, cuneate-attenuate at base, long acuminate oracuminate-caudate at apex, entire at margin, coriaceous, glabrous above, midrib straight, channeled above, raised beneath; lateral nerves 16-23 pairs, almost perpendicular to the midrib; petiole 1 -1.5 cm long, glabrous. Flowers solitary, axillary; pedicels subclavate, 4-5 mm long, pubescent with deciduous hairs; bracts 2-3 at base, 1-1.5 mm long, densely brown pubescent on both surfaces. Sepals 3, broadly ovate, 4 -4.5 x 3.8-4 mm, shortly acuminate at apex, pubescent abaxially. Petals 3+3 in two rows, greenish-yellow to dark ivorycoloured; outer petals broadly ovate, 1.5-1.8 x 0.8-1.2 cm, obtuse-acute at apex, densely pubescent; inner petalsforming a cone, rhombic, unguiculate, 0.9-1.2 x 0.7-0.8 cm, densely pubescent with appressed fulvous hairs oneither surfaces except the glabrous region at base of the adaxial surface. Stamens many, to 1.5 mm long, linear-oblong, connective convex, glabrous. Carpels 4-8, ovoid-oblongoid, to 3 mm long. Ovary densely hairy; stigma funnel shaped. Monocarps 4-6, almost sessile, ellipsoid, 2-2.5 x 1-1.5 cm, acuminate at apex, yellowish-brown, glabrous. Seeds 2 in each carpel, each 0.6 x 0.5 cm, dark brown, glabrous Fl. & Fr.: Agust-December

Western Ghats: Evergreen forestrs above 1100 m in Kerala (Endemic).

IUCN: NE (IUCN, 2024).

Goniothalamus rhynchantherus Dunn., Bull. Misc. Inform. Kew 1914: 182. 1914.

Trees to 5–8 m tall; bark smooth, black. Leaves simple, alternate, oblong, elliptic-oblong or ovate, 9–22 x 3.5–6.5 cm, base acute, obtusely acuminate or cuspidate-acuminate, margin entire, recurved, coriaceous, glabrous; lateral nerves 12–15 pairs, pinnate, faint, intercostae reticulate; petiole 6–10 mm long, slender, glabrous, grooved above. Flowers axillary, solitary, fragrant, 2.5 cm across; petiole stout, 0.5-1 cm long, glabrous. Sepals 3, 8 x 9 mm, broadly ovate, connate at base, acute at apex, appressed golden brown pubescent outside, glabrous inside. Petals 3+3 in two rows, outer petals 2-3 cm long, obovate, adnate to 2/3 from apex, acute, densely silvery white or golden sericeous pubescent outside, glabrous inside except the margin; inner petals smaller, shortly clawed, cohering in a veluted cap over the stamens and ovary. Stamens many, closely packed, 1.5-2 mm long; anthers beaked, remote, hidden by the connective, produced into a beak. Carpels many, awl-shaped, densely pubescent, 2-3 mm long, ribbed; style 3-4 mm long, subulate; stigma pointed, entire, ovules 1 or 2. Fruit an aggregate of berry, monocarps many, 15 x 8 mm, obovoid, sometimes constricted, glabrous. Seed one, ovoid, shiny, reddish-brown. Fl. & Fr.: April – June.

Western Ghats: Evergreen forests up to 1000 m in Kerala and Tamil Nadu (Endemic).

IUCN: VU (Low risk/near threatened) (IUCN, 2024), endangered (Narasimhan and Irwin, 2021).

Goniothalamus salicinus Hook.f. & Thoms., Fl. Brit. Ind. 1: 106. 1855. Vern.: *Pulithalai* (Tam.)

Small trees to 5 m tall; bark smooth; branchlets terete, densely strigose with antrorsely purplishbrown or blackish hairs. Leaves elliptic-lanceolate, 7-10 x 3-4 cm, obtuse at both ends, glabrous above, glacous beneath; midrib prominent, lateral nerves faint; petiole to 1.5 cm long, slender, canaliculate above, glabrous. Flowers solitary axillary or sometimes on old leaf scar; pedicel short, slender, densely strigose. Petals 3+3 in two rows, fleshy, lobes triangular; outer petals linear-triangular, 1.2–1.6 x 0.4–0.6 cm, long-acuminate atapex; inner petals slightly shortern than the outer one. Fruits red. Fl. & Fr.: April.

Western Ghats: Evergreen forests of Kerala. World: Native to Sri Lanka.

IUCN: VU (Low risk/near threatened) (IUCN, 2024).

Goniothalamus sericeus Sujana & Vadhyar, Taiwania 65: 176. 2020.

Small trees to 8 m tall; bark surface rugulose, greyish-brown with greenish-white blotches; mature branchlets lenticellate, glabrous, blackish. Leaves linear-lanceolate, 8–11 x 1.8– 2.1 cm, cuneate-attenuate at base, acute or shortly acuminate at apex, margin entire, recurved, glabrous adaxially, puberulent abaxially, midrib straight, channelled above, lateral veins 10–12 pairs, slightly distinct abaxially; petiole 0.4–0.5 cm long, grooved above. Flowers solitary, axillary or on older branches, pendent, fragrant at maturity; pedicels subclavate, 0.6–0.8 cm long, pubescent; bracts 6, oblong, 1–1.3 mm long, sericeous, brown. Sepals 3, broadly ovate, acute at apex, 5–5.2 x 6–6.3 mm, basally connate, adaxially glabrous, white at base, green towards apex, deciduous in fruits. Petals 3+3 in two rows; outer petals ovate, unguiculate, 15–22 x 15–16 mm, obtuse at apex, densely sericeous abaxially, glabrous at base, fleshy; inner petals rhomboidal, formed mitriform dome, unguiculate, 8–9 x 6.5–7 mm, very densely sericeous abaxially, glabrous adaxially, fleshy, yellowish-white. Stamens many, 0.9–1 mm long, linear-oblong, connective produced in to a round apex, papillate. Carpels 20–24, ovoid, 2.5 mm long; ovary 1.8 x 0.3 mm, densely hairy with long golden-brown hairs; style indistinct,

0.7 mm long, curved, greenish white; stigma bifid, black. Monocarps 6 –13, sessile, ovoid, 1.7–1.8 x 1–1.1 cm, yellow when mature, black strigose hairs at base. Seeds 1 or 2 in each carpel, embedded in mucilage, 1.1–1.2 x 0.6–0.7 cm, flattened-ovoid, reddish brown. Fl. & Fr.: January - August

Western Ghats: Evergreen forests of Tamil Nadu (Endemic).

IUCN: CR -B1ab (ii, iii) + B2ab (i, ii) (Sujana and Vadhyar, 2020) (NE, IUCN, 2024).

Notes: This tree species is recently described from Kanyakumari Wildlife Sanctury of southern most Western Ghats. It is characteristically distinguished from its allied species *Goniothalamus wightii* Hook.f. & Thomson. by the presence of smaller leaves, deciduous sepals and densely sericeous abaxial inner petals.

Goniothalamus thwaitesii Hook.f. & Thoms., Fl. Ind. 1: 106. 1855.

Small trees to 8 m tall; branchlets subglabrous. Leaves simple, alternate, oblong to narrow elliptic-oblong, 10-15 x 4-5.5 cm long, apex bluntly acuminate, base acute to subattenuate, chartaceous, glabrous; midrib stout, prominent beneath; secondary nerves 10-12 pairs; tertiary nerves very slender, coarsely reticulate; petiole 0.8-1 cm long, terete, petiole articulate at base, glabrous. Flowers axillary, solitary; pedicels 1-2 cm long, glabrous. Sepals 3, 0.5-0.6 cm long, triangular ovate. Petals 3+3 in two rows; outer larger, 2.5-2.8 x 0.8-1 cm, ovate-lanceolate; inner smaller, 0.8-1 x 0.5-0.6 cm, ovate-lanceolate, clawed. Stamens many, connective capitate. Carpels many, 0.6-0.8 cm long; style subulate. Monocarps up to 20, subsesile or very shortly pedicelled, obtuse or rounded at apex. Seed 1. Fl. & Fr.: August – March.

Western Ghats: Evergreen forests above 800 m in Kerala and Tamil Nadu. World: Sri Lanka.

IUCN: NE (IUCN, 2024).

Goniothalamus wightii Hook.f. & Thoms., Fl. Ind. 1: 106. 1855.

Small trees to 5-7 m tall; bark black, smooth; branchlets glabrous, young branchlets puberulous. Leaves simple, alternate, linear-lanceolate, 8-22 x 4-4.5 cm, apex shortly acuminate, base cuneate, coriaceous, glabrous, often black dotted beneath; midrib canaliculate above; secondary nerves 13-25 pairs, slender, looped at margin; tertiary nerves coarsely reticulate, not prominent; petioles 0.6-1 cm long, canaliculate, glabrous. Flowers solitary, axillary; pedicel 7 mm long, stout, pubescent with 4 distichous bracts at base. Sepals broadly ovate, pubescent. Outer petals 12 x 7 mm, ovate, acute, pubescent; inner petals 6 x 5 mm, ovate, pubescent. Stamens 1.5 mm long; anthers septate; ovaries densely hairy. Monocarps ellipsoid, 2 x 1 cm, glabrous. Fl. & Fr.: February – September.

Western Ghats: Evergreen forests at 1000 to 1500 m, Kerala and Tamil Nadu (Endemic).

IUCN: EN - B1ab(iii)+2ab(iii) (IUCN, 2024).

Goniothalamus wynaadensis (Bedd.) Bedd., Pl. Ind. Or.: 13. 1870. Synonym: *Atrutegia wynaadensis* Bedd.

Small trees to 3-5 m tall; young parts minutely fusco-pubescent, becoming glabrous. Leaves alternate, oblong to oblong-lanceolate, attenuate at base, abruptly acuminate at apex, 10-24 x 3-5.5 cm; petioles 6-8 mm long. Flowers yellowish-green, 1.5 cm long, solitary, axillary or from small tubercles on old wood; pedicels 5 mm long, with 2-3 minute basal bracts. Sepals 3, almost orbicular, obtuse or sometimes acute, 4-5 mm across. Petals 3+3 in two rows; outer petals broadly ovate, subacute, 1.5-2 x 0.5 cm long, prominently veined; inner ones 7 x 3 mm, closely connivent, glabrous inside, densely hairy at the tips inside. Stamens many; connectives prolonged beyond the anthers into blunt tips. Carpels many, oblong, 4 mm long, hairy; style linear-elongate; stigma bifid; ovule solitary. Monocarps many, elliptic to oblong, apiculate, 1.2-1.5 cm long, shortly stalked, 1-seeded. Fl. & Fr.: December – March.

Western Ghats: Occasional in evergreen forests of Karnataka, Kerala and Tamil Nadu (Endemic).

IUCN: NT (Low risk/near threatened) (IUCN, 2024).

Huberantha Chaowasku

1a. Petiole to 5 mm long, lamina glabrous above, pubescent beneath; fruitlets 15-18 mm long .... **H. ceasoides**

1b. Petiole to 10 mm long, lamina glabrous; fruitlets 5-6 mm long .... **H. korinti**

Huberantha cerasoides (Roxb.) Chaowasku., Kew Bull. 70 (2)-23: 2 (2015). Synonyms: *Polyalthia cerasoides* (Roxb.) Benth. ex Hook.f. & Thoms., *Uvaria cerasoides* Roxb., *Hubera cerasoides* (Roxb.) Chaowasku Vern.: *Narela* (Mal.); *Kothukala* (Tam.).

Evergreen trees to 15 m tall; bark black or pale grey, smooth; branches horizontal, young branches minutely tomentose, old glabrous. Leaves simple, alternate, oblong-lanceolate or ovate, 4-20 x 2-4 cm, base obtuse or acute, apex acute or acuminate, margin entire, coriaceous, glabrous above and pubescent beneath; lateral veins 4-12 pairs, ascending, slender, pinnate, slightly raised beneath, intercostae reticulate, obscure; petiole 3-5 mm, slender, tomentose, grooved above. Flowers axillary or extra axillary, solitary or 2-3 fascicled, 8-13 mm across. Sepals 3, free, ovate, 5-8 x 3-4 mm, appressed hairy outside, glabrous inside. Petals 6 (3+3), 6-8 x 4 mm, puberulous, outer ones oblong or ovate, inner ones linear or lanceolate, slightly inflexed; torus convex. Stamens numerous, cuneate, anther thecae extrorse, cells remote, connectives almost flat at top, concealing the anthers. Carpels many, oblong-linear, stalked, 1.5-2 mm long, dark red, covered with stiff hairs; ovule 1, basal, ascending; stigma clavate, slightly curved. Fruit an aggregate of berries, red, monocarps 15-18 mm long, ellipsoid to broadly ovoid, puberulent. Seed 1, globose. Fl. & Fr.: March – December.

Westen Ghats: Evergreen forests at 600 – 1000 m in Maharashtra, Karnataka, Kerala and Tamil Nadu. India: Andhra Pradesh, Arunachal Pradesh, Assam, Bihar, Odisha and West Bengal. World: Sri Lanka, Myanmar, China, Laoz, Cambodia, Vietnam and Thailand.

IUCN: NE (IUCN, 2024).

Huberantha korinti (Dunal) Chaowasku., Kew Bull. 70(2)-23: 2 (2015). Synonyms: *Polyalthia korintii* (Dunal) Thw., *Guatteria korintii* Dunal Vern.: *Karuvall, Kurintipanel* (Mal.); *Sirunetti, Uluvindai* (Tam.).

Small trees to 3-5 m tall; branchlets glabrous. Leaves alternate, ovate, lanceolate or elliptic-lanceolate, 5-14 x 2-5 cm, acute at base, acuminate at apex, coriaceous, glabrous; petioles to 1 cm long. Flowers solitary, axillary, rarely 1-3 together from tubercles on old wood; pedicels 2.5-2.8 cm long, pubescent; bracts in middle of pedicel, ovate, 2 x 1 mm, glabrous inside, pubescent outside. Sepals ovate, 4 x 2 mm, pubescent. Outer petals ovate, acute or subobtuse, 5-7 x 2-3 mm, pubescent on both sides less so on inside; the inner longer and broader. Stamens to 1 mm long. Carpels 7-14, oblong, 1.5 mm long, pubescent; ovule 1; stigmas capitate, sessile. Monocarps ellipsoid or globose, 6 x 5 mm, succulent, yellowish orange; stalks to 7 mm long. Fl. & Fr.: January – December.

Western Ghats: Occasional in evergreen forsts of Kerala and Tamil Nadu. India: Andhra Pradesh. World: Sri Lanka.

IUCN: NE (IUCN, 2024).

Meiogyne Miq.

Meiogyne pannosa (Dalz.) Sinclair., Sarawak Mus. J. 5: 604. 1951. Synonyms: *Unona pannosa* Dalz., *Desmos pannosa* (Dalz.) Saff., *Meiogynera marowii* (Dunn) Gandhi., *Unonara marowii* Dunn., *Desmos ramarowii* (Dunn.) D.Das

Small trees to 5-9 m tall; bark smooth, greyish, mottled with green, young shoots puberulous. Leaves simple, alternate, estipulate, elliptic ovate, ovate or ovate lanceolate, 2.5-7 x 1.5-3 cm, base obtuse or rounded, apex acuminate or caudate acuminate, margin entire, coriaceous, glabrous and glossy above, slender, prominent, intercostae reticulate; petiole 5-9 mm long, slender, pubescent. Flowers subsessile, axillary or terminal, solitary, dull yellowish-green or dirty white. Sepals 3, 4-6 x 4-5 mm, ovate, tomentose outside, glabrous inside, connate at base. Petals 3+3 in two rows, 2.5-5 x 0.5-1.5 cm, outer spreading, clawed, inner erect and smaller, oblong-lanceolate or ovate-lanceolate, densely silky tomentose. Stamens numerous, cuneate, connective broad, concealing the anthers, capitate. Carpels many, sessile or subsessile, oblong, 2-2.5 mm long, strigose; ovules 2-8 in one row; stigma capitate. Fruit aggregate, fruitlets 12 x 8 mm, obovoid, fulvous tomentose, apiculate. Seeds 1-3, globose, shiny. Fl. & Fr.: August – January.

Western Ghats: Evergreen forests above 1000 m in Maharashtra, Karnataka, Kerala and Tamil Nadu (Endemic).

IUCN: NT (Low risk/near threatened) (IUCN, 2024)

Uses: Bark is yielding a strong fibre.

Miliusa Lesch. ex A. DC.

1a. Tall tree up to 7-20 m tall 2
1b. Small shrubs up to 5 m tall 4
2a. Inner petal pouched below; ovules 6 or more .... **M. tomentosa**
2b. Inner petals flat or saccate below; ovules 1 or 2 3
3a. Lateral nerves to 16 on each sides; cymes with 3-6-flowered; peduncle to 2.5 cm long; monocarps pilose .... **M. velutina**
3b. Lateral nerves 10 on each sides; flowers solitary; peduncle to 4.5 cm long; monocarps globose .... **M. paithalmalayana**
4a. Stamens 6-12, uni or biseriate 5
4b. Stamens more than 15, many-seriate 6
5a. Flowers purplish brown; stamens 8-12, uniseriate .... **M. nilagirica**

5b. Flowers green; stamens up to 9, biseriate.......................... **M. wayanadica**
6a. Fruiting pedicel 2-4 cm long; monocarp pisiform 7
6b. Fruiting pedicel 0.3-1.5 cm long; monocarp globose.................................8
7a. Young branchlets glabrous; monocarp apex apiculate........ **M. wightiana**
7b. Young branchlets hairy; monocarp apex obtuse ....**M. tirunelvelica**
8a. Sepals glabrous, subequal; fruiting pedicel greenish 9
8b. Sepals pubescent, equal; fruiting pedicel purplish 13
9a. Young branches glabrous; flushing leaves pale greenish 10
9b. Young branches pubescent; flushing leaves coppery red.......................11
10a. Petal lobes broadly ovate-lanceolate, margin recurved ...**M. malnadense**
10b. Petals lobes broadly ovate-triangular, margin not recurved ....**M. manickamiana**
11a. Flowers in axillary white; carpel oblong................................. **M. ammaii**
11b. Flowers in pseudo axillary or on fallen leaf scar; greenish yellow; carpel linear-oblong 12
12a. Young parts densely hairy; monocarps with apicualte apex ....**M. gokhalaei**
12b. Young parts sparse villous; monocarps globose with appressed apex...... ....**M. horsfieldii**
13a. Leaves with marginal looping nerves; pedicel glabrous .... **M. indica**
13b. Leaves without marginal looping nerves; pedicel pubescent..................... ....**M. montana**

Miliusa ammaii Karup. & Richard, J. of Biolog. Rec. 97-105. 2016.

Small tree up to 5 m tall; branches parallel, young stem cylindrical, glabrous, mature stem lenticellate. Leaves elliptic-lanceolate, 5-10 x 3-4 cm, glabrous on both surface, base round to cuneate, minutely subequal, margin entire, recurved; lateral nerves up to 10, vein lets obscurely reticulate, ending with oil glands, tip obtuse. Flowers solitary, axillary; pedicel pendulous, filiform, cylindrical, glabrous, up to 1.5 cm long. Sepals 3, minute, ovate-lanceolate, 0.2 mm long, succulent, lobesequal. Petals 3+3 in two rows, outer 3 small, slightly longer than sepals, glabrous, up to 1.5 mm long, inner petals large, 1 cm long, base transparent, glabrous outside, succulent, white, lobes curved back. Stamens numerous in many series, anthers 4-lobed, connective thick, included; filament short or absent. Pistils 10, glabrous, 0.6 mm long; stigma oblique, grey coloured, lobepubescent. Fruiting pedicel elongate up to 1.8 cm long, stipe elongate, succulent, up to 1.2 cmlong, stout when ripe, orange-red coloured, grooved above, ripened monocarps purplish-red, globose to subglobose, glabrous. Seeds single to two in each monocarp, smooth. Fl. & Fr.: December – February.

Western Ghats: Mid-evergreen forests of Tirunelveli range in Tamil Nadu (Endemic).

IUCN: NE (IUCN, 2024).

Miliusa gokhalaei Ratheesh, Sujanapal, Anil Kumar & Sivad., Phytotaxa 42: 27. 2012.

Evergreen trees to 5 m tall; bark black, branches terete, drooping, young parts densely hairy. Leaves simple, alternate, 6.0–12. 0 x 2.5–4. 0 cm, oblong or obovate, glabrous, midrib pubescent below, slightly unequally rounded at base, apex caudate acuminate, subsessile or petiole to 3 mm long, membranous; lateral nerves 10–14 pairs, intramarginal nerves sub-marginal, looping; margins entire, slightly curved, thickened. Flowers solitary, pseudo-terminal, greenish-yellow; pedicel 2–4mm long, glabrous; bracts two, small, ovate. Sepals 3, 1.5 x 2.0 mm, broadly ovate, acute, glabrous. Petals 3+3 in two rows, outer petals 3, broadly ovate, acute, hairy on margins, slightly hooded, 2.0 x 1.5 mm; inner petals 3, ovate-lanceolate, 2.0–2.4 x 0.8–1.3 cm, thick, fleshy, glabrous, thickly hooded on the lower half, greenish-yellow with brown streaks inside. Stamens many, 40–50, anthers in pairs, 1 mm, connective included; staminodes absent; anthers extrorse. Carpels 10–15, linear-oblong in outline, slightly curved, 2 mm long, glabrous; stigma ovoid-acute, about half the height of the ovary, with viscous exudate; ovules 1 or 2. Fruiting stalk terete, 8–11 mm long. Monocarps usually 8–12, each 10–12 mm across, obovoid, mammose, apex projecting, more or less obturbinate, glabrous, deep pink; stipe 5 mm long. Seeds 1-2. Fl. & Fr.: September – February.

Western Ghats: Evergreen forest of Kerala (Endemic).

IUCN: NE (IUCN, 2024).

Miliusa horsfieldii (Benn.) Pierre, Fl. Forest. Cochinch.: t. 38. 1881. Synonyms: *Saccopetalum horsfieldii* Benn., *Miliusa tectona* Hutch. Ex C.E. Parkinson, *Saccopetalum tectona* (Hutch. Ex C.E. Parkinson) Chatterjee., *Miliusa arborea* (Elmer) Sinclair., *Saccopetlaum arboreum* Elmer., *Saccopetalum unguiculatum* C.E.C. Fisch., *Miliusa unguiculata* (C.E.C. Fisch.) Sinclair

Trees to 15-35 m tall; bark yellowish-brown colored; branches villous. Leaves elliptic to oblong, 4-13 x 1.8-4 cm, papery, abaxially villous, adaxially sparsely pubescent but densely so on veins, secondary veins 10-14 on each side of midvein, base rounded and slightly oblique, apex acute to acuminate; petiole 2 mm long, villous. Inflorescences axillary, cymose, 1-30- flowered; pedicel 1.5-3 cm long, recurved. Sepals broadly triangular, 3-5 x 1-2 mm. Petals 3+3 in two rows, dark red; outer petals lanceolate, 7-11 x 1-2 mm, reflexed at anthesis; inner petals ovate-lanceolate, 20-26 x 5-10 mm, incurved, with a conspicuous midvein. Stamens 1 mm long. Carpels incurved, densely sericeous; ovules 8 per carpel, in 2 series; stigmas ovoid, glabrous. Monocarp globose to subglobose, stipes 1-1.5 cm long; 5-20, dark red at maturity, puberulent. Seeds 2-8 per monocarp, reniform, 11 x 5 mm. Fl. & Fr.: May – June.

Western Ghats: Evergreen forests of Karnataka and Tamil Nadu. World: Native to Andaman & Nicobar Islands, Myanmar, China, Thailand, Laos, Peninsular Malaysia, Indonesia, Philippine and Australia.

IUCN: NE (IUCN, 2024).

Miliusa indica Lesch. ex A.DC., Mem. Soc. Phys. Genève 5: 212. 1832. Synonyms: *Miliusa montana* Gardner ex Hook.f. &Thoms., *Miliusa eriocarpa* Dunn., *Miliusa indica* var. *tomentosa* (Thwaites) Hook.f. & Thomson., *Uvaria ciliata* B.Heyne ex Wight & Arn. Vern.: *Sakkada maram* (Tam.).

Small trees up to 7 m tall; branchlets pubescent. Leaves simple, alternate, Leaves ovate, 5-7 x 2-2.7 cm, glabrous above, pubescent beneath when young, apex acute to obtuse, base rounded; secondary nerves 12 pairs, faint, tertiary nerves obscure; petioles to 0.2 cm long, pubescent. Flowers solitary or paired; pedicel 7 mm long, slender. Sepals 2 x 1.5 mm, ovate, ciliate. Outer petals 2.5 x 1.5 mm, ovate, acute, glabrous; inner petals ovate, acute, saccate at base, dotted with red glands. Torus hairy. Anthers curved; filaments pubescent, glandular. Stigma ovoid. Monocarps many, globose to ellipsoid, slightly constricted at middle, glabrous. Fl. & Fr.: May – September.

Western Ghats: Evergreen forestr of Karnataka, Kerala and Tamil Nadu. India: Andhra Pradesh. World: Sri Lanka.

IUCN: NE (IUCN, 2024).

Miliusa malnadensis N.V. Page & Nerlekar, Phytotaxa 245: 79. 2016.

Small evergreen trees to 5 m tall; bark grayish brown; branches terete, lenticelled and pubescent when young. Leaves simple, alternate, oblong to lanceolate, 5.0–11.5 x 1.5–3.5 cm, bases slightly unequal, cuneate or rounded, apices acute to acuminate; secondary nerves 6-9 pairs, prominent beneath and faintly looping; petioles 2–3 mm long, pubescent in young, glabrous when mature. Flowers axillary and solitary, bisexual; pedicels 2.5–3.0 cm long, glabrous; bracts 3, minute, triangular, to 0.5 mm long, ovate-lanceolate, densely tomentose. Sepals 3, triangular, acute, ciliate, reflexed. Petals 6, 3+3 in two rows, outer 3 reduced, sepaloid, ovate, 1.0 mm, alternate to sepals; inner 3 elliptic-lanceolate, acute, purple with faint maroon streaks inside, 2.0 x 1.5 cm, tightly appressed at base, veins 7, glabrous outside, apices and margins densely puberulous, basal saccate portion glabrous inside, recurved alon margins; torus ovoid. Stamens numerous, paired, in 5 whorls, 1 x 1 mm, intermixed with numerous 0.5 mm long hairs that are present on the torus; staminodes absent; connectives, not extending at apex, obtuse at top; anthers extrose. Carpels up to 10, elongated, curved, pubescent throughout the outer ovary wall, 2 mm long, with 1–2 ovules; styles short orabsent; stigmas ovoid to oblong, slightly tilted. Fruiting pedicel terete, 3 cm long; monocarps 6–10, each to 10 mm across, globose, glabrous, mucronate; stipes 8–10 mm long. Seeds one or two. Fl. & Fr.: November – May.

Western Ghats: Rare in evergreen forests of Karnataka (Endemic).

IUCN: DD (IUCN, 2024).

Miliusa manickamiana Murugan, '*M. manickamii*'. Indian J. Forest. 39: 371. 2016. Vern.: *Thirukurungudi Kothukala* (Tam.).

Small trees to 4 m tall; branchlets pubescent, terete. Leaves simple, alternate, ovate to elliptic, 2-8 x 2-3.4 cm, base subcordate, margin ciliate, apex acute; petioles 2-4 mm long. Inflorescence axillary or terminal. Flowers 1-1.5 cm across; bracts ovate, to 1 mm long, densely pubescent. Sepals 3, 2 mm long, base connate. Petals 3+3 in two rows, outer lanceolate 2.5 x 1.5 mm, reflexed, inner lanceolate, 1-1.8 x 1-1.4 cm erect, fleshy. Stamens 5-6 whorls, 30-35, 1 mm long. Carpel 5-10; stigma sessile, 0.5 mm long, globose. Monocarps subglobose, 5-8, 10 x 7 mm, glabrous. Seeds 1-2, 8 mm long. Fl. & Fr.: July – October.

Western Ghats: Evergreen forests of Tamil Nadu (Endemic).

IUCN: NE (IUCN, 2024).

Miliusa montana Gardner ex Hook.f. & Thomson, Fl. Brit. India 1: 148. 1855. Synonyms: *Miliusa indica* var. *strigosa* Hook.f. & Thomson, *M. montana* var. *major* Hook.f. & Thomson,

Small trees to 6 m tall; branchlets drooping, terete, new foliage coppery. Leaves oblong or ovate-lanceolate, 3-6 x 2-3 cm, base subcordate, margin entire.

Flowers solitary, axillary or on old leaf scar, flesh coloured, pendulous. Petals 3+3. Fruitlets cluster of 5-6 monocarps, globose, subsessile, reddish, orange when ripe, glabrous. Seed solitary. Fl. & Fr.: July – September.

Western Ghats: Common in deciduous forests of Kerala and Tamil Nadu. World: Sri Lanka.

IUCN: NE (IUCN, 2024).

Miliusa nilagirica Bedd., Icon. Pl. Ind. Or.: 18. 1874.

Small trees to 5 m tall; young branches rugose. Leaves alternate, ovate, lanceolate or elliptic-lanceolate, 5-10 x 2-4 cm, subcuneate at base, shortly acuminate at apex, glabrous; petioles 3-7 mm long. Flowers purplish green, bisexual, solitary, axillary or extra-axillary; pedicels 5-10 mm long, glabrous; bracts 2-3, basal, minute, glabrous. Sepals 3, minute, ovate, ciliate along margins towards apex, otherwise glabrous. Petals 3+3 in tow rows; outer petals like the sepals; inner oblong, acute, 10 x 2 mm, thick, glabrous except the thickened margins towards the apex. Stamens 8-12, uniseriate, 1 mm long; connectives blunt at apex. Carpels 13, oblong, 2 mm long; ovules 1 or 2; style very short; stigma sessile, oblong or globose. Monocarps many, globose or subglobose, apiculate, red, glabrous; stalks 5-7 mm long. Fl. & Fr.: March – December.

Western Ghats: Evergreen forests above 1500 m in Kerala and Tamil Nadu (Endemic).

IUCN: VU-B1+2c. (IUCN, 2024).

Miliusa paithalmalayana Josekutty, Phytotaxa 255: 287. 2015.

Small trees to 8 m tall; bark grayish; branches spreading, pubescent when young. Leaves simple, alternate, elliptic-lanceolate, 5.0–20.0 x 1.8–6.0 cm, base round, slightly oblique, apex caudate acuminate-obtuse or acuminate-obtuse, glabrous both sides, shining, margin entire; lateral veins 8–10 pairs, obscure above, prominent below, intra marginal veins looping, sub marginal; petiole 3–4 mm long. Flowers solitary, axillary or pseudo-terminal, deep purplish; pedicel 4–5 cm long, glabrous; bract 1, rudimentary; bracteoles 2, minute, subulate. Sepals 3, 1.3 x 1.1 mm, broadly ovate, acute at apex, hairy on margins. Petals 3 + in two rows, outer petals 3, broadly ovate, apex acute, hairy on margins, 1.5 x 1.1 mm; inner petals 3, elliptic-lanceolate, involute along margins, 20–30 x 6–7 mm, glabrous, base hooded, cohering when young along margins, deep purplish or brown. Torus ovoid, hispid with white silky hairs. Stamens many, up to 40; filaments short, 0.1 mm long, glabrous; anthers in pairs, 0.7 x 1.0 mm, glabrous, connectives included. Carpels 12–15, purplish, linear-oblong in outline, slightly curved, 2.0 x 0.5 mm, hispid with silky white hairs, style short; stigma ovoid, black, up to 0.6 mm long; ovules 1–2. Fruiting stalk terete, glabrous, 5–7 cm. Monocarps 3–7, each 8–10 mm across, glabrous, crimson red, oblong, apex and base oblique; stipe crimson red, 20 mm long. Seeds 1–2 per monocarp. Fl. & Fr.: August – December.

Western Ghats: Rare in wet-evergreen fostes of Agasthiyamalai environ in Kerala (Endemic).

IUCN: NE (IUCN, 2024).

Miliusa tirunelvelica Murugan, Manickam, Sundaresan & Jothi, Novon 14: 102. 2004. Vern.: *Nellai Kothukala* (Tam.).

Small trees to 4 m tall; branches and branchlets terete, glabrous. Leaves simple, opposite, elliptic to lanceolate, 3.5 x 1.3-3.5 cm, base cuneate or attenuate, apex acute to obtusely acuminate, margin entire; petioles 5-7 mm, glabrous. Inflorescence axillary branched cymes, up to 4 cm long; peduncles 1.5 - 2.5 cm long, quadrangular. Flowers 1 cm diam, pedicels 5 mm long, terete, glabrous; bracts lanceolate 2-4 mm long, persistent. bisexual 4-merous epigynous. Sepals 3, lobes 3 mm long, lobes 4 suborbicular, 1.5 x 3 mm, campanulate, pink, scarious along margin. Petals 3-4, ovate, 5 x 5 mm, base slightly clawed, abtusely acute at apex. Stamens 8, 2 mm long; anthers 3 mm long, oblong, 1.25 x 0.5 mm. Ovary subglobose, 3 mm across; styles linear-subulate, 3-5 mm, glabrous. Fruit berry; monocarps up to 12, globose to ellipsoid, glabrous, redish. Seeds 1. Fl. & Fr.: March – December.

Western Ghats: Evergreen forsts of Kerala and Tamil Nadu (Endemic).

IUCN: NE (IUCN, 2024).

Miliusa tomentosa (Roxb.) Finet & Gagnep., Mem. Soc. Bot. France 4: 153. 1906. Synonyms: *Saccopetalum tomentosum* (Roxb.) Hook.f. & Thomson, *Uvaria tomentosa* Roxb.

Deciduous trees to 15-20 m tall; bark thick, brownish-black, longitudinally grooved; young parts tomentose. Leaves simple, alternate, ovate or oblong-ovate, 8-17 x 2-10 cm, base cuneate to round or subacute; apex acute, margin entire chartaceous, tomentose above when young, glabrous with age except midrib; lateral nerves 8-10 pairs, pinnate, slender, prominent beneath, intercostae reticulate; petiole 2-6 mm long, slender, pubescent. Flowers greenish-yellow, solitary or a few together, leaf opposed or subterminal cymes;

pedicels 3-7 cm long, very slender, tomentose. Sepals 3, 5 x 1.5 mm, linear-lanceolate, tomentose. Petals 3+3 in two rows, outer petals sepalloid, linear-lanceolate, 4-7 mm long, tomentose, inner petals ovate-oblong, saccate at base, obtuse, 1.5-2 x 0.5 cm, tomentose on both sides, outer petal shorter than inner. Stamens many, connectives apiculate, not concealing the anthers. Carpels many, stalked, broadly ovate, brown tomentose; ovules 4-6, in two rows; style absent; stigma capitate. Fruit aggregate of berries, monocarps subglobose, 2.5 cm across, tapering to base, succulent, purple, stalk stout, 1-1.5 cm long. Seeds 4-5. Fl. & Fr.: February – April.

Western Ghats: Evergreen and semievergreen forests at 1000 m in Maharashtra, Goa, Karnataka, Kerala and Tamil Nadu. World: Sri Lanka and Nepal.

IUCN: LC (IUCN, 2024).

Miliusa velutina (Dunal) Hook.f. &Thoms., Fl. Ind. 1: 151. 1855 var. **velutina**. Synonyms: *Uvaria velutina* DC., *Guatteria velutina* (DC.) A. DC. Vern.: *Kana-kaitha, Vilum* (Mal.).

Trees to 20 m tall; bark brown, deeply cracked; branches tomentose when young. Leaves elliptic, ovate or oblong, 8-15 x 5-10 cm, secondary veins 16 on each side of midvein, base obliquely rounded to cordate, apex acute to shortly acuminate; petiole 2-7 mm, tomentose. Inflorescence leaf-opposed or on short few-leaved branches, cymose, 3-6-flowered; peduncle 6-25 mm, tomentose; bracts minute, caducous; pedicel 5-10 cm, slender; bracteoles similar to bracts. Torus glabrous. Sepals and outer petals ovate-triangular, 2-8 x 1.5-3.5 mm; inner petals dark brown, ovate-lanceolate to lanceolate-rhombic, 10 x 6 mm, outside with very dense appressed golden hairs, inside glabrous. Carpels many, pubescent; ovules 2 per carpel; stigma clavate. Monocarp globose to ovoid, stipes 5-8 mm long; 20 x 16 mm, pilose. Fl. & Fr.: April – July.

Western Ghats: Deciduous forests of Maharashtra, Karnataka, Kerala and Tamil Nadu. India: Andhra Pradesh. World: Myanmar, Laos, Cambodia, Vietnam and Thailand.

IUCN: NE (IUCN, 2024).

Key to the varieties

1a. Leaves velutinous; fruit pedicel 4-8 cm long ....var. **velutina**

1b. Leaves glabrescent; fruit pedicel up to 4 cm long ....var. **deviyarina**

Miliusa velutina (Dunal) Hook. F. & Thomas var. **deviyarina** S.M. Rajendran *et al.* Indian. J. Forest. 26(3): 220. 2003.

Fl. & Fr.: May – October.

Western Ghats: Moist deciduous forest of Virudhunagar district in Tamil Nadu and Chandrapur district of Maharashtra (Endemic).

IUCN: NE (IUCN, 2024).

Miliusa wayanadica Sujanapal, Ratheesh & Sasidh., J. Bot. Res. Inst. Texas 4: 64 (2010)

Small trees to 7 m tall; bark black, branches terete, branching horizontal, young parts densely pubescent. Leaves simple, alternate, lanceolate or elliptic-lanceolate, 6–9 x 2–3 cm, base rounded, slightly oblique, apex caudate acuminate, margin entire; petiole to 3 mm or subsessile, coriaceous, lateral nerves 5–8 pairs, intramarginal nerves sub-marginal, tomentose on both sides. Flowers yellow or yellowish green, solitary, axillary or slightly above the axil; pedicel 4–7 mm long, glabrous; bracteoles two at the base, unequal, triangular, very small. Sepals 3, 4 x 3 mm, broadly ovate. Petals 3 + 3 in two rows, outer petals triangular, 3 mm long, slightly curved on mature flower, prominently hairy on margins; inner petals broadly ovate, 7–9 x 2–4 mm, hairy on the mouth. Torus triangular, long hairy. Stamens 6+3, anthers in pairs; staminodes 3, arranged alternate with two stamens, one pair slightly above the staminode, connective slightly prolonged above the anther, connective truncate, anthers extrorse. Carpels 7–12, flask shaped, 2 mm long, glabrous; stigma club-shaped, with viscous exudation on the torus in mature flower; ovules 1 or 2. Monocarps 6–8, subglobose, glabrous, brownish or blackish when dry, apiculate; stipe terete, 10–13 mm long, 8–10 mm across. Seeds 1 or 2. Fl. & Fr.: December – March.

Western Ghats: Evergreen forests of Kerala (Endemic).

IUCN: NE (IUCN, 2024).

Miliusa wightiana Hook.f. & Thoms., Fl. Ind. 1: 149. 1855. Synonym: *Miliusa flaviviridis* N.V. Page, M. Poti & K. Ravikumar, *Miliusa agasthyamalana* V.S.A. Kumar & Sindhu Arya

Ergreen trees to 10-15 m tall; bark black; branches glabrous; young parts pubescent. Leaves simple, alternate, oblong-lanceolate, 2.5-9.5 x 1-4 cm, base obtuse, apex obtusely acuminate, margin entire, coriaceous, glabrous except nerves beneath, lateral nerves 5-8 pairs, prominent and slightly pubescent beneath, intercostae reticulate; petiole 2-6 mm, slender, pubescent. Flowers green, solitary, axillary; pedicels 1.5-2.5 cm long, slender, glabrous. Sepals 3, minute, valvate, reflexed. Petals 3+3 in two rows; outer petals like the sepals; inner ones ovate-oblong, 10-15 x 6-8 mm, connate at base, puberulous outside, inner petal longer than outer, glabrous within. Stamens many, much broader, connectives hardly apiculate; anther subdidymous, cells not concealed. Carpels many, oblong, glabrous, 1-2 ovuled; style short, stigma oblong. Fruit an aggregate, monocarps pisiform, glabrous, granulate. Seeds 1-2. Fl. & Fr.: May – December.

Western Ghats: Common in evergreen forests of Karnataka, Kerala and Tamil Nadu (Endemic).

IUCN: NE (IUCN, 2024).

Mitrephora (Blume) Hook.f. & Thoms.

1a. Outer petals to 3 cm long; styles prominent, stigma club-shaped; leaves with minutely pubescent on nerves .... **M. grandiflora**

1b. Outer petals less than 1.5 cm long; styles obscure, stigma globose; leaves glabrous throughout .... **M. heyneana**

Mitrephora grandiflora Bedd., Icon. Pl. Ind. Or.: 21. 1870. Synonym: *Kinginda grandiflora* Kuntze

Trees to 12 m tall; bark black, smooth; branchlets horizontal; young shoots golden-brown, tomentose. Leaves simple, alternate, elliptic, elliptic-lanceolate, ovate or ovate-lanceolate, 4-15 x 3-7.5 cm, base obtuse, round or truncate, apex acute or slightly acuminate, margin entire, glabrous, coriaceous, glabrous and shining above, glabrescent beneath with hairy glands in the axil; lateral veins 6-10 pairs, intercostae reticulate, intramarginal nerves present, nerves pubescent and prominent beneath; petiole 1 mm, pubescent when young, stout, slightly grooved above. Flowers white, turning yellow, solitary or in groups of 2-3 in leaf opposed cymes; pedicels 1 cm long. Sepals 3, densely tomentose, connate at base, broadly ovate or orbicular 4-5 mm across, yellowish-brown pubescent. Petals 3+3 in two rows, outer petals 3 x 2 cm, obovate-acute, white, turning yellow, pubescent outside, subglabrous inside; inner petals 1-1.5 x 0.3-0.5 cm, triangular at apex with long broad claw, white with red streaks, tomentose on both sides, inner petal dissimilar to the outer, forming a mitriform cap, cauducous. Torus convex. Stamens many, oblong, 1 mm long, connectives flat at top, hiding the anther cells. Carpels many, oblong-ovate, 2 mm long, strigose; ovules 4-6; stigma club shaped. Fruit aggregate of berries, monocarps globose, tomentose. Seeds 1-2. Fl. & Fr.: December - April.

Western Ghats: Evergreen forests of Karnataka, Kerala and Tamil Nadu (Endemic).

IUCN: VU-B1+2c. (IUCN, 2024).

Mitrephora heyneana (Hook.f. & Thoms.) Thw., Enum. Pl. Zeyl.: 8. 1858. Synonyms: *Orophea heyneana* Hook.f. &Thoms., *Kinginda heyneana* Kuntze, *Uvaria heyneana* Wall.

Trees to 7-12 m tall; bark smooth, pale brown; branches ascending, puberulous when young, later glabrous. Leaves simple, alternate, ovate-lanceolate or ovate, 5-12 x 2.5-6 cm, base obtuse or round, apex acuminate; margin entire, glabrous above, pubescent beneath when young, glabrous when mature, coriaceous; lateral nerves 6-10 pairs, intercostae reticulate; petiole 4-7 mm, canaliculate above. Flowers solitary or 2-3 together in axillary or leaf-opposed cymes, yellowish; pedicels 3-5 mm long, tomentose. Sepals 3, minute, ovate or orbicular, brown pubescent. Petals 3+3 in two rows, outer petals 10-15 x 5-8 mm, elliptic obovate or ovate-lanceolate, yellow, inner petals 5-8 x 3-5 mm, clawed, cohering, dissimilar to outer, yellow streaked with pink. Stamens many, oblong-cuneate; anthers concealed by flat connectives. Carpels 6-9, oblong, strigose, style shor, stigma globular, sessile, ovules in two rows. Fruit aggregate of berries, monocarps obovoid to subglobose, 1.5 x 1.2 cm, sessile or subsessile, apiculate, sometimes constructed between seeds; pedicels thick. Seeds 1-3. Fl. & Fr.: March – December.

Western Ghats: Evergreen forests of Karnataka, Kerala and Tamil Nadu. World: Sri Lanka.

IUCN: NE (IUCN, 2024).

Monoon Miq.

1a. Flowers on extra-axillary fascicle on trunk 2

1b. Flowers axillary on fallen leaves 3

2a. Sepals tomentose; monocarps ovoid .... **M. coffeoides**

2b. Sepals glabrous; monocarps oblong-ovoid .... **M. shendurunii**

3a. Flowers fragrant; fruitinmg pedicel more than 3.5 cm long... **M. frgarans**

3b. Flowers not fragrant; fruiting pedicel less than 2 cm long .... **M. tirunelveliense**

Monoon coffeoides (Thw. ex Hook.f. &Thoms.) B. Xue & R.M.K. Saunders, Taxon 61: 1030. 2012. Synonyms: *Guatteria coffeoides* Thw. ex Hook.f. &Thoms., *Poyalthia coffeoides* (Thw. ex Hook.f. &Thoms.) Benth. &Hook.f. ex Bedd. Vern.: *Villa* (Mal.); *Nedunarai* (Tam.).

Trees to 15-25 m tall; bark smooth, dark green, often mottled with greyish-white; young shoots slender, minutely puberulous; branches horizontal. Leaves simple, alternate, young leaves red flushed, oblong, oblong-lanceolate or elliptic-lanceolate, 10-27 x 3.5-8.5 cm, base obtuse, cuneate-round or acute, apex acute or acuminate, margin entire, slightly undulate, coriaceous, glabrous and shiny above; lateral nerves 8-14 pairs, prominent beneath, intercostae scalariform, obscure; petiole 10-13 mm, stout, glabrous, grooved above. Flowers greenish-yellow, solitary, extra axillary or some times fascicled on woody tubercles on trunk; pedicels 1-4 cm long, slender, pubescent. Sepals 3, 2 x 5 mm, suborbicular, tomentose. Petals 3+3 in two rows, 3-5 x 0.5-1 cm, greenish-yellow, flat and strap shaped, subequal. Torus convex. Stamens numerous, 1 mm long, cuneate, connectives subglobose at top, concealing the anthers, anther cells remote. Carpels numerous, 3-4 mm long, densely pubescent, oblong, stalked; stigma capitate; ovule one, basal and ascending. Fruit aggregate of berries, monocarps ovoid, 2.5 x 1.5 cm, apiculate, dark purple. Seed one, broadly ovoid, grooved, smooth and shiny. Fl. & Fr.: December – October.

Western Ghats: Evergreen forest at 500 – 1500 m in Karnataka, Kerala and Tamil Nadu. World: Sri Lanka.

IUCN: NE (IUCN, 2024).

Uses: Bark is used for making ropes.

Monoon fragrans (Dalz.) B. Xue& R.M.K. Saunders, Taxon 61: 1031. 2012. Synonyms: *Guatteria fragrans* Dalz., *Polyalthia fragrans* (Dalz.) Bedd. Vern.: *Cehlla, Kodangi, Neddunar, Pullarrei* (Mal.). Vern.: *Nedunarai, Pullarrei, Chella, Kodangi, Udambilai* (Mal. & Tam.).

Trees to 18 m tall; branches hoary when young. Leaves ovate-oblong to oblong-lanceolate, 10-24 x 5-12.5 cm, membranous, abaxially minutely puberulent on major veins but glabrescent, adaxially glabrous or puberulent, base rounded, apex rounded to long acuminate; lateral veins 14-16 pairs, oblique, reticulate veins inconspicuous; petiole 1-1.8 cm long. Inflorescences in axils of fallen leaves or on woody tubercles on branches, cymose, pedunculate, few to many flowered, hoary-tomentose. Flowers fragrant, pedicel to 2.5 cm long, bracteoles inserted below middle of pedicel, orbicular. Sepals orbicular, 3 mm in diam., recurved at anthesis. Petals greenish yellow to pale yellow, linear, 3.8 x 0.5 cm, subequal, densely minutely hairy. Stamens many; connectives apically truncate. Carpels many, hairy. Fruiting pedicel to 3.5 cm long, stipes to 5 cm long, densely pubescent; monocarps broadly ovoid, 2.5-4 cm, walls thin and brittle, surface hoary. Seed 1 per monocarp. Fl. & Fr.: October – May.

Western Ghats: Evergreen forests at 1000 m in Maharashtra, Goa, Karnataka, Kerala and Tamil Nadu (Endemic).

IUCN: NE (IUCN, 2024).

Monoon shendurunii (Basha & Sasidh.) B. Xue & R.M. K. Saunders, Taxon 61: 1033. 2012. Synonyms: *Polyalthia shendurunii* Basha &Sasidh.

Trees to 15 m tall; bark smooth, brownish-black; bole surface with warty tubercles. Leaves simple, alternate elliptic, obovate, oblong or elliptic-lanceolate,15-26 x 6-8.5 cm, base rounded or truncate, apex acuminate, acumen to 3 cm long, margin entire, glabrous, midrib impressed and channelled above, lateral veins 10-14 pairs, raised beneath; intercostae subscalariform; petiole 5-10 mm long, stout, pubescent. Flowers greenish to yellow, on twigs solitary and or up to 10 in fascicles on tubercles on the trunk; pedicels 3-5 cm long, tomentose; bract 3-4 x 2.5-3 mm towards the lower half of the pedicel, ovate-acute, tomentose. Sepals 3, deltoid-ovate, 1.5-2 x 1-1.3 cm, acuminate, reflexed, connate at base, persistent. Petals 3+3 in two rows, subequal; outer slightly smaller, oblong, acute or acuminate, coriaceous, minutely rugose and appressed-tomentose on both sides, greenish-yellow to yellow; inner 8-9 x 1.5-2 cm. Stamens numerous, closely arranged, 3 mm long; anthers and connectives hirsute, more on the ventral surface, appendage glabrous. Carpels many, each 2.5 mm long; ovule solitary, basal; style indistinct; stigma orbicular, depressed at middle, margins shortly lobed, glabrous. Fruit an aggregate of berries; stipes 4-5 mm across; monocarps to 7, 3 x 2.5 cm, oblong-ovoid, glabrous; stalk 1.8 cm long. Seed oblong, 2.8 x 1.8 cm. Fl. & Fr.: April – May.

Western Ghats: Occasional in evergreen forests of Kerala and Tamil Nadu (Endemic).

IUCN: EN- B1+2c. (IUCN, 2024).

Monoon tirunelveliense (M.B.Viswan. & Manik.) B. Xue & R. M. K. Saunders, Taxon 61: 1034. 2012. Synopsis: *Polyalthia tirunelveliensis* M.B. Viswan. & Manik. Vern.: *Nellai Nedunarai* (Tam.).

Trees to 25 m tall; bark blackish grey, lenticellate; young branches rusty-pubescent, glabrescent. Leaves elliptic-lanceolate, oblong-elliptic or oblong-oblanceolate, 8- 38.5 x 4 - 14.4 cm, dark greenish brown above, paler beneath, mostly truncate, rarely acute at base, subacute or subacuminate at apex, margin undulate; midrib grooved above, raised beneath, rusty-pubescent when young, rusty-puberulous later; lateral nerves 10-19 pairs, slightly impressed above, prominent beneath; petioles canaliculate, rusty-pubescent, to 1.7 cm long. Flowers 6.4 - 9.1 cm across, together, solitary or in fascicles of 1 - 3 in axils of fallen leaves of older branches; pedicels slender, broadened upwards, flattened. Fl. & Fr.: October – Decemberr.

Western Ghats: Rare in evergreen foress of Tamil Nadu (Endemic).

IUCN: Critically Endangered (Narasimhan and Irwin, 2021) (NE-IUCN, 2024).

Orophea Blume

1a. Outer petals and young parts glabrous; carpels oblong .... **O. erythrocarpa**

1b. Outer petals and young parts pubescent or glabrous; carpels globose ...2

2a. Pedicels upto 5 mm long .... **O. thomsonii**

2b. Pedicles more than 5 mm long 3

3a. Buds and young shoots shortly tomentose; stamens 6 .... **O. zeylanica**

3b. Buds and young shoots glabrous; stamens 12 4

4a. Outer petals tomentose along the margin; inner petals 7 mm long .... **O sivarajanii**

4b. Outer petals ciliate along the margin; inner petals 9 mm long .... **O. uniflora**

Orophea erythrocarpa Bedd., Madras J. Lit. Sci., ser. 2, 22: 71. 1861. Vern.: *Milagu naari* (Tam.).

Trees to 8 m tall; bark black, smooth; young plants softly pubescent. Leaves simple, alternate, ovate, elliptic or oblong, 4-13 x 2-4.5 cm, base round or truncate, apex acuminate or acute, margin entire, glabrous above, pubescent beneath, membranous, lateral nerves 6-8 pairs, pinnate, intercostae scalariform; petiole 3-4 mm, pubescent, slightly grooved above. Flowers yellowish 3-4 together, axillary or supra-axillary, tomentose racemes, 7 mm across; pedicels 1-2 cm long, villous. Sepals 3, broadly ovate, tomentose, 1.5 mm long. Petals 3+3 in two rows; outer petals 4 x 3 mm, orbicular, brown tomentose outside, glabrous inside; inner rhomboid, clawed, curved inward, pubescent at tips, nectaries of inner petals free, oblong. Stamens 12 in 2 rows, outer 6 imperfect, very small, connectives slightly apiculate at top. Carpels 6, red, ovate-oblong, densely strigose; ovules 2; style short, stigma sessile. Fruit aggregate, monocarps oblong, sessile, reddish, 2.5 cm long. Seed one, scrobiculate. Fl. & Fr.: January – July.

Western Ghats: Rare in moist deciduous to evergreen forests of Kerala and Tamil Nadu (Endemic).

IUCN: VU- B1ab(iii)+2ab(iii) (IUCN, 2024).

Orophea sivarajanii Sasidh., Nordic J. Bot. 19: 301. 1999.

Small trees to 4 m tall; bark smooth; young shoots glabrous. Leaves alternate, oblong or elliptic-oblong, 7-15 x 3-6 cm, acuminate at apex, obtuse or rounded at base, glabrous, lateral nerves 7-11 pairs, prominent below and faint above, intercostae widely reticulate; petiole 2-3 mm long, glabrous. Flowers in 2 or 3-flowered, racemes or solitary, axillary or lateral, 1.5-2 cm long, glabrous; pedicels 8-12 mm long, glabrous; bract 0.5 mm long, ovate-acute, towards the upper half of the pedicel. Sepals 2 x 2.5 mm, jointed at base, ovate-deltoid, acute, ciliolate along the margins. Petals 3+3 in two rows; outer petals 4 mm across, ovate-orbicular, tomentose along the margins; inner petals 7 mm long, greenish towards the upper half and white in the lower half, tomentose along the margins; nectaries 2, distinct. Stamens 12 in 2 rows of 6 each. Carpels 6, glabrous; ovules 2, style hairy. Monocarps globose, 8 mm across. Fl. & Fr.: November – February.

Western Ghats: Rare in evergreen forests of Kerala (Endemic).

IUCN: NE (IUCN, 2020).

Orophea thomsonii Bedd., Madras J. Lit. Sci., ser. 2, 22: 72. 1861.

Trees up to 8 m tall; branchlets pubescent when young, glabrouw at maturity. Leaves ovate-oblong, 5-11 x 3-4 cm, apex obtusely acuminate, base acute, glabrous except on midrib beneath; petiole very short, to 5 mm long. Flowers creamy white, axillary, solitary or in few flowered cymes, peduncle very short. Sepals 3, ovate, strogose. Petals 3+3 in two rows; outer petals strigose, 4 x 5 mm, inner pubescent inside, acute, short-calwed, nectaries oblong, horizontal. Stamens 10-12 in 2 series; staminodes absent. Carpels 5 or 6, ovoid, strigose; ovules 2; style short; stigma produced. Monocarps globose-pisiform, 8 mm across, stalked. Fl. & Fr.: February – March.

Western Ghats: Semi-evergreen to evergreen forests of Kerala and Tamil Nadu (Endemic).

IUCN: NE (IUCN, 2024).

Orophea uniflora Hook.f. & Thoms., Fl. Ind. 1: 110. 1855. Vern.: *Otraipoo, Milagunaari* (Tam.).

Small trees to 7 m tall; bark rough, black; branchlets pubescent when young, glabrous at maturity. Leaves ovate, elliptic-oblong, 8-10 x 3 cm, apex acuminate, base acute, glabrous, nerves faint; petiole to 4 mm long, tomentose. Flowers 8-10 mm across, solitary or in axillary clusters; pedicels 3-5 mm long, tomentose. Sepals 1.5 mm across, orbicular, margin ciliate. Petals 3+3 in two rows; outer petals 5 mm across, orbicular, ciliate, inner petals 9 x 4 mm, rhomboid, glabrous, nectaries oblong. Stamens 9-12, glabrous; staminodes absent. Carpels 6, glabrous; ovules 2. Monocarps globose, 8 mm across, glabrous. Seed one. Fl. & Fr.: February – August.

Western Ghats: Rare in evergreen forests of Karnataka, Kerala and Tamil Nadu (Endemic).

IUCN: NT- B1ab(iii)+2ab(iii) (IUCN, 2024), vulnerabl (Narasimhan and Irwin, 2021).

Orophea zeylanica Hook.f. & Thoms., Fl. Ind. 1: 111. 1855. Synonym: *Orophea malabarica* Sasi. & Sivar.

Small trees to 5 m tall; young branches yellow pubescent. Leaves alternate, ovate or elliptic-oblong, 6.5-8 x 3-4.5 cm, almost rounded at base, shortly and obtusely acuminate at apex, dark green above, pale beneath, puberulous beneath when young, later glabrous, lateral nerves 4-7 pairs; petioles 2 mm long, glabrous. Flowers solitary or 2-3 together, axillary or extra-axillary; pedicels 1-2 cm long, pubescent when young, bracts small, at the base of pedicels. Sepals 3, ovate-orbicular, acute, 1 x 1 mm, connate at base, pubescent. Petals 3+3 in two rows; outer ones almost orbicular, 3-4 mm across, veined, downy; inner ones clawed, 4-5 mm long, glabrous, except the pubescent margin. Stamens 6, in one row, 0.5 mm long; connectives slightly pointed at the apex. Carpels 9-15, obovate, 1.5 mm long, glabrous, 2-ovuled; stigma capitate, subsessile. Monocarps globose, 7.5 mm in diam.; stalks 2-3 mm long. Seeds 1-2. Fl. & Fr.: May – September.

Western Ghats: Fairly common in evergreen forests of Kerala, Karnataka and Maharashtra. World: Sri Lanka.

IUCN: NE (IUCN, 2024).

Polyalthia Blume

1a. Leaves attenuate or acuminate at apex 2
1b. Leaves rounded or very obtuse at apex; bark corky .... **P. suberosa**
2a. Leaves acuminate at apex; twigs rufous-pubescent .... **P. rufescens**
2b. Leaves gradually acuminate at apex; twigs rusty-tomentose ....
....**P. malabarica**

Polyalthia malabarica (Bedd.) I.M. Turner, Nordic J. Bot. 33: 284. 2015. Synonym: *Phaeanthus malabaricus* Bedd.

Small trees to.5 m tall; young branches puberulous on flushing, glabrous. Leaves, ovate to broadly elliptic - oblong, 4-25 x 1.5-7 cm, base round, unequal or cordate, apex acute to abruptly acuminate, subcoriaceous, glabrous above, sparsely hairy along the nerves, lateral nerves 6-12 pairs, reticulate; petiole 2 - 4 mm long, glabrous. Flowers axillary or extra axillary, solitary or rarely paired; pedicel, slender, 2.5 cm long, puberulous; bracts 2, deltoid to ovate, 0.5 mm long, puberulous above, glabrous below. Sepals 3, deltoid, 3-4 x 4 - 5 mm, apex acute, pubescent above, glabrous within, margin ciliate. Petals 3+3 in two rows; outer petals 3, ovate rhomboid, 6-8 x 4-6 mm, 1 mm thick, apex acute, pubescent above, glabrous within, spreading, greenish yellow; inner petals 3, ovate oblong, 1.2-2 × 0.8-1 cm, cucullate, apex obtuse, pubescent above, glabrous within. Torus hemispherical, 5 x 8 mm. Stamens numerous, 1.5 x 0.8 mm, connective prolongation slightly raised, apex densely ciliate, creamy yellow. Carpels many, cylindrical or ovate oblong, 2 x 1 mm, strigose; style indistinct; stigma obovate, capitate, apex sparsely pilose; ovules 2. Monocarps up to 20, globose, 0.7-1 cm across, apex beaked, puberulous to pubescent, red; stipitate 5 mm long. Seed one or rarely two, spherical or hemispherical, slightly grooved, pitted. Fl. & Fr.: February – September.

Western Ghats: Rare in evergreen forests of Kerala (endemic).

IUCN: NT- B1ab(iii)+2ab(iii) (IUCN, 2024).

Notes: In India, the genus *Phaeanthus* was represented by *P. malabaricus* Bedd. (Gamble 1914; Mitra 1993; Karthikeyan *et al.*, 2009). As per the revisionary study by Mols & Kebler (2000) the species was treated as *Polyalthia suberosa*. Later, Turner (2015) critically studied both the specimens that are distinctly different with unique characteristic features of their own and made a combination as *Polyalthia malabarica*. Hence, based on above studies it can be concluded that the genus *Phaeanthus* does not exists in India

Polyalthia rufescens Hook.f. & Thoms., Fl. Brit. India 1: 66. 1872. Synonyms: *Phaeanthus moulmeinensis* Craib Vern.: *Nellai Kothukala* (Tam.).

Small trees to 5 m tall; branchlets densely tomentose when young, puberulous when old. Leaves alternate, oblong-lanceolate, 8-13 x 3.5 -5.5 cm, obtuse or slightly cordate at base, shorlyt acuminate at apex, glabrous on both surfaces, shining above, pale beneath; lateral nerves to 16 pairs, faint above, prominent beneath; petioles 2 mm long, swollen, glabrous. Flowers solitary or a few together, leaf-opposed, 1 cm across, brown tomentose; pedicels to 3 cm long, puberulous; bracts basal, minute. Sepals small, broadly ovate, acute, 2-3 x 2 mm, pubescent. Petals 3+3 in two rows; outer petals ovate-oblong, subacute, 8-12 x 5-7 mm, tomentose; the inner smaller. Carpels many, linear, 1 mm long, densely silky; ovules 2; styles short; stigmas capitate. Monocarps globose, 6 mm across. Fl. & Fr.: February – May.

Western Ghats: Evergreen forests of Kerala, Tamil Nadu including Andaman & Nicobar Islands (Endemic).

IUCN: EN- B1+2c (IUCN, 2024).

Polyalthia suberosa (Roxb.) Thw., Enum. Pl. Zeyl.: 398. 1864. Synonyms: *Phaeanthus malabaricus* Bedd., *Uvaria suberosa* Roxb. Vern.: *Kothukala* (Tam.).

Small trees to 4-6 m tall; bark fissured, corky, rough; young branches rufous tomentose. Leaves simple, alternate, oblong, elliptic-oblong or oblong-lanceolate, 3-12 x 1-3.5 cm, base slightly narrowed, oblique or obtuse, apex obtuse or obtusely acute, margin entire, undulate, glabrous and shiny above, submembranous; lateral veins 9-12 pairs, pinnate, faint, intercostae reticulate; petiole 2-3 mm long, brown pubescent, slender. Flowers yellowish-green, mostly suffused with purple, solitary, rarely in pairs, extra-axillary, 8-10 mm long; pedicels slender, 1.5-2.5 cm long. Sepals 3, spreading, ovate, acute, 2 x 1 mm, pubescent outside, glabrous inside. Petals 3+3 in two rows; outer petals ovate to oblong-lanceolate, acute, slightly reflexed, 6 x 4-5 mm, thickly coriaceous, silky pubescent outside, glabrous inside; inner ones slightly longer. Torus convex. Stamens numerous, 1 mm long, connectives slightly convex at top concealing the anther cells. Carpels many, 2 mm long, pubescent; ovule one, style oblong, stigma triangular, flat. Fruit an aggregate of berries; monocarps subglobose, purple 5 mm across, puberulous, stalks slender, 0.5-1 cm long. Seeds 1, globose, smooth. Fl. & Fr.: March – January.

Western Ghats: Common in evergreen forests of Karnataka, Kerala and Tamil Nadu. India: Andhra Pradesh, Assam, Madhya Pradesh, Meghalaya,

Odisha and West Bengal. World: Sri Lanka, China, Laoz, Cambodia, Vietnam, Thailand, Myanmar, Malaysia, Philippines and Java.

IUCN: NE (IUCN, 2024).

Popowia Endl.

Popowia beddomeana Hook.f. & Thoms., Fl. Brit. India 1: 68. 1872. Synonym: *Popowia ramosissima* Bedd.

Small trees to 5 m tall; branches parallel, brown tomentose when young. Leaves ovate, tomentose, to 8 x 2.5 cm, smooth above, acute or rounded at base, glabrous above, tomentose beneath; secondary nerves prominent beneath, brown tomentose. Flowers solitary, yellow, sometimes paired, extra-axillary, 0.8 cm across, globose, villous outside, glabrous inside, inner one concave, inflexed at tip. Sepals connate below, ovate, acute, 4 mm long, densely pubescent. Petals ovate, 5 x 4 mm, densely pubescent, inner slightly longer, concave, inflexed at tip, thick. Stamens about 24; connective truncate or capitate. Ovaries 4-7; ovules solitary. Carpels 4-7, oblong, tomentose; ovule 1, on ventral sature; style large; stigma obovate, recurved. Monocarps globose 1 cm across, red, stiped. Fl. & Fr.: February – April.

Western Ghats: Evergreen forests above 1000 m in Kerala and Tamil Nadu (Endemic).

IUCN: EN- B1+2c (IUCN, 2024).

Sageraea Dalzell

1a. Leaf base acute; flowers fascicled on base of pedicels of young branches .... **S. laurina**

1b. Leaf base rounded; flowers on fascicled on tuberculate old branches .... **S. thwaitesii**

Sageraea laurina Dalz., Hooker's J. Bot. Kew Gard. Misc. 3: 207. 1851. *Sagerea laurifolia* (J.Grah.) Blatt. & McCann., *Guatteria laurifolia* J. Grah., *Sageraea dalzellii* Bedd., *Bocagea dalzellii* (Bedd.) Hook.f. Vern.: *Kanakatha, Manja nara, Manyarei* (Mal.).

Trees to 20 m tall; bark surface black, smooth, often blotched with grey; young shoots glabrous. Leaves simple, alternate, elliptic or broadly oblong, 10-30 x 3.5-10 cm, base acute, attenuate or obtuse; apex acute, margin entire, glabrous, coriaceous, shining above; lateral nerves 10-15 pairs, intercostae reticulate; petiole 10-12 mm, stout, glabrous, grooved above. Flowers yellow,

15-18 mm across, in axillary fascicles with a few scaly bracts at the base of the pedicels. Sepals 3, 3 x 5 mm orbicular, ciliate, slightly connate at base; outer petals yellow, 13 x 8 mm, orbicular, concave, ciliate along margin, inner petals similar but smaller than the outer. Torus scarcely elevated. Stamens upto 25, in 2 or more rows; outer sometimes sterile and flattened, connectives produced, not concealing the anthers, anthers oblong. Carpels 3-6, hairy; ovules 1-8, style absent; stigma capitate. Fruit an aggregate of berries, monocarps globose, 7-7.5 x 3.5 cm. Seed single. Fl. & Fr.: October – February.

Western Ghats: Deciduous forests of Maharashtra, Goa, Karnataka, Kerala and Tamil Nadu (Endemic).

IUCN: NE (IUCN, 2024).

Uses: Wood is useful in building construction.

Sageraea thwaitesii Hook.f & Thoms., Fl. Ind. 1: 93. 1855. Synonyms: *Bocagea thwaitesii* (Hook.f. &Thoms.) Hook.f. & Thoms., *Sageraea grandiflora* Dunn., *Sageraea dalzellii* Bedd. Vern.: *Maja uara, Kana kaitha, Manyarei* (Mal.).

Trees to 20 m tall; bark black, smooth. Leaves simple, alternate, broadly oblong,18-35 x 7-12.5 cm, base round, acute or obtuse, apex acute or short-acuminate, margin entire, glabrous, coriaceous, shining above; lateral nerves 10-16 pairs, slender, faint, intercostae reticulate; petiole 8-20 mm long, stout, glabrous, grooved above. Flowers yellow, fascicled axillary or on woody tubercles; pedicles to 2.5 cm long. Sepals 3, 2-3 x 5 mm, orbicular, ciliate, connate at base. Petals 3+3 in two rows, 13 x 8 mm, orbicular, concave, ciliate along margins; inner petals similar but smaller than outer, glabrous. Stamens 12, anthers oblong, connective prolonged, not concealing the anthers. Carpels 3-6, free, strigose; style absent, stigma globose. Fruit an aggregate of berries; monocarps globose, 3.5 cm across, smooth. Seed single. Fl. & Fr.: November – February.

Western Ghats: Rare in evergreen forests of Karnataka, Kerala and Maharashtra. World: Sri Lanka.

IUCN: NT- B2b(iii) (IUCN, 2024), endangered (Narasimhan and Irwin, 2021).

Xylopia L.

Xylopia patoniae I.M. Turner., Edinburgh J. Bot. 68: 340. 2011. Synonyms: *Xylopia parvifolia* (Wight) Hook.f. & Thoms., *Patonia parvifolia* Wight Vern.: *Kalpottan, Santhu* (Mal.); *Kalpotta* (Tam.).

Trees to 15-25 m tall; bark 5-6 mm thick, grey; young branches brown pubescent. Leaves simple, alternate, oblong-lanceolate, 6-10 x 2.5-3 cm, base acute, apex acuminate, margin entire, coriaceous, glabrous, shiny above; lateral nerves 8-9 pairs, pinnate, very slender, faint, intercostae reticulate; petiole 5-8 mm, slender, grooved above. Flowers yellowish-brown, 2-5 together in axillary fascicles, 15-18 mm across; pedicels very short, pubescent; bracts longer than the sepals. Sepals 3, connate up to middle, ovate, acute, 1.5-2 x 2.5-3 mm, pubescent out side, glabrous inside. Petals 3+3 in two rows; outer petals linear, slightly clawed and dilated at base, 10-13 x 1.5-2 mm, brown pubescent outside, hairy inside; inner ones a little shorter and narrower, linear, concave and spoon-shaped at base, densely tomentose. Stamens numerous, 2 mm long, connectives flat, round, anthers transversely septate. Carpels 5, linear, 4 mm long, slightly pubescent; ovules 4-6; style linear, stigma fusiform. Fruit an aggregate of berries, monocarps obovoid, 1-4 together, subsessile, blunt at apex, tapering to base, pubescent, stalk 1-2 mm long, woody. Seeds 4-6, oblong. Fl. & Fr.: August – February.

Western Ghats: Kerala and Tamil Nadu. India: Gujarat, Haryana, Jammu & Kashmir, Madhya Pradesh, Punjab, Rajasthan and Uttar Pradesh. World: Sri Lanka.

IUCN: NE (IUCN, 2024).

Uses: Fibre is obtained from the bark.

Cultivated:

Annona reticulata L., Vern.: *Ramphal* (Hindi), *Amrytaphala* (Kan.), *Athamaram* (Mal.), *Aant* (Mar.), *Atta* (Tam.); Exotic: Native to Tropical America

Cananga odorata (Lam.) Hook. f. & Thomson.; Native: India, Tropical Asia, Australia.

Monoon longifolium (Sonn.) B. Xue & R. M. K. Saunders. Vern.: Ashoka (Hin.), Putranjiva (Kan.), Arana (Mal.), Ashok (Mar.), Asokamaram (Tam.).

APOCYNACEAE Juss.

Key to the genera

1a. Anthers free from stigma, cells rounded at base 2

1b. Anthrers conniving in a cone around stigma, cells produced downward .... **Wrightia**

2a. Fruit indehiscent, 1-8-seeded (rarely many) 3

2b. Fruit dehiscent, follicles many seeded 5

3a. Leaves opposite; fruit fleshy 4

3b. Leaves scattered; fruits fibrous-woody .... **Cerbera**

4a. Corolla tube and lobes white; fruit a drupe or mericarp 7

4b. Corolla tube and lobes greenish yellow; fruit capsule .... **Hunteria**

5a. Leaves whorled; coma present 6

5b. Leaves opposite; seeds without hairs .... **Tabernaemonatana**

6a. Seeds with a deciduous silky coma at the top .... **Hollarhena**

6b. Seeds with a deciduous siky comma at both ends .... **Alstonia**

7a. Seeds 2-8 in drupe, not arillated .... **Kopsia**

7b. Seeds many in mericarp, arillated, aril orange red .... **Voacanga**

Alstonia R.Br.

1a. Corolla hairy outside, with suborbicular lobes; tertiary venation distinct on upper surface of driedleaves .... **A. scholaris**

1b. Corolla glabrous outside, with ovate to oblong lobes; tertiary venation indistinct on upper surface of dried leaves 2

2a. Follicle terete, glabrous, to 45 cm long .... **A. macrophylla**

2b. Follicle 2-angled, slightly hairy, to 15 cm long .... **A. venenata**

Alstonia macrophylla Wall. ex G. Don., Gen. Hist. 4: 78. 1837. Vern.: *Periya Yezhilai paalai* (Tam.).

Large trees to 30 m tall; bark whitish shiny, rough with glabrescent. Leaves simple, in whorls of three or mostly of four, 10-40 x 3-12 cm, widest in or above the middle, cuneate at the base, abruptly shortly acuminate or more rarely obtuse at the apex, membranous, glabrous above, densely pubescent beneath when young, lateral nerves 18-30 rather distant 0.6-2 cm apart; lateral veins scarcely raised above, slightly so beneath; petiole to 1-3 cm long. Flowers white, in terminal cymes, arranged in a sessile umbel; pedicels much longer than the calyx. Seplas 5, lobes up to 1 mm long, semicircular, puberlous to almost glabrous without. Petals glabrous lobes 5, tube 0.5-0.6 cm long, oblong, in bud overlapping to the right. Stamens attached to the inflated portion near the mouth, not exserted; anthers ovate, acute. Follicles pendulous, terete, linear, 30-45 cm long, 0.25-0.35 cm in diameter, glabrous when ripe. Seeds many, acute but not acuminate at one end, obtuse at the other. Fl. & Fr.: October-April

Western Ghats: Moist deciduous forests of Kerala, Maharashtra and Tamil Nadu (cultivated as social forestry tree and avenue tree). World: South East Asia

IUCN: LC (IUCN, 2024).

Alstonia scholaris (L.) R.Br., Asclepiadeae: 64. 1810. Synonyms: *Echites scholaris* L., *Pala scholaris* (L.) Roberty, *Alstonia kurzii* Hook.f. Vern.: *Paala* (Mal.); *Mukkam palei, Elelaipalei* (Tam.).

Trees to 40 m tall; bark gray; branchlets copiously lenticellate. Leaves in whorls of 3-10, blade narrowly obovate to very narrowly spatulate, 7-28 x 2-11 cm, leathery, base cuneate, apex usually rounded; lateral veins 25-50 pairs; petiole 1-3 cm. Cymes dense, pubescent; peduncle 4-7 cm; pedicel usually as long as or shorter than calyx. Petals white, tube 6-10 mm, lobes broadly ovate or obovate, 2-4.5 mm, overlapping to left. Ovary distinct, pubescent. Follicles linear, to 57 x 2-5 mm. Seeds oblong, margin ciliate, ends with tufts of hairs 1.5-2 cm. Fl. & Fr.: December – March.

Western Ghats: Deciduous to evergreen forests of Maharashtra, Goa, Karnataka, Kerala and Tamil Nadu. World: Sri Lanka, Nepal, China, Myanmar, South East Asia and Australia.

IUCN: LC (IUCN, 2024).

Uses: The wood is used to make packaging cases, tea-boxes and rarely furnitures. The bark is source of fibre and latex.

Alstonia venenata R.Br., Asclepiadeae: 64. 1810. Synonym: *Blaberopus venenatus* (R.Br.) A.DC., *Alstonia venenata* var. *pubescens* Monach., *Echites venenatus* Roxb. ex A.DC. Vern.: *Paalai, Elaipaalai* (Tam.).

Small trees to 10 m tall; bark smooth; latex milky. Leaves 12-20 x 1.2-3 cm, linear-oblong, acuminate at both ends, 5 at each node, lateral nerves many, very close. Cymes terminal, to 20 cm across, corymbose. Flowers many or corymbose cyme, white; pedicels stout. Sepals 1 mm long, ovate, acute, ciliate. Petals white, tube slender to 1.5 cm long, hairy inside, glabrous outside, lobes

13 mm long, oblong, acute. Follicle 2-angled, strongly compressed, falcate, acute, to 15 cm long. Seeds papery, smooth. Fl. & Fr.: February – August.

Western Ghats: Common in mid-elevation to 1000 m in Maharashtra, Karnataka, Kerala and Tamil Nadu (Endemic). India: Cultivated elsewhere.

IUCN: NE (IUCN, 2024).

Cerbera Lour.

Cerbera odollam Gaertn., Fruct. Sem. Pl. 2: 193. 1791. Synonyms: *Tanghinia odollam* (Gaertn.) G.Don, *Cerbera dilatata* Markgr., *Cerbera forsteri* Seem. Vern.: *Othalam, Chatthaukai* (Mal.); *Kadama, Kaatrali* (Tam.).

Evergreen trees, to 6 m tall; bark greenish-brown; branchlets stout with prominent leaf scars, latex milky. Leaves simple, alternate, crowned at the end of branches, lanceolate or oblanceolate,10-25 x 2.5-6.5 cm, base cuneate or attenuate, apex acuminate or acute, margin entire, glabrous, fleshy, lateral nerves many, close, slender, looped near the margin forming intramarginal nerve, intercostae reticulate; petiole 8-35 mm long, slender, glabrous. Pseudoterminal cymes, flowers many. Flowers white, 5 cm across. Sepals 5, lobes 5, linear, recurved, eglandular. Petals 5, tube funnel-shaped above the throat with 5 villous scales. Stamens 5, small, included; anthers lanceolate, apiculate. Disc absent. Carpels 2, free, ovules 4 in each cell on both sides of thick placenta. Drupe globose or ellipsoid, pericarp green turning rose. Seeds 1 or 2 compressed. Fl. & Fr.: June – January.

Western Ghats: Moist deciduous forests of Maharashtra, Goa, Karnataka and Kerala. World: Sri Lanka, Bangladesh, Myanmar, Thailand, Vietnam, Malaysia, Indonesia, Philippines and Pacific Islands.

IUCN: LC (IUCN, 2024).

Holarrhena R.Br.

Holarrhena pubcescens Wall. ex Don., Gen. Hist. 4: 78. 1837. Synonyms: *Echites pubescens* Buch.-Ham., *Chonemorpha pubescens* (Wall. ex G.Don) G. Don, *Chonemorpha antidysenterica* G.Don, *Holarrhena antidysenterica* var. *pubescens* (Wall. ex G. Don) J.L. Stewart & Brandis Vern.: *Kodasapalai, Kudasam* (Mal. & Tam.).

Trees to 10 m tall; branchlets with whitish, lenticels doted. Leaves ovate or elliptic, 10-24 x 4-11.5 cm, membranous, pubescent, base rounded, apex acute or obtuse, lateral veins 10-15 pairs; petiole 1-5 mm, grooved, glandular inside groove. Cymes 5-8 cm; peduncle 1-2 cm; pedicel 0.3-3 cm. Sepals elliptic to linear, 2-12 mm. Petals white, pubescent, tube 0.9-1.9 cm, lobes oblong, 1-3 cm. Stamens 5, anthers included, narrowly ovate, base rounded. Follicles linear, 20-43 x 0.5-1.5 cm, with whitish, lenticellate. Seeds 0.9-1.6 cm, coma 2.5-4.5 cm. Fl. & Fr.: February – June.

Western Ghats: Deciduous forests of Maharashtra, Goa, Karnataka, Kerala and Tamil Nadu. World: China, Nepal, Bangladesh, Myanmar, Bhutan, Laos, Cambodia, Vietnam and Africa.

IUCN: NE (IUCN, 2024).

Uses: The wood is used for making toys, furniture and plates. The leaves and fruits are used medicinally as tonic and febrifuge.

Hunteria Roxb.

Hunteria zeylanica (Retz.) Gard. ex Thw., Enum. Pl. Zeyl.: 191. 1860. Synonyms: *Cameraria zeylanica* Retz., *Hunteria corymbosa* Roxb., *Hunteria corymbosa* var. *roxburghiana* (Wight) Trimen., *Hunteria roxburghiana* Wight

Trees to 15 m tall; branches slender, glabrous. Leaves oblong, elliptic, or narrowly ovate, 5-18 x1-9 cm, base broadly cuneate to rounded, apex acuminate, lateral veins more than 30 pairs, subparallel, joining marginal veins; petiole 1-1.5 cm. Flowers strongly fragrant, white. Pedicel usually longer than calyx. Sepals ovate, 1.5-1.7 mm, acute. Petals 5, tubular, tube 7-10 mm long, pubescent inside. Berries yellow, globose, usually paired, 1-2 cm in diam. Seeds brownish, ovoid, 1.2 cm x 8 mm. Fl. & Fr.: May – December.

Western Ghats: Moist deciduous forests to evergreen forests above 900 m in Kerala and Tamil Nadu. World: Sri Lanka, Myanmar, China, Indonesia, Malaysia, Laos, Cambodia, Vietnam and Africa.

IUCN: LC (IUCN, 2024).

Kopsia Blume

Kopsia arborea Blume, Catalogus: 13. 1823. Synonyms: *Kopsia longifolia* Merr. *Kopsia scortechinii* King & Gambe, *Kopsia jasminiflora* Pit.

Trees to 15 m tall; branches greenish, terete, slightly compressed, puberulent when young. Leaves elliptic, narrowly elliptic or narrowly ovate, 8-24 x 3.5-8.5 cm, glabrous, apex acute or short acuminate, lateral veins 10-20 pairs; petiole 5-15 cm. Cymes corymbose, many flowered; peduncle to 14 cm, puberulent or glabrous; bracteoles narrowly oblong, puberulent or glabrous; pedicel 3-4 mm. Sepals narrowly oblong, 4-6 mm, puberulent or glabrous. Petals white, tube 2.5

cm, lobes narrowly oblong, 1.5-2 cm. Disc scales narrowly oblong, longer than ovary, fleshy. Ovary ovoid, puberulent. Drupes black or blue-black, ellipsoid, 2.5-3.5 x 1.5-2 cm. Fl. & Fr.: March – August.

Western Ghats: Evergreen forests of Kerala. World: China, Indonesia, Malesia, Thailand, Vietnam and Australia.

IUCN: LC (IUCN, 2024).

Tabernaemontana L.

1a. Leaves opposite, capsules not curved ....**T. gamblei**

1b. Leaves alternate, capsules curved .... **T. alternifolia**

Tabernaemontana alternifolia L., Sp. Pl.: 211. 1753. Synonyms: *Tabernaemontana crispa* Roxb. *Tabernaemontana heyneana* Wall., *Ervatamia heyneana* (Wall.) T. Cooke Vern.: *Thongupiraikai Nanthiyavattai* (Tam.).

Small trees to 8 m tall; bark corky with shallow fissures, latex milky. Leaves elliptic-oblong, 13-18 x 4-7 cm, apex acuminate, base acute, lateral nerves 12-15 pairs, intercostae reticulate; petiole 2 cm long. Cymes terminal or axillary; peduncles to 4 cm long. Flowers white, pedicellate. Sepals 4 mm long, lobes short, oblong, obtuse. Petals white, tubular, tube 25 mm long; lobes 15 mm long, oblong, obtuse, apex crisped. Capsule yellow, 5 x 1.5 cm, slightly curved, beak acuminate; aril red. Fl. & Fr.: November – June.

Western Ghats: Moist deciduous forests of Maharashtra, Goa, Karnataka, Kerala and Tamil Nadu (Endemic).

IUCN: NE (IUCN, 2024).

Tabernaemontana gamblei Subr. & A.N. Henry, Bull. Bot. Surv. India 12: 5. 1972. Synonym: *Ervatamia caudata* Gamble Vern.: *Kaattu Nanthiayavattai* (Tam.).

Small tree to 7 m tall; branchlets white. Leaves simple, opposite, often unequal, 7-10 x 4 cm, elliptic, apex caudate-acuminate, base acute; lateral nerves 7 pairs; petiole 1 to cm long. Cymes terminal. Flowers white; pedicels stout, 1 cm long. Sepals ovate, obtuse. Petals tubular, tube to 17 mm long, cylindrical; lobes 15 mm long, oblong, obtuse. Capsule 3 x 1.25 cm, not curved; aril red. Fl. & Fr.: December– September.

Western Ghats: Evergreen forests of Karnataka, Kerala and Tamil Nadu (Endemic).

IUCN: NT (Low risk/near threatened) (IUCN, 2024).

Voacanga Thours.

Voacanga grandifolia (Miq.) Rolfe, J. Bot. 21: 202. 1883. Synonyms: *Voacanga grandifolia* (Miq.) Rolfe var. *glabrifolia* Bakh.f., *Pootia grandifolia* Miq., *Orchipeda grandifolia* (Miq.) Miq., *Pootia exauriculata* Teijsm. & Binn. Vern.: *Nanjukai maram* (Tam.).

Trees to 2-6 m tall; bark greyish or yellowish, smooth. Leaves opposite, often in unequal pairs, sessile or subsessile, amplexicaule, 24-47 x 8-16 cm, broadly elliptic to oblanceolate, base rounded or auriculate, apex acute or acuminate; secondary veins 15-25 pairs, glabrous except on nerves below. Inflorescence axillary, long-peduncled, erect or drooping, laxly-flowered; peduncle stout, 9-20 cm long. Flowers 5-merous; pedicels to 2.5 cm long; bracteoles to 8 mm long, ovate, deciduous. Sepals 5, companulate, divided to middle, with many basal glands within; lobes 5, 4-7 mm across, suborbicular. Petals white, rotate or salver-shaped; tube to 1.7 cm long, twisted; lobes 5, overlapping to the left in bud, 1.5-2 cm across, obovate-rounded. Stamens 5, inserted at the mouth of the corolla tube; filaments short; anthers to 5 mm long, sagittate, cohering with the stigma. Disc annular, 5-lobed, yellow. Ovary 2-carpellary, many-ovuled; style to 1.6 cm long; stigma acute with a ring at base. Fruits of two, recurved or divaricate mericarps, 7-8 x 4-6 cm, somewhat elongate-reniform or obovoid-oblong, yellow or green with many pale grey or brownish warts. Seeds many, to 8 mm long, ovoid, dark brown, rugose; aril orange red. Fl. & Fr.: Throughout.

Western Ghats: Moist deciduous forests of Goa, Kerala and Tamil Nadu (Naturalized). World: Native to Malesia, Queensland, Australia, Philippines, and cultivated elswhere.

IUCN: NE (IUCN, 2024).

Wrightia R.Br.

1a. Petals yellowish, pinkish or redihs; corona scales 10 .... **W. arborea**

1b. Petals white; corona scales many in 2-series 2

2a. Flowers few in terminal cymes; corona orange redish ....**W. indica**

2b. Flwoers many in terminal cymes; corona white 3

3a. Base of corolla lobes clawed; seed coma silky white...... **W. dolichocarpa**

3b. Base of corolla lobes nto clawed; seed coma dull brown ....**W. tinctoria**

Wrightia arborea (Dennst.) Mabber., Taxon 26: 533. 1977. Synonyms: *Periploca arborea*Dennst., *Wrightia tomentosa* Roem. & Schult., *Beaumontia wallichii* (A.DC.) Walp. Vern.: *Mailam pala* (Mal.); *Karuppalai, Thonthampalei* (Tam.).

Trees to 20 m tall; branches gray or brown, pubescent, lenticellate. Leaves elliptic to broadly ovate or obovate, 6-18 x 3-8.5 cm, pubescent to glabrescent adaxially, tomentose abaxially; lateral veins 10-15 pairs; petiole 2-10 mm. Cymes pubescent; peduncle to 2 cm; pedicel 1-1.5 cm long. Sepals ovate, 3 mm long. Petals yellowish, pinkish or salmon redish, rotate or subrotate, corolla tube 3-7 mm long, glabrous, lobes narrowly elliptic to ovate, 0.8-1.6 cm, papillate; corona scales 10, shorter than anthers, glabrous inside, apex dentate. Ovary connate. Follicles connate, cylindric, 14-21 x 3-4 cm, lenticellate. Seeds linear-fusiform, 2 cm long, coma to 3.5 cm long. Fl. & Fr.: March – December.

Western Ghats: Deciduous forests of Gujarat, Maharashtra, Karnataka, Kerala and Tamil Nadu. World: Sri Lanka, China, Nepal, Myanmar, Laos, Vietnam, Thailand and Malaysia.

IUCN: LC (IUCN, 2024).

Wrightia dolichocarpa Bahadur & Bennet, Indian Forester 104: 32. 1978.

Small trees up to 5 m tall; bark dull white. Leaves yellowish green, glabrous, elliptic or elliptic oblong, apex acuminate to caudate acuminate, 8–19 x 3–7 cm, base acute, midrib prominent beneath, secondary veins 8–12 pairs; petiole 3–5 mm long, glabrous. Follicles 2, 60–95 cm long, dark green, coherent at the tip, slender, terete, lenticellate. Seeds 3–3.4 cm long, linear; coma white, about 5 cm long. Fl. & Fr.: May – Agust.

Western Ghats: Dry deciduous forests of Dadra and Nagar Haveli of Maharashtra (Endemic).

IUCN: NE (IUCN, 2024).

Note: Description adapted from original protologue

Wrightia indica Nagn., Ann. Missouri Bot. Gard. 52: 140. 1965. Vern.: *Paalai* (Tam.).

Deciduous trees upto 15 m tall; bark grey, smooth; outer layer thin, inner layer thick, brittle, creamy; latex milky white. Leaves simple, opposite, distichous, 3.5-11 x 2-4 cm, oblong-lanceolate or elliptic-ovate, apex acuminate or caudate acuminate, base acute, margin entire, glabrous, glaucous beneath, chartaceous, lateral nerves 5-10 pairs, slender, arched, intercostae reticulate; petiole 2-5 mm long, stout, glabrous; Flower bisexual, white, scented, in terminal cymes appearing along with new leaves; pedicel 1.5 mm long; bracts 2, minute. Sepals 5, 5-lobed, with membranous margin, 2 mm long, ovate, obtuse, ciliate, glandular inside. Petals in salver shaped, lobes 5, broad lobes 12 x 5 mm, oblong, obtuse, throat with 1-2 series of erect, fimbriate corona scales, tube 1.5-2 mm. Stamens 5, inserted on the mouth of the tube, anthers sagitate, spurred at base, conniving and adhering to the stigma. Disc absent. Carpels 2, free; ovules many; style filiform, stigma ovoid, usually with a toothed basal ring. Follicle 2, 15-45 cm long, slender, smooth, green, cohering at the tip. Seeds 16 mm long, black, commate at one end. Fl. & Fr.: March – July.

Western Ghats: Dry deciduous forests of Tamil Nadu and extended its distribution in Andhra Pradesh andTelangana State (Endemic)

IUCN: NE (IUCN, 2024).

Wrightia tinctoria (Roxb.) R.Br., Asclepiadeae: 59. 1810 ssp. **tinctoria.** Synonyms: *Nerium tinctorium* Roxb., *Wrightia tinctoria* var. *typica* Pichon Vern.: *Vetpalai, Velpalai, Paalamaram, Neela palei, Irum palei, Thontha palei* (Tam.); *Aiya pala* (Mal.).

Deciduous trees to 10 m tall; brak whitish, smooth. Leaves simple, opposite, 5–11 x 2-4 cm, oblong-lanceolate or elliptic-ovate, apex acuminate, base acute, chartaceous. Flower white, in terminal cymes appearing along with new leaves; pedicel 1– 1.5 mm long; bracts 2, minute. Sepals 5, 5-lobed, with membranous margin, 2 mm long, ovate, obtuse at apex, ciliate, glandular inside. Petals in salver shaped, 5-lobed, lobes 12 x 5 mm, obtuse, corolla tube 1.5-2 mm long. Stamens 5, inserted on the mouth of the tube, anthers sagitate, spurred at base. Carpels 2, free; ovules many; style filiform, stigma ovoid, usually with a toothed basal ring. Follicle 2, 15-45 cm long, slender, smooth, green, cohering at the tip. Seed 14 mm long, grey-brown, silky commate at one end. Fl. & Fr.: March – May.

Western Ghats: Common in deciduous forests of Maharashtra, Goa, Karnataka, Kerala and Tamil Nadu. World: Sri Lanka, Myanmar and Timor.

IUCN: LC (IUCN, 2024).

Key to the subspecies

1a. Leaves narrowly elliptic or narrowly obovate to elliptic, glabrous; inflorescence glabrous; antipetalous corona segments glabrous near the base within, the fimbriae usually simple ....ssp. **tinctoria**

1b. Leaves elliptic to broadly elliptic, puberulent throughout; inflorescence puberulent; antipetalous corona segments puberulent near the base within, the fimbriae compound ....ssp. **rothii**

Wrightia tinctoria (Roxb.) R.Br., ssp. **rothii** (G. Don) Ngan. Ann. Missouri Bot. Gard. 52: 144. 1965. Synonyms: *Wrightia rothii* G. Don, *Wrightia tinctoria* var. *rothii* (G. Don) Hook.f.

Fl. & Fr.: March – July.

Western Ghats: Rare in deciduous forests of Tamil Nadu (Endemic).

IUCN: LC (IUCN, 2024).

Planted:

Funtumia africana (Benth.) Stapf.; Exotic: Native to Africa

Plumeria alba L.; Exotic: Native to Tropical America.

Plumaria obtusa L.; Exotic: Native to Central America.

Plumeria rubra L.; Exotic: Native to Tropical America

Wrightia coccinea (Roxb.) Sims; Native: India, Bangladesh, Myanmar, Philippines.

AQUIFOLIACEAE Bercht. & J. Presl

Ilex Tourn. ex L.

1a. Leaves serrate; lateral nerves 6-9 pairs; flwoers 4-merous ..**I. denticulata**

1b. Leaves entire; lateral nerves 10-12 pairs; flowers 5-6-merous 2

2a Leaves acute at apex; flowers 6-merous; drupe with persistent stigma ... . **L. malabarica**

2b. Leaves mucronate or emarginate at atpex; flowers 5-merous; drupe not as above 3

3a. Large tree with subsessile umbellules; leaf midrib slightly impressed ....**I. wightiana**

3b. Small trees with sessile umbellules; leaf modrip strongly impressed 4

4a. Leaves ovate-lanceolate, long acuminate at apex ....**I. gardneriana**

4b. Leaves ovate-elliptic, minutely mucronate or emarginate 5

5a. Tree with densely branched, small leaves ....**I. walkeri**

5b. Tree with sparsely branched, large leaves ....**I. tadiandamolensis**

Ilex denticulata Wall. ex Wight, Ill. Ind. Bot. 2: 147. 1850. Synonym: *Ilex nilagirica* Miq. ex Hook.f.

Trees to 15 m tall; bark thick, grey, smooth; branchlets glabrous. Leaves simple, alternate, elliptic-oblong, obovate or elliptic, 5-16 x 2-3 cm, base cuneate or subacute, apex acuminate, margin dentate or serrate, coriaceous, glabrous; lateral nerves 6-9 pairs, prominent, ascending, intercostae reticulate; petiole 10-13 mm, slender, glabrous, grooved above. Flowers unisexual, white, 4-merous, 6 mm, across; male flowers in branched short axillary cymes; sepals 4, cupular, 0.5 mm long, lobes 4, persistent broad, subacute. Petals 4, 4 x 2 mm, white, oblong, connate at the very base. Stamens 4, adhering to the base of the corolla; filaments subulate; anthers oblong, dorsifixed. Disc absent. Pistillode globose. Female flowers in fascicles. Sepals as in male flowers. Petals free. Staminodes hypogynous. Ovary ovoid, syncarpous; style very short. Drupe globose, 6-7 mm across, brown, smooth, with persistent calyx, stigma. Pyrenes 4, oblong-elliptic, 4-5 x 3 mm, striate or rugose; fruiting pedicel stout. Fl. & Fr.: February – March.

Western Ghats: Evergreen forests above 1500 m in Karnataka, Kerala and Tamil Nadu. World: Africa and Australia, China and Sri Lanka.

IUCN: NE (IUCN, 2024).

Ilex gardneriana Wight, Icon. Pl. Ind. Orient. 4: t. 1217. 1848.

Trees to 15 m tall; branchlets glabrous, terete. Leaves simple, alternate, elliptic-oblong, ovate or ovate-lanceolate, 5-7.5 x 2-2.5 cm, base obtuse or round, apex acuminate, margin entire, revolute, coriaceous, glabrous; lateral nerves 7-8 pairs, pinnate, obscure, intercostae reticulate; petiole 10-20 mm long, slender, glabrous, grooved above. Flowers unisexual, 5-merous. Male flowers solitary or in fascicles; peduncle 1.2 cm long; pedicel 0.6 cm long, slender. Sepals 5-lobed, 1.5 x 1 mm, lobes triangular, ciliolate. Petals 5, oblong, obtuse, connate at base. Stamens as long as petals. Berry globose, 4.5 x 3 mm, yellowish, glabrous. Fl. & Fr.: March – April.

Western Ghats: Evergreen forests above 1200 m of Kerala and Tamil Nadu (Endemic).

IUCN: CR-D (IUCN, 2024). (Not rediscovered after the discovery of this tree species from the Sispara ghats of Nilgiris, possibly extinct).

Ilex malabarica Bedd., Fl. Sylv. S. India: t. 143. 1871. Synonym: *Ilex wightiana* Dalzell & A. Gibson

Trees to 25 m high; bark greyish, branches glabrous. Leaves simple, alternate, elliptic or elliptic-oblong, 4.5-11.5 x 1.5-3.5 cm, base acute or cuneate, apex acute or acuminate, often mucronate, margin entire, glabrous, shiny above, coriaceous; lateral nerves 4-11 pairs, slender, faint, intercostae reticulate, obscure; petiole 6-15 mm, slender, grooved above. Flowers unisexual, white, 6-merous. Male flowers in short axillary panicled umbellules; pedicels 6-8 mm long, slender. Sepals 6, lobes ovate, puberulous outside, ciliate, persistent. Petals 4-6, white, orbicular, connate at base. Disc 0. Stamens 4-6, filaments short, inserted at the base of the corolla, anthers broadly oblong; pistillode with 4 minute stigmas. Female flowers solitary or in fascicles, very rarely pedunculate; pedicels 6-8 mm long, slender. Sepals as in males. Petals 4-6, orbicular, ciliate, connation lesser than that in male; staminodes hypogynous. Ovary globose; style short, thick; stigma large, absolutely 6-lobed. Drupe depressed globose, small, puberulous, red, 6-celled, crowned by the thick style and supported by the persistent calyx. Fl. & Fr.: February – June.

Western Ghats: Evergreen forsts of Maharashtra, Goa, Karnataka, Kerala and Tamil Nadu (Endemic).

IUCN: LC (IUCN, 2024).

Uses: Wood is useful in making cabinets, tea cups, bowls and packing cases.

Ilex tadiandamolensis Kesh.Murthy, Yogan. & Vasud. Nair Curr. Sci. 56: 270. 1987.

Trees to 10 m tall; branches terete, minutely pubescent when young, glabrous when mature. Leaves elliptic-lanceolate, 3.5–9 x 1.2–2.8 cm, glabrous on both sides, acute or emarginate at apex, attenuate at base, midrib impressed above, lateral nerves 5 to 7; petiole 3 – 8 mm long, channeled, puberulous. Flowers 5–6 mm across, in dense axillary cymes; pedicels 2–5 mm long. Sepals 6, lobed, each up to 1 mm long, triangular-ovate, acute or blunt at apex, toothed along margin. Petal lobes 6, connate below, 1–2 mm long, oblong, margins wavy. Stamens 6, 1–2 mm long, alternating with corolla lobe; anther dorsifixed. Ovary ovoid, 1.5–2 mm across, glabrous; ovules 6. Fl. & Fr.: March – April.

Western Ghats: Rare in evergreen forests of Karnataka (Endemic).

IUCN: NE (IUCN, 2024).

Note: So far known only from type locality

Ilex walkeri Wight & Gard. ex Thw., Enum. Pl. Zeyl.: 184. 1860. Synonym: *Ilex emarginella* Turcz., *Ilex walkeri* var. *emarginella* (Turcz.) Loes., *Ilex walkeri* var. *major* Thwaites ex Trimen

Small trees to 10 m tall; branchlets glabrous. Leaves alternate, elliptic-obovate, cuneate to obtuse at base, subentire and recurved at margin, mucronate or retuse at apex, coriaceous or membranous and glabrous; lateral nerves 2 or 3 pairs; petioles to 2.5 mm long; stipules subulate-deltoid, to 7.5 mm long. Inflorescences umbels; peduncles and pedicels short in both sexes. Flowers 4 or 5-merous. Sepals 4-lobed, lobes subulate, deltoid and glabrous. Petals 4 or 5, connate at base, orbicular or ovate, 2.5 x 1.25-4.7 mm. Stamens 4 or 5, included. Ovary 5-celled; stigma 5-lobed. Drupes globose, 4 mm, red; pyrenes 4 or 5, 3-gonous, hard. Fl. & Fr.: February – May.

Western Ghats: Evergreen forests of Kerala, Maharashtra and Tamil Nadu. World: Sri Lanka.

IUCN: NE (IUCN, 2024), vulnerable (Narasimhan and Irwin, 2021).

Ilex wightiana Wall. ex Wight, Icon. Pl. Ind. Orient. 4: t. 1216. 1848 var. **wightiana**. Vern.: *Parasal, Vellodi* (Mal.); *Badaga, Vellodai* (Tam.).

Trees to 15 m tall; branchlets glabrous. Leaves elliptic, ovate, oblong or oblanceolate, acute or rounded at base, subentire or serrulate at margin, acute to acuminate, mucronate-apiculate or rarely obtuse at apex, glabrescent; petioles to 1.4 cm long. Inflorescences in umbellules; peduncles of male inflorescences 0.8-1.25 cm long. Flowers 5-merous, 5-6 mm across. Sepals cupular, 1-2 x 2-2.5 mm, lobes 5 or 6, oblong-ovate, ovate or oblong, acute-apiculate, sometimes obtuse at apex, ciliate at margin, puberulous. Petals 5 or 6, connate below, white, tube 1 mm long, lobes oblong, oblong-obovate or ovate-elliptic, entire at margin, obtuse at apex, 2-2.5 x 1.5-2 mm. Stamens 5 or 6, shorter than petals, epipetalous; filaments 1 mm long; anthers oblong, 0.8-0.9 mm long. Ovary rudimentary, puberulent; stigma 4-6 lobed. Female umbels lax; peduncles 1.2 cm long. Sepals and petals as in male flowers. Ovary globose, 2 mm long, 6-loculed; ovules 1 or 2 in each locule, pendulous; style 0.5 mm long; stigma capitate. Drupe globose, apiculate, smooth, dark red or blue; pyrenes 5 or 6, trigonous. Fl. & Fr.: February – June.

Western Ghats: Everegreen forests above 1500 m in Kerala and Tamil Nadu. World: Sri Lanka.

IUCN: LC (IUCN, 2024).

Key to the varieties

1a. Leaves to 5 cm long, nerves obscure beneath; umbellules pedicellate .... var. **wightiana**

1b. Leaves to 6 cm long, nerves distinct beneath; umbellules subsessile .... var. **peninsularis**

Ilex wightiana Wall. ex Wight, Hook.f. Fl. Brit. India 1: 603. 1875 var. **peninsularis**.

Fl. & Fr.: March – September.

Western Ghats: Evergreen forests of Kerala and Tamil Nadu (Endemic).

IUCN: LC (IUCN, 2024).

Uses: Wood is useful for building construction.

Planted:

Ilex aquifolium L.; Exotic: Native to China, Europe, West Asia

Ilex cornuta Lindl. & Paxton; Exotic: Native to China, Korea

Ilex opaca Aiton; Exotic: Native to America

ARALIACEAE Juss.

Key to the genera

1a. Leaves pinnately compound 2

1b. Leaves palmately compound ....**Heptapleurum**

2a. Leaflets in unipennate; flwoers greenish yellow............................**Polycias**

2b. Leaflets in bipinnate; flowers white .... **Aralia**

Aralia L.

1a. Leaves palmately or digitatly compound; leaflets entire............................ .... **A. leschenaultia**

1b. Leaves bipinnately compound; leaflets sharply serrate.... **A. malabarica**

Aralia leschenaultii (DC.) J. Wen, Brittonia 45: 53. 1993. Synonyms: *Aralia fragrans* (D. Don) Jebb& J. Wen, *Hedra fragrans* (DC.) Seem., *Pentapanax fragrans* (D. Don) Ha, *Pentapanax leschenaultii* (DC.) Seem., *Panax leschenaultii* DC., *Hedera leschenaultia* (DC.) Wight, *Hedera trifolia* Wight & Arn., *Pentapanax leschenaultii* Seem., *Pentapanax longipedunculatus* N.S. Bui, *Pentapanax fragrans* var. *longipedunculatus* (N.S. Bui) Ha, *Pentapanax leschenaultii* var. *simplex* K.M. Feng.

Trees to 10 m tall; bark thin, smooth, greening. Leaves palmate or digitate; leaflets 3-5, to 15 x 10 cm, ovate-acuminate, glabrous, highly reticulate. Umbels 2-3 cm across. Flowers 3-4 mm across, numerous. Sepal lobes 5, acute. Petals 5, yellow. Stamens 5. Ovary inferior; styles connate. Drupe 4-5 celled, obscurely angled. Fl. & Fr.: June – November.

Western Ghats: Evergreen forests of Karnataka, Kerala and Tamil Nadu. World: Sri Lanka, Nepal, Myanmar, Bangladesh, Bhutan, Western China, Thailand and Vietnam.

IUCN: LC (IUCN, 2024).

Aralia malabarica Bedd., Fl. Sylv. S. India: cxxi. 1872.

Small trees to 5 m tall; bark greyish brown, prickly. Leaves to 1 m long, bipinnately compound; leaflets opposite, 14-18 x 5-7 cm, ovate, apex acuminate, base oblique, sharply serrate, penninerved, sessile; rachis prickly, sheathing at base. Inflorescence large panicled umbels, 30 cm across; bracts ovate, acuminate. Flowers 3-4 mm across, 12-16 together; pedicels 1 cm long. Sepals 5, ovate, acute. Petals ovate-oblong, obtuse, spreading, yellow. Stamens 5, inflexed in bud. Ovary inferior, 5-celled, globose, cells 1-ovuled; styles 5, free. Berry globose, 7 x 7 mm, 5-angled, rugose. Seeds five, 3 mm across, 3-angled, light brown. Fl. & Fr.: March – June.

Western Ghats: Evergreen forests above 1000 m in Kerala and Tamil Nadu (Endemic).

IUCN: VU- B1+2c. (IUCN, 2024).

Heptapleurum Gaertn.

1a. Flowers pedicellate 2
1b. Flowers sessile .... **H. capitatum**
2a. Style long, connate in a cylindric column; large erect trees 3
2b. Style short, not connate in a column, usually straggleing trees 4
3a. Flowers in branched umbels .... **H. rostratum**
3b. Flowers in racemes .... **H. racemosa**
4a. Leaflets prominently reticulated 6
4b. Leaflets not prominently reticulated 5
5a. Petilule to 15 mm long; leaflets 1-2; peducle to 2 cm long............................
.... **H. chandrasekharanii**

5b. Petiolule to 5 cm long; leaflets 7-9; peduncle to 10 cm long .... **H. wallichianum**

6a. Leaflets below 7 cm long; umbels in small panicles .... **H. bourdillonii**

6b. Leaflets more than 7 cm long; umbels in large branched panicles 7

7a. Umbels in terminal; leaflets obovate, rounded or emarginate. **H. stellata**

7b. Umbels in terminal and axillary; leaflets ovate or elliptic, shortly pointed at apex..8

8a. Peduncle rachins much longer than 60 cm ....**H. digitatum**

8b. Peducle rachis shorter than 30 cm 9

9a. Leaves and bark silvery shining; young branches pubsescent .... **H. venulosa**

9b. Leaves and bark greenish not shining; your branches glabrous ....**H. agathiyamalayana**

Heptapleurum agasthiyamalayanum (Manickam, Murugan, Sundaresan & Jothi) Lowry & G.M.Plunkett, Novon 28: 146 (2020). Synonyms: *Schefflera agasthiyamalayana* Manickam, Murugan, Sundaresan & Jothi Vern.: *Agasthiar Ettilai maram* (Tam.).

Small trees to 5 m tall; branchlets cylindric, greenish, glabrous. Leaves alternate, digitately compound; leaflets 5, elliptic- oblong, 8-17 x 4-6 cm, obtusely acute base, sub-revolute margin, obtusely acuminate apex; petioles 10-17 cm long; stipules connate; petiolules 2.5-5 cm long, glabrous. Inflorescence terminal, panicles umbells, 5-8 cm long; peduncles 3-5 cm long, terete. bracts minute. Sepals 5, campaulate, tube to 2 mm long, adanate to ovary; lobes indistinct. Petals 5, 3 mm long, valvate. Stamens 5, equal, reflexed; anthers 1 mm long. Ovary 2 mm long; styles and stigma indistinct. Berry globose, dark purple, 6 mm long. Seeds 5, compressed. Fl. & Fr.: August – February.

Western Ghats: Evergreen sholas of Agasthiyamalai hills in Tamil Nadu (Endemic).

IUCN: NE (IUCN, 2024), endangered (Narasimhan and Irwin, 2021).

Heptapleurum bourdillonii (Gamble) Lowry & G.M. Plunkett, Novon 28: 148. 2020. Synonyms: *Schefflera bourdillonii* Gamble

Trees to 8 m tall. Leaves digitately compound. Leaves alternate, leaflets 5-8, leaflets 4-9 x 0.6-2 cm, obovate, spathulate, elliptic-obovate or rhomboid, base attenuate or cuneate, apex obtuse, emarginate, margin entire, reflexed, glabrous, coriaceous; lateral nerves 2-7 pairs, pinnate, prominent, intercostae reticulate; stipules intrapetiolar, connate within the petiole; rachis 4-10 cm long; petiolule 0.5-1.5 cm, slender, articulate, sheathing at the base, glabrous. Flowers bisexual often polygamous, 5 mm across, in subterminal stellate villous umbellate racemes upto 4 cm long; pedicels 5-10 mm long. Sepals truncate, lobes 0. Petals 5, ovate, acuminate, inflexed. Staminodes in females 5, subulate. Stamens 5. Ovary inferior, 5-celled; styles short, not connate in to a cylindrical column. Berry globose, 3 mm across, 5-angled. Seeds compressed. Fl. & Fr.: March – May.

Western Ghats: Moist evergreen forests of Kerala and Tamil Nadu (Endemic).

IUCN: EN- B1+2c. (IUCN, 2024).

Heptapleurum capitatum (Wight & Arn.) Seem., J. Bot. 3: 81. 1865. Synonyms: *Schefflera capitata* (Wight &Arn.) Harms, *Brassaia capitata* (Wight & Arn.) C.B. Clarke, *Paratropia capitata* Wight & Arn.

Small trees to 8 m tall. Leaves alternate, palmately compound; leaflets seven, their stalks 2 inches, 15 x 5 cm, coriaceous, glossy on the upper side, dull and light green on lower side; midrib very stout, brown; lateral neves slender, prominent on the upper side, the smaller reticulations distinct on the lower sides; leaf-base sheathing with a triangular intra-petiolar stipule, covering the axillary bud; petiole to 15 cm long. Inflorescens a terminal panicle up to 35 cm long, dark purple. Flowers in dense sessile heads of 6 or 7, at the ends of the stout branches of the panicle; heads 1.5-2.5 cm wide, of about ten flowers. Bracts concave, tomentose. Petals 6, valvate. Stamens longer than the petals abruptly bent inwards just below the anther; anther opening inwards. Ovary bluntly conical, continuous with the disc at the base; 6-celled, ovules one in each cell; styles fused in to a column. Drupes sub globose, 6-angled. Fl. & Fr.: February – May.

Western Ghats: Evergreen forests of Karnataka, Kerala and Tamil Nadu (Endemic).

IUCN: NE (IUCN, 2024).

Heptapleurum chandrasekharanii (Ramam. & Rajan) Lowry & G.M. Plunkett, Novon 28: 149. 2020. Synonyms: *Schefflera chandrasekharanii* Ramam. & Rajan

Small trees to 7 m tall; bark papery, branchlets terete, lenticellate. Leaves digitately compound, alternate, leaflets 1-2; 5-10 x 3-4 cm, obovate, base cuneate or oblique, apex apiculate, margin crenate, wavy, coriaceous, glabrous; lateral nerves 5-7 pairs, pinnate, prominent, intercostae reticulate; stipules adnate within the petiole; rachis cylindric glabrous, base sheathing; petiolule 5-15 mm long, terete. Flower in terminal umbellate panicles; peduncle 2 cm long; pedicel 1-2 mm long, tomentose. Sepals 1 mm long, cup shaped, adnate to the ovary. Petals 5, 1 mm long, ovate, glabrous. Stamens 5, inserted around the disc; filaments incurved; anthers oblong. Ovary 5-celled, ovule 1 in each cell; style short. Fl. & Fr.: March – May.

Western Gahts: Rare in evergreen forests of Kerala (Endemic).

IUCN: NE (IUCN, 2024).

Heptapleurum digitatum (G.Don ex Loudon) Lowry & G.M.Plunkett, Novon 28: 151. 2020. Synonyms: *Schefflera roxburghii* Gamble, *Schefflera venulosa* var. *roxburghii* (Gamble) Khanna, *Sciodaphyllum digitatum* G.Don ex Loudon, *Aralia digitata* Roxb.

Large climbing trees to 10-12 m tall. Leaves alternate, digitate, leaflets chartaceous, up to 15 x 6 cm, ovate-elliptic, apex acute, main nerves scarcely more prominent than the secondary and reticulations. Flowers in panicles of umbels or compound racemes, usually terminal; bracts usually deciduous; bracteoles 0 or few. Sepals 5, cupular, mouth truncate or toothed. Petals 5 or 6 or many, valvate. Stamens as many as the petals. Ovary-cells as many as the petals; styles short, free. Drupes 5-6 angled. Seeds compressed. Fl. & Fr.: March – May.

Western Ghats: Evergreen forests of Maharashtra, Karnataka, Kerala and Tamil Nadu (Endemic).

IUCN: NE (IUCN, 2024).

Heptapleurum racemosa (Wight) Bedd Fl. Sylv. S. India: 214. 1872. Synonyms: *Schefflera racemosa* (Wight) Harms, *Hedera racemosa* Wight, *Heptapleurum racemosum* (Wight) Bedd. Vern.: *Ettilai maram* (Tam.).

Trees to 10 m tall; young parts rusty tomentose. Leaves digitately compound, alternate, stipulate, leaflets 4-8, elliptic or elliptic-lanceolate, 9.5-18 x 3.8-6 cm, base acute, apex acuminate, margin entire or wavy, glabrous, glaucous beneath, chartaceous; lateral nerves 7-14 pairs, ascending, slender, raised, intercostae reticulate, obscure; petiolule 2.5-5 cm long, slender, grooved above, glabrous; stipules adnate within the petiole; rachis 12-23 cm long, stout, sheathing at base, glabrous. Flowers bisexual, pale green, in paniculate racemes from stout older branches; bracts, bracteoles and pedicels rusty pubescent, bracts persistent. Sepals cupular, 5-toothed, obtuse. Petals 5, ovate, acute, valvate. Stamens as many as petals. Ovary cells as many as petals; styles 5, short, connate in a cylindric column; stigma distinct. Berry ovoid, 5-6 mm across, 5-celled, 5 angled, glabrous with persistent style. Fl. & Fr.: March – June.

Western Ghats: Evergreen forests above 1000 m in Kerala and Tamil Nadu (Endemic).

IUCN: NE (IUCN, 2024).

Heptapleurum rostratum (Wight) Bedd., Fl. Sylv. S. India, Forest. Man.: cxxii. 1872. Synonyms: *Schefflera rostrata* (Wight) Harms, *Hedear rostrata* Wight,

Trees to 8-12 m tall; sometimes epiphytic on trees or rocks. Leaves digitate, leaflets 7-11, 5-12 x 3-8 cm, oblong lanceolate, acute at apex, round at base, rusty pubescent when young; petioles 10-20 cm long; petiolule 0.5-2.5 cm long, both rusty pubescent when young. Inflorescence at tip of old branches or axillary; umbels 2-2.5 cm diam; peduncles 2-3 cm long; pedicels 0.5-1 cm long; bracts 0.2-0.3 cm long, ovate, rusty pubescent when young. Flowers 0.4-0.5 cm across. Sepals cupular, tube 0.2-0.25 cm long, 0.25 cm diam, minutely 5-6 toothed. Petals 0.25-0.3 cm long, ovate, valvate. Filaments 0.5 cm long; anthers 0.1 cm long. Ovary 0.4-0.5 cm long; style short, connate in into a cylindric column. Fl. & Fr.: March – June.

Western Gahts: Semi-evergreen to evergreen forests of Karnataka, Kerala and Tamil Nadu (Endemic).

IUCN: NE (IUCN, 2024).

Heptapleurum stellata Gaertn., Fruct. Sem. Pl. 2: t. 178, f. 3. 1791. Synonyms: *Schefflera stellata* (Gaertn.) Harms, *Heptapleurum acutangulum* Gaertn., *Hedera obovata* Wight Vern.: *Peimiratti* (Tam.).

Small trees to 6 m tall; often epiphytic on large trees and rocks; bark smooth, papery, greenish. Leaves digitately compound, leaflets 7-8, 3-5, 12 x 8 cm, obovate, apex emarginate or slightly obtuse; petiole 10-15 cm long, sheathing, petiolule to 5 cm. Panicles to 35 cm long; pedicels to 0.7 cm long. Seplas cupular, calys tube united with ovary forming a rim. Petals 7-8, 3 x 1.5 cm, triangular. Stamens 7-8; filaments 1.5 mm; anthers 1.5 mm. Ovary seated in the calyx cup, 7-8 celled. Drupe globose. Fl. & Fr.: April – May.

Western Ghats: Deciduous forest of Karnataka, Kerala and Tamil Nadu. India: Andhra Pradesh. World: Sri Lanka.

IUCN: NE (IUCN, 2024).

Heptapleurum venulosa (Wight & Arn.) Seem., J. Bot. 3: 80. 1865. Synonyms: Schefflera venulosa (Wight &Arn.) Harms, *Paratropia venulosa* Wight &Arn., *Schefflera elliptica* (Blume) Harms Vern.: *Aithilai mara, Modakama* (Tam.).

Small trees to 7 m tall; branchlets 0.5-0.7 cm thick. Leaves digitately compound, 5 or 6 foliolate, leaflets 10-14 x 5-8 cm, ovate-oblong, apex abruptly acuminate, base rounded; nervules prominently reticulate; petiolule 7-10 cm long; stipule 5 mm broad, black. Panicles 20 cm across, glabrous; umbels 2 cm across, 3-12 flowered. Flowers brown; pedicel 8-12 mm long; Sepals cupular, apex truncate. Petals 2 x 2.5 mm, triangular. Drupe globose, 4 x 4 mm, 5-angled. Fl. & Fr.: September – June.

Western Ghats: Moist deciduous to evergreen forests of Maharashtra, Goa, Karnataka, Kerala and Tamil Nadu. India: Andhra Pradesh and Odisha. World: Myanmar, China, Laos, Cambodia, Vietnam, Malaysia and Singapore.

IUCN: NE (IUCN, 2024).

Heptapleurum wallichianum (Wight & Arn.) Seem., J. Bot. 3: 80. 1865. Synonyms: *Schefflera wallichiana* (Wight &Arn.) Harms, *Paratropia wallichiana* Wight &Arn. *Aralia wallichiana* (Wight & Arn.) D.Dietr.

Trees to 12 m tall; sometimes epiphytic on other trees; bark 5-8 mm thick, surface grey, rough; branchlets 2 cm thick. Leaves digitately compound, leaflets 5-8, 14-24 x 5-8 cm, oblong or ovate, apex obtusely acuminate or acute, base subcordate, truncate or round margin entire, coriaceous, glabrous; lateral nerves 10-21 pairs, very slender, nearly horizontal, intercostae obscure; stipules adnate within the petiole, 1.5 cm broad; rachis 17-37 cm long, stout, cylindric, glabrous, sheathing at base; petiolule 4-6.5 cm long, stout. Flowers bisexual, pale green, 6-8 mm across, 12-20 together in umbels on elongate racemes arranged in panicles on short axis; pedicel 7 mm long. Seplas cupular, apex truncate. Petals 5, valvate, cohering in to a cap above the stamens. Stamens 5. Ovary inferior, cells as many as petals; style absent; stigma sessile. Berry globose, 5-6 mm across, 5 angled, crowned with disc; pyrenes 5, 1-seeded. Fl. & Fr.: March – August.

Western Ghats: Evergreen forests of Karnataka, Kerala and Tamil Nadu (Endemic).

IUCN: NE (IUCN, 2024).

Polycias J.R. Forst. & G. Forst.

Polyscias acuminata (Wight) Seem., J. Bot. 3: 181. 1865. Synonyms: *Eupteron acuminatum* (Wight) Miq., *Hedera acuminata* Wight Vern.: *Croton chedi* (Tam.).

Small trees to 8 m tall; bark smooth, lenticellate. Leaves to 30 cm long, 1-pinnate, leaflets opposite, 5-6 paired, 12 x 4 cm, elliptic-oblong, acuminate, entire, petiolulate; petiole sheathing at base. Umbels in axillary or terminal panicles, 30-50 cm long; bracts lanceolate. Flowers 4-5 mm across, 6-12 together, pedicellate. Seplas 5, 5-lobed, ovate, acute. Petals brown, 3 x 2 mm, ovate, acute, valvate. Stamens 5; filaments shorter than anthers. Ovary 5-celled, inferior; styles 5, free. Berry globose, 5-angled, glabrous. Fl. & Fr.: February – March.

Western Ghats: Evergreen forests above 1000 m in Kerala and Tamil Nadu. World: Sri Lanka.

IUCN: NE (IUCN, 2024).

Planted:

Polyscias austaliana (F. Muell.) Philipson; Exotic: Native to New Guinea to North Australia.

Polyscias scutellaria (Burm. f.) Fosberg; Exotic: Native to South west Pacific Islands

Schefflera actinophylla (Endl.) Harms; Exotic: Native to Australia

Schefflera elegantissima (H.J. Veitch ex Mast.) Lowry & Frodin; Exotic: Native to New Caledonia, Polynsenia

Tetrapanax papyrifer (Hook.) K. Koch; Exotic: Native to South China and Taiwan

Trevesia palmata (Roxb. ex Lindl.) Vis.; Native: India, South West China, Bangladesh, Nepal, Laos, Cambodia, Vietnam, Thailand.

ARECACEAE Bercht. & J. Presl

1a. Leaves bipinnate; leaflets broadly cuneate ....**Caryota**
1b. Leaves unipennate; leaflets linear-lanceolate 2
2a. Base of leaflet auricled ....**Arenga**
2b. Base of leaflet not auricled 3
3a. Male flowers small, sunken in the rachis ....**Bentinckia**
3b. Male flowers large, not sunken in the rachis ....**Pinanga**

Arenga Labill. ex DC.

Arenga wightii Griff., Calcutta J. Nat. Hist. 5: 475. 1845. Synonyms: *Saguerus wightii* (Griff.) H.Wendl. & Drude Vern.: *Alatthil Thenga, Aazham, Aazhuthumpana, Alathilthenga, Alathiltenna, Karampana, Kaattuthengu, Malam Tengu, Njettipana, Panei* (Mal.); *Alampanei, Ala panei, Cheeni panai, Kaatuthengai* (Tam.).

Monoecious palms to 6 m tall; trunk cylindrical, covered with persistent leafsheaths, caudex 10-15 m across. Leaves pinnatisect, to 8 m long, leaflets to 100 x 5 cm, linear, base auricled, glaucous below; petiole to 3 m long. Spadices to 1 m long, basipetal. Male flowers orbicular; stamens many, free. Female flowers with 6 tepals, biseriate, similar, accrescent. Ovary 3-celled. Berry depressed-globose, smooth, hard, pale green. Seeds 2-3. Fl. & Fr.: April – June.

Western Ghats: Evergreen forests above 900 m in ravines and valleys of Tamil Nadu and Kerala (Endemic).

IUCN: VU- B1+2c. (IUCN, 2024).

Uses: From the flowerstalk obtained toddy, which is making local plam jaggery by Kani tiribes.

Bentinckia Berry ex Roxb.

Bentinckia condapanna Berry ex Roxb., Fl. Ind., ed. 1832. 3: 621. 1832. Vern.: *Kanthal, Kanthakamugu* (Mal.); *Konda panai, Varei-kamugu* (Tam.).

Monoecious palms to 20 m tall; trunk 20 cm thick, annulate, caudex 2-3 m across. Leaves pinnately compound, leaflets 50-70 x 3-5 cm, linear-lanceolate, evenly scattered, sessile; petiole 30-60 cm long, sheath rounded, 55 cm long, 26 cm broad, encircling the stem. Spadix axillary, below the leaves, branched, 30-50 cm across; spathes many. Male flowers small or sunken in the spirally arranged pits towards the tips of branches. Perianth lobes 6, biseriate, outer smaller, 5 mm across, orbicular, inner longer. Stamens 6, free. Female flowers larger, towards the base of the branches of spadix. Ovary 3-celled, 1 cell larger with an ovule. Fruit subglobose, 1.2 x 0.9 cm, dark, pink. Fl. & Fr.: May-December

Western Ghats: Slopes of Evergreen forests above 1000 m in Tamil Nadu and Kerala (Endemic).

IUCN: VU- A1c. (IUCN, 2024).

Uses: Terminal apical bud and pith is edible either raw or cooked by local tribes.

Caryota L.

Caryota urens L., Sp. Pl.: 1189. 1753. Vern.: *Ohunda-pana, Kala-pana* (Mal.); *Kunthal-panei, Thippili-panei* (Tam.).

Monoecious palms to 16-20 m tall; trunk 30-50 cm diam, smooth with prominent annular leaf-scars. Leaves bipinnate, 4-6 m long, pinna 5-7 pairs, to 1.5 m long, leaflets broadly cuneate, fan-shaped, 12-20 cm long, 7-10 cm wide at wider portion, raemorse at apex, many ribbed. Spadix interfoliar, shortly peduncled, much branched, pendulous, to 4 m long; spathes few, 40-50 cm long. Flowers many, in triads with female flower in the middle. Sepals 3, rounded, imbricate. Petals linear-oblong, valvate. Stamens many. Ovary 3-celled, 3-gonous; ovule 1-per locule. Fruit globose, 2 cm across, reddish purple. Seeds plano-convex, subreniform. Fl. & Fr.: Round the Year

Western Ghats: Deciduous to evergreen forests of Kerala, Tamil Nadu, Karnataka, Goa and Maharashtra. World: Sri Lanka and cultivated elsewhere.

IUCN: LC (IUCN, 2024).

Uses: Fresh inflorescence used for decorating festive occasions.

Pinanga Blume

Pinanga dicksonii (Roxb.) Blume, Rumphia 2: 77. 1839. Synonyms: *Areca dicksonii* Roxb., *Ptychosperma dicksonii* (Roxb.) Miq., *Seaforthia dicksonii* (Roxb.) Mart. Vern.: *Kana-kamugu* (Mal.); *Iragu panai* (Tam.).

Monoecious palms to 4-6 m tall; trunk 6-8 cm diam. Leaves pinnate, 1-1.3 cm long, leaflets 2-3 united, 30-50 cm long, 2-3 cm wide, toothed margins. Spadix small, at axil below the crown, 4-6 branched, deflexed and drooping. Spathe 1. Flowers unisexual, 3 at a node, middle female and side ones male. Male flowers: sepals keeled. Petals valvate, 0.2-0.3 cm, ovate lanceolate. Stamens 6. Female flowers: sepals and petals orbicular, 0.1 cm; staminodes 6. Ovary 1-celled; ovule 1. Fruit ellipsoid, 0.6-1 x 0.2-0.3 cm. Seed solitary, ellipsoid. Fl. & Fr.: February – July.

Western Ghats: Valleys and forest slopes above 900 m in Kerala, Tamil Nadu, Karnataka and Maharashtra (Endemic).

IUCN: NE (IUCN, 2024).

Uses: The nut is eaten by the hill tribes substitute for bettle nut. Stems used as walking sticks.

ASTERACEAE Bercht. & J. Presl

Key to the genera

1a. Trees above 15 m tall; anthers not auricled at base....................................2

1b. Trees below 10 m tall; anthers auricled at base ....**Monosis**

2a. Head peduncled; achenes 3-4-angled ....**Strobilocalyx**

2b. Head subsessile; achenes subcylindric ....**Vernonia**

Monosis DC.

1a. Heads with 4-6 florets, subpyramidal .... **M. travancorica**

1b. Jeads with 20-25 florets, pyramidal .... **M. kannikattiensis**

Monosis kannikattiensis (Rajkumar, Selvak., S. Murug. & Chellap.) Kottaim., Int. J. Curr. Res. Biosci. Pl. Biol. 6(10): 37. 2019. Synonym: *Vernonia kannikattiensis* Rajkumar, Selvak., S. Murug. & Chellap.

Trees to 15 m tall. Leaves oblanceolate to obovate, cuneate at base, margin coarsely toothed, pubescent and glabrous below; petiole to 3-8 mm long. Heads in terminal paniculate, homogamous, sessile or pedunculate, pyramidal, each head surrounded by 4-5 imbricate scaly bracts. Flowers 20-25 per head. Fl. & Fr.: January – March.

Western Ghats: Everegreen forests of KMTR, Kannikatti area in Tamil Nadu (Endemic).

IUCN: NE (IUCN, 2024).

Monosis travancorica (Hook.f.) H. rob. & Skvarla., Proc. Biol. Soc. Washington 119: 606. 2006. Synonym: *Vernonia travancorica* Hook.f.

Trees to 6 m tall; bark grey or greyish-brown with narrow vertical fissures; branchlets pubescent. Leaves simple, alternate, obovate or elliptic-oblanceolate, 5-24 x 2.5-10 cm, base cuneate or attenuate, apex obtuse or obtusely acute, margin entire, undulate or distantly dentate towards distal end, glabrous, chartaceous; lateral nerves 4-8 pairs, ascending, prominent, intercostae reticulate; petiole 6-25 mm, stout, pubescent, grooved above. Head homogamous of 4-6 florets, in terminal subpyramidal panicles; outer bracts orbicular, obtuse, hairy; inner bracts 7 x 2.5 mm, oblong, obtuse, glabrous. Flowers bisexual, pink. Seplas hairy pappus, adnate to the ovary. Peals 5, tubular, 10 mm long, lobes 5, glabrous. Stamens 5, inserted within the corolla tube; filaments free; anthers basifixed, connate, connective produced upwards, anthers obtusely auricled at base. Ovary 1-celled, ovule solitary, erect, style arms yellowish-brown, puberulous. Achene 10-ribbed, ribs prominent with yellow glands between; pappus similar, 9-10 mm long, setaceous, white, bent at base. Fl. & Fr.: December – March.

Western Ghats: Evergreen forests of Karnataka, Kerala and Tamil Nadu (Endemic).

IUCN: NE (IUCN, 2024).

Strobocalyx (Blume ex DC.) Spach

Strobocalyx arborea (Buch.-Hamil.) Sch., Bip. Jahresber. Pollichia 18-19: 171. 1861. Synonyms: *Vernonia arborea* Buch.-Ham., *Decaneurum javanicum* (Blume) Miq., *Gymnanthemum arboretum* (Buch.-Hamil.) H. Rob., *Cacalia arboera* (Buch.-Hamil.) Kuntze

Trees to 20 m tall; bBranches densely fulvous tomentulose or glabrescent, rarely glabrous. Leaves ovate, elliptic-ovate, or oblong, 8-25 x 4-10 cm, thinly leathery, abaxially white pubescent, adaxially fulvous pubescent along veins or glabrous, glandular, lateral veins 8-12-paired, adaxially distinctly raised, veinlets reticulate, base obtusely rounded or attenuate, margin entire, apex acuminate; petiole 10-30 mm, densely pubescent. Synflorescences terminal, panicles broader than long, shallowly rounded, flat-topped. Heads very many; peduncles to 10 mm, mostly much shorter, densely pubescent; bracteoles absent. Involucre cup-shaped, 2-3 x 6-8 mm; phyllaries 5-seriate, ovate or ovate-oblong, 0.5-2 mm, densely pilose, rarely subglabrous, apex obtuse or subacute, outer very short, ovate. Receptacle flat, areolate. Florets 5 or 6. Petals pink, pale purplish, or white, tubular, 5-7 mm, lobes acute, glabrous. Style branches long. Achenes compressed, 2-3 mm, 3- or 4-angled and 8-10-ribbed, inconspicuously pilose and glandular. Pappus 1- or 2-seriate, dirty white; outer setae few, almost absent, 1 mm; inner setae 5-7 mm. Fl. & Fr.: November – April.

Western Ghats: Semi-evergreen forests of Tamil Nadu (Endemic).

IUCN: LC (IUCN, 2024).

Vernonia Schreb

Vernonia vivekanathanii Uniyal, Fl. India 13: 391. 1995. Synonyms: *Monosis wightiana* DC., *Vernonia shevaroyensis* Gamble, *Monosis shevaroyensis* (Gamble) H. Rob. & Skvarla

Evergreen trees to 15 m tall. Leaves simple, oblanceolate to obovate, rarely elliptic lanceolate, 9–26 x 4–10 cm, round to cuneate at base, subacute to acute at apex, entire to repand-crenate at margins, abaxially scabrid-pubescent to brown tomentose, adaxially glabrous except the veins; lateral veins 9 -15 pairs; petioles 0.8–1.2 cm long. Synflorescence in axillary or terminal panicles, thyrse-like, homogamous heads at the end of the rigid divaricating branches of the panicle; peduncle 0.8–12 cm long. Capitulum subsessile, medium sized, 1 cm long. Involucre tubular, imbricate; phyllaries 5–6-seriate, outermost ovate or suborbicular, 1–2 mm long, subacute or acuminate at apex, villous almost throughout, innermost linear, 4.5–5 mm long, subacute or obtuse at apex. Floret single per capitulum, bisexual, 0.8–1.0 cm long, purplish. Petals tubular, 5-lobed, tube slender, 4 mm long, with free apices, apical part funnel shaped; free lobes linear-lanceolate, acute at apex, glabrous. Stamens 5, syngenesious, 5 mm long; anthers linear-sagittate, 2.5 mm long. Carpels 2, 1.3–2.5 mm long, unilocular with single ovule in basal placentation; style slender, bifid at apex, branches equal, subulate, pilose. Achenes subcylindric-clavate, 2–3 x 0.5–1 mm, light brown to brown, 10-ribbed, glabrous or with a few hairs. Pappus numerous, plumose, in two rows, 0.7–1.0 cm long, white, barbed, persistent, often slightly bent at base. Fl. & Fr.: November – April.

Western Ghats: Deciduous to evergreen forests of Karnataka, Kerala and Tamil Nadu (Endemic).

IUCN: NE (IUCN, 2024).

BERBERIDACEAE Juss.

Berberis L.

Berberis napaulensis (DC.) Spreng., Syst. Ved. ed. 16.2: 120. 1825. Synonyms: *Mahonia leschenaultii* (Wall. ex Wight & Arn.) Takeda ex Dunn., *Berberis leschenaultii* Wall. ex Wight & Arn., *Berberis nepalensis* var. *leschenaultia* Hook.f. & Thoms. Vern.: *Nepala Maramanjal* (Tam.).

Small trees to 6 m tall; bark rough, yellowish inside. Leaves alternate, imparipinnate, membranous, herbaceous, coriaceous, petiolate or sessile, exstipulate; rachis articulated at each leaflet node, leaflets few to 41, sessile to petiolulate. Inflorescences solitary, fascicled racemose or panicle, terminal cymes. Flowers bisexual, yellow or greenish yellow, actinomorphic, pedicellate.

Sepals 9, usually in 3 series; bracts longer or shorter the pedicels. Petals usually in one series, base with or without 2 oblong glands. Stamens 6, slightly adnate, anther opening by valves. Ovary superior, ellipsoid; ovules few, basal, erect; style short or sessile, persistent in fruit. Berry ywllow turn into purple or black when ripe. Seeds about 1-7, with conspicuous endosperm. Fl. & Fr.: June – December.

Western Ghats: Occasional in evergreen forest of Kerala (Idukki high ranges) and Tamil Nadu (Nilgiris, Palnis and Anamalais). World: Nepal and China.

IUCN: NE (IUCN, 2024).

BETULACEAE Gray

Key to the genera

1a. Branchlets yellow pubescent; female inflorescence numerpoous in panicle .... **Alnus**

1b. Branchlets villous resinous; female inflorescence 3-5 in racemes...**Betula**

Alnus Mill.

Alnus nepalensis D.Don, Prodr. Fl. Nepal 58. 1825. Synonyms: *Clethropsis nepalensis* (D.Don) Spach, *Betula leptophylla* Regel, *Betula leptostachya* Wall. *Nom.nud*.

Trees to 15 m tall; bark gray or dark gray, smooth; branchlets dark brown, sparsely yellow pubescent when young, glabrescent at maturity. Leaves simple, alternate, obovate-lanceolate, obovate-oblong, ovate or elliptic, 4-16 x 2.5-10 cm, abaxially with dense, resinous glands, yellow pubescent along veins, bearded in axils of lateral veins, adaxially glabrous, base cuneate or broadly cuneate, margin entire or remotely minutely serrate, apex abrupt or acute, rarely acuminate; lateral veins 8-16 pairs; petiole stout, 1-2.5 cm long, subglabrous. Female inflorescences numerous, in a panicle, ellipsoid, 2-2.2 cm x 7-8 mm; peduncle stout, 2-8 mm, glabrous; bracts 4 mm, woody, persistent, base cuneate, apex rounded, 5-lobed. Nutlet oblong, 2 mm, with membranous winged. Fl. & Fr.: May – September.

Western Ghats: Cultivated in estate borders and road sides in high altitudes of Kerala, Karnataka, Maharashtra and Tamil Nadu. India: Arunachal Pradesh,

Himachal Pradesh, Meghalaya, Nagaland, Sikkim and Uttar Pradesh. World: Native of Himalaya and widely distributed in Bangladesh, China, Myanmar, Tibet, Thailand, Vietnam, and Nepal.

IUCN: LC (IUCN, 2024).

Uses: Alder trees is useful for making fire-wood and charcoal. In Western Gahts, wood choaps used for fuel wood in Tea processing units.

Betula L.

Betula alnoides (Buch.-Ham.) ex D.Don, Prodr. Fl. Nepal 58. 1825. Synonyms: *Betula nitida* D.Don, *Betula acuminata* Wall. *nom.illeg.*, *Betulaster acuminata* Spach, *Betulaster affinis* Spach, *Betulaster nitida* (D.Don) Spach

Trees to 30 m tall; bark gray, smooth; branchlets densely white villous and resinous glandular. Leaves lanceolate, ovate-lanceolate or ovate-elliptic, 4-12 x 2.5-5.5 cm, papery, abaxially densely glandular punctate, sparsely villous along veins, bearded in axils of lateral veins, adaxially glabrous, base cuneate, broadly cuneate, or subrounded, rarely subcordate, margin irregularly incurved setiform serrate, apex acuminate or caudate-acuminate; lateral veins 10-13 pairs; petiole to 1.5-3 cm long, densely villous and glandular punctate. Female inflorescences 3-5 in a raceme, pendulous, narrowly cylindric, 5-10 x 4-6 mm when mature; peduncle 2-3 mm, densely yellow villous; bracts 3 mm long, densely pubescent and ciliolate, 3-lobed, middle lobe oblong or obtuse, lateral lobes reduced, auriculate. Nutlet obovate, 1.5-2 mm, sparsely pubescent at apex, with membranous winged. Fl. & Fr.: October – May.

Western Ghats: Cultivated in high altitudes of Kerala, Karnataka, Maharashtra and Tamil Nadu. India: Assam, Manipur, Nagaland and Uttar Pradesh. World: China, Bhutan, Myanmar, Nepal, Thailand and Vietnam.

IUCN: NE (IUCN, 2024).

Uses: The bark is used for paper like sheets to write in olden days. The bark is also used to cure hysteria, snake bites and an antiseptic.

BIGNONIACEAE Juss.

Key to the genera

1a. Leaves pinnate or bipinnate; stamens 4 2

1b. Leaves tripinnate; stamens 5 .... **Oroxylum**

2a. Calyx spathaceous, split one side .... **Dolichandron**

2b. Calyx not spathaceous, divided into lobes 3

3a. Capsule winged; septum flat .... **Panjanelia**

3b. Capsule not winged; septum cylindrical 4

4a. Capsule cylindrical, woody, lenticellate to 80 cm long .... **Radermachera**

4b. Capsule compressed or subcylindric, not woody to 30 cm long 5

5a. Leaves unipennate; pods oblong-elliptic, straight .... **Heterophragma**

5b. Leaves bipinnate; pods linear curved .... **Stereospermum**

Dolichandrone (Fenzl) Seem.

1a. Corolla longer than 10 cm, mouth broad; seeds with corky wings .... **D. spathacea**

1b. Corolla shorter than 10 cm, mouth narrow; seeds with membranous wings.. 2

2a. Calyx-spathe with a recurved point; capsule arcute with conspicuous speckles 3

2b. Calyx-spathe obtuse or emarginate; capsules smooth falcately curved ... . **D. falcata**

3a. Leaflets ovate, glabrous or nearly so .... **D. atroviresns**

3b. Leaflests oblong-ovate, pubescent or subfloccose .... **D. arcuata**

Dolichandrone arcuata (Wight) Clarke, in J.D. Hooker, Fl. Brit. India 4: 380. 1884. Synonym: *Spathodea arcuata* Wight Vern.: *Ram palai* (Tam.).

Trees to 15 m tall; bark brown, peeling off in thin flakes. Leaves compound, imparipinnate, opposite, leaflets 5-11, oblong-ovate, orbicular or elliptic-ovate, 4-7.5 x 2-4 cm, base oblique or acute, apex acute or obtuse, margin entire, pubescent, coriaceous; lateral nerves 5-6 pairs, pinnate, intercostae reticulate, faint; rachis 7-20 cm long, slender, tomentose; petiolule to 5-30 mm, slender, tomentose. Flowers white, few in terminal corymbs or panicles. Seplas spathaceous, split on one side and recurved to 2.5 cm, pubescent. Petals tubular, 4 cm across, widened at mouth, tube 6 cm, lobes 5, fringed. Stamens 4, didydynamous, divaricate, included; filaments 5 and 7 mm long. Disc annular. Ovary 5 mm, sessile; ovules numerous, style 2 mm; stigma 2 lobed. Capsule 2 valved, upto 45 x 1.5 cm, linear, terete, pubescent, speckled with white dots, curved. Seeds 0.6 mm, winged. Fl. & Fr.: October – February.

Western Ghats: Deciduous forests above 700 m in Karnataka, Kerala and Tamil Nadu (Endemic).

IUCN: NE (IUCN, 2024).

Dolichandrone atrovirens (Roth) K. Schum. H.G.A. Engler & K.A.E. Prantl, Nat. Pflanzenfam. 4(3a): 240. 1894. Synonyms: *Bignonia atrovirens* B. Heyne ex Roth, *Spathodea atrovirens* (Roth) Spreng., *Dolichandrone crispa* Seem. Vern.: *Poompadhiri* (Tam.).

Deciduous trees to 8-18 m tall; young parts velvety pubescent. Leaves opposite, imparipinnate, 15-30 cm long, leaflets 2 or 3 pairs, elliptic, 3-7.7 x 2-3.5 cm, acuminate or obtuse, tapering and often unequal sided at base; lateral nerves 4-6 pairs; petiolules 6-10 mm. Flowers in few flowered corymbose racemes; pedicels 1-3 cm long. Sepals spathaceous, 2- 2.5 cm long, split on one side, puberulous when young. Petals tubular, white, tube 2,5-3 cm long, lobes 5, rounded, crisped crenate. Stamens 4, didynamous, included. Ovary sessile; ovules many; stigma 2-lobed. Capsule 2-valved, 25-40 x 1.5-2 cm. speckled smooth; seedswiged on both ends, 3 x 0.6 cm. Fl. & Fr.: February – October.

Western Ghats: Common in deciduous forests of Karnataka, Kerala and Tamil Nadu (Endemic).

IUCN: NE (IUCN, 2024).

Dolichandrone falcata (Wall. ex DC.) Seem., J. Bot. 8: 340. 1870. Synonyms: *Spathodea falcata* Wall. ex DC., *Bignonia falcata* J.Koenig ex DC., *Dolichandrone lawii* Seem. Vern.: *Kaattuvarsana, Kalar Aathi, Kaliyakka, Sithaathi* (Tam.).

Small deciduous trees to 10 m tall; bark whitish, smooth; branchlets terete, white, tomentose. Leave pinnately compound, terminal leaflets large, foliolate 3-6 x 2-4 cm, obovate-rounded, base obtuse to cuneate, rounded and emarginate at apex, tomentose beneath; rachis to 10 cm long, terete, tomentose; minutely petiolule. Flowers in axillary racemes, large, whote. Sepals 5, connate, spathaceous, eamrginate apex, leafy, green, to 4 cm long, tomentose out side. Petals 5, tubular, tubes to 6 cm long, 5-lobes, apex of the lobes crumbled and fimbriate. Stamens 4; staminode present. Ovary oblong-linear, glabrous above; style to 4 cm long; stigma simple. Capsule linear, pubescent when young, glabrous at maturity, many seeded. Seeds papery, winged, brown. Fl. & Fr.: March – July.

Western Ghats: Deciduous forests of Maharashtra, Karnataka, Kerala and Tamil Nadu (Endemic). India: Andhra Pradesh and Odisha.

IUCN: NE (IUCN, 2024).

Dolichandrone spathacea (L.f.) K. Schum., Fl. Kais. Wilh. Land: 123. 1889. Synonyms: *Bignonia spathacea* L.f., *Dolichandrone rheedei* (Spreng.) Seem., *Spathodea rheedei* Spreng. Vern.: *Paadhiri, Vil pathiri* (Tam.).

Deciduous trees to 18 m tall; bark greyish-brown, smooth. Leaves compound, imparipinnate opposite, leaflets 5-9, opposite, rhomboid or ovate, elliptic, ovate-lanceolate, 7.5-15.2 x 2.5-7.5, base oblique; apex acuminate; margin entire, glabrous and shiny; rachis 23-30 cm long, stout, glabrous; petiolule upto 6 mm, stout, glabrous. Flowers white, 3-4 together on terminal short erect corymbs. Sepals spathaceous, cleft to the base on one side, spathe acuminate at tip. Petals tubular, tube 10-18 cm long, slender below, widening half way to a funnel; lobes 5, spreading, crisped. Stamens 4, didynamous, included. Disc annular. Ovary sessile; ovules many; style long; stigma 2 lobed. Capsule to 45 x 2.5 cm, purplish-brown. Seeds winged, flat, rectangular, 1.6 cm across. Fl. & Fr.: June – July.

Western Ghats: Deciduous forests of Goa, Karnataka and Kerala. India: Andhra Pradesh and Odisha. World: Sri Lanka, Myanmar and Malay Archipelago.

IUCN: LC (IUCN, 2024).

Heterophragma DC.

Heterophragma quadriloculare (Roxb.) K. Schum., H.G.A. Engler & K.A.E. Prantl, Nat. Pflanzenfam. 4(3b): 243. 1895. Synonyms: *Bignonia quadriloculare* Roxb., *Heterophragma roxburgghii* (Spreng.) DC., *Spathodea roxburghii* Spreng.

Trees to 15 m tall; bark thick, brown or grey, exfoliating in small angular scales. Leaves compound, imparipinnate, opposite, clustered at apex, leaflets 7-11, opposite or subopposite, elliptic, elliptic-oblong, 4-5 x 1.8-2.5 cm, base acute or oblique, apex obtuse or acute, margin entire or crenate, puberulent, coriaceous; lateral nerves, 4-6 pairs, slender, puberulent, intercostae reticulate; rachis 14-23 cm long, slender, pulvinate, puberulent; petiolule 3-15 mm long, slender, puberulent. Flowers rose coloured, in terminal woolly tomentose panicle; pedicels to 5 mm, stout. Sepals to 2 cm long, tubular, irregularly 5 lobed, densely tomentose. Petals to 3.5 cm, floccose, lobes 5, subequal, rounded, crisped. Stamens 4, didydynamous, included; anthers divaricate. Disc cushion shaped. Ovary sessile, ovate; ovule many; style to 3 cm; stigma 2 lobed. Capsule 25-28 x 5-6.5 cm, oblong, 2 valved, compressed, dissepiments spongy. Seeds 2 x 6 cm, membranous, winged at one end. Fl. & Fr.: January – June.

Western Ghats: Evergreen forests of Gujarat, Maharashtra, Goa, Karnataka, Kerala and Tamil Nadu (Endemic).

IUCN: LC (IUCN, 2024).

Oroxylum Vent.

Oroxylum indicum (L.) Vent., Forest Fl. Burma 2: 237. 1877. Synonyms: *Bignonia indica* L., *Calosanthes indica* (L.) Blume, *Oroxylum flavum* Rehder Vern.: *Achi, Arandhai, Panna, Vangam* (Tam.).

Trees to 12 m tall; bark thick, surface brownish-grey, smooth. Leaves compound, 2-3 pinnate, pinnae 5-9; imparipinnate, opposite, leaflets 3-5 in each pinnae, opposite, 7.5-15 x 4-9 cm, ovate, apex acuminate, base cordate, oblique or truncate, margin entire, glabrous, chartaceous; lateral nerves, 4-6 pairs, pinnate, prominent, puberulent beneath, intercostae reticulate; rachis 60-100 cm, stout, glabrous. Flowers in lax terminal racemes, reddish-purple outside, pinkish-yellow within, racemes to 30-50 cm long; pedicels 1-2 cm long. Sepals 5, 2.5 cm long, campanulate, coriaceous, glabrous, limb truncate or obscurely toothed. Petals to 10 cm long, fleshy, campanulate, 5-lobed, lobes 3 cm long. Stamens 5, perfect, the posterior one shortest; filaments inserted below the base of the tube, filaments hairy at base; connective with a short mucrone. Ovary subsessile, contracted at the base; ovule many; style slender; stigma 2-lobed. Capsule 2-valved, 40-75 x 5-8 cm, compressed, tapering at both ends. Seeds thin, papery, 5-6 cm long, winged all around except at base. Fl. & Fr.: June – December.

Western Ghats: Semi-evergreen forests of Gujarat, Goa, Karnataka, Kerala and Tamil Nadu (Endemic).

IUCN: NE (IUCN, 2024).

Uses: Roots are important ingredients in the ayurvedic drug Dasamoolaristam which is traded in the drug market.

Panjanelia DC.

Pajanelia longifolia (Willd.) K. Schum., H.G.A. Engler & K.A.E. Prantl, Nat. Pflanzenfam. 4(3b): 244. 1895. Synonyms: *Bignonia longifolia* Willd., *Pajanelia rheedii* Wight, *Pajanelia multijuga* DC. Vern.: *Arandhal* (Tam.).

Deciduous trees to 20 m tall; trunk unbranched to a considerable height; bark thick, pale grey, shining, pustular lenticels in vertical rows. Leaves compound, imparipinnate, leaflets 7-17, opposite, 17-25 x 7-12 cm, ovate, apex acuminate, base oblique, margin entire, glabrous, chartaceous; lateral nerves 6-10 pairs, pinnate, prominent, intercostae reticulate; rachis to 100 cm, glabrous, stout; petiolule 3-7 mm, channelled above, glabrous. Flowers bisexual, crimson-purple, white within, in large terminal recemose panicles. Sepals to 4.5 cm long, campanulate, lobes 5, rusty tomentose. Petals tubular, 9 x 8 cm, tube 6-8 cm long, base narrow, broadly ventricose, lobes 5, subequal, crisped. Stamens 4, free, didydynamous with a rudimentary fifth, filaments arching, slender; anthers equal, oblong. Ovary 0.5-1 cm long, 2-celled, oblong; ovules many; style 6 cm long; stigma 2 lobed. Capsule 2 valved, 30-45 x 5-7.5 cm, brown, winged on both margins, compressed, smooth. Seeds 2.5 x 1.5 cm, flat, papery, winged on both sides. Fl. & Fr.: December – June.

Western Ghats: Evergreen forests of Maharashtra, Goa, Karnataka and Kerala. India: Assam and Meghalaya. World: Myanmar.

IUCN: LC (IUCN, 2024).

Radermachera Zoll. & Moritzi

Radermachera xylocarpa (Roxb.) K. Schum., H.G.A. Engler & K.A.E. Prantl, Nat. Pflanzenfam. 4(3b): 243. 1895. Synonyms: *Bignonia xylocarpa* Roxb. *Stereosprmum xylocarpum* (Roxb.) Benth. & Hook.f. Vern.: *Kadalathi, Maan kombu, Pathiri, Vedankanni* (Tam.).

Deciduous trees to 20 m tall; bark light greyish-brown, rough, corky, flaking off in thin irregular scales. Leaves compound, bipinnate or tripinnate, opposite, pinnae 7-9, leaflets 3-9 in each pinnae, opposite, elliptic-lanceolate, ovate-lanceolate or ovate, 3.5-15 x 2-7 cm, base oblique or acute, apex acute or acuminate, margin entire, membranous, glabrous; lateral nerves 4-7 pairs, glabrous, raised beneath, prominent, slender; intercostae reticulate; rachis 24-50 cm long, stout, swollen at base and each nodes; petiolule 1.5-5 mm long. Flowers white, tinged with yellow, fragrant, in terminal panicled cymes. Seplas 5, campanulate, 1 cm long, irregularly 5 lobed. Petals 5, campanulate, 4.5 x 4 cm, bilabiate, lobes subequal, obtuse, crisped, glabrous. Stamens 4, didynamous, included; anthers divaricate, 4 mm; staminodes filiform. Ovary oblong, glabrous, superior; ovules many, multiseriate; style slender; stigma spoon shaped. Capsule bivalved, elongate, 30-75 x 1.5-3.7 cm, woody, warty tubercled, slightly curved. Seeds winged, 1.5 cm long, with membranous wings at both ends. Fl. & Fr.: February – October.

Western Ghats: Semi-evergreen to evergreen forests of Gujarat, Maharashtra, Goa, Karnataka, Kerala and Tamil Nadu (Endemic).

IUCN: LC (IUCN, 2024).

Stereospermum Cham.

1a. Filaments of the stamens with a tuft of woolly hair at base; leaves and panicles nearly or quite glabrous .... **S. colais**

1b. Filaments of the stamens without a tuft of woolly hair at base; leaves and panicles more or less pubescent .... **S. chelonoides**

Stereospermum chelonoides (L.f.) DC., Biblioth. Universelle Genève, n.s., 17: 125. 1838. Synonyms: *Bignonia chelonoides* L.f. *Stereospermum suaveolens* (Roxb.) DC., *Bignonia suaveolens* Roxb. Vern.: *Pungaali* (Tam.).

Deciduous trees to 25 m tall; bark grey, exfoliating in flat scales. Leaves compound, imparipinnate, opposite, leaflets 5-11, elliptic or ovate, 7.5-16 x 5-7.5 cm, base oblique, acute or unequally rounded, apex acuminate, margin entire or serrulate on young trees, shiny and glabrous above and pubescent beneath, coriaceous; lateral nerves 6-8 pairs, pinnate, prominent, puberulent beneath, intercostae reticulate; rachis 150-450 mm long, slender, swollen at base; petiolules 10-20 mm, slender, pubescent. Flowers dull crimson, in drooping panicles, pubescent, 18 mm long. Sepals campanulate, hairy, lobes 3-5, short, broad. Petals funnel shaped, 3-4 cm long, pubescent outside, lobes 5 unequal, rounded, crisped. Stamens 4, didynamous, included; staminodes present, anther divaricate. Disc cupular, fleshy. Ovary sessile, superior, 2-celled, syncarpous; ovules many; style slender; stigma 2. Capsule terete, 30-60 x 0.5 cm, grey, lenticellate. Seeds 3 cm long, trigonus, wedge shaped, with a membranous wing with obtuse ends. Fl. & Fr.: January – September.

Western Ghats: Deciduous forests of Maharashtra, Karnataka, Kerala and Tamil Nadu. World: Sri Lanka, Myanmar.

IUCN: LC (IUCN, 2024).

Stereospermum colais (Buch.-Ham. ex Dillw.) Mabber., Taxon 27: 553. 1978 publ. 1979. Synonyms: *Bignonia colais* Buch.-Ham. ex Dillw., *Stereospermum tetragonum* DC., *Stereospermum personatum* (Hassk.) Chatterjee, *Dipterospermum personatum* Hassk. Vern.: *Ambu, Pathiri, Poompathiri, Pungali* (Tam.).

Large trees to 30 m tall; bark thick, pale greyish-brown, transversely rugose, exfoliations irregular. Leaves compound, imparipinnate, opposite, leaflets 7-13, elliptic-lanceolate, elliptic-oblong, ovate or obovate, 4.5-13 x 2-4.5 cm, apex acuminate or caudate acuminate, base oblique, acute or obtuse, margin entire or serrate, glabrous above and puberulent beneath, chartaceous; lateral nerves 6-10 pairs, pinnate, puberulent beneath, prominent; intercostae reticulate; rachis 10-21 cm long, slender, puberulent, swollen at base; petiolule 7-12 mm, slender, grooved above. Flowers in terminal lax panicles. Sepals campanulate, 6 mm long, shallowly lobed, lobes 3-5, dorsal lobe much larger than the others, puberulent, tube 1.2-1.5 cm, obtuse. Petals infundibuliformis, 2 x 1.5 cm across, yellow with red streaks, pubescent inside, base glandular hairy, lobes 5, subequal, crisped. Stamens 4, didymous, divaricate; filaments unequal, smaller ones 1-1.2 cm, larger to 2.2-2.5 cm, hairy below; anthers divergent, 5 mm long; staminodes present. Ovary sessile, oblong, superior, 2-celled, syncarpous; many ovuled, 1 seriate; style slender; stigma 2, spoon shaped. Capsule subtetragonus, to 35 x 0.7 cm, tapering at apex and base, epicarp thin, spirally splitting. Seeds 8 mm long, wings obtuse at both ends. Fl. & Fr.: February – October.

Western Ghats: Semi-evergreen forests of Maharashtra, Goa, Karnataka, Kerala and Tamil Nadu. World: Sri Lanka, Myanmar, China, Bhutan, Nepal, Bangladesh, Laos, Cambodia, Vietnam, Thailand and Malesia.

IUCN: NE (IUCN, 2024).

Planted:

Catalpa bignonioides (L.) Walter; Exotic: Native to North America

Crescentia alata Kunth; Exotic: Native to Mexico, South America

Crescentia cujete L.; Exotic: Native to North America

Dolichandrone stipulata (Wall.) Seem. ex Gamble; Exotic: Native to China, Myanmar, Laos, Cambodia, Vietnam, Thailand.

Fernandoa adenophylla (Wall. ex G.Don) Steenis; Exotic: Native to Myanmar

Handroanthus serratifolius (Vahl) S.O.Grose; Exotic: Native to Tropical America

Jacaranda mimosifolia D. Don; Exotic: Native to Peru

Kigelia africana (Lam.) Benth.; Exotic: Native to Tropical Africa

Markhamia lutea (Benth.) K.Schum.; Exotic: Native to Tropical Africa

Millingtonia hortensis L.f.; Exotic: Native to South East Asia

Parmentiera cereifera Seem; Exotic: Native to Panama

Phyllarthron bojerianum DC.; Exotic: Native to Comor Island

Phyllarthron comorensa DC.; Exotic: Native to Madagascar, Comoro Island

Spathodea campanulata P. Beauv.; Exotic: Native to Tropical Africa

Tabebuia aurea (Silva Manso) Benth. & Hook.f. ex S. Moore; Exotic: Native to South America

Tecoma stans (L.) Juss. ex Kunth; Exotic: Native to America

Tecomella undulata (Sm.) Seem.; Native: India, Pakistan, South West Asia

BIXACEAE Kunth

Key to the genera

1a. Leaves entire; petals pink; ripe seeds with thick red fleshy testa ......**Bixa**

1b. Leaves palmately lobed; pental yellow; ripe seeds with long white wool ....**.Cochlospermum**

Bixa L.

Bixa orellana L., Sp. Pl.: 512. 1753. Synonyms: *Bixa tinctaria* Salisb., *Bixa acuminata* Bojer, *Orellana orellana* (L.) Kuntze, *Bixa upatensis* Ram.Goyena Vern.: *Kumkumachedi, Kuppa-majal, Kurangaka* (Mal.); *Japra, Packyavathi poo, Chindura maram, Maramanjal* (Tam.).

Small evergreen trees to 2-9 m tall; young branches densely dark scaly. Leaves alternate, ovate, subcordate or truncate at base, long-acuminate at apex, 7-24 x 4-16 cm, shining above, densely red- dotted, initially scaly beneath, glabrous later; lateral nerves 4 or 5 pairs; petioles slender, 4-10 cm long, stipules 5-6 mm long. Inflorescence a terminal corymb or panicle, 8-50 flowered, scaly; pedicels 7-10 mm long. Flowers 3-5 cm across. Sepals broadly ovate to suborbicular, concave, scaly, purple. Petals 5-7, unequal, obovate, rounded at apex, 2-3 x 1-2 cm, pink. Stamens many; filaments slender, yellow at base, red at apex; anthers purple. Ovary densely covered with thick red bristles; style 12-15 mm long, swollen above, red. Capsules globose or broadly ovoid, nearly truncate at base, abruptly short-acuminate at apex or elongate-ovate with narrowly long-acuminate apex, 2-4 x 1.5-3.5 cm, with dense long stiff but soft and flexible bristles, green when young, brownish red with maturity. Seeds obpyramidal, 5 x 4-5 mm, orange-red. Fl. & Fr.: Throughout the year.

Western Ghats: Cultivated throughout and naturalized in open wetlands. India: Most of the states. World: Native to America.

IUCN: LC (IUCN, 2024).

Uses: Seeds yielding red dye which is used in colouring food stuffs, butter, cheese, chocolate, cosmetics, oil, soap, varnishes and silk. There are two cultivars usally cultivated in estates of Western Ghats such as white flowers with green capsules and pink flowers with red capsules.

Cochlospermum Kunth ex DC.

Cochlospermum religiosum (L.) Alston, in H. Trimen, Handb. Fl. Ceylon 6(Suppl.): 14. 1931. Synonyms: *Bombax religiosum* L., *Cochlospermum gossypium* DC., *Bombax congo* Burm.f. Vern.: *Appakadukka, Cheemappanni, Chempanni* (Mal.); *Kaatu paruthi, Kongilavu, Malamparuthi, Maraparuthi* (Tam.).

Deciduous trees to 10 m tall; bark thick, surface dark grey, fibrous. Leaves simple, palmately 3-5 lobed, alternate, lobes, elliptic or elliptic-lanceolate, 5-15 x 7-20 cm, base cordate, tips acute or acuminate, margin entire or crenate, serrate, glabrous above, densely white tomentose below, coriaceous; 5-7 nerved from the base, lateral nerves 7-10 in each lobe, parallel, prominent, intercostae scalariform, faint; petiole 6-20 cm long, stout, swollen at base, pubescent. Flowers 7.5-10 cm across, bright yellow, in grey tomentose terminal panicles. Sepals 5, 2.5 x 1.5 cm, unequal, obovate, densely tomentose, deciduous. Petals 5, 5 x 3 cm, obovate, deeply emarginate, contorted. Stamens many, inserted on an eglandular disc, shortly connate at base into 8-10 clusters; filaments 1.5-2.5 cm long; anthers linear. Ovary superior, globose, glabrate, 3-5-celled, ovules many; style 1, to 1 cm, filiform; stigma somewhat lobed. Capsule obovoid, 5-valved, 6-8 x 4.5-5.5 cm, pear shaped, straight, leathery, brown. Seeds many, 6 mm long, black, curved, embedded in white cotton. Fl. & Fr.: January - April

Western Ghats: Common in deciduous forests of Gujarat, Maharashtra, Karnataka, Kerala and Tamil Nadu. India: Andhra Pradesh, Bihar, Madhya Pradesh, Odisha and Uttar Pradesh. World: Sri Lanka, Myanmar, Vietnam, Cambodia, Thailand and Malesia.

IUCN: NE (IUCN, 2024).

Uses: Bark yields an edible gum. Seed fiber is used for stuffing mattresses.

BORAGINACEAE Juss.

Key to the genera

1a. Flowers polygamous; drupe with one stone .... **Cordia**

1b. Flowers bisexual; drupe with 2-4 pyrenes .... **Ehretia**

Cordia L.

1a. Corolla tube equal or shorter than calyx; stamens 5 2

1b. Corolla tube longer than calyx; stamens 8 ....**C. subcordata**

2a. Leaves braoadly ovate-rouended or elliptic, basally palminerved 3

2b. Leaves obovate-ellipitc, usually penninerved 6

3a. Branches and leaves brown tomentum; inflorescence much condensed terminal panicle ....**C. macleodii**

3b. Branhces and leaves simple pubescence; inflorescence lax or few-flowered 4

4a. Panicle very lax, terminal and axillary; ripened fruit white.........**C. myxa**

4b. Panicle aggregare, terminal; ripened fruit orange yellow 5

5a. Leaves silvery beneath, reticulate prominent; fruit globose with depressed apx...**C. monoica**

5b Leaves glabrous beneath, reticulate faint; fruit subglobose apiculate apex.. **C. dichotoma**

6a. Leaves glabrous or faint pubescent; style more 1 cm long ....**C. quercifolia**

6b. Leaves dense pubescent; style less than 0.5 cm long ....**C. sinensis**

Cordia dichotoma G. Forst., Fl. Ins. Austr.: 18. 1786. Synonyms: *Gerascanthus dichotomus* (G.Forst.) Borhidi, *Cordia griffithii* C.B.Clarke, *Cordia latifolia* Roxb., *Cordia tremula* Griseb., *Corida obliqua* Willd., *Cordia obliqua* Willd. var. *tomentosa* (Wall.) Kazmi, *Cordia wallichii* G. Don, Vern.: *Mookuchali pazham, Narivili* (Tam.).

Trees to 20 m tall; bark whitish, rough. Leaves ovate to broadly ovate or elliptic, 6-13 x 4-9 cm, sparsely pubescent or glabrous, base rounded to broadly cuneate, margin usually subundulate to undulate dentate, rarely entire, apex obtuse to mucronate; petiole 2-5 cm long. Inflorescences terminating leafy lateral branches, dichotomously branched into corymbose cymes, widely spaced, 5-8 cm wide. Flowers dimorphic, sessile. Sepals 5, campanulate, 5-6 mm long, 5-lobed; lobes unequal, triangular. Petals white, as long as calyx; lobes shorter than tube, margin somewhat undulate. Filaments of staminate flowers. 3.5 mm long, filaments of bisexual flowers 1-2 mm long. Rudimentary pistil globose. Style united portion 1-1.5 mm, first branches to. 1 mm long, second branches 2-3 mm long; stigma spatulate. Drupes yellow or reddish, subglobose, 1-1.5 cm in diam, with sticky mesocarp, surrounded by persistent calyx. Fl. & Fr.: January – June.

Western Ghats: Deciduous forests of Gjarat, Maharashtra, Goa, Karnataka, Kerala and Tamil Nadu. World: Sri Lanka, Pakistan, China, Laos, Cambodia, Vietnam, Taiwan, Australia and New Caledonia.

IUCN: LC (IUCN, 2024).

Uses: Fruits locally edible.

Cordia macleodii (Griff.) Hook.f. & Thoms., J. Proc. Linn. Soc., Bot. 2: 128. 1858. Synonym: *Hemigymnia macleodii* Griff.

Small trees to 6 m tall; bark thin, whitish, pubescent. Leaves alternate, broad ovate, base cordate, tomentose; basal pair of nerves opposite, 3-5 nerved. Flowers white, fragrant, terminal cymes. Calyx ribbed, densely tomentose. 5-6 lobes. Corolla white, 5, oblong-ellptic lobes, recurved. Fruiting calyx ribbed tomentose. Drupes yellow when ripe, conical, style persistent. Fl. & Fr.: February – August.

Western Ghats: Deciduous to semi-evergreen forests of Gujarat, Maharashtra, Karnataka, Kerala and Tamil Nadu. India: Andhra Pradesh, Odisha, Punjab, Telangana and West Bengal. World: Pakistan.

IUCN: NE (IUCN, 2024).

Cordia monoica Roxb., Pl. Coromandel 1: 43. 1796. Synonyms: *Gerascanthus monoicus* (Roxb.) Borhidi Vern.: *Mookalapazham, Mookkattakai, Pasakaimaram, Virimaram* (Mal.); *Kaatu koyya, Viriampazham* (Tam.)

Trees to 10 m tall; branchlets lenticellate, tomentose. Leaves simple, alternate, ovate- elliptic, 4-10 x 1.5-7 cm, base round, truncate or acute, apex acute or obtuse; upper surface scabrous with prominent whitish points, villous or tomentose beneath, coriaceous; 3-5 nerves from base, palmate, lateral nerves 4-5 pairs, intercostae scalariform, prominent; petiole 5-20 mm, slender, tomentose when young, glabrous at maturity. Flowers polygamous, white, subsessile, in small dense tomentose paniculate cymes. Seplas 4-5, tubular, tube 8 mm, obconic, densely tomentose outside, glabrous inside, unevenly 3-4 lobed, lobes 5 mm, triangular, accrescent. Petals funnel shaped, lobes 4-5, longer than the tube, oblong. Stamens 4-5; filaments attached to the mouth of corolla tube, 3 mm long, glabrous; anthers 1.5 mm long. Ovary superior, 4-celled, hairy; ovule one in each cell; stigma clavate. Drupe ovoid with a mucronate tip, 2.5 x 1.5 cm, yellow, fulvous-pubescent, stone 4-celled. Seed one. Fl. & Fr.: December–April.

Western Ghats: Deciduous forests of Karnataka, Tamil Nadu and Kerala India: Andhra Pradesh. World: Saudi Arabia, Somalia, Sri Lanka, Sudan and Swaziland.

IUCN: LC (IUCN, 2024).

Cordia myxa L., Sp. Pl.: 199. 1753. Synonym: *Cordia domestica* Roth, *Corida myxa* var. *doemstica* (Roth) Clarke, *Cordia fulvosa* Wight, *Cordia evolutior* (Clarke) Gamble, *Cordia fulvosa* var. *evolutior* Clarke Vern.: *Narivili* (Tam.).

Trees to 17 m tall; bark thick, peeling off; smooth. Leaves ovate-oblanceolate, 7-13 x 4-6 cm, acute, cuneate with numerous peltate scales, whitish pubescent; nerves palmate, 3-5 from the base; lateral nerves prominent, reticulate. Flowers white. Calys ribbed, tubular, margin recurved in fruit. Corolla lobes narrow, whote. Drupe ovoid-conical, turning yellow when ripe. Seed one. Fl. & Fr.: April – July.

Western Gahts: Common in foot hills to deciduous forests of Maharashtra, Karnataka and Tamil Nadu (Endemic). India: Himachal Pradesh, Punjab and Rajasthan.

IUCN: LC (IUCN, 2024).

Cordia quercifolia Klotzsch. W.C.H. Peters, Naturw. Reise Mossambique 6(Bot., 1): 247 (1861). Synonyms: *Cordia nevillii* Alston, *Cordia perrottetii* sensu Wight, non DC. *Cordia rithii* sensu Balfour, non Roem. & Schult., *Cordia gharaf* sensu Chiovenda, non Ehrenb. ex Asch. Vern.: *Narivirian* (Mal.)

Small trees to 10 m tall; bark grey or brownish-grey with deep longitudinal furrows, smooth; young branches brown tomentose. Leaves simple, subopposite or alternate, oblanceolate, orbicular or ovate, 4-10 x 2.5-5 cm, base acute or obtuse, apex obtuse or rounded; margin entire, tomentose beneath, chartaceous; nerves 3-5 from base, lateral nerves 3-4 pairs, intercostae scalariform, faint; petiole 7-20 mm, slender, tomentose. Flowers polygamous, creamy white, in axillary or terminal cymes to 7 cm long. Sepals 5, tubular, tube 7 mm, tomentose, lobes 5. Petals 5, tube 8 mm across, lobes 5, recurved. Stamens 5; filaments 5 mm, hairy at the base, anthers 2 mm. Ovary superior, 1.5 mm, ovoid, 4-celled; ovules one in each cell; style 6 mm long. Drupe ovoid, 1-1.2 cm long, acute at apex. Seed one. Fl. & Fr.: November – March.

Western Ghats: Deciduous forests of Tamil Nadu. World: Sri Lanka, Pakistan, Oman, Socotra, Mauritiana and Tropical Africa.

IUCN: NE (IUCN, 2024).

Cordia subcordata Lam., Tabl. Encycl.: n. 1899. 1792. Synonyms: *Cordia octandra* A.DC., *Lithocardium subcordatum* (Lam.) Kuntze, *Cordia moluccana* Roxb., *Cordia serrata* Roxb.

Small trees to 10 m tall; bark pale brown or dark grey, stringy; branchlets sparsely hairy. Leaves simple, subopposite and alternate, ovate, 5-13 x 3-8 cm, base acute or rounded, apex acuminate, margin entire or serrate, scabrous, membranous; 3 nerved from base, lateral nerves 3 pairs, pinnate, prominent, intercostae scalariform, prominent; petiole 20-50 mm long, slender, glabrous. Flowers polygamous, white, 8.5 mm across, in terminal dichotomous, long peduncled corymbose cymes; pedicel pilose. Petals 14 mm long, cupular, densely woolly within, sparsely hairy outside, accrescent. Petals 20 mm long, campanulate, lobes 7-9, oblong, obtuse. Stamens 8, alternately long and short; filaments shorter than corolla lobes, hairy at base. Ovary superior, obovoid, 4-celled; ovule one in each cell; style short or absent; stigma sessile, 2 lobed. Drupe ellipsoid, 3 x 1.5 cm, glabrous, acute. Seed one. Fl. & Fr.: August – December.

Western Ghats: Cultivated in gardens and avenue trees in Kerala and Tamil Nadu. World: Bangladesh, Western Australia, Madagascar, Sri Lanka and South China.

IUCN: LC (IUCN, 2024).

Cordia sinensis Lam., Tabl. Encycl. 1: 423. 1792. Synonyms: *Cordia gharaf* (Forssk.) Ehrenb. ex Asch., *Cornus gharaf* Forssk., *Corida rothii* Roem. & Schult., *Cordia reticulata* Roth ex Roem. & Schult., *Cordia angustifolia* Roxb., *Corida senegalensis* var. *pelida* DC., *Cordia quercifolia* Klotzsch, *Cordia ovalis* R.Br. Vern.: *Narivizhi* (Tam.).

Small trees to 10 m tall; bark grey or brownish-grey with deep longitudinal furrows, smooth, peeling of in thin linear strips. Leaves simple, subopposite or alternate, oblanceolate, orbicular or ovate, 4-10 x 2.5-5 cm, base acute or obtuse, apex obtuse or rounded, margin entire, scabrous with white cystoliths above and tomentose beneath, chartaceous; nerves 3-5 from base, palmate, lateral nerves 3-4 pairs, intercostae scalariform, faint; petiole 7-20 mm, slender, tomentose. Flowers polygamous, creamy white, in axillary or terminal cymes to 7 cm long. Sepals tubular, tube 7 mm, tomentose, lobes 5. Petals tubular, 8 mm across, lobes 5, recurved. Stamens 5; filaments 5 mm, hairy at the base, anthers 2 mm. Ovary superior, 1.5 mm, ovoid, 4-celled, ovules one in each cell; style 6 mm. Drupe ovoid, 1-1.2 cm long, acute. Seed one. Fl. & Fr.: April – October.

Western Ghats: Cultivated as an avenue tree in Maharashtra, Goa, Karnataka, Kerala and Tamil Nadu. World: Sri Lanka, Pakistan, South West Asia and Africa.

IUCN: LC (IUCN, 2024).

Ehretia R.Br.

1a. Leaves more than 7 cm lomng; flowers in axillary or terminal lax cymes of unilateral spikes or racemes .... **E. aspera**

1b. Leaves usually less than 7 cm long; flowers in close terminal or axillary corymbs 2

2a. Calyx as longer or as the corolla tube .... **E. matthewii**

2b. Calyx shorter than the corolla tube .... **E. wightiana**

Ehretia aspera Willd., Phytographia 1: 4. 1794. Synonyms: *Bruxanelia indica* Dennst. ex Kostel. *Ehretia laevis* var. *aspera* (Willd.) Clarke, *Ehretia obtusifolia* Hochst., *Ehretia championii* Wight & Gardner ex C.B.Clarke, *Ehretia cutranga* C.B.Clarke, *Ehretia laevis* var. *canarensis* C.B.Clarke, *Ehretia laevis* var. *floribunda* (Royle ex Benth.) Brandis, *Ehretia laevis* var. *pubescens* (Royle ex Benth.) C.B. Clarke, *Ehretia canarensis* (Clarke) Gamble, *Ehretia laevis* Roxb., *Ehretia pubescens* Benth. Vern.: *Kalvirasu* (Tam.).

Small trees to 6 m tall; bark grey, smooth. Leaves simple, alternate, elliptic, 6-10 x 3-4 cm, apex acuminate, base cuneate, thin coriaceous, glabrous; lateral nerves 5-7 pairs; petiole to 2 cm long. Flowers 2.5-3 mm across, in axillary and terminal unilateral corymbose cymes. Seplas tubular, minute, 5-toothed. Petals white, 2.5 mm long, lobes 5, oblong, obtuse. Stamens 5, exserted. Ovary minute, 2-celled; ovules 2 in each cell; style 2-fid above middle. Drupe subglobose, 4-5 mm across, red. Seeds 4. Fl. & Fr.: March – June.

Western Ghats: Common in deciduous forests of Maharashtra, Goa, Karnataka, Kerala and Tamil Nadu. World: Sri Lanka, Nepal, Pakistan, China, Myanmar, Iran, Laos, Cambodia, Vietnam, Australia and Polynesia.

IUCN: DD (IUCN, 2024).

Ehretia matthewii Kottaim., Webbia 71: 239. 2016. Synonym: *Ehretia ovalifolia* Wight

Small trees to 5 tall; bark thin, whitish, rough; branchlets lenticellate, terete. Leaves to 7 x 3.5 cm, oblong or elliptic, apex obtuse, nerves 6-8 pairs, pubescent in the nerve axils; petiole 1-1.5 cm. Corymbs terminal and axillary, 8-10 cm. Flowers white, slightly pubescent; pedicel 1-2 mm. Seplas tubular, tube to 1 mm long, lobes 1 mm long, pubescent. Petals rotate, tube 3 mm long, lobes 2 mm long. Stamens 5, filaments 4 mm. Ovary 4 celled; styles 2, fused to 3 mm. Drupe globose, 6.5 mm diam, glabrous, black when ripe. Fl. & Fr.: August – November.

Western Ghats: Deciduous forests of Kerala and Tamil Nadu. World: Sri Lanka.

IUCN: NE (IUCN, 2024).

Ehretia wightiana Wall. ex G. Don, Gen. Hist. 4: 388. 1837. Synonyms: *Ehretia lanceolata* B. Heyne ex C.B. Clarke

Small trees to 7 m tall; branchlets pale. Leaves clustered, to 4 x 1.5 cm, obovate, elliptic, apex acute, base cuneate; nerves 4-6 pairs; petiole 1 cm long. Corymbs terminal, to 4 cm long, slightly pubescent. Fruit 0.5 cm across, fruiting calyx tube 1 mm long, lobes 2 mm, pubescent. Fl. & Fr.: Februayr – April.

Western Ghats: Rare in dry deciduous forests of Karnataka, Kerala and Tamil Nadu (Endemic).

IUCN: NE (IUCN, 2024).

Planted :

Cordia sebastina L.; Exotic: Native to Tropical America

BURSERACEAE Kunth

Key to the genera

1a. Flowers in cymes; stamens 8-10 .... **Commiphora**

1b. Flowers in panicles; stamens 5-6 2

2a. Fruit dehiscent ....**Boswellia**

2b. Fruit indehiscent 3

3a. Flowers 3-merous.. **Canarium**

3b. Flowers 4-6-merous .... **Garuga**

Boswellia Roxb. ex Colebr.

Boswellia serrata Roxb. ex Colebr., Asiat. Res. 9: 379. 1807. Synonyms: *Boswellia glabra* Roxb., *Boswellia serrata* var. *glabra* (Roxb.) Bennet, *Chloroxylon dupada* Buch. -Ham. Vern.: *Adimar, Chitta, Gugula, Dhupa, Chilakdhupa, Madimar, Tallaki, Maddi* (Kan.); *Parangi, Chaambaraani* (Mal.); *Salai chad dink* (Mar.); *Kungiliyam, Sambranai, Kunthiricum, Parangi chambrani, Vaalugam* (Tam.).

Deciduous trees to 20 m tall; bark yellowish-white with dark blotches, exfoliations thin, papery, smooth flakes; exudation white gum-resin. Leaves imparipinnate, alternate, apically clustered, leaflets 15-31, sessile or subsessile, opposite or subopposite, elliptic-oblong, oblong-lanceolate, oblong-ovate, 0.8-9.5 x 0.5-3.5 cm, base oblique, acute, apex obtuse, margin entire or crenate, chartaceous, glabrous; lateral nerves 8-14 pairs, intercostae reticulate, faint; rachis 11-44 cm, slender, pubescent, swollen at base. Flowers bisexual, small, white, in axillary or subterminal fascicled racemes. Sepals campanulate, pubescent, short; lobes 5-7, persistent. Petals 5-7, 7 x 2.5-4 mm, white, ovate-oblong, shortly clawed, inflexed at apex pubescent out side except margin. Disc annular, crenate, free from calyx. Stamens 10, free, filaments alternately longer and shorter connective produced beyond the anther lobe. Ovary sessile, superior, ovoid, 3-celled, ovules 2 in each cell; style to 3 mm, grooved; stigma 3-lobed. Drupe ovoid, trigonous; pyrenes 3. Seeds 3. Fl. & Fr.: December – June.

Western Ghats: Semi-evergreen to evergreen forests of Gujarat, Maharashtra, Karnataka, Kerala and Tamil Nadu (Endemic). India: Andhra Pradesh, Assam, Bihar, Madhya Pradesh, Punjab, Rajasthan and Uttar Pradesh.

IUCN: NE (IUCN, 2024).

Uses: It is a multipurpose tree which is used for timber and good source of oleo-gum resin and is having several medicinal properties.

Canarium L.

Canarium strictum Roxb., Fl. Ind., ed. 1832. 3: 138. 1832. Synonyms: *Cordia tiliifolia* Warb., *Rumphia tiliifolia* Poir., *Pimela stricta* Blume Vern.: *Kari dhup, Halemaddu, Raala dhupa* (Kan.); *Karuthukungilyam, Thellim* (Mal.); *Attam, Karumkungiliyam, Dhoopam* (Tam.).

Trees to 30 m tall; bark thick, surface grey-brown or yellow, smooth, exfoliation small, irregular; aromatic, exudation brownish-black, resinous. Leaves imparipinnate, alternate, leaflets 7-11, opposite, ovate, ovate-lanceolate, oblong-ovate or elliptic-ovate, 7-16 x 3.5-7 cm, base acute or oblique, apex acuminate, margin serrulate or crenulate, glabrous above, rusty tomentose beneath, coriaceous, lateral nerves 10-20 pairs, parallel, prominent; intercostae scalariform, prominent; petioule 3-10 mm long, stout, tomentose; rachis 22-36 cm long, stout, tomentose, swollen at base. Flowers polygamous, 5-6 mm across, bright yellow; in large axillary panicles; bracts caudcous. Male flowers shortly pedicellate. Sepals campanulate, tube pubescent without, 5 mm long; lobes 3, 1 mm long. Petals 3, oblong, concave, apiculate. Disc annular, to 6 lobed, apically pilose. Staminal tube to 3 mm long; stamens 6, free from the disc; filaments 1 to 2 mm long; anthers oblong, subequal; pistillode short. Bisexual flowers pedicellate to 3 mm long. Sepals in urceolate, tube 4 mm long, pubescent; lobes 3, valvate. Petals 3, white, 1 cm long, oblong, pubescent without. Disc obscurely lobed, pilose above; staminal tube to 3 mm; filaments 0.5 mm; anthers subequal. Ovary superior, to 3.5 mm, 3-celled; ovules 1 in each cell; style 1, stout; stigma capitate, 2-3 lobed. Drupe ellipsoid, 3.5 x 1.5 cm, dark blue, 1-3-celled. Seeds 1-3. Fl. & Fr.: February – May.

Western Ghats: Semi-evergreen to evergreen forests above 1000 m iin Maharashtra, Goa, Karnataka, Kerala and Tamil Nadu. India: Arunachal Pradesh, Assam, Meghalaya, Odisha and Sikkim. World: Myanmar, Thailand and China.

IUCN: NE (IUCN, 2024).

Uses: Black dammer is oleo-gum resin extracted from the bark and it can be used for manufacturing varnishes, bottling wax, and substitute for medicinal plaster.

Commiphora Jacq.

1a. Leaves uni or trifoliate; flowers deep red, pink or cream color, fruit oblong or ovoid-subg-lobose 2

1b. Leaves pinnately 5-foliate; flowers greenish yellow, fruit globose to ellipsoid. ....**C. caudata**

2a. Flowers pedicellate, disc 6-8-lobed; stamonodes absent ....**C. berryi**

2b. Flwoers sessile, disc lobe inconspicuous; staminiodes 8 .... **C. wightii**

Commiphora berryi (Arn.) Engl. A.L.P.P.de Candolle & A.C.P.de Candolle, Monogr. Phan. 4: 17. 1883. Synonyms: *Balsmodendrum berryi* Arn., *Commiphora gileadensis* (Roxb.) M.R. Almeida, *Protium gileadense* Wight & Arn. Vern.: *Mullu kilivai, Paarai kiluvai, Vaelikiluvai* (Tam.).

Trees to 10 m tall; bark reddish-brown or grey; branchlets spine-tipped. Leaves 3-foliate, alternate, clustered leaflets opposite, sessile, laterals smaller, terminal larger, ovate-obovate, obovate, ovate or suborbicular, 0.6-2 x 1.2-2 cm, base cuneate or attenuate, apex obtuse, margin crenate or entire, glabrous, chartaceous; lateral nerves 3-5 pairs, intercostae reticulate, faint; rachis 6-14 mm, slender, grooved above, flattened at base, pubescent. Flowers small, polygamous, deep red, pink or cream coloured, solitary or fascicled, axillary.

Petals in narrowly campanulate, lobes 4, shorter than the tube, glabrous. Petals 4, oblanceolate, apiculate. Stamens 8, free, alternately long and short, inserted on the margin of the disc; filaments subulate. Disc small, 6-8 lobed. Ovary superior, ovoid, sessile, 2-celled; ovules 2 in each cell; style short; stigma 3-4 lobed. Drupe oblong or ovoid-subglobose, 0.8-1.2 x 0.6-0.9 cm, beaked. Seeds 2. Fl. & Fr.: February – June.

Western Ghats: Common in foot hills and scrub forests of Maharashtra, Karnataka, Kerala and Tamil Nadu. India: Andhra Pradesh. World: Sri Lanka.

IUCN: NE (IUCN, 2024).

Commiphora caudata (Wight &Arn.) Engl. A.L.P.P.de Candolle & A.C.P.de Candolle, Monogr. Phan. 4: 27. 1883 var. **caudata**. Synonyms: *Protium caudatum* Wight &Arn., *Balsmodendron caudatum* Marchand, *Protium caudatum* var. *roxburghianum* (Wight &Arn.) A.W. Benn., *Protium roxburghianum* Wight &Arn. Vern.: *Kiluvai, Karpura kiluvai, Malaikiluvai, Pachaikiluvai* (Tam.).

Trees to 15 m tall; bark green with reddish-brown stripes, peeling off in thin scales. Leaves imparipinnate, alternate, leaflets 3-7, opposite, terminal large, ovate or ovate-lanceolate, 2.7-10.5 x 1-5 cm, base cuneate, attenuate or acute, apex caudate-acuminate, margin entire, glabrous, chartaceous; lateral nerves 6-10 pairs, parallel, intercostae reticulate, prominent; petiolule 5-10 mm, slender, glabrous; rachis 5-14 cm long. Flowers polygamous, small, greenish-yellow, in lax dichotomous axillary panicles; bracts 2, opposite, glandular-hairy. Sepal tubular, tube narrowly campanulate, fused with disc, glandular-hairy; lobes 4, as long as tube, deltoid. Petals 4, broadly linear, reflexed at apex. Disc cupular. Stamens 8, free, inserted on the margins of disc, alternately long and short; anthers oblong. Ovary superior, oblong or ovoid, attenuate into style, 2-celled; ovules 2 in each cell; stigma 2-lobed. Drupe globose or ellipsoid, red when ripe, pyrenes ovoid. Seeds solitary. Fl. & Fr.: February – May.

Western Ghats: Deciduous forests of Karnataka, Kerala and Tamil Nadu. India: Andhra Pradesh. World: Sri Lanka.

IUCN: NE (IUCN, 2024).

Key to the varieties

1a. Leaves glabrous benath, lateral nerves 6-10 pairs; fruit globose or ellipsoid. .... var. **caudata**

1b. Leaves pubescent beneath, lateral nerves 8-12 pairs; fruit ovoid or obovid .... var. **pubescens**

Commiphora caudata (Wight & Arn.) Engl. var. **pubescens** (Wight & Arn.) K.M. Matthew. Kew Bull. 46: 540. 1991. Synonyms: *Protium pubescens* Wight & Arn., *Commiphora pubescens* (Wight &Arn.) Engl., *Balsamodendrum arnottianum* Marchand, *Balsamea pubescens* (Wight & Arn.) Engl. Vern.: *Pachaikiluvai* (Tam.).

Western Ghats: Deciduous forests of Kerala and Tamil Nadu. World: Pakistan.

IUCN: NE (IUCN, 2024).

Commiphora wightii (Arn.) Bhandari, Bull. Bot. Surv. India 6: 327. 1965. Synonyms: *Balsmodendron wightii* Arn., *Balsmodendron roxburghii* Stocks, *Commiphora mukul* (Hook. ex Stocks) Engl. *Balsmodendron mukul* Hook. ex Stocks. Vern.: *Antu guggula, Guggula, Kaushika, Mahishaaksha* (Kan.); *Guggala, Gulag, Mukul* (Mar.); *Malaikiluvai* (Tam.).

Small trees to 4 m tall; branches aromatic, thorny and knotty with papery bark. Leaves sessile, alternate or fascicled, 1-3-foliolate; leaflets glabrous, the terminal sessile or subsessile, obovate, serrate, 1-5 cm long, 0.5-2.5 cm broad, lateral when present sessile, serrate or entire, less than half the size of the terminal leaflet. Plants dimorphic, one having bisexual and male flowers, and the other having female flowers with staminodes. Bisexual and male flowers sessile, 3-5 mm long, usually red, sometimes pinkish white; bracts 2, opposite, glandular hairy. Sepals connate basally with the disc; tubular or urceolate, 1-2 mm long; lobes usually triangular, valvate, glandular hairy outside. Petals reflexed, acute, 3-5 mm long, 1 mm broad. Stamens 8, rarely 10, 3-5 mm long, free, alternately short and long, included. Disc conspicuous, toothed; shorter stamens inserted alternately in deeper sinuses. Ovary 2-loculed with sessile 2-lobed stigma. Fruit up to 1 cm long, red when ripe, marked with 2 white longitudinal lines, mucronate, 4 lined and fused at the base; epicarp dehiscing from the base upwards on maturation. In female flowers, sepals 2 mm long, petals 3-4 mm long, 1 mm broad; staminodes 8, alternately short and long, l-1.5 mm long. Ovary and fruit the same as in bisexual flowers. Fl. & Fr.: March-May.

Western Ghats: Rare busy tree in scrub and deciduous forests of Karnataka, Maharashtra and Tamil Nadu. World: Pakistan.

IUCN: CR- A2cd (IUCN, 2024).

Garuga Roxb.

1a. Flowers yellow; petals abaxially pubescent.. **G. floribunda** var. **gamblei**

1b. Flowers white, pale yellow or yellowish green........................ **G. pinnata**

Garuga floribunda Decne var. **gamblei** (King ex W.W. Sm.) Kalkman Blumea 7: 466. 1953. Synonyms: *Garuga gamblei* King ex W.W.Sm. Vern.: *Kaarai* (Tam.).

Trees to 8-26 m tall; branchlets glabrous except young parts, conspicuously lenticellate. Leaflets 9-19, without stipels, elliptic or oblong-lanceolate, membranous or papery, base rounded, oblique, margin serrate, lowest pair stipulelike, 5-8 mm, middle pair 9-11 x 3-5 cm, terminal leaflet with petiolule, 6-25 mm long; rachis and midrib sparsely pubescent. Panicles lateral or axillary, 25-35 cm, fascicled toward ends of branches. Flowers yellow, 4-6 mm diam; receptacle cupular, abaxially pubescent. Sepals deltoid, abaxially pubescent. Petals oblong-lanceolate, 3 mm, both surfaces villous. Stamens pubescent at base; disk lobes deltoid or quadrangular. Ovary globose, with short stipe, tomentose; style tomentose or pubescent; stigma shallowly 5-lobed. Drupe nearly globose, 5-9 x 5-12 mm, base without persistent calyx. Fl. & Fr.: February – June.

Western Ghats: Evergreen forests of Kerala and Tamil Nadu. India: Arunachal Pradesh, Assam, Meghalaya, Mizoram, Nagaland, Sikkim and Tripura. World: Bangladesh, China.

IUCN: LC (IUCN, 2024).

Uses: Wood is used for manufacturing furniture and agreiculture implements.

Garuga pinnata Roxb., Pl. Coromandel 3: 5. 1811. Synonyms: *Kunthia pinnata* (Roxb.) Kuntze, *Garuga pinnata* var. *obtusa* Engl., *Garuga pinnata* var. *sikkimensis* Engl. Vern.: *Ennai Kaarai, Arunelli, Karuvaembu, Karuvilaangan* (Tam.).

Trees 4-10 m tall; bark gray-brown, rough; branchlets glabrous except very young parts, lenticellate with leaf scars. Leaves with 9-23 leaflets; lowest leaflet blades stipulelike, spoon-shaped or linear, 5-10 mm, early deciduous, middle blades elliptic, oblong, or lanceolate, 5-11 x 2-3 cm, base rounded, sometimes obliquely cuneate, margin sparsely serrate, apex usually narrowly acuminate; lateral veins 10-15 pairs; rachis and leaflets pubescent with long hairs, especially when young; petiolules absent to 4 mm, terminal petiolule to 5-10 mm long. Panicles lateral or axillary, 7.5-19 cm, densely pubescent with long hairs when young; peduncle 2-6 cm long. Flowers white, yellowish white, or greenish yellow; pedicel 1-3 mm, long pubescent. Sepals deltoid, 2.5-3.5 mm, pubescent on both surfaces. Petals oblong, 5-5.5 x 1.5-2 mm, pubescent with short somewhat curved hairs. Stamens slightly unequal; filaments with long hairs at base. Disc lobes trapeziform or deltoid. Ovary oblong, with short stipe, sparsely pilose, especially when young; style pilose; stigma shallowly 5-lobed. Drupe globose, yellow when ripe, 11-15 x 11-18 mm, sometimes pubescent. Fl. & Fr.: January – May.

Western Ghats: Evergreen forests of Gujarat, Goa, Maharashtra, Karnataka, Kerala and Tamil Nadu. India: Most of the states. World: South China, Myanmar, Laos, Cambodia, Vietnam, Thailand, Philippines and Malaysia.

IUCN: NE (IUCN, 2024).

Cultivated:

Commiphora myrrha (Nees) Engl.; Exotic: Native to Arabian Peninsula, Tropical Africa

Bursera penicillata (Sessé & Moc. ex DC.) Engl.; Exotic: Native to Mexico

Canarium vulgare Leenh.; Exotic: Native to Malaysia, New Guinea

Commiphora stocksiana (Engl.) Engl.; Exotic: Native to Pakistan

CACTACEAE Juss.

Piptanthocereus Riccob

Piptanthocereus forbesii (Hort. Berol ex Forest) Riccob., Boll. Reale Orto. Bot. Palermo. 8: 288. 1909. Synonym: *Cereus forbsii* C.F. Forst.

Woody columnar catoid trees to 5 m tall; bark rough, forming areoles with undulated surface; aerial branches succulent, rib forming leaf leass spiny, glaucous.

IUCN: NE (IUCN, 2024).

Planted:

Pereskia grandiflora Haw.; Exotic: Native to Brazil

CALOPHYLLACEAE J.Agardh

Key to the genera

1a. Sepals 2 ....**Mammea**
1b. Sepals 4 or 5 2
2a. Leaves with pellucid dots; fruits enclosed by enlarged sepals ..... .**Mesua**
2b. Leaves without pellucid dots; fruits not enclosed by sepals 3
3a. Stamens more than 40; fruit a druoe .... **Calophyllum**
3b. Stamens less than 25; fruits a capsule 4
4a. Flowers in axillary racemes; 2.3 cm diam, petals white ....
.... **Agasthiyamalaia**
4b. Flowers in terminal panicles; to 6 mm diam, petals whitish yellow..........
....**Poeciloneuron**

Agasthiyamalaia S. Rajkumar & Janarth.

Agasthiyamalaia pauciflora (Bedd.) S. Rajkumar & Janarth., J. Bot. Res. Inst. Texas 1: 131. 2007. Synonym: *Poeciloneuron pauciflorum* Bedd. Vern.: *Pudangalli, Puli-vayila* (Mal.)

Trees to 15 m tall; branchlets minutely hairy. Leaves opposite, elliptic, lanceolate or oblong-elliptic, attenuate at base, entire or slightly recurved at margin, bluntly caudate acuminate at apex, 5.5-13 x 1.5-4 cm, glabrous; petioles 7-10 mm long. Flowers solitary, paired or 4 in fascicles, terminal or in axilary leaf scars axils, 1-1.2 cm across, white; bracts 3-angled, 3 mm long. Sepals 4, unequal; inner pair 6 mm long; the outer 2 mm long, puberulous outside. Petals 6 imbricate, glabrous. Stamens 16-24 in 2 whorls, subsessile, inserted on disc below ovary; filaments to 0.2 mm long; anthers linear-oblong, acute or cuneate at base, 2.5-3.5 mm long with a transparent 1.5 mm long flap with a distinct nerve and mucronate apex, smooth, 2-loculed. Ovary 2 mm long, 2-loculed; ovules 2 in each locule, erect; styles 2, 3.5 mm long. Capsules ovoid or globose, 2.5 cm, with persistent sepals and 3 stylar bases, dehiscing into 2 coriaceous valves, 1-loculed. Seed 1, globose, 1 cm across. Fl. & Fr.: February – March.

Western Ghats: River banks of evergreen forests above 1200 m in Kerala and Tamil Nadu (Endemic).

IUCN: CR- B1+2c. (IUCN, 2024).

Uses: Timebr is used for construction purposes, making walking sticks.

Calophyllum L.

1a. Perianth segments 4 only .... **C. apetalum**

1b. Sepals and petals 4 each 2

2a. Petiole >1 cm long 3

2b. Petiole <1 cm long 4

3a. Leaves >5 cm broad, elliptic-oblong or obovate; fruit globular .... **C. inophyllum**

3b. Leaves <5 cm broad, elliptic-lanceolate or oblong-lanceolate; fruit ovoid. .... **C. polyanthum**

4a. Leaves obovate .... **C. austroindicum**

4b. Leaves linear oblong to linear-lanceolate .... **C. pascalianum**

Calophyllum apetalum Willd., Mag. Neuesten Entdeck. Gesammten Naturk. Ges. Naturf. Freunde Berlin 5: 79. 1811. Synonyms: *Calophyllum calaboides* G.Don, *Calophyllum decipiens* Wight, *Calophyllum wightianum* Wall. ex Planch. & Triana, *C. calaba* L. Vern.: *Attupunna, Cherupunna, Kattapunna, Manjapunna, Porapunna* (Mal.).

Evergreen trees to 10-30 m tall; bark smooth, with longitudinal diamond or boat shaped fissures, inner bark reddish with usually with resinous milky or yellow latex; branches quadrangular or compressed, glabrous when young, oil glands present in the branches. Leaves simple, opposite decussate, oblong-obovate to elliptic, 5-18 x 2.5-4.5 cm, base rounded or subtruncate, margins entire, slightly recurved, apex obtuse, glabrous, shining on both the surfaces; lateral veins numerous, slender 5 mm apart, translucent latex canals present between the veins; petiole canaliculated, glabrous, to 6 mm long. Inflorescence axillary, 5-13-flowered, racemes about 4-7 cm long; peduncles slender, 1 cm long. Flowers white, 1-2.5 cm across; pedicel slender, glabrous, 7-15 mm long; bracts ovate, deciduous, about 4 mm long. Perianth lobes 4-8, imbricate; sepals 4 rarely 3, decussate in pairs, obovate-ovate, deflexed, biseriate, veined prominently, membranous, outer 2, broadly ovate-orbicular, 3.5-6 x 3.3-5.5 mm, inner 2, obovate-elliptic, petaloid, 6-9 x 3.5-5 mm, petals absent, rarely 2-5, but deciduous before anthesis. Stamens 45-85, obscurely fascicled, free; filaments slender, connate at the base, 5 mm long, yellow or reddish, 1.5-2.5 mm across. Ovary superior, globose, unilocular, 1.5 mm across; ovule single, anatropus, style slender, longer than the stamens; stigma peltate, margins entire or crenulate. Fruit indehiscent drupe, ovoid-subglobose, 1-2 cm across, red when ripe. Seeds 1, ovoid-globose, 7.5-12 x 5-7 mm, smooth with large cotyledons, with oil content. Fl. & Fr.: October – April

Western Ghats: Evergreen forests of Maharashtra, Goa, Karnataka, Kerala and Tamil Nadu (Endemic).

IUCN: VU- A2cd (IUCN, 2024).

Uses: Wood is used for construction of bridges, boats, cabinet work, buildings, etc. Seed oil is used in treatment of rheumatism and leprosy. Fruit is said to be edible.

Calophyllum austroindicum Kostrem. ex P.F. Stevens., J. Arnold Arbor. 61: 250. 1980. Vern.: *Chiru punnai* (Tam.).

Trees to 15-30 m tall; bark pale brown to yellowish; branchlets usually 4-angled; young branches minutely rusty-pubescent. Leaves simple, opposite, elliptic, subrhomboid to obovate or oblanceolate, 3-8.5 x 1.2-4.2 cm, subacute or shortly and bluntly acuminate at apex, cuneate to acute at base, slightly undulate, recurved and margined, rounded, retuse; lateral nerves 5-10 pairs, more or less prominent on both surfaces; petioles 5-8 mm long, grooved above, glabrous or minutely rusty-pubescent. Racemes axillary, 3-8 cm long, 7-17-flowered, glabrous; peduncles 3-7.5 cm long, glabrous. Flowers 2 cm across; pedicels 6-15 mm long, glabrous. Sepals 4, subequal; outer pair suborbicular, 3.7-5 x 3.5-6 mm; the inner broadly elliptic, 6 x 4 mm. Petals 4, rarely 6, elliptic or oblong to obovate, obtuse, 4-9 x 1.5-6 mm. Stamens 100-125; filaments 4.5 mm long; anthers elliptic to oblong, cordate, truncate or oblique at base, 0.7-1.3 x 0.4-0.8 mm. Ovary globose, 2-3 mm; styles 1.5-3.5 mm long; stigmas peltate, 3-radiate, 0.7-1 mm across. Berries ovoid, beaked at apex, 3-4.3 x 1.6-2.7 cm; stone ellipsoid, 2.5 x 1.5 cm, smooth; stalks 1.2-1.7 cm long. Fl. & Fr.: January – May.

Western Ghats: Evergreen forests above 1000 m in Karnataka, Kerala and Tamil Nadu (Endemic).

IUCN: NE (IUCN, 2024).

Calophyllum inophyllum L., Sp. Pl.: 513. 1753. Vern.: *Pinna, Punna* (Mal.); *Punnai, Pinnai, Nagam* (Tam.).

Trees to 20 m tall; bark smooth or with boat-shaped fissures, brown to pale grey; exudate milky or yellow; branchlets slightly 4-angled, glabrous. Leaves broadly elliptic-oblong or obovate, cuneate to rounded at base, retuse or sub-acute at apex, 6-15 x 3-10.5 cm, coriaceous, glabrous; lateral nerves 4-20 pairs; petioles flat, 1-1.7 cm long, glabrous. Racemes axillary, 5-12 cm long, 7-12-flowered. Flowers polygamous, 2-2.5 cm across, white, fragrant; bracts ovate, 3-4 mm long, caducous; pedicels 1-3 cm long, glabrous. Sepals 4, reflexed; outer pair ovate to suborbicular, concave, 5.5-6.5 x 4.5-6.5 mm, nerved; the inner subelliptic, 9-15 x 7.5-9.5 mm, petaloid, nerved. Petals usually 4, rarely 3 or 5, obovate to elliptic or oblong, reflexed, 9-12 x 5-9 mm, rarely glabrous, nerved. Stamens 210-360; filaments 3.5-5.5 mm long, connate for 1.5 mm into 4-6 bundles, cream coloured; anthers rounded or retuse at apex, 0.7-2 mm long. Ovary globose-depressed, 2.5-3 x 2.8-3 mm, stipitate; style 4.5-6.5 mm long; stigmas peltate, 3-5 radiate. Drupes globose to obovoid, short apiculate, 2.5-5 x 2.5-4 cm, smooth; stone subglobose. Fl. & Fr.: December – February.

Western Ghats: Low level forest of Maharashtra, Goa, Karnataka, Kerala, Tamil Nadu. India: Andaman & Nicobar Islands and Odisha. World: Sri Lanka, Old and New World Tropics.

IUCN: LC (IUCN, 2024).

Uses: Wood is used for making knees for boats. The tree is cultivated for ornamental purposes and sacred tree. Seed oil extracted for using flame.

Calophyllum pascalianum B.R.Ramesh, Ayyappan & De Franceschi, Kew Bull. 67: 115. 2012. Vern.: *Malai punnai* (Tam.).

Trees to 18 m tall; bark yellowish, longitudinally fissured, sap yellow. Leaves opposite, linear-oblong to linear-lanceolate, 6.5 - 10.5 0.9 - 1.3 cm, cuneate at base, obtuse at apex, margin slighdy undulate, dark greenish above, paler beneath, coriaceous, glabrous; midrib slightly impressed above, prominent beneath; venation closely parallel, 9 pairs, prominent on both sides; petiole 0.5 - 1 cm long. Inflorescence a panicle, slender, axillary, shorter than leaves, 3.5 - 5 cm long. Flowers 6 mm long, polygamous; pedicels 0.5 - 1.2 cm long. Sepals 4, unequal, the outer pair smaller, ovate to suborbicular 2.5 - 3.5 x 2 - 2.5 mm; inner pair obovate, 5-6 x 2 3 mm. Petals 4, oblong-obovate, 4 - 5 x 2 - 2.5 mm. Stamens 50, filaments 2.5 - 3 mm long; anther oblong, 1 - 1.2 mm long. Ovary 1- 1.5 mm long; style 2.5 - 3 mm long; stigma peltate. Fruits drupaceous, ellipsoid, 7 x 3.5 mm, apiculate, wrinkled. Seed 1. Fl. & Fr.: March – May.

Western Ghats: River banks of evergreen forests in Tirunelveli range of Tamil Nadu (Endemic).

IUCN: NE (IUCN, 2024).

Calophyllum polyanthum Wall. ex Choisy, Mem. Soc. Phys. Geneve 12: 423. 1849. *Calophyllum elatum* Bedd., *Calophyllum smilesianum* Craib, *Calophyllum* *smilesianum* var. *luteum* Craib, *Calophyllum williamsianum* Craib Vern.: *Katta-pinna, Pinnapai, Viri* (Mal.); *Kaattu punnai* (Tam.).

Lofty trees to 35 m tall; bark 15-16 mm thick, surface yellowish-brown, deeply fissured; branchlets subquadrangular, minutely brown tomentose. Leaves simple, opposite, decussate, elliptic, elliptic-lanceolate, 8-12.5 x 3-8 cm, apex acute or acuminate, base acute, margin entire, glabrous, shining, coriaceous; lateral nerves many, slender, parallel, close, intercostae obscure; petiole 12-25 mm long, slender, grooved above, glabrescent. Flowers white, in panicles or racemes, terminal and from the axils of the upper leaves; peduncle slender; pedicel 1-1.5 cm long, slender, white. Sepals 4, concave, elliptic, outer ones reflexed, ciliolate, inner 2 petalloid, 7.5 mm long. Petals 4, 5 mm across, white, orbicular, early cauducous, inner ones smaller. Stamens numerous, 4 mm long; filaments united at the base. Ovary superior, slender, 1-celled, ovule one, erect; style 2 mm long, flexuous; stigma peltate, concave, obscurely lobed. Drupe obliquely ovoid, 2-2.5 cm long, dark purple, smooth. Fl. & Fr.: January – July.

Western Ghats: Evergreen foersts above 800 m in Maharashtra, Goa, Karnataka, Kerala and Tamil Nadu. India: Assam, Meghalaya, Sikkim and West Bengal. World: Sri Lanka, Bangladesh, China, Myanmar, Thailand, Vietnam and Malesia.

IUCN: NE (IUCN, 2024).

Uses: Wood is used for building ships, construction purposes and paper pulp. Fruit is edible.

Mammea L.

Mammea suriga (Buch. -Ham. ex Roxb.) Kostrem, Pengum. Lemb. Pusat Penjel. Kehut. 72: 23. 1961. Synonyms: *Calophyllum suriga* Buch. -Ham. ex Roxb., *Ochrocarpus longifolius* (Wight) Benth. ex Hook.f. & T. Anderson, *Calysaccion longifolium* Wight, *Mammea longifolia* (Wight) Planch. &Triana Vern.: *Seraya, Suran punna* (Mal.); *Naga poo* (Tam.).

Evergreen trees to 18 m tall; bark rough, defoliating in irregular pieces, latex milky; branchlets obscurely 4-angled. Leaves opposite or verticillate, clustred at ends of branchlets, oblong, oblong-lanceolate or elliptic, 12-20 x 4.5-7.8 cm, obtuse, acute or shortly cuneate at base, entire at margin, emarginate, retuse or short-acuminate at apex, glabrous; midnerve stout, channelled above, raised beneath, lateral nerves 15-20 pairs, faint; petioles stout, 6-10 mm long, channelled above. Fascicles in axils of fallen leaves. Flowers uni-and bisexual, 1 cm across, white or pinkish; bracteoles 8, subulate, 1-1.5 mm long; pedicels slender, thick above, 1.5-2.5 cm long. Sepals in cupular, opening into 2 valves; lobes obtuse with a minute cusp at apex, concave, 5-7 mm long, reddish. Petals 4, ovate or oblong-obovate, 4-5 x 8.5-10 mm, white with red streaks, caducous. Stamens 60-100; filaments 4-5 mm long; anthers oblong or linear, retuse at apex. Ovary depressed-subglobose, 2-2.5 mm, 2-4 loculed; styles subulate, 1.5-3 mm long; stigmas large, umbonate, peltate, obscurely sinuate or 2-lobed. Berries obliquely ovoid, 2.2-3 x 1-1.5 cm, with a hard-pointed style at apex. Seeds 1-4, 2 x 1 cm; stalks 1.7-2.5 cm long. Fl. & Fr.: February – March.

Western Ghats: Deciduous and evergreen forests of Maharashtra, Goa, Karnataka, Kerala and Tamil Nadu (Endemic).

IUCN: NE (IUCN, 2024).

Uses: Cultivated as an avenue tree for its fragrant flowers. Flower is used to extract perfumes and medicinally important dye. Fruit is said to be edible.

Mesua L.

1a. Flowers and fruits distinctly stalked; bracts and bracteoles absent .... **M. ferrea**

1b. Flowers and fruits sessile; bracts and bracteoles absent 2

2a. Petals 2 cm long; fruit ovoid conical .... **M. thwaitesii**

2b. Petals 9 mm long; fruits ovoid-globose .... **M. pulchella**

Mesua ferrea L. Sp. Pl. 515. 1753 var. **ferrea**. *Mesua nagassarium* (Burm.f.) Kostrem., *Calophyllum nagasarium* Burm.f., *Mesua ferrea* ssp. *vera* Vesque, *Mesua ferrea* ssp. *thwaitesii* (Planch. &Triana) Vesque., *Mesua thwaitesii* Planch. & Triana. Vern.: *Chruli, Elipongu, Nagachempakam, Nanga, Peri, Vayanaavu, Veluthapala* (Mal.); Nagasampige (Kan.); *Nagchampa, Nagechapha, Thorachmpa* (Mar.); *Nangu, Pongu, Sirunaga poo* (Tam.).

Trees to 12 m tall; bark ash or grey coloured, reddish when cut. Leaves red and pendulous when young, simple, opposite, elliptic to oblong- lanceolate, 8-12 x 2-4.5 cm, base broadly attenuate or rounded, apex acuminate or cuspidate, coriaceous, shiny above, glaucous beneath and generally covered with a wax-like white powder beneath; nerves many, close and parallel; petiole to 1 cm long. Flowers white, solitary or in pairs at axils, 4.5-5.5 cm across. Sepals 4 in two pairs, 1.3-1.6 cm long, broadly ovate, persistent. Petals 4, sometimes 5, white with brown veins, 2.5-3.5 x 1.3-1.5 cm, obovate or obcordae, margins curled. Stamens numerous; anthers yellow. Ovary 4-6 mm long, ovoid; ovules 2 in each locule; stigma peltate. Fruits ovoid, 2-2.5 x 1.5-2 cm, glabrous. Seeds 1-4, pyriform, dark brown. Fl. & Fr.: March – May.

Western Ghats: Maharashtra, Goa, Karnataka, Kerala and Tamil Nadu. World: Sri Lanka, Bangladesh, Nepal, Myanmar, Thailand, Cambdia, Vietnam, Indonesia, Singapore and Malaysia.

IUCN: NE (IUCN, 2024).

Key to the varieties

1a. Inflorescence pubescent; flowers and fruits pedicelled; outer 2-3 teplals smaller than the inner .... .var. **ferrea**

1b. Inflorescence glabrous; flowers and fruits sessile; outer and inner petals subequal ....var. **coromandeliana**

Mesua ferrea var. **coromandeliana** (Wight) N.P.Singh. J. Econ. Taxon. Bot. 10: 203. 1988. Synonyms: *Mesua coromandeliana* Wight, *Mesua nagassarium* var. *coromandeliana* (Wight) K.K.N.Nair Vern.: *Nangu* (Tam.).

Fl. & Fr.: December – February

Western Ghats: Evergreen forests of Kerala and Tamil Nadu (Endemic).

IUCN: NE (IUCN, 2024).

Mesua pulchella Planch. & Triana. Ann. Sci. Nat., Bot., sér. 4, 15: 307. 1861. Synonyms: *Mesua ferrea* var. *pulchella* (Planch. & Triana) Vasque, *Mesua nagasarium* var. *pulchella* (Planch. &Triana) Trimen Vern.: *Malai nangu* (Tam.).

Evergreen trees to 20 tall; bark 8-10 mm thick, brown rough, irregularly flaking, fibrous, exudation scanty, colourless; branchlets slender. Leaves simple, opposite, decussate, elliptic or narrowly lanceolate, 5-10 x 1-3.5 cm, base acute, apex acuminate, margin entire, glabrous, shining, coriaceous; lateral nerves many, parallel, intercostae reticulate, obscure; petiole 8-10 mm long, slender, slightly grooved above, glabrous. Flowers white, 3-3.5 cm across, sessile, solitary or in pairs, axillary, fragrant. Sepals 4, in pairs, outer pair 7-9 mm long, suborbicular, inner 10-11 mm long. Petals 4, cuneate-oblong. Stamens numerous, yellow; filaments variable; anthers large. Ovary superior, 2-celled; ovules 2 in each cell; styles as long as ovary; stigma small, peltate. Capsule ovoid to globose, 2.5-3 x 3.4 cm with persistent accrescent sepals. Seeds smooth. Fl. & Fr.: April – December.

Western Ghats: Evergreen forests of Maharashtra, Karnataka, Kerala and Tamil Nadu. World: Sri Lanka, Myanmar.

IUCN: NE (IUCN, 2024).

Mesua thwaitesii Planch. & Triana in Ann. Sci. Nat. Bot. 4, 15: 305. 1861. Synonym: *M. ferrea* L. var. *thwaitesii* (Planch. & Triana) Vesque. Vern.: *Nangu* (Tam.).

Trees to 20 m tall; bark smooth, dark brown. Leaves opposite, elliptic, oblong- elliptic or lanceolate, 10-23 x 2-5 cm, acute to shortly cuneate or rounded at base, entire and slightly recurved at margin, acute or acuminate at apex, coriaceous, faintly nerved. Flowers solitary, paired or rarely in 3, axillary; peduncles short, pubescent; bracts and bracteoles broadly ovate or suborbicular, obtuse, thick. Sepals 4, outer pair smaller than the inner and petals, slightly pubescent, persistent. Petals 4, orbicular, 2 cm long. Stamens numerous; anthers oblong, cordate, unequal, oblique at base, connective thick. Ovary 2-celled; stigma peltate. Fruits globose-depressed or obovoid, apiculate or conic, 3-5 cm long, woody, surrounded by enlarged sepals and bracts. Seeds 2. Fl. & Fr.: February – December.

Western Ghats: Evergreen and semi-evergreen forests of Kerala and Tamil Nadu (Endemic).

IUCN: NE (IUCN, 2024).

Poeciloneuron Bedd.

Poeciloneuron indicum Bedd. J. Linn. Soc., Bot. 8: 267. 1865. Vern.: *Vayala, Vayila* (Mal.)

Evergreen trees to 30 m tall; bark thick, grey to brown, rough, exudation milky yellow, scanty; branchlets terete. Leaves simple, opposite, elliptic, elliptic-oblong or linear-oblong, 11-25 x 2.5-6.5 cm, apex caudate acuminate, base acute or obtuse, margin entire, glossy, glabrous, coriaceous; lateral nerves many, close, parallel, slender, obscure, intercostae reticulate; petiole 10-30 mm, glabrous, grooved above. Flowers bisexual, yellowish-white, fragrant, in axillary and terminal panicles; pedicels 1-2 cm long, puberulous; bracteoles triangular. Sepals 5, 3 mm long, ovate, puberulous without. Petals 5, 5-6 mm, elliptic to obovate. Stamens numerous, free or slightly connate; filaments 0.5 mm long; anthers 3.5 mm, erect. Ovary 1-2 mm, bilocular; ovules 2 in each locule; styles 2, subulate. Capsule globose, 2-2.5 cm across, beaked, pink, 1-celled. Seed one, fleshy. Fl. & Fr.: April – June.

Western Ghats: Evergreen sholas above 1800 m in Karnataka, Kerala and Tamil Nadu (Endemic).

IUCN: NE (IUCN, 2024).

Uses: Timber is useful in making railway sleepers, bridges, house construction materials, and agriculture implements.

CANNABACEAE Martinov

Key to the genera

1a. Leaf base oblique; fruit leass than 1 cm diam .... **Trema**

1b. Leaf base rounded; fruits moer than 1 cm diam 2

2a. Leaves with entire margin; drupe ellipsoid glabrous............ **Aphananthe**

2b. Leaves minutely serrate or serrulate; drupe ovoid glabrous .... **Celtis**

Aphananthe Planch.

Aphananthe cuspidata (Bl.) Planch., A.P.de Candolle, Prodr. 17: 209. 1873. Synonyms: *Cyclostemon cuspidatum* Blume, *Gironniera cuspidata* (Blume) Planch. ex Kurz, *Gironniera reticulata* Thw., *Gironniera thompsonii* King ex A.M. Cowan & Cowan Vern.: *Kodi thani* (Tam.).

Trees to 25 m tall; bark brownish, exfoliating in irregular, projecting flakes. Leaves simple, alternate, stipules sheathing, lamina elliptic, ovate-lanceolate or oblong-ovate, 6-18 x 2.5-7.5 cm, base rounded, apex acuminate, margin entire, glabrous, coriaceous; lateral nerves 10-14 pairs, intercostae reticulate, prominent; petiole 6-10 mm, slender, grooved above, glabrous. Flowers unisexual, male in short axillary panicles; female axillary, solitary. Tepals 5, concave, obtuse or acute. Stamens 5; filaments 2 mm long; pistillode woolly. Ovary sessile, superior, glabrous; stigmas 2, spreading, filiform. Drupe obovoid, 1.5-2 cm long, glabrous, stoutly beaked, slightly 2-angled. Fl. & Fr.: October - March.

Western Ghats: Evergreen forests of Karnataka, Kerala and Tamil Nadu. World: Myanmar, Laos, Cambodia, Vietnam, Hong-Kong, Thailand, Hainan and Malesia.

IUCN: NE (IUCN, 2024).

Celtis L.

1a. Leaves prominently 3-nerved; fruiting pedicel more than 2 cm long 2
1b. Leaves except basal nerve, penninerved; fruiting pedicel less than 2 cm long 3
2a. Fruit globose-ellipsoid, orange red when ripe ....**C. philippenesis**
2b. Fruit subglobose-ellipsoid, blackish when ripe ....**C. timorensis**
3a. Leaf tip caudate acuminte, scabrous beneath .... **C. australis**
3b. Leaf tip obtuse, pubescent beneath .... **C. tetrandra**

Celtis australis L., Sp. Pl.: 1043. 1753.

Deciduous trees to 20 m tall; bark thin, grey or pale brown, smooth, with horizontal wrinkles similar; sometimes becoming more rugged with warty excrescences when old. Leaves simple, alternate, lanceolate or oval-anceolate, 5-15 x 3-6 cm, serrate with regular jagged teeth, usually wavy, acuminate or twisted apex, base cuneate or rounded, slightly asymmetric, tomentose beneath; petiole to 1.5 cm long, groovd above, tomentose. Flowers andromonoecious, small, greenish, solitary or grouped in 3-5, produced on new branhces. Drupe ovoid or spherical, 10-15 mm in diam, with a scant sweetish fleshy part, blackish when ripe. Fl. & Fr.: Setember.

Western Ghats: Deciduous forests of Tamil Nadu. World: Native to Western Asia, North Africa and Southen Europe.

IUCN: LC (IUCN, 2024).

Celtis philippensis Blanco., Fl. Filip.: 197. 1837. Synonyms: *Bosea trinervia* Roxb., *Celtis brevinervis* (Blume) Planch., *Sponia strychnifolia* Teijsm. & Binn., *Celtis philippensis* var. *wightii* (Planch.) Soepadmo Vern.: *Vakkanai, Elumbiritti* (Tam.).

Trees to 25 m tall; bark pale green or grey with black streaks inside, covered with lenticels; branchlets pubescent. Leaves simple, alternate, stipules lateral, elliptic-ovate, 6-12 x 2-6 cm, ovate, elliptic or elliptic-oblong, base round or acute, apex acuminate, margin entire, glabrous, coriaceous; 3-nerved from the base, prominent; petiole 4-15 mm, slender, pubescent. Flowers polygamous, yellow, in axillary puberulous cymes, usually on new shoots. Male flowers usually at the basal part and female flowers on the upper side. Tepals 5, 2 x 1 mm, ovate, concave, ciliate, membranous. Stamens 5, free, inserted round a woolly torus. Ovary superior, sessile, ovoid, 1 mm, 1-celled; style 8 mm long, pubescent. Drupe ovoid, 7 x 5 mm, smooth. Fl. & Fr.: January - March.

Western Ghats: Semi-evergreen forests of Karnataka and Tamil Nadu. World: Tropical Africa, Solomon Islands.

IUCN: LC (IUCN, 2024).

Celtis tetrandra Roxb., Fl. Ind. 2: 63. 1824. Synonyms: *Celtis trinervia* Roxb., *Celtis xizangensis* E.W.Ma, *Sponia tetrandra* Voigt Vern.: *Ada, Kuviya, Kona, Murukkan* (Tam.).

Trees to 30 m tall; bark grayish white, lenticels few or none. Stipules narrowly lanceolate, caducous, to 7 mm long, never enclosing buds. Leaves ovate-elliptic, ovate-lanceolate or rhombic, 5-13 x 2.5-5.5 cm, thickly papery, abaxially puberulent when young, with hairs scattered on major veins and sometimes tufted in vein axils with age, base oblique with one side or rounded, margin entire or obtusely serrate, apex acuminate to shortly caudate-obtuse; secondary veins 3 or 4 on each side of midvein; petiole 6-13 mm, brown, broadly sulcate, puberulous at least near adaxial furrow. Flowers fascicled, occasionally in racemes of 2 or 3 flowers. Infructescences unbranched or seldom forked, rarely showing a small scar from a fallen male flower, 1-3 in leaf axil, rather slender, glabrous or pubescent, 0.8-1.5 cm long; fruiting pedicel 1 or 2or as long as subtending petiole. Drupe yellow to orange when mature, globose, 7-8 mm in diam, 5 mm in diam., 4-ribbed. Fl. & Fr.: February - June.

Western Ghats: Semi-evergreen forests of Maharashtra, Karnataka, Kerala and Tamil Nadu. World: Bangladesh, Myanmar, Laos, Cambodia, Vietnam, Thailand and Malesia.

IUCN: LC (IUCN, 2024).

Celtis timorensis Spanoghe, Linnaea 15: 343. 1841. Synonyms: *Celtis cinnamomea* Lindl. ex Planch., *Celtis dysodoxylon* Thwaites, *Sponia pendula* Span. Vern.: *Kalluviri, Kodalimurukki, Peenari* (Tam.).

Trees to 20 m tall; bark grayish white, gray, or grayish brown, with scattered short linear lenticels. Stipules linear-lanceolate to linear, 2-7 mm long, not enclosing bud, caducous. Leaves ovate-elliptic to ovate-oblong, 5-15 x 2.5-7.5 cm, coriaceous, base broadly cuneate to rounded and distinctly asymmetric, margin entire or weakly serrate, apex acuminate to caudate-acuminate; secondary veins 1 or 2 on each side of midvein; petiole 3-12 mm, pubescent, glabrescent. Inflorescence a branched cyme, 10-flowered, golden brown pubescent when young; proximal inflorescences with male flowers, distal inflorescences with male and female flowers. Infructescences branched,

2-3.5 cm, pubescent or glabrous, with 3 or 4 drupes and several prominent scars from fallen flowers. Drupe yellow, becoming red to orange-red when mature, broadly ovoid, 7-9 mm, base rounded, apex conic-acute; stone white, ovoid, 6 mm, reticulately foveolate, conspicuously 4-ribbed. Fl. & Fr.: January -Septermber.

Western Ghats: Evergreen forests of Maharashtra, Goa, Karnataka, Kerala and Tamil Nadu. World: Nepal, Bangladesh, Myanmar, Laos, Cambodia, Vietnam and Malesia.

IUCN: LC (IUCN, 2024).

Trema Lour.

Trema orientale (L.) Blume, Mus. Bot. 2: 62. 1856. *Celtis orientalis* L., *Celtis rigida* Blume, *Ceanothus leschenaultii* DC., *Sponia wightii* Planch. Vern.: *Ambarathai, Nudalai* (Tam.).

Trees to 20 m tall; bark gray, smooth, irregularly fissured on old branchlets. Stipules linear-lanceolate, 5-9 mm long. leaf blade 10-18 x 5-9 cm, leathery and fragile, abaxially grayish white to grayish green when dry, pubescent, and surface completely hidden by hairs, adaxially green to grayish green when dry, scabrous, base cordate and oblique, margin denticulate, apex acuminate to acute; basally 3-veined, secondary veins 4-6 on each side of midvein; petiole 0.8-2 cm, pubescent. Male inflorescences 1.8-2.5 cm long, pubescent. Female inflorescences 1-2.5 cm long. Male flowers: 1.5-2 mm in diam. Tepals 5. Ovary rudimentary, obovate-conic, compressed. Female flowers: pedicellate; tepals 4 or 5, triangular-ovate, 1-1.5 mm long. Drupes black when mature, globose to ovoid-globose, compressed, 3-5 x 2.5-3.5 mm, rugose, with perianth persistent. Seed broadly ovoid, compressed, 2-5 mm in diam. Fl. & Fr.: Septemebr - March.

Western Ghats: Common in stream sides and forests above 400 m in Gujarat, Maharashta, Goa, Karnataka, Kerala and Tamil Nadu. World: Sri Lanka, Nepal, Bhutan, Myanmar, Philippines, Australia, Taiwan, Thailand, Vietnam, Polynesia, Malaysia, Indonesia, Phlippines, Australia and Tropical Africa.

IUCN: LC (IUCN, 2024).

CAPPARIDACEAE Juss.

Key to the genera

1a. Leaves simple .... **Capparis**

1b. Leaves trifoliate 2

2a. Stamens 6-12; fruit lomentum .... **Maerua**

2b. Stamens 12-50; berry globose or ellipsoid .... **Crataeva**

Capparis L.

1a. Upper petals with yellow or purplish blotch; ovary densely hairy .... **C. brevispina**

1b. Upper petals not blotched as above; ovary glabrous 2

2a. Outer sepals connate in bud with free tips; outer pericarps in fruit sculptured .... **C. divaricata**

2b. Outer sepals not connate in bud; outer pericarp ribbed 3

3a. Leaves with olive green; nerves close, more or less parallel ... **C. grandis**

3b. Leaves with pale green; nerves distant, not parallel .... **C. decidua**

Capparis brevispina DC. Prodr. 1: 246. 1824. Synonyms: *Capparis acuminata* Roxb., *Capparis wallichiana* Wight & Arn. Vern.: *Kaattu adhondai, Puthar adhondai* (Tam.).

Large shrubs or small tree to 2-4 m tall; branchlets hairy; bark rough with longitudinal fissures. Leaves elliptic or oblong, 5-7 x 2-4 cm, tip acute, coriaceous; stipular spines very small, cataphylls linear. Flowers axillary, solitary, 4-5 cm across; pedicels 4 cm long. Sepals 4 in two whorls, greenish, ovate, rounded apex. Petals 4, oblong, obovate, white with brown patches. Stamens numerous, exserted; gynandrophore as long as or longer than the filaments; anthers dorsifixed. Ovary 6 mm, ellipsoid, densely hairy; ovules few to many. Fruits subglobose, glabrous, beaked, redish when ripe, seeds many. Fl. & Fr.: November – April.

Western Ghats: Common in scrub jungles and foot hills of Maharashtra, Karnataka, Kerala and Tamil Nadu. India: Andhra Pradesh, Odisha and Pondicherry. World: India, Sri Lanka.

IUCN: NE (IUCN, 2024)

Capparis decidua (Forssk.) Edgew. J. Proc. Linn. Soc., Bot. 6: 184. 1862. Synonyms: *Sodada deciduas* Forssk., *Capparis aphylla* Roth Vern.: *Ilai adhondai* (Tam.).

Small trees to 10 m tall; branches with leafless green crooked spiny; bark rough, thick. Leaves present on young twigs, caducous, linear, 4-20 mm long, 1-3 mm broad, often spine-tipped, subsessile; stipular spines 1-6 mm long, straight or slightly curved, yellow or brown. Inflorescence few to many flowered, ebracteate corymbs on short lateral shoots. Flowers 1-2 cm across, pedicel 1-1.5 cm long, usually brick red. Sepals petaloid, usually 5-8 mm long, 3-5 mm broad, ovate-oblong, upper one distinctly saccate, often with floccose-ciliate margins. Petals about as long as the sepals, puberulous, upper pair slightly larger and hidden in the saccate sepal. Stamens 10-15, 10-20 mm long, often red in colour. Gynophore 10-15 mm long; ovary to 2 mm in diam. with a beak about 1 mm long. Fruit globose, 10-15 mm in diam, slightly beaked, glabrous smooth, deep red when ripe. Seeds reniform, 2-5 mm in diam. Fl. & Fr.: July – November.

Western Ghats: Scrub forests of Maharashtra, Karnataka and Tamil Nadu. India: Andhra Pradesh, Gujarat and Odisha. World: Pakistan, West Asia and Tropical Africa.

IUCN: LC (IUCN, 2024).

Capparis divaricata Lam. Encycl. 1: 606. 1785. Synonyms: *Capparis stylosa* DC., *Capparis horrida* Banks ex Wight & Arn. Vern.: *Neela ilia adhondai, Koratti* (Tam.).

Small trees to 5 m tall; bark rough, fissured, greyish. Leaves elliptic-lanceolate to linear-lanceolate, 4-6 x 1.5-1.7 cm, mucronate, thick, coriaceous; petioles 0.5-0.7 cm, stipular spines 3 mm long, straight, stout. Flowers axillary, solitary. Sepals 4, two series, greenish, elliptic-rounded, apex rounded, concave. Seplas greenish white, 4, obovate-lanceolate, basally tinged, margin finely serrated. Berry globose, 5-6 cm across, rugose, ribbed and beaked. Seeeds angular, black. Fl. & Fr.: January – September.

Western Ghats: Scrub forests of Maharashtra, Karnataka, Kerala and Tamil Nadu. India: Andhra Pradesh and Pondicherry. World: Sri Lanka.

IUCN: NE (IUCN, 2024).

Capparis grandis L.f. Suppl. Pl.: 263. 1782. Synonyms: *Capparis auricans* (Kurz) Craib, *Capparis bisperma* Roxb., *Capparis grandis* var. *auricans* Kurz. Vern.: *Mutkondai* (Tam.).

Small trees to 5 m tall; branchlets densely tomentose; bark whitish, smooth. Leaves ovate, 4-6 x 3-4 cm, apex emarginate, base acuminate, tomentose on both sides; petiole to 1 cm, densely tomentose. Corymbs terminal, lower with leaves, upper with small bracts. Flowers to 2 cm across, white; peduncles to 3 cm, tomentose; pedicels 2.5 cm, buds tomentose. Berry globose, 2.5 cm across, glabrous. Fl. & Fr.: April – January.

Western Ghats: Common in scrub forests and deciduous forests of Maharashtra, Goa, Karnataka, Kerala and Tamil Nadu. India: Andhra Pradesh, Gujarat, Pondicherry and Rajasthan. World: Sri Lanka, Myanmar, Thailand, Laos and Vietnam.

IUCN: NE (IUCN, 2024).

Crateva L.

1a. Leaflets oblong-laceolate; berry oblong-ellipsoid ....**C. magna**

1b. Leaflets ovate-lanceolate; berry globose ....**C. adansonii**

Crateva adansonii ssp. **odora** (Buch. -Ham.) M. Jacobs. Blumea 12: 198. 1964. *Crateva odora* Buch. - Ham. *Crateva roxburghii* R.Br. *Crateva religiosa* sensu Dunn, *Crateva religiosa* var. *roxburghii* (R.Br.) Hook.f. &Thoms. Vern.: *Mavalingam* (Tam.).

Deciduous tree to 6-15 m tall; Bark smooth, whitish, glabrous. Leaves 3-foliolate, leaflets shortly petiolulate, elliptic, elliptic-lanceolate, ovate-lanceolate, 5-12 x 2.5-6 cm, base slightlyn oblique, acuminate, narrowed towards the base, subcoriaceous, often turning brownish when dry; petiole 4-8 cm long. Inflorescence-axis 2-5 cm long, often extending into a leafy extension after flowering. Flowers 3-6 cm across, greenish-white turning yellowish after opening, appearing before or just after the flush of new leaves, pedicellate; pedicels 2-4 cm long. Sepals 4-5 mm long, ovate-deltoid, greenish-white, distant. Petals ovate, or suborbicular, reticulately veined, eventually 20-30 mm long, 10-17 mm wide; 2 adaxial ones generally somewhat larger. Stamens 20-25 numbers, exceeding the petals, 3-5 cm long, inserted at base of the gynophore. Gynophore 3-5 cm long, slender, thickened in fruit. Fruit globose, 3-5 cm in diam, yellowish. Seeds reniform, about 3-4 mm in diam., smooth, brown. Fl. & Fr.: March – August.

Western Ghats: Scrub forests of Maharashtra, Karnataka, Kerala and Tamil Nadu. India: Andhra Pradesh, Gujarat, Madhya Pradesh and Odisha. World: Sri Lanka, Pakistan, Myanmar, Tropical Africa.

IUCN: LC (IUCN, 2024).

Crateva magna (Lour.) DC. Prodr. 1: 243. 1824. Synonyms: *Capparis magna* Lour., *Crateva religiosa* var. *nurvala* (Buch. - Ham.) Hook.f. &Thoms., *Crateva nurvala* Buch. - Ham., *Crateva religiosa* Dunn. Vern.: *Malai mavalingam* (Tam.).

Small trees to 4–20 m tall; bark grayish brown, smooth or verrucose, with circular lenticels. Leaves trifoliolate, leaflets oblong-lanceolate, 7-18 x 3-8 cm, papery to thinly leathery, abaxially dusty gray, adaxially brownish green and glossy; secondary veins 10-15 pairs, reticulate veins obvious, apex acuminate to long acuminate; Petiole 5-12 cm, adaxially with several pale glands toward rachis; petiolules 2-6mm. Inflorescence corymbose racemes, 20-30-flowered, subtended by several leaves; peduncle 4-6 cm; pedicel 3-4 cm long. Sepals lanceolate, 3-5 x 1.2-1.5 mm. Petals white, claw 5-10 mm long, blade 1.5-2.5 cm with an abruptly acuminate apex. Stamens 11-17; filaments 2.5-5 cm long; anthers 2-3 mm. Gynophore 4-6 cm; ovary oblong-ellipsoid, terete, 5-6 x 2 mm. Fruit oblong-ellipsoid to oblong-ovate, 2.5-5 cm diam; stipe 1-3 mm wide. Seeds 30-50 per fruit, dark brown, 6-15 x 6-10 mm, slightly flattened, dorsally irregularly crested. Fl. & Fr.: January – July.

Western Ghats: Forests up to 400 m in Gujarat, Maharashtra, Goa, Karnataka, Kerala and Tamil Nadu. India: Andhra Pradesh, Gujarat, Himachal Pradesh, Manipur, Odisha, Punjab, Tripura and West Bengal. World: Bangladesh, China, Myanmar, Laos, Cambodia, Vietnam, Malaysia, Indonesia and Tropical Africa.

IUCN: NE (IUCN, 2024).

Maerua Forssk.

Maerua apetala (B.Heyne ex Roth) M.Jacobs. Blumea 12: 207. 1964. Synonyms: *Capparis apetala* B.Heyne ex Roth, *Niebuhria apetala* (B.Heyne ex Roth) Dunn, *Capparis linifolia* Roxb.

Deciduous tree upto 5 m tall; bark pale, smooth. Leaves thick, sub coriaceous, 3-foliolate, leaflets linear-oblong and lanceolate, 2–6 x 1.7–3.5 cm, base rounded, margin entire, apex acute to acuminate; secondary nerves 3–6 pairs, obscure; petiole to 7.5 cm long, terete, slightly grooved above. Flowers fragrent, creemish tinged with purple, axillary or terminal corymbs. Stamens many. Berry oblong-ovoid, 1-2 ridged, 1.3 cm long. Seeds muricate, 6 mm across, white. Fl. & Fr.:

Western Ghats: Common in scrub jungles of Tamil Nadu, Kerala, Karnataka and Maharashtra. India: Andhra Pradesh, Telangana and Pondicherry.

IUCN: NE (IUCN, 2024).

CASUARINACEAE R.Br.

Planted:

Allocasuarina littoralis (Salisb.) L. A. S. Johnson; Exotic: Native to Tasmania, Queensland, Victoria

Allocasuarina torulosa (Aiton) L. A. S. Johnson; Exotic: Native to Australia

Allocasuarina verticillata (Aiton) L. A. S. Johnson; Exotic: Native to Australia

Casuarina equisetifolia L.; Exotic: Native to Australia, Malaysia, Pasifica Island.

Casuarina glauca Sieber ex Spreng.; Exotic: Native to Australia.

CELASTRACEAE R.Br.

Key to the genera

1a. Leaves spiral or alternate; seeds partly enveloped by aril .**Gymnosporia**

1b. Leaves dicussate or opposite; seeds basally enveloped by aril 2

2a. Petals connate at base; disc inconspicuous; fruits 1-seeded .... **Microtropis**

2b. Petals free; disc conspicuous; fruits 3-many-seeded 3

3a. Ovary 1-2-loculed, ovule 2 per locule; fruit drupaceous indehiscen t.....4

3b. Ovary 3-5-loculed, ovule 1-18 in each locule; fruit capsular dehiscent ..6

4a. Disc not cupular; fruit with persistent termial style 5

4b. Disc cupular; fruit withour persistent lateral style ....**Pleurostylia**

5a. Fruit two seeded .... **Elaeodendron**

5b. Fruit one-seeded .... **Cassine**

6a. Ovary 3-loculed, ovule 8-18 in each locule; fruit 3-angled; seeds winged ....**Lophopetalum**

6b. Ovary 4-5-loculed, ovule 1-2 in each locule; fruit 4-5-angled; seeds not wined 7

7a. Frits dehisced by 5-valves; seeds 2 in each valves .... **Euonymus**

7b. Fruits dehisced by 3-valves; seeds one in each valve ......**Glyptopetalum**

Cassine L.

1a. Leaves obtuse rounded at apex; sepals orbicular 2

1b. Leaves caudate-acuminate at apex; seplas broadly ovate .... **C. kedarnathii**

2a. Petiole to 1 cm long; sepals and petals 4-5 .... **C. glauca**

2b. Petiole to 3 cm long; sepals and petals 5 .... **C. albens**

Cassine albens (Retz.) Kostrem., Gard. Bull. Singapore 39: 178. 1987. Synonyms: *Schrebera albens* Retz., *Elaeodendron glaucum* sensu Hook.f., *Cassine dichotoma* (Roxb.) M.R. Almeida, *Neerija dichotoma* Roxb. Vern.: *Mukarti, Kannira* (Kan.); *Karumiraka* (Mal.); *Arn* (Mar.); *Eerkolli, Irgoli, Kannimaram, Karuvali, Pirai* (Tam.).

Trees to 20 m tall; bark grey to greyish-brown; exudation watery; branchlets glabrous. Leaves simple, opposite, elliptic, ovate, elliptic-ovate or ovate-oblong, 5-10 x 2-6 cm, base acute, attenuate, obtuse or round, apex acute or acuminate, margin entire or shallowly crenate or serrate, glabrous, coriaceous; lateral nerves 4-7 pairs, pinnate, intercostae reticulate, prominent; petiole 15-20 mm long, slender, glabrous. Flowers bisexual, to 8 mm across, greenish, in axillary or extra axillary dichotomously branched cymes to 5 cm long; bracts ovate; pedicel 3 mm long, glabrous. Sepals 5, 5-lobed; lobes unequal. Petals 5, 3-4 mm long, oblong, spreading, reflexed. Stamens 5, inserted on the disc; filaments 1.5 mm long; anthers subglobose. Disc cushion-like, sinuate. Ovary embedded in the disc, 2-3-celled, ovules 2 in each cell; style thick; stigma not lobed. Drupe oblong, 1 x 0. 8 cm, yellowish-green, smooth, apiculate. Seed one, curstaceous. Fl. & Fr.: July – October.

Western Ghats: Deciduous forests of Karnataka, Kerala, Maharashtra and Tamil Nadu. World: Sri Lanka, Bhutan, Cambodia, Indonesia, Nepal and Pakistan.

IUCN: NE (IUCN, 2024).

Uses: Wood used for making furniture and household implements.

Cassine glauca (Rottb.) O. Kuntze., Revis. Gen. Pl. 1: 114. 1891. Synonyms: *Mangifera glauca* Rottb., *Elaeodendron glaucum* (Rottb.) Pers., *Elaeodendron roxburghii* Wight & Arn. Vern.: *Karuvaali, Khiri* (Tam.).

Trees to 20 m tall; bark red inside, exuding a watery sap when cut; branchlets redish. Leaves opposite or sub-opposite, elliptic, ovate-oblong or obovate, 3-16 x 1.5-7.5 cm, glabrous, dark green, shining above, glaucous beneath; petiole

0.5-1 cm long, grooved above. Cymes axillary, dichotomous, 7.5-12 cm long; peduncle 2.5-6.5 cm long, often red. Flower 5 mm in diam, whitish. Sepals 4-5, lobed, lobes 4-5, rounded, margin membranous. Stamens included; filaments recurved; anthers with a 2-lobed connective. Ovary adnate to the disc; style very short, persistent. Drupe ovoid, 1-15 cm long, 1-celled, 1-seeded. Fl. & Fr.: September – April.

Western Ghats: Deciduous to semi-evergreen forests of Gujarat, Maharashtra, Goa, Karnataka, Kerala and Tamil Nadu. India: Andhra Pradesh and Odisha. World: Sri Lanka and Malaysia.

IUCN: NE (IUCN, 2024).

Cassine kedarnathii Sasidh. & Swarupan., Reinwardtia 11: 30. 1992.

Trees to 30 m tall; bark greyish brown, reddish inside. Leaves opposite, elliptic-oblong, acute at base, crenate at margin, caudate-acuminate at apex, coriaceous and glabrous; lateral nerves 5-9 pairs; petioles 2-2.5 cm long. Cymes dichasial, terminal and axillary, glabrous. Flowers 12 mm across, greenish yellow; pedicels 4-5 mm long. Sepals 5, imbricate, oblong, broadly ovate, obtuse, 1.5 x 3 mm, glabrous. Petals imbricate, oblong, obtuse, slightly revolute at margin, 5 x 2-2.5 mm. Stamens 5, inserted at margin of wavy annular disc; filaments 2 mm long. Ovary within disc, 2-loculed with 2 basal ovules in each; sytles short; stigmas entire. Drupes ellipsoid, apiculate, green, vertically grooved on both sides, brown inside. Seed 1, elliptic, planoconvex, shallowly grooved above, brown. Fl. & Fr.: April – June.

Western Ghats: Evergreen forests of Kerala (Endemic).

IUCN: NE (IUCN, 2024).

Elaeodendron Jacq.

Elaeodendron paniculatum Wight & Arn., Prodr. Fl. Ind. Orient. 1: 157. 1834. Synonym: *Cassine paniculata* (Wight &Arn.) Lobr.- Callen. Vern.: *Kanneermaram* (Tam.).

Trees to 20 m tall; bark grey, smooth, exudation watery; branchlets slender, terete. Leaves simple, opposite, decussate, elliptic-oblong, obovate, oblanceolate, lanceolate or ovate lanceolate, 7-9 x 2.5-4 cm, base acute or cuneate, apex acute or obtusely-acuminate, margin crenate, glabrous, coriaceous; lateral nerves 6-8 pairs, intercostae reticulate, faint; stipules minute, cauducous; petiole 8-15 mm, slender, grooved above, glabrous. Flowers greenish-yellow, 1 cm across, in axillary stout dichasial corymbose cymes. Seplas 5, lobed, lobes 5, unequal, imbricate. Petals 5, obovate, spreading. Stamens 5, inserted into the margin of the disc; filaments slender at length recurved; anthers subglobose. Disc thick, fleshy, margin sinuate. Ovary immersed in the disc, conical, 2-5-celled; ovules 2 in each cell; style short; stigma small. Drupe globose to ellipsoid, 2.5 x 1.5 cm, apiculate. Seed one, erect. Fl. & Fr.: March – July.

Western Ghats: Deciduous to semi-evergreen forests of Maharashtra, Karnataka, Kerala and Tamil Nadu. World: Sri Lanka, Malay Archipelago, Philippines and Singapore.

IUCN: NE (IUCN, 2024).

Euonymus L.

1a. Cymes less than 7-flowered 2
1b. Cymes more than 7-flwoered 3
2a. Leaves with crenulate margin; flowers in dichotomous cymes .... **E. crenulatus**
2b. Leaves with entire or serrate margin; flowers in umblled cymes .... **E. indicus**
3a. Branchlets acutely 4-angeld .... **E. angulatus**
3b. Branchlets terete 4
4a. Leaves oblong-lanceolate to ovate-laneolate, entire or crenate 5
4b. Leaves elliptic, crenate-serrate .... **E. laxiflorus**
5a. Sepals margin dentate; petals orbicular .... **E. barberi**
5b. Seplas margin entire or crispate; petals ovate 7
6a. Leaves linear-lanceolate, prominently serrate .... **E. serratifolius**
6b. Leaves elliptic-lanceolate; margin entire or crenulate 8
7a. Petals margin entire; seeds 5 in capsule .... **E. dichotomus**
7b. Petals margin crispate; seeds 10 in capsule .... **E. kanyakumariensis**

Euonymus angulatus Wight, Icon. Pl. Ind. Orient. 3: t. 1053. 1846. Synonyms: *Euonymus acutangulus* Wight, *Euonymus pterocladus* Hohen. ex G.Lawson.

Small trees to 6 m tall; branchlets acutely 4-angled, glabrous. Leaves opposite, oblong-lanceolate to elliptic, 8-12 x 4-8 cm, tapering at base, entire and thick at margin, acuminate at apex, coriaceous, glabrous, shining above, pale beneath; lateral nerves not conspicous. Cymes lax, usually drooping, more than 7-flowered; peduncles slender. Sepals 5, entire or fimbriate at margin.

Petals 5, free, orbicular, purplish. Stamens 5, alternating with petals, inserted at margin of disc. Disc 5-lobed, fleshy. Ovary within disc; styles short, thick. Capsule 5-loculed, turbinate, attenuate below, obtusely winged above. Fl. & Fr.: December – April.

Western Ghats: Evergreen forests at 1500 m in Karnataka, Kerala and Tamil Nadu (Endemic).

IUCN: VU- B1+2c. (IUCN, 2024).

Euonymus barberi Murugan& Manickam, Indian J. Forest. 29: 199. 2006.

Trees, up to 5 m high; internodes 3-5 cm long, branchlets flat. Leaves simple, deccussate, 5-8.5 x 3-4.5 cm elliptic-ovate, base acute, margin entire, apex acuminate; petioles 5-8 mm long; stipules linear, 3 x 1 mm. Inflorescence axillary or terminal cymes, up to 5 cm long. Flowers 7-15, 1.2 cm across, bisexual, pedicels 5 mm long; bracts minute. Sepals 5, imbricate, ovate, 2-3 x 2-3 cm, persistent. Petals 5, free,5 mm in diam, margin dentate. Stamens 5, alternate to petals, filaments reduced, anther 2-celled. Ovary inferior 5-celled, stigma simple. Capsule obcordate, 1.2 x 1.5 cm. Seeds 5. Fl. & Fr.: October – December.

Western Ghats: Rare in evergreen forests of Agasthiyamalai hills of Tamil Nadu (Endemic).

IUCN: NE (IUCN, 2024), endangered (Narasimhan and Irwin, 2021).

Euonymus crenulatus Wall. ex Wight &Arn., Prodr. Fl. Ind. Orient. 1: 161. 1834.

Evergreen trees to 6 m tall; bark grey, smooth, young branches slightly compressed. Leaves simple, opposite, decussate, elliptic, ovate, elliptic-obovate or oblong-lanceolate, 3-10 x 1.5-4 cm, base acute or attenuate, apex obtuse or emarginate, margin crenate-serrate towards apex, glabrous, coriaceous; lateral nerves 4-5 pair, faint, intercostae reticulate, obscure; petiole 2-6 mm long, slender, glabrous. Flowers reddish, 0.8-1 cm across, 3-7-flowered axillary dichasial cyme; peduncles and pedicels moderately stout. Seplas 5, deeply 5 lobed, 3 x 3 mm, ovate. Petals 5, 5 x 5 mm, free, orbicular, not fimbriate but some times crispate. Stamens 5, attached on the disc; filaments short; anthers broad. Disc fleshy, 5-lobed. Ovary 5-celled, sunk in the disc; ovules 2 in each cell; style very short; stigma blunt. Capsule turbinate, 1-1.5 x 0.8-1 cm, drooping, scarlet, 5-valved, valves obcordate. Seeds solitary in each cell; aril small, orange-red. Fl. & Fr.: March – September.

Western Ghats: Evergreen forests above 1000 m in Karnataka, Kerala and Tamil Nadu (Endemic).

IUCN: NE (IUCN, 2024).

Euonymus dichotomus B. Heyne ex Wall. W. Roxburgh, Fl. Ind. 2: 410. 1824. Vern.: *Vini* (Tam.)

Small trees to 10 m tall; branchlets terete, very slender, tender portions quadrangular. Leaves oblong-lanceolate to ovate-lanceolate, 4-6 x 2-3 mm, apex acuminate, coriaceous, thin, base tapering, entire or crenate, wavy; petioles grooved above. Inflorescence more than 7-flowered cymes. Sepals 5-lobed. Petals 5, orbicular, entire, flat, greenish. Stamens 5, inserted on the margins of the disc. Ovary sunk on the disc; stigma blunt. Disc lobed. Capsules clavate, 1.5 cm long, tapering downwards. Seeds oblong, brownish, glabrous, enclosed in fleshy aril. Fl. & Fr.: April – September.

Western Ghats: Evergreen forest above 1000 m in Karnataka, Kerala and Tamil Nadu (Endemic). World: Extended in Sri Lanka.

IUCN: NE (IUCN, 2024).

Euonymus indicus B. Heyne ex Wall. W. Roxburgh, Fl. Ind. 2: 409. 1824. Synonyms: *Euonymus goughii* Wight, *Euonymus javanicus* var. *talungensis* Pierre, *Euonymus javanicus* var. *coriaceus* (Ridl.) Blakelock, *Euonymus javanicus* var. *elmeri* (Merr.) Blakelock. Vern.: *Kanparchi* (Kan.); *Vini* (Tam.).

Trees to 8 m tall; bark greyish-brown, rough; branchlets terete, yellow, twigs quadrangular. Leaves simple, opposite decussate, elliptic, lanceolate, elliptic-lanceolate, oblong-lanceolate, 7-11 x 2-4 cm, base cuneate or acute, apex acuminate, margin entire, glabrous, shining above, coriaceous; lateral nerves 4-5 pairs, faint, intercostae reticulate, obscure; stipules linear-subulate, cauducous; petiole 5-10 mm, slender, glabrous. Flowers yellowish-red, 1 cm across, in 1-3-flowered axillary cymes; peduncles to 3 cm, stout; pedicels to 6 mm long, stout. Seplas 5, deeply divided, 5-lobed, to 1.5 mm long, fleshy, glabrous. Petals 5, 4 mm long, orbicular-oblong, fimbriate. Stamens 5, inserted on the disc near its margin; filaments short; anthers broad. Disc thick, cupular, 5-lobed, fleshy. Oovary angular, sunk in the disc, 5-celled; ovules 2 per cell; style short, thick; stigma minutely lobed. Capsule obovoid, 5-angled, orange-yellow, loculicidal, valves deeply obcordate. Seeds reddish, oblong with red aril. Fl. & Fr.: January – October.

Western Ghats: Moist deciduous to evergreen forests of Maharashtra, Goa, Karnataka, Kerala and Tamil Nadu. India: Andaman Islands. World: Myanmar, Bangladesh and New Guinea.

IUCN: LC (IUCN, 2024).

Euonymus kanyakumariensis Murugan & Manickam, J. Bombay Nat. Hist. Soc. 102: 198. 2006.

Small trees to 3–5 m tall; bark smooth, brownish; branchlets obscurely 4-angular. Leaves decussate, rarely ternate, elliptic-oblong, 3-7 x 1.5-3 cm, membranous, base acute, margin entire, recurved, apex obtusely acute; nerves 5-7 pairs, obscure above, prominent beneath; petioles up to 5 mm long, hemispherical; stipules glabrous. Inflorescence axillary, of dichotomously branched cymes, to 6 cm long; peduncles 3-4 cm long. Flowers 7-9, 1.5 cm diam, 5-merous; pedicels 5-10 mm long, slender; bracts subulate, 1-2 mm long. Sepals 5, imbricate in bud, subequal, sub-orbicular, 2-3 mm diam, persistent, entire at margin. Petals 5, orbicular, 3-4 mm diam, greenish-purple, base slightly clawed, margin crispate, apex rounded. Stamens 5, opposite to sepals, shorter than petals; filaments obscure; anthers 2-celled, deltoid. Disc flat, 5-lobed. Ovary 1-2 mm long, 5-celled, immersed in disc; ovules 2 in each; style subulate; stigma simple, acute. Capsule obcordate, 2 x 1.5 cm, pink, apex retuse, base acute; lobes 5, towards base, obtuse. Seeds 5-10, 7.5 mm long, brown to black, yellow arillate. Fl. & Fr.: March.

Western Ghats: Evergreen forest above 1000 in Tamil Nadu (Endemic).

IUCN: NE (IUCN, 2024), endangered (Narasimhan and Irwin, 2021).

Euonymus laxiflorus Champ. ex Benth., Hooker's J. Bot. Kew Gard. Misc. 3: 333 (1851). Synonyms: *Euonymus paniculatus* Wight ex G. Lawson., *Euonymus longifolius* Champ. ex Benth., *Euonymus incertus* Pit.

Small trees to 4 m tall; bark brown, rough, glabrous. Leaves 5-8 x 2-2.5 cm, lanceolate, apex acuminate, base attenuate, upper half serrulate; petiole 7 mm long, grey. Cymes broad, on short axillary peduncle; secondary peduncle to 2 cm long, slender, trichotomously branched, spreading, 6-12-flowered. Flowers 6 mm across. Sepals 5, entire, orbicular, green. Petals 5, entire, orbicular, deep purple, glabrous. Capsule to 1.5 x 2 cm. Seeds 5, orange-yellow, aril white. Fl. & Fr.: January – June.

Western Ghats: Evergreen forests of Kerala and Tamil Nadu (Endemic).

IUCN: LC (IUCN, 2024), endangered (Narasimhan and Irwin, 2021).

Euonymus serratifolius Bedd., Fl. Sylv. S. India: lxiv. 1871.

Stunted trees to 3 m tall; branches terete, sturdy, twigs green to light green. Leaves papery, linear-lanceolate, 15-22 x 3-5.5 cm, base subrounded, margin variously serrate, denticles ciliate, apex acuminate or caudate; lateral veins 12-15 pairs, reticulate, disappearing before reaching margin; petiole sturdy, 6-10 mm long. Peduncle slender, 2-5 cm long, several flowered; pedicel 6-8 mm. Flowers 5-merous, 7-9 mm in diam, purple. Capsule obpyramidal, 5-angled with deep grooves, brown or yellow-brown to red-brown, 3 x 2.5 cm. Seeds ellipsoid, 1-2, dark brown; aril orange-red. Fl. & Fr.: January – July.

Western Ghats: Evergreen forests of Kerala and Tamil Nadu. World: China South-Central and China Southeast

IUCN: EN- B1+2c. (IUCN, 2024).

Glyptopetalum Thw.

1a. Leaves elliptic-oblong or elliptic-obovate, margin entire, serrate towards apex 2

1b. Leaves ovate-lanceolate, margin sharply serrate except at base.. **G. zeylanicum**

2a. Cymes 3-flowered; capsules globose .... **G. grandiflorum**

2b. Cymes 6–7-flowered; capsules subglobose .... **G. lawsonii**

Glyptopetalum grandiflorum Bedd., Icon. Pl. Ind. Or. 1: 21. 1871.

Small trees to 5 m tall; branchlets terete, angled when young. Leaves elliptic-oblong, 10-15.5 x 5-6.5 cm, rounded or cuneate at base, entire, rarely obscurely serrate towards apex, acute or acuminate at apex, coriaceous. Cymes 3-flowered; peduncles filiform, slender, 10-16 cm long. Pedicels to 2.5 cm long, inserted with scale-like bracts. Sepals 4-lobed. Petals 4, obovate, 8-10 mm long, inconspicuously pitted or not, yellow. Stamens 4, sessile. Disc more or less rounded. Ovary within disc. Capsules globose, 2 cm diam, smooth, pale, 4-seeded. Seeds ovate, 1.5 cm across, arillate. Fl. & Fr.: February – June.

Western Ghats: Evergreen forests above 1500 in Karnataka, Kerala and Tamil Nadu. World: Sri Lanka

IUCN: NE (IUCN, 2024).

Glyptopetalum lawsonii Gamble, Bull. Misc. Inform. Kew 1916: 131. 1916. Vern.: *Kodikurunthu* (Tam.).

Small trees to 7 m tall; bark rough, brownish; branchlets terete, young shoots flattened. Leaves elliptic-obovate, 5-9 x 3-6 cm, obtuse, base cuneate, coriaceous, entire or with a few teeth at apex; nerves 7-9 pairs, distinct. Cymes axillary or supra-axillary, 6-7-flowered. Flowers bisexual. Sepals 4-lobed, short, scarious at apex. Petals 4, ovate, 3 mm long, obtuse, 2 oblique pits towards apex. Stamens 4; filaments short, thick. Ovary immersed in the disc; styles short. Capsular, subglobose, 1-1.6 cm across. Seeds subglobose, 10 x 5 mm, covered with fleshy aril. Fl. & Fr.: August – February.

Western Ghats: Dry deciduous to evergreen forests in Kerala and Tamil Nadu (Endemic).

IUCN: VU- B1+2c. (IUCN, 2024).

Glyptopetalum zeylanicum Thw., Hooker's J. Bot. Kew Gard. Misc. 8: 268. 1856.

Small trees 5 m tall; branchlets terete, flat at tip. Leaves opposite, ovate-lanceolate, 10-15.5 x 3.5-6.5 cm, cuneate at base, sharply serrate at margin except at base, acute at apex, membranous and glabrous. Lateral nerves conspicuous; petioles 8 mm long, grooved. Cymes axillary or supra-axillary, 5-7-flowered. Sepals 4, small. Petals 4, oblong, 5 mm long, with a pit near apex, yellow. Stamens 4, inserted at angles of disc; connectives enlarged. Ovary within disc, 4-lobed; styles conical. Capsules rounded, 1-4-loculed, 1-4-seeded. Seeds 1.5 cm across; aril partially covering seed, red. Fl. & Fr.: March – December.

Western Ghats: Deciduous forests abive 800 m in Kerala and Tamil Nadu. World: Sri Lanka.

IUCN: NE (IUCN, 2024), vulnerable (Narasimhan and Irwin, 2021)

Gymnosporia (Wight & Arn.) Hook.f.

1a. Leaves and flowers bearing on spines .... **G. ovata**

1b. Leaves and flowers not bearing on spines 2

2a. Branchlets zigzag; stigma capitate; aril not fleshy .... **G. wallichiana**

2b. Branchlets not zigzag; stigma not capitate; aril fleshy 3

3a. Aril of the seed thin, memberanous 4

3b. Aril of the seed fleshy, enveloped the base of the seed 5

4a. Leaves obovate, minutely serrate, grey when dry; capsule globose .... **G. emarginata**

4b. Leaves ovate-lanceoalte, crenate-serrate, reddish when dry; capsule obovaoid .... **G. rufa**

5a. Flowers large in panicles; leaves cunate at base .... **G. heyneana**

5b. Flwoers smallin axillary cymes or fascicle; leaves obtuse at base 6

6a. Leaves tip rounded; capsule 6-lobed ....**G. puberula**

6b. Leaves top acuminate; capsule 3-lobed ....**G. rothiana**

Gymnosporia emarginata (Willd.) Thw., Enum. Pl. Zeyl.: 409. 1864. Synonyms: *Maytenus emarginata* (Willd.) Ding Hou, *Celastrus emarginatus* Willd., *Celastrus montanus* Roth, *Gymnosporia montana* (Roth) Benth., *Maytenus senegalensis* (Lam.) Excell., *Celastrus senegalensis* Lam., *Catha montana* G. Don Vern.: *Halumanike* (Kan.); *Kattanji, Mulmaram* (Tam.).

Small unarmed tree up to 5 m tall; branches terete, bark smooth. Leaves obovate-spathulate, 3.5-14.5 x 2-9 cm, apex rounded or emarginate, base attenuate, chartaceous to coriaceous, margin crenate to entire; petioles 2-12 mm long; nerves 5-6 pairs. Flowers in axillary fascicles; bracts deltoid, fimbriate. Sepals lobed. Petals oblong or ovate-oblong, 2-3 x 1 mm. Stamens 2-3 mm long, inserted beneath the margin of the disc. Ovary partially immersed, 3-loculed; stigmas 3, distinct. Disc fleshy, rounded. Capsule obovoid, 10-12 x 8-9 mm. Seeds ellipsoid, 3x 2 mm, arillate. Fl. & Fr.: September – February.

Western Ghats: Common in deciduous forests of Gujarat, Maharashtra, Karnataka, Kerala and Tamil Nadu. India: Andhra Pradesh, Uttar Pradesh. Odisha, World: Sri Lanka, Afghanistan, China, Malaysia, Indonesia, Australia and Africa.

IUCN: NE (IUCN, 2024).

Gymnosporia heyneana (Roth) M.A. Lawson, J.D. Hooker, Fl. Brit. India 1: 620. 1875. Synonyms: *Maytenus heyneana* (Roth) D.C.S. Raju & Babu *Celastrus heyneanus* Roth. Vern.: *Nandunari* (Tam.).

Small throny tree to 3 m tall; spines terminating short shoots, also axillary. Leaves alternate, ovate to lanceolate, to 7 x 4 cm, base obtuse, margin serrate, apex acute, retuse; petiole to 1 cm long. Cymes to 5 cm long; peduncle 1 cm. Flowers 3 mm wide; pedicel to 4 mm. Sepals 5-lobed, lobes ovate. Petals 5, cream, oblong to obovate, 2 mm. Disc lobed. Stamens 5. Ovary immersed in disc; ovules 2 per cell; stigma 3-fid. Capsule obovoid, to 1 cm diam. Seeds 2 per cell; aril fleshy. Fl. & Fr.: January – April.

Western Ghats: In deciduous forests of Maharashtra, Karnataka, Kerala and Tamil Nadu (Endemic). India: Andhra Pradesh.

IUCN: NE (IUCN, 2024).

Gymnosporia ovata (Wall. ex Wight & Arn.) M.A. Lawson, J.D. Hooker, Fl. Brit. India 1: 620. 1875. Synonyms: *Maytenus ovata* (Wall. ex Wight & Arn.) Loes, *Cleastrus ovatus* Wall. ex Wight & Arn.

Small trees to 5 m tall; branchlets with axillary spines, mostly confined to basal portion. Leaves broadly ovate, 8 x 6 cm, obtuse or emarginate, base cuneate, crenate, thinly coriaceous, nerves indistinct; petiole to 1 cm. Cymes axillary, clustered, 2.5 cm long; bracts ovate, 1 mm long; pedicels slender, 3 mm long. Sepals 1 mm long, ovate. Petals 3.5 mm long, obovate, spreading. Stamens 0.8 mm long, inserted on disc. Ovary 1.5 mm, ovoid; style 0.7 mm, stout; stigma 3-fid, recurved; ovary 3-celled, ovules 2 in each cell. Capsule obovoid, 1.5 cm long, dehiscent, 3-lobed at apex. Fl. & Fr.: December – June.

Western Ghats: Scrub to deciduous forests of Maharashtra, Karnataka, Kerala and Tamil Nadu (Endemic).

IUCN: NE (IUCN, 2024).

Gymnosporia puberula M.A. Lawson, J.D.Hooker, Fl. Brit. India 1: 619. 1875. Synonyms: *Maytenus puberula* (M.a. Lawson) Loes., *Gymnosporia konkanensis* Talbot

Small unarmed trees to 5 m tall; branchlets terete, smooth, glabrous. Leaves obovate, 5-6.5 x 2.5-4 cm, apex shortly acuminate to acute, base tapeing, margin serrate-crenate, membranous, glabrous above, hispid near nerves beneath; veins conspicuous, reticulated; petioles 8 mm long. Inflorescence a dichotomous cyme, axillary or at the end of the branchlets, 3-7 cm long; bracts lanceolate. Calyx deeply lobed; lobes suborbicular. Petals, obtuse, elliptic. Capsule obovoid, 1 cm long. Fl. & Fr.: July – September.

Western Ghats: Mid elevation forests above 700 m in Maharashtra and Karnataka (Endemic).

IUCN: NE (IUCN, 2024).

Gymnosporia rothiana (Walp.) M.A. Lawson, J.D.Hooker, Fl. Brit. India 1: 620. 1875. Synonyms: *Maytenus rothiana* (Walp.) Lobr.-Callen, *Catha rothiana* Walp., *Maytenus wightiana* Babu, *Celastrus rothianus* Wight & Arn., *Celastrus serrulatus* Roth, *Gymnosporia wightiana* (Babu) R.S. Rao. Vern.: *Yenkli* (Kan.).

Small trees to 3 m tall; branches terete. Leaves alternate, simple, broadly ovate, 10.5–15.5 x 6–10 cm, membranous, green above, pale beneath, acute at base, margins serrate, acute at apex; midrib greenish, turning brown later; lateral veins 8–12 pairs; petioles 0.8–2.2 cm long, canaliculate above, young petioles covered with whitish powder withering when matured. Inflorescence axillary, fascicled cymes to 1 cm long; peduncles 3–4 mm long. Flowers pale green, 3–5 mm diam; pedicel 3–10 mm long; bracteole 0.5 × 0.3 mm, tridentate, entire at margin; bract 0.4–0.5 mm long, reddish-brown, sub-deltoid, minutely ciliate at margins. Sepals 5, imbricate, greenish, 5 mm long, deltoid, ciliate at margins. Petals 5, 4–6 mm long, imbricate, oblong, obtuse, tip entire, inserted below the yellow disc. Stamens 5, inserted below the disc, alternate to petals; filaments 7–8 mm long; anther ovate, 2 mm long; Disc 5-lobed. Ovary 3-loculed, sunk in the fleshy disc; ovules 2 in each locule; style obscure, 0.5–1.0 mm long; stigma indistinct. Capsules loculicidal, 3-lobed, obtriangular, 12 mm long; locules 3 seed1 in only onelocule. Seeds ovoid, glossy, 10 x 6 mm, pale white in fresh specimens; aril white, fleshy rim at base of seed. Fl. & Fr.: April – November.

Western Ghats: Semi-evergreen forests of Maharashtra, Goa, Karnataka and Kerala (Endemic).

IUCN: NE (IUCN, 2024).

Gymnosporia rufa (Wall.) Hook.f., Cat. Pl. Kew Herb. Griffith, Falconer & Helfer: 17. 1865. Synonyms: *Maytenus rufa* (Wall.) Hara, *Celastrus rufa* Wall., *Gymnosporia rufa* var. *latifolia* Haines, *Maytenus rufa* var. *latifolia* (Haines) H.B.Naithani & S.Biswas

Small trees to 8 m tall; often branchlets with spines. Leaves oblong-lanceolate to ellipticlanceolate, 5-13 x 1-4 cm, apex acuminate, base narrowed, sub-coriaceous, glabrous, margin serrate; nerves 7-10 pairs. Cymes axillary or below the leaves; peduncles up to 2 cm long. Calyx 5-toothed, triangular. Petals suborbicular. Capsule turbinate, 8 mm in diameter, 1-3-loculed, each locule with one seed. Seeds black, arillate. Fl. & Fr.: March – December.

Western Ghats: Deciduous forests in Tamil Nadu. India: Andhra Pradesh, Assam, Himachal Pradesh, Manipur, Meghalaya, Odisha and Uttar Pradesh. World: Bhutan.

IUCN: NE (IUCN, 2024).

Gymnosporia wallichiana (Spreng.) M.A.Lawson, J.D.Hooker, Fl. Brit. India 1: 621. 1875. Synonyms: *Maytenus wallichiana* (Spreng.) D.C.S. Raju & Babu, *Celastrus wallichianus* Spreng., *Maytenus wallichii* (G.Don) Bennet & Sahni, *Catha wallichii* G.Don

Small tree to 4 m tall. Leaves simple, alternate, obovate, 2.5-4 x 2-3 cm, obtuse or emarginate at apex, cuneate at base, coriceous and glabrous; nerves distinct, reticulate. Flowers in cymes or fascicles, axillary or on spines. Sepals 5, 3-angled. Petals 5, oblong, thin. Stamens 5, interpetaloid, inserted below disc. Disc rounded and fleshy. Ovary within the disc; stigmas capitate. Capsular slightly obcordate, 12 mm long, 3-loculed, 6-seeded; seeds flat, arillate; aril attached only to the base of seed. Fl. & Fr.: June – October.

Western Ghats: Deciduous forests of Karnataka, Kerala and Tamil Nadu. India: West Bengal. World: Sri Lanka, Pakistan and Bangladesh.

IUCN: NE (IUCN, 2024).

Lophopetalum Wight

Lophopetalum wightianum Arn., Ann. Nat. Hist. 3: 151. 1839. Synonyms: *Euonymus fimbriatus* Baill. ex Laness., *Lophopetalum fimbriatum* Wight, *Lophopetalum winkleri* Loes. Vern.: *Balpale* (Kan.); *Venkotta* (Mal.); *Venkottai* (Tam.).

Evergreen trees to 30 m tall; bark yellowish-brown, mottled with white, smooth, brittle; branchlets terete, brown. Leaves simple, opposite or subopposite, elliptic, elliptic-oblong, oblong, ovate or ovate-oblong, 7.5-25 x 3.7-10 cm, base obtuse or round, apex obtuse or acute, margin entire, coriaceous, glabrous; lateral nerves 6-12 pairs, intercostae reticulate, prominent; petiole 10-25 mm long, stout, grooved above, glabrous. Flowers to 15-18 mm across, pinkish to dull red, in lax axillary or terminal dichasial cyme. Sepals 5, cupular, broadly flattened at base, 0.5-0.6 cm across; lobes 5, broad, obtuse, very short. Petals 5, 0.5-0.6 cm long, continuous with the disc, orbicular, uneven, clawed, persistent, spreading. Disc large, flat, 5 lobed. Stamens 5, inserted on the disc; filaments subulate; anthers oblong. Ovary small, immersed in the disc and continuous with it, triquetrous, 3-celled; ovules 4 or more in each cell; style short, stigma capitate. Capsule trigonous, 8-10 x 2.5-3.5 cm, 3-angled, 3 valved, brown, smooth, pointed at both ends, loculicidal. Seeds thin, surrounded by long linear wing, 5 x 1.2 cm, brown, compressed, arillate. Fl. & Fr.: December – May.

Western Ghats: Evergreen forests of Maharashtra, Goa, Karnataka, Kerala and Tamil Nadu. India: Arunachal Pradesh, Assam and Meghalaya. World: Bangladesh, Myanmar, Singapore, Indonesia and Malaysia.

IUCN: LC (IUCN, 2024).

Uses: Wood is useful in carpentary work, for making tea cups.

Microtropis Wall. ex Meisn.

1a. Flowers sessile, clustered on axile or fallen leaf scars 2

1b. Flowers in axillary or extra-axillary peduncled cymes..... .**M. microcarpa**

2a. Branchlets yellowish, youngest darker; leaves long attenuate at base .... **M. wallichiana**

2b. Branchlets black or purple; leaves cuneate or rounded at base 3

3a. Flowers small, clusters with a few flowers ....**M. stocksii**

3b. Flowers large, clusters with many flowers 4

4a. Leaves with very thick rugulose, obscurely nerved .... **M. latifolia**

4b. Leaves subcoriaceous, prominently nerved ....**M. ramiflora**

Microtropis latifolia Wight ex M.A. Lawson., J.D. Hooker, Fl. Brit. India 1: 613. 1875.

Small trees to 7 m tall. Leaves simple, subopposite, ovate-elliptic to oblong-lanceolate, 7-15 x 2.5-6.5 cm, acuminate at apex, coriaceous, glabrous, rugulose above; lateral nerves obscure on both surfaces; petioles 6-7 mm long. Flowers sessile, clustered in axils of fallen leaves or solitary, yellowish white. Sepals 5, 5-lobed, membranous at margin above, rugose outside. Petals obovate-cuneate, connate at base forming a ring. Stamens inserted at ring; filaments short. Ovary flask-shaped, furrowed. Capsules ellipsoid, narrow at base, 1.5-2 cm long, 1-seeded. Seed oblong, rugose. Fl. & Fr.: October – July.

Western Ghats: Evergreen forsts at 700 – 2000 m in Maharashtra, Karnataka, Kerala and Tamil Nadu (Endemic).

IUCN: NE (IUCN, 2024).

Microtropis microcarpa Wight, Icon. Pl. Ind. Orient. 3: t.975. 1845. var. **macrocarpa** Synonyms: *Elaeodendron nilghirense* Wall., *Microtropis ovalifolia* Wight

Small trees to 5 m tall. Leaves opposite, elliptic or oblong-lanceolate, 3.5-5 x 2-2.5 cm, attenuate at base, coriaceous, glabrous above, pale beneath; petioles 5 mm long. Flowers in cymes; peduncles shorter than leaves. Sepals 5, 5-lobed; imbricate, thin and fleshy below, membranous above, persistent in fruit. Petals obovate, clawed, nerved. Stamens inserted at base of petals; filaments short; anthers large. Ovary ovoid-conic, ribbed; styles short. Capsule ellipsoid, pointed and rugose. Seed 1, shining. Fl. & Fr.: February – May.

Western Ghats: Evergreen forests of Maharashtra, Kerala and Tamil Nadu (Endemic).

IUCN: NE (IUCN, 2024).

Key to the varieties

1a. Cymes with lax flowers, 4-8 cm long; capsule oblong with pointed apex ....var. **microcarpa**

1b. Cymes with dense flowers, 1-2 cm long; capsule ellipsoid with pointed ends .... var. **densiflora**

Microtropis microcarpa Wight var. **densiflora** (Wight) Meisn., Proc. Amer. Acad. Arts 73: 287. 1940. Synonyms: *Microtropis denisflora* Wight, *Paracelastrus densiflorus* (Wight) F.N.Williams

Fl. & Fr.: February – May.

Western Ghats: Tamil Nadu (Endemic).

IUCN: NE (IUCN, 2024).

Microtropis ramiflora Wight, Icon. Pl. Ind. Orient. 3: t. 977. 1845. Synonym: *Paracelastrus ramiflorus* (Wight) F.N.Williams

Evergreen trees to 25 m tall; bark greenish-yellow, smooth with bluish-green blotchesbranchlets black purple, rough. Leaves simple, opposite, decussate, elliptic, elliptic-ovate or obovate, base round or subcordate, 5-13 x 2.5-6.5 cm, apex rounded or slightly emarginate, margin revolute, coriaceous, glabrous; lateral nerves 5-7 pairs, intercostae reticulate, prominent; petiole 3-10 mm long, stout, glabrous. Flowers yellowish, sessile, 6-10 in axillary condensed clusters. Sepals 5, ovate to suborbicular, unequal, margin laciniate, imbricate. Petals 5, fleshy, joined at the base with disc. Disc annular. Stamens 5, inserted on the disc; filaments subulate, basally connate. Ovary conical, seated on the disc, 2-3-celled, ovules 2 in each cell; style short, stigma 2-3 lobed. Capsule ellipsoid, 2 x 0.8 cm, woody, apiculate, grey to rusty, 2-valved, surrounded at the base by persistent calyx. Seed one, 4 x 6 mm, oblong, stipitate, erect. Fl. & Fr.: December – May.

Western Ghats: Evergreen forests above 1000 m in Kerala and Tamil Nadu. World: Sri Lanka and Sumatra.

IUCN: NE (IUCN, 2024).

Microtropis stocksii Gamble, Bull. Misc. Inform. Kew 1916: 132. 1916.

Small tree to 4 m tall; branchlets reddish-brown. Leaves elliptic, 7-10 x 3-5 cm, apex obtusely acuminate, margins folding, green or yellow when dried; petiole to 8 mm long, yellow. Flowers 3 mm across, few in axillary clusters. Sepals ovate, obtuse. Petals 2 x 1 mm, ovate, obtuse, white, glabrous. Capsule oblong, to 13 x 8 mm, brown, aril red. Fl. & Fr.: February – Sepember.

Western Ghats: Evergreen forests of Maharashtra, Karnataka, Kerala and Tamil Nadu (Endemic).

IUCN: NE (IUCN, 2024).

Microtropis wallichiana Wight ex Thw., Enum. Pl. Zeyl.: 71. 1858. Synonyms: *Paracelastrus wallichianus* (Wight ex Thwaites) F.N.Williams, *Microtropis suborbiculata* Merr. & F.L.Freeman

Trees to 15 m tall; branchlets yellowish, youngest darker. Leaves simple, opposite, decussate, ovate, elliptic, oblanceolate or elliptic-oblanceolate, 5-10.5 x 1.5-5 cm, glabrous, coriaceous; lateral nerves 5 or 6 pairs, intercostae reticulate, slender, prominent; petiole 7-10 mm, slender, grooved above, glabrous. Flowers pale yellow, 2 mm across, sessile, in axillary fascicles. Sepals 5, subreniform, margin fringed, imbricate. Petals 5, obovate-oblong, obtuse, connate at base. Stamens 5, inserted on the upper edge of the disc. Disc annular, slightly lobed. Ovary ovoid, seated on the disc, 2-celled, ovules 2 in each cell; style short; stigma minutely lobed. Capsule 2-valved, 1-seeded, surrounded at the base by the persistent calyx, tapering at the base, round at apex with persistent style. Seeds erect, oblong. Fl. & Fr.: February – September.

Western Ghats: Evergreen forests of Karnataka, Kerala and Tamil Nadu. World: Sri Lanka, Indonesia and Malaysia.

IUCN: NE (IUCN, 2024).

Pleurostylia Wight &Arn.

Pleurostylia opposita (Wall.) Alston., H. Trimen, Handb. Fl. Ceylon 6 (Suppl.): 48. 1931. Synonyms: *Celastrus opposita* Wall., *Pleurostylia wightii* Wight &Arn., *Euonymus godaverensis* Haines, *Pleurostylia wightii* Wight & Arn. Vern.: *Chirupiyari, Karuvali* (Tam.).

Trees to 15 m tall; branchlets subterete or obscurely 4-angled. Leaves ovate to obovate-oblong, rarely suborbicular, 3-8 x 1.5-5.2 cm, papery, base cuneate, margin entire, sometimes slightly recurved, apex obtuse, acute, rarely acuminate; veins 5 or 6 pairs, thin and obvious; petiole short, 2-5 mm. Cymes of 1 or 2 branches, few flowered; peduncle short, 2-3 mm; pedicel 1-2 mm. Flowers green. Sepals 5-lobed, lobes rounded or subreniform, 0.5 x 0.75-1 mm. Petals elliptic or broadly ovate, 1.25-1.5 x 1 mm, reflexed at anthesis, obtuse or slightly acute. Disc margin slightly crenate. Stamens attached just abaxially on disk margin; filament subulate, fleshy, slightly flattened; anther slightly ovoid. Ovary adnate to disc at base; style very short; stigma capitate. Nut ellipsoid or sometimes slightly obovoid, 5-7 x 4-5 mm. Seed sustained by persistent floral parts. Fl. & Fr.: July – February.

Western Ghats: Evergreen forests above 900 m in Maharashtra, Karntaka, Kerala and Tamil Nadu. World: Sri Lanka, Mauritius and Madagascar.

IUCN: LC (IUCN, 2024).

Planted:

Catha edulis (Vahl) Forssk. Ex. Endl.; Exotic: Native to Africa

Euonymus japonicus Thunb.; Exotic: Native to Japan, Korea

CENTROPLACACEAE Doweld & Reveal

Bhesa Buch.-Ham. ex Arn.

1a. Ripened fruit brown; seeds minutely covered the basal aril ......**B. indica**

1b. Ripened fruit pinkihs-red; seeds half way covered the basal aril ....**B. paniculata**

Bhesa indica (Bedd.) Ding Hou, Blumea, Suppl. 4: 152. 1958. Synonyms: *Trochisandra indica* Bedd., *Kurrimia indica* (Bedd.) Gamble Vern.: *Kadalpala, Perumgarandam* (Tam.).

Evergreen trees to 30 m tall; bark 5-6 mm thick, grey mottled with greenish-white; brittle. Leaves simple, alternate, stipule large, deciduos; blade 7-22 x 1.6-12 cm, broadly elliptic-ovate, oblong, ovate to lanceolate, base obtuse or subtruncate, apex acute, mucronate, margin entire glabrous, coriaceous; lateral nerves 15-20 pairs, parallel, intercostae scalariform; petiole 16-60 mm, stout, swollen at base and tip, glabrous. Flowers in terminal panicles, 6-8 mm across, pale yellow. Sepals 5, cupular, deeply 5-fid, lobes imbricate. Petals 5, deflexed. Disc fleshy, 5-lobed. Stamens 5, inserted below the margin of disc, filaments subulate. Ovary ovoid, seated on the disc, 2-celled, ovules 2 in each cell; styles two, free, stigma small, capitate. Capsule two-lobed, 3-3.7 cm long, two valved, lobes flat, divergent, reddish-brown. Seeds 1-3, fleshy with aril. Fl. & Fr.: March – May.

Wstern Ghats: Common in evergreen sholas above 1200 m in Kerala and Tamil Nadu. World: Malesia, Myanmar and Thailand.

IUCN: NE (IUCN, 2024).

Bhesa paniculata Arn., Edinb. Phil. Journ. 16: 315. 1834. Synonyms: *Kurrimia paniculata* (Arn.) Wall. ex M.A. Lawson, *Pyrospermum calophyllum* Miq.

Trees to 60 m tall; bark grey-brown, smooth; branchlets terete, glabrus with prominent leaf scars. Leaves elliptic to oblong or obovate, 5.5-20 x 2.2-12 cm, cuneate at base or rounded, apex pointed or blunt; lateral veins 15-20 pairs, prominently raised below, intercostal scalariform, faint above; petioles to 9.5 cm long, swollen at both ends. Inflorescnce panicles, 10-37 cm long ro racemes to 20 cm long on leafy twigs. Flowers yellowish-cram or dark purplish-red, 5 mm wide; pedicel to 2-3 cm long. Sepals deeply 5-lobed, lobes hairy outside. Petals 5, oblong or ovate, hairy inside. Stamens 1.5-2 mm long; anthers triangular. Ovary ellipsoid, styles free. Fruits 2-lobed, flattened, heart-shaped, 1-2 x 0.8-1.3 cm, yellow then rose-red on clusters. Seeds 2-4, lower half covered with fleshy pink aril. Fl. & Fr.: November – January.

Westren Ghats: Anamalai hills of Tamil Nadu. World: Malaysia, Philippines, Myanmar, Sumatera and Thailand.

IUCN: NE (IUCN, 2024).

CHRYSOBALANACEAE R.Br.

Atuna Raf.

1a. Leaves oblong to elliptic-lanceolate, 5–7 cm broad; racemes lax, stout; petals lanceolate, acuminate at apex **A. indica**

1b. Leaves lanceolate, less than 3.5 cm broad; racemes dense, slender; petals broadly ovate, rounded at apex **A. travancorica**

Atuna indica (Bedd.) Kostrem., Reinwardtia 7: 422. 1969. Synonyms: *Cyclandrophora indica* (Bedd.) Prance ex Kosterm., *Entosiphon indicum* Bedd., *Entosiphon indicus* Bedd. *Parinari indica* (Bedd.) Bedd. '*Parniarium indicum*'

Evergreen trees to 15 m tall; bark brown, smooth, thin. Leaves simple, alternate, stipulate, stipules lanceolate; blade 17-21 x 5.5-7.5 cm, oblong, elliptic-lanceolate, elliptic-oblong or elliptic-ovate, base acute, apex acuminate or obtusely acuminate, margin entire, glabrous, chartaceous; lateral nerves 12-18 pairs, pinnate, prominent, intercostae reticulate, prominent; petiole 6-12 mm long, stout, glabrous. Flowers white, in axillary or terminal corymbose racemes; bracts brown, hairy. Sepals 5, tubular, tube funnel-shaped, pubescent; lobes 5, imbricate. Petals 5, white, inserted to the mouth of calyx tube. Stamens many; filaments basally connate, anthers small. Ovary adnate to the side of calyx tube, 2-celled, ovules 2, erect; style filiform, basal; stigma truncate. Drupe ovoid, 3.5-4 x 2.5-3 cm, smooth. Fl. & Fr.: November – February.

Western Ghats: Evergreen forest above 1500 m in Kerala and Tamil Nadu (Endemic).

IUCN: EN- B1+2c. (IUCN, 2024).

Atuna travancorica (Bedd.) Kostrem., Reinwardtia 7: 422. 1969. Synonyms: *Parinari travancorica* Bedd, *Cyclandrophora travancorica* (Bedd.) Prance ex Kosterm., *Ferolia travancorica* (Bedd.) Kuntze Vern.: *Kallan Kai maram* (Tam.).

Evergreen trees to 25 m tall; bark greyish-brown mottled with white, smooth, thin; branchlets pubescent. Leaves simple, alternate; stipules free, subulate; blade elliptic, elliptic-oblong or elliptic-lanceolate, 4-16.5 x 1.7-4.8 cm, base acute, apex acuminate or acute, margin entire, glabrous and glossy above except silky pubescent midrib, coriaceous; lateral nerves 11-17 pairs, intercostae reticulate, prominent; petiole 2-6 mm, stout, grooved above, silky tomentose when young and glabrous at maturity. Flowers pale lavender, 1.5 cm across, usually in axillary silky-villous, slender drooping racemes; bracts ovate-lanceolate, villous. Sepals 5, tube funnel-shaped, lobes ovate-cuspidate, villous outside. Petals 5, 7 mm long, inserted on the mouth of calyx tube, deciduous. Stamens numerous, purple; filaments connate at base; anthers small. Ovary superior, adnate to the side of the calyx tube, 2-celled, ovules 2, erect; style filiform, basal; stigma truncate. Drupe suborbicular, 2-3.5 x 2-3 cm, compressed brown, smooth. Fl. & Fr.: November – June.

Western Ghats: Evergreen forests above 1800 m in Kerala and Tamil Nadu (Endemic).

IUCN: EN- B1+2c. (IUCN, 2024).

Planted:

Chrysobalanus icaco L. Exotic: Native to America, Africa

Parinari excelsa Sabine, Exotic: Native to South America, Africa

CLUSIACEAE Lindl.

Garcinia L.

1a. Flowers pentamerous 2

1b. Flowers tetramerous 6

2a. Ovary 6–8-locular; berries top-shaped, irregularly wrinkled .... **G. pushpangadaniana**

2b. Ovary usually 3-5-locular; berries globose, ovoid or discoid, smooth or irregularly ridged 3

3a. Staminodes free or bundles of 2–3; berries discoid, irregularly ridged ..4

3b. Staminodes 5-phalangiate; berries globose, obliquely globose or ovoid, smooth 5

4a. Leaves linear-oblong, lateral nerves 15-20 pairs; male flowers 4-10 in fascicle .... **G. xanthochymus**

4b Leaves ovate-elliptic, lateral nerves 25 to 40 pairs; male flowers 5-15 in fascicle. .... **G. gamblei**

5a. Staminate fowers always fascicled; stigma 3-lobed, berries globose, smooth .... **G. talbotii**

5b. Staminate fowers spicate; stigma 5-lobed; berries depressed globose, slightly ridged .... **G. spicata**

6a. Flowers sessile or subsessile, pedicels to 0.5 cm long 7

6b. Flowers pedicellate, pedicels >0.5 cm long .... **G. gummi-gutta**

7a. Flowers always axillary 8

7b. Flowers terminal or sub-terminal 9

8a. Leaves linear-lanceolate, attenuate towards tip .... **G. wightii**

8b. Leaves elliptic-ovate, acute at tip .... **G. cambogioides**

9a. Ovary 2-loculed 10

9b. Ovary many-loculed 11

10a. Leaves oblanceolate, caudat-cuminate at apex; berries laterally compressed; stigma not persistent .... **G. imberti**

10b. Leaves linear-oblong to obovate, berries ovoid-oblong with persistent discoid stigma .... **G. travancorica**

11a. Berry surface rough 12

11b. Berry surface smooth.......... 13

12a. Berries sub-globose or ellipsoid, echinate .... **G. rubro-echinata**

12b. Berries ovoid-conical, 3–5 grooved .... **G. conicarpa**

13a. Leaves elliptic-oblong; sepals yellow; berry yellow .... **G. cowa**

13b. Leaves ovate-oblong; sepals pinkish-orange; berry purplish pink .... **G. indica**

Garcinia cambogioides (Murray) Headland, Mat. Med. Ind., ed. 3: 839. 1837. var. **cambogioides** Synonyms: *Garcinia morella* (Gaertn.) Desr., *Mangostana morella* Gaertn., *Stalagmitis cambogia* Pers. Vern.: *Chigiri, Daramba, Karukkampuli, Pinnarpuli* (Mal.).

Evergreen trees to 18 m tall; bark thick, brownish-grey or brown to dark brown, smooth; exudation dark yellow or orange yellow, sticky; branchlets quadrangular, glabrous. Leaves simple, opposite decussate, elliptic-obovate or elliptic-oblanceolate, 6-16 x 2.5-9 cm, apex obtuse, obtusely acuminate or caudate-acuminate, base acute or cuneate, margin entire, glabrous, coriaceous; lateral nerves 8-16 pairs, intercostae reticulate, faint; petiole 6-15 mm, stout, glabrous, grooved above, very shortly ligulate at base. Flowers polygamodioecious, reddish, sessile. Male flowers: 2-4 in axillary fascicles or on old wood, Sepals 4 orbicular, decussate, outer pairs smaller than the inner, glabrous. Petals 4, little larger than sepals, orbicular, veined, concave. Stamens 10-12, monadelphous, the filaments combined in to a subquadrangular central column, anthers red, orbicular, plurilocular. Pistillode absent. Female flowers: axillary, solitary, larger than male flowers. Staminodes 10-12 in a ring round the ovary, connate at the base. Ovary superior, greenish, globose, smooth, 4-celled, ovule one in each cell; stigma peltate, sessile, irregularly lobed, persistent, margin dentate. Berry subglobose or globose, 1.5-2.5 cm long, smooth, yellowish or light pink, surrounded at the base by persistent sepals, crowned by flat tuberculate round stigmas, pulp sweet. Seeds 2-4, kidney shaped, laterally compressed, dark brown. Fl. & Fr.: January – May.

Western Ghats: River banks of evergreen forests above 1100 m in Goa, Karnataka, Kerala and Tamil Nadu. World: Sri Lanka, Bangladesh, Myanmar, Thailand, Malaysia and Singapore.

IUCN: NE (IUCN, 2024).

Garcinia cowa Roxb. ex Choisy., Mem. Nouv. Gen. Guttif.: 17. 1823. Synonyms: *Stalagmitis cowa* (Roxb. ex Choisy) G.Don, *Garcinia cambogia* Roxb., *Garcinia roxburghii* Wight Vern.: *Kodam puli* (Tam.).

Evergreen trees to 15 m tall; bark smooth, surface greyish-brown; exudation yellow, sticky, scanty; branchlets quadrangular, drooping. Leaves simple, opposite decussate, elliptic-oblong, oblanceolate or broadly lanceolate, 8-17 x 2.5-7 cm, base acute, attenuate or cuneate, apex acute or obtuse, margin entire, glabrous, thickly coriaceous; lateral nerves 15-21 pairs, looped along the margin forming intramarginal nerve, intercostae reticulate, obscure; petiole 8-13 mm long, stout, glabrous. Flowers dioecious, small, yellow. Male flowers: to 1 cm across, 3-8 in axillary or terminal fascicles; pedicels about 6 mm long. Sepals 4, 4-6 mm long, unequal, broadly ovate, fleshy, yellow. Petals 4, 8-10 mm long, oblong, yellow flushed with pink or red. Stamens numerous on a convex fleshy receptacle; anthers oblong; filaments short; rudimentary pistil absent. Female flowers: upto 1.5 cm across, 2-5 in terminal fascicles, longer than male flowers, yellow; pedicel short. Staminodes in a ring of 4 bundles of 3-8 around the ovary; filaments unequal. Ovary superior, subglobose, 6-8 locular, ovules one in each cell; stigma sessile, flat, deeply divided into 6-8, papillose, wedge-shaped rays. Berry depressed globose, 2-4 cm across, with 4-8 vertical grooves, smooth, yellow, ridid, beaked. Seeds 4-8, oblong with a soft aril. Fl. & Fr.: January – May.

Western Ghats: Evergreen forest at 1200 m in Kerala and Tamil Nadu. India: Assam, Bihar, Meghalaya, Nagaland, Odisha, Sikkim, Tripura and West Bengal. World: Bangladesh, China, Myanmar, Laos, Thailand, Malaysia and Vietnam.

IUCN: LC (IUNC, 2024).

Garcinia conicarpa Wight., Icon. Pl. Ind. Orient. 1: t. 121. 1839. Synonyms: *Garcinia gummi-gutta* (L.) Robson var. *conicarpa* (Wight) N.P.Singh, *Carcinia cambogia* var. *conicarpa* (Wight) T.Anderson, *Garcinia darwiniana* Kesh.Murty *et al.*

Evergreen trees to 20 m tall; bark grayish, smooth, dark brown; branchlets drooping, terete, glabrous; latex usually resinous, yellow. Leaves simple, opposite or rarely ternate, linear lanceolate-oblong, 8-15 x 3-5.5, usually broadest in the middle, base cuneate, margins entire, apex shallow acute to obtusely acuminate, dark glossy green, glabrous on both sides, with translucent glands; lateral veins indefinite, irregular with oblique parallel short veins between, reticulate; petiole slender with raised margins, 0.5-1.2 cm long. Inflorescence axillary solitary, cymes or fascicles. Flowers pseudobisexual. Male flowers: axillary, 3-5 fascicled, pale yellow, greenish or white, 10-20 mm long; pedicels to 15 mm long, thick towards tip. Sepals 4, imbricate, ovate-obovate, unequal, decussate in pairs, inner pair 7-8 mm long, outer pair 5-6 mm long. Petals 4, imbricate, obovate-oblong, alternating with sepals. Stamens 20 or more, inserted on short convex torus; rudimentary pistil absent; filaments to 0.5 mm long; anthers basifixed, 2 loculed, apex obtuse, to 0.5 mm long, dehiscing vertically. Bisexual flowers: in axillary fascicles, bigger than male flowers, globose. Stamens 20 or more, filaments arranged in a ring, unequal, connate at the base, in unequal bundles; anthers 2 locular, few sterile, otherwise fertile. Ovary globose-ovoid, 4 locular; ovules lateral or erect, connate at the base; style short; stigma rays 4. Berry globose-conical, 5-7 cm across, sulcate or smooth, yellow or red, 4 grooved, with angular furrows. Seeds 4, usually equal to number of grooves, oblong ovoid, to 3 cm long, compressed, pale brown, covered with juicy cream white or red pulp. Fl. & Fr.: March – January.

Western Ghats: Rare in evergreen forests of Maharashtra, Karnataka, Kerala and Tamil Nadu (Endemic).

IUCN: NE (IUCN, 2024).

Garcinia gamblei Shameer, Sabu et Mohanan, in Phytotaxa 297 (1): 71-76, 2017.

Dioecious evergreen trees to 15 m tall; bark grayish white with white mottle, old ones flaked; exudation milky; branchlets tetrangular. Leaves opposite, ovate or broadly elliptic, 20–28 x 10–12 cm, obtuse–cuneate at base, acute at apex, margin sub revolute, glabrous; midribs raised on both sides, secondary veins conspicuous, 25–40 pairs; petioles 1–1.5 cm long, stout, pale greenish, adaxially ligulate at the base. Male flowers 5–15, sessile, fascicled, axillary or on axils of fallen leaves. Sepals 5, free, 5 x 3 mm, imbricate, orbicular, convex; margin membraneous and ciliate. Petals 5, free, 8×6 mm, imbricate, pale greenish, obovate; margin membraneous, glabrous. Stamen 5-phalangiate, glabrous; phalanges antipetalus, alternating with 5 intra staminal discs. 7–17 in each phalanges; filament 0.5 mm long, anther lobes brown; rudimentary pistil absent. Female flowers usually solitary, rarely 2–4 in fascicles, axillary or on axills of fallen leaves, sub-sessile. Sepals 5, free, unequal, inner ones 6×5 mm, outer 3×2 mm, imbricate, orbicular, convex; margin ciliate. Petals 5, free, 0.9–1 x 0.8 cm, imbricate, orbicular obovate; margin membraneous.

Staminodes in loose bundles, filaments free or 2–3-united. Ovary globose, 5 mm diam, 5-locular; style sessile; stigma 5-lobed, peltafid. Berry globose, 8 x 7 cm, irregularly ridged on surface, depressed on distal end. Seeds1–3, rarely 5, oblong, 2.3 x 1.7 cm. Fl. & Fr.: December – August.

Western Ghats: Rare in Ponmudi hills of Kerala (Endemic).

IUCN: DD (IUCN, 2024).

Garcinia gummi-gutta (L.) Robson, Brittonia 20: 103. 1968. var. **gummi-gutta** Synonyms: *Cambogia gummi-gutta* L. *Garcinia cambogia* (Gaertn.) Desr., *Mangostana cambogia* Gaertn. Vern.: *Gorakkapuli, Kodampuli, Kodapuli, Pinaru, Penanga* (Mal.); *Kodukkapuli, Kodam puli* (Tam.).

Evergreen trees to 20 m tall; bark black, rough, thinly scaly; exudation pale yellow, sticky; branchlets terete, glabrous. Leaves simple, opposite decussate, elliptic or elliptic-oblanceolate or lanceolate, 7-15 x 2-8 cm, base cuneate, or attenuate, apex acute or shortly obtusely acuminate, margin entire, glabrous, coriaceous; lateral 10-16 pairs, arched towards the margin forming intramarginal nerve, intercostae reticulate, faint; petiole 5-20 mm long, stout, channelled above, ligulate at the base small, glabrous. Flowers polygamodioecious, yellowish-white. Male flowers: 3-5 in short axillary fascicles; pedicels 7-15 mm long, thickened towards tip, often reflexed. Sepals 4, fleshy with membranous margin, coriaceous, ovate or obovate, outer 2 smaller than inner. Petals 4, twice as long as the setals, concave with narrow membranous margin. Stamens 10-20 or more, inserted on a short convex receptacle, monadelphous, forming a globular head; anthers 0.5 mm long. Pistillode absent or minute. Female flowers: 1-3 in terminal and axillary fascicles, larger than male flowers; pedicels 4-6 mm long. Sepals and petals as in male. Staminodes 10-20, filaments unequal, connate at base in a ring around the ovary or grouped in unequal bundles; anthers bilocular. Ovary globular, 8-10 locular, ovule one in each locule; stigmatic rays 8-10, margin tuberculate. Berry ovoid to subglobose, 4-5 cm diam, yellow or red, 8-10 grooved, fleshy. Seeds 6-8, ovoid, compressed, pale brown, veined; aril succulent, white or red. Fl. & Fr.: January – May.

Western Ghats: Common in evergreen forests above 1200 m in Maharashtra, Goa, Karnataka, Kerala and Tamil Nadu (Endemic). World: Sri Lanka.

IUCN: LC (IUCN, 2024).

Uses: Dry fruit pulp is used as a substitute for tamarind to make fish curry, mostly in Kerala an southern Tamil Nadu.

Key to the varieties

1a. Outer seplas smaller than the inner; berries ovoid to subglobose .... var. **gummi-guttta**

1b. Outer sepals subequal to the inner; berries globose to grooved................ .... var. **papilla**

Garcinia gummi-gutta (L.) Robson var. **papilla** (Wight) N.P. Singh. Fl. India 3: 111. 1993. Synonyms: *Garcinia papilla* Wight, *Garcinia cambogia* var. *papilla* (Wight) T. Anderson Vern.: *Kodampuli* (Mal.).

Fl. & Fr.: Throught the year.

Western Ghats: Evergreen forests of Karnataka, Kerala and Tamil Nadu (Endemic).

IUCN: LC (IUCN, 2024).

Uses: The fruit is used for making pickle. Ripen fruit is very acidic.

Garcinia imbertii Bourd., J. Bombay Nat. Hist. Soc. 12: 349. 1899. Vern.: *Manja-kanji* (Tam.).

Evergreen trees to 15 m high; bark brown mottled with white, rough, vertically fissured; exudation sweet-scented; branchlets quadrangular, glabrous. Leaves simple, opposite decussate, elliptic or oblanceolate, 4-10 x 1.5-4 cm, base narrow or cuneate, apex acuminate or obtusely acuminate, margin entire, glabrous, shining, coriaceous; lateral nerves 15-25, looped near the margin forming intramarginal nerve, intercostae obscure; petiole 2-5 mm long, grooved above. Flowers yellow, polygamodioecious, sessile. Male flowers: 3-9 in terminal fascicles, 4-5 mm across. Sepals 4, in 2 rows, orbicular or suborbicular, concave, outer 2 smaller, 1.5-2 mm long, inner ones larger, 2.5-3 mm long, greenish-yellow. Petals 4, 3-4 mm long, imbricate, broadly ovate or orbicular, concave, yellow. Stamens 16-20, monadelphous in a central mass round the rudimentary pistil, 1-2 mm in diam, stalk 0.5-1 mm long. Female flowers: usually solitary, terminal, rarely in pairs. Sepals and peals as in male flowers. Ovary superior, globose, 1.5-2 mm across, 2-loculed, ovule 1 in each locule; stigma sessile, capitate, convex. Staminodes many, united and surrounding the ovary in a ring. Berry subglobose, 2.5 x 2.5 cm, green. Seeds 1 or 2, enclosed in a thick leathery covering. Fl. & Fr.: April – September.

Western Ghats: Rare in evergreen shola of Chemunji rage of Agathiyamalai hills in Kerala and Tamil Nadu (Endemic).

IUCN: CR- C2a(i). (IUCN, 2024).

Garcinia indica (Thouars) Choisy., Mem. Nouv. Gen. Guttif.: 17. 1823. Synonyms: *Brindonia indica* Thouars, *Oxycarpus indica* (Thouars) Poir, *Garcinia purpurea* (G.Don) Roxb, Vern.: *Penampuli* (Mal.); *Goa Kodam puli* (Tam.).

Trees to 15 m tall; bark lpale brown, very thin, smooth, rather shiny; young branches subterete, irregularly striate. Leaves simple, opposite decussate, lanceolate or obovate-oblong, 6.5-11 x 1.5-4 cm, base attenuate, apex acute or acuminate, margin entire, glabrous, shining, membranous; lateral nerves 7-18 pairs, parallel, intercostae reticulate; petiole 5-12 mm long, slender, glabrous. Flowers polygamodioecious. Male flowers: 4-8 in axillary and terminal fascicles; pedicels 6 mm long. Sepals 4, yellowish-orange to pinkish-orange, coriaceous, ovate-rotundate; outer ones 3-4.5 mm long, inner ones 4.5-5 mm long. Petals 4, 5-6 mm long, thick. Stamens many, inserted on a hemispheric subquardate torus; filaments short; anthers oblong, truncate; rudimentary pistil absent or a few equalling the stamens. Female flowers: solitary, terminal; pedicels 3 mm long. Sepals and petals as in male flowers; staminodes 10-18, in 4 unequal, 2 to 3 seriate phalanges alternating with petals, 1-3 mm long. Ovary superior, 4-8 locular, subglobose; stigma 4-8 rayed, convex, coronate, rays tuberculate, often 2-seriate. Berry depressed globose, 2.5-4 cm across, 4-8 loculed, purple or wine brown, surrounded by persistent calyx. Seeds 5-8, compressed in acidic pulp. Fl. & Fr.: November – February.

Western Ghats: Evergreen forests of Maharashtra, Karnataka, Kerala and Tamil Nadu (Endemic).

IUCN: VU- A2cd (IUCN, 2024).

Uses: Fruit is edible. Seed yields an edible fat known as kokum butter which is used for manufacturing soap and candle. Wood is useful in pulp industry.

Garcinia rubro-echinata Kostrem., Ceylon J. Sci., Biol. Sci. 12: 128. 1977. Synonyms: *Garcinia echinocarpa* sensu Gamble, non Thw., *Garcinia echinocarpa* var. *monticola* Maheshw Vern.: *Para, Pura* (Mal.); *Kalmanickam, Malamkongu* (Tam.).

Evergreen trees to 20 m tall; bark surface green, mottled with yellow, red and white, smooth, with numerous small lenticels; exudation yellowish-white, sticky; branchlets angular, glabrous. Leaves simple, opposite decussate, obovate, elliptic or oblong-ovate, 8-15 x 3-7.5 cm, base cuneate or acute, apex obtuse or subretuse, margin entire, revolute, glabrous, coriaceous; lateral nerves numerous, parallel, looped near the margin forming intramarginal nerve, intercostae obscure; petiole 10-30 mm, stout, glabrous, ligulate projections at the base prominent, clasping the branch. Flowers polygamodioecious, reddish-green, sessile. Male flowers: 2-12 in axillary or terminal clusters. Sepals 4, red, orbicular, concave. Petals 4, pale green, dark green towards tip, almost twice as long as sepals. Stamens 12-40, monadelphous, staminal column short, very thick, anthers red; pistillode absent. Female flowers: solitary, terminal, slightly larger than male. Staminodes uniseriate, connate in to a ring at base. Ovary superior, globose, covered with imbricate fleshy scales, 3-4 locular, ovules one in each cell; style short; stigma massive, white, circular, papillate. Berry subglobose or ellipsoid, 3-6 x 2.5-4 cm, green to yellow green, covered with pyramidal spines or broad tubercles, crowned by short style and thick stigma. Seeds 1-3 large, oblong, veined, light brown. Fl. & Fr.: April – June.

Western Ghats: Evergreen forests above 1300 m in Kerala and Tamil Nadu (Endemic).

IUCN: EN-D (IUCN, 2024).

Uses: Seed oil is used for illuminating lamps by local tribes. It is also used in soap and candle manufacture.

Garcinia pushpangadaniana T. Sabu, N. Mohanan, Krishnaraj & Shareef, in Phytotaxa 116 (2): 52, 2013.

Trees to 20 m tall; branchlets tetragonous, dark brown, glabrous; latex remains milky. Leaves elliptic-oblong, 14–20 x 6–8 cm, coriaceous, base rounded, margin subrepand, acute or obtuse at apex, dark green above, light green below; lateral nerves 30–34 pairs, laxly reticulated; petiole 1.5–2 cm long, quadrangular; stipule 6 x 3 mm, intrapetiolar. Inflorescence axillary or from the axils of fallen leaves, fascicled, 2–10-flowered. Flowers pinkish-white. Male flower: pedicel 7–9 mm long. Sepals 5, ciliate, membranous at margin, unequal, 6 x 5 mm, orbicular–suborbicular. Petals 5, 1 x 0.8 cm, orbicular, glabrous, shortly clawed at base. Stamens 12–15 in phalanges, alternating with fleshy intrastaminal 5-parted disc; filaments 1 mm long; anthers brown, 0.5 mm long; rudimentary pistil present. Female flower: pedicel 1–1.5 cm long. Sepal 6 x 5 mm, orbicular–suborbicular. Petals 1–1.2 x 1.2 cm, membranous and ciliate at margin, fleshy at middle. Staminodes 3-5, slender, peltate. Ovary globose, 6 x 6 mm, locules 6–8, greenish yellow; style very short; stigmatic rays 6–8, oblong, stellate, brownish, lobes glabrous. Berry globose, 12 x 11 cm, arranged perpendicular to the axis at maturity, fleshy, without any pulp, irregularly ridged on the surface, symmetric. Seeds plano-convex, smooth or rugose, whitish-yellow, 2 x 1 cm; seed coat light brown, papery. Fl. & Fr.: May – December.

Western Ghats: Semi-evergreen forests of Kerala and Tamil Nadu (Endemic).

IUCN: NE (IUCN, 2024).

Garcinia spicata (Wight & Arn.) Hook.f., J. Linn. Soc., Bot. 14: 486. 1875. Synonyms: *Xanthochymus spicatus* Wight & Arn., *Garcinia ovalifolius* Hook.f., *Garcinia ovalifolia* var. *spicata* (Wight &Arn.) T. Anderson Vern.: *Majanangu* (Mal.); *Kaattupinnai, Kokottai, Manjal naangu* (Tam.).

Evergreen trees to 25 m tall; bark thick, yellowish-brown mottled with white; exudation milky, white to yellowish-green, sticky; branchlets 6-ribbed. Leaves simple, opposite decussate, ovate-elliptic, elliptic-oblong or oblong, 7-22 x 4-8 cm, base acute or obtuse, apex obtuse, emarginate or obtusely acute, margin entire, glabrous, coriaceous; lateral nerves 12-20 pairs, parallel, arched towards the margin forming intramarginal nerve, intercostae laxly reticulate, prominent; petiole 10-20 mm, stout, grooved above. Flowers polygamodioecious, white, in the axils of leaves or fallen leaves. Male flowers: in axillary fascicles or pseudo-spikes; bracts minute, pilose; pedicels slender, 6-12 mm long. Sepals 4 or 5, 3 x 4 mm, pale green, suborbicular, concave, the outer two coriaceous, the inner larger, thinner, margin ciliate. Petals 5, white, obovate, concave, membranous. Stamens in 5 erect, long clawed spathulate fascicles, opposite to petals; anthers didymous; pistillode clavate, smooth. Female flowers: solitary, axillary or paired, usually on much more longer pedicels than the male. Staminoides 5, small with weak anthers. Ovary superior, globose, terete, 4-celled, ovule one in each cell; style 1 mm long; stigma peltate, lobes cuneiform. Berry broadly oblong, yellowish, smooth, 3-4 cm across. Seeds 1-3, up to 2.5 cm long, oblong, latterly flattened, testa brown, veined. Fl. & Fr.: March – May.

Western Ghats: Evergreen forests of Maharashtra, Goa, Karnataka, Kerala and Tamil Nadu. India: Andhra Pradehs. World: Sri Lanka.

IUCN: NE (IUCN, 2024).

Uses: The wood is useful for general construction purposes. Seed oil has been reported for effective in skin treatment.

Garcinia talbotii Raiz. ex Sant., Rec. Bot. Surv. India 16: 14. 1960. Synonyms: *Garcinia malabarica* Talbot., *Garcinia ovalifolia* var. *macrantha*Hook.f. ex T.

Anderson, *Garcinia spicata* var. *macrantha*sensu T. Cooke, *Xanthichymus ovalifolius* Gaertn. Vern.: *Malam pongu* (Tam.).

Trees to 15 m tall; branchlets angled, swollen at nodes; latex brownish. Leaves simple, opposite, elliptic-oblong or ovate, 7-18 x 4-10 cm,, obtuse or retuse at apex, rounded at base, entire at margin, coriaceous; lateral nerves 16-18 pairs or more, distinct; petioles 12-18 mm long, rugose. Inflorescences congested, often an old wood. Flowers 1.8- 2.7 cm across, creamy white or white to greenish yellow; pedicels 5-12 mm long; bracteoles many, minute, at base of pedicel. Sepals 5, orbicular, 3 mm, green. Petals 5, orbicular to suborbicular, concave, 5-10 mm across, white. Male flowers: stamens in 5 phalanges, opposite petals; anthers 8-12 in each phalange, didymous, reddish; pistillode absent. Female flowers: staminodes in 5 thin flat phalanges; anthers 4 or 5 in each phalange, flat, didymous. Ovary globose, 3 or 4 loculed; stigmatic lobes 3, thick, spreading. Berries broadly oblong, 4-6 cm across, with abundant yellow latex. Fl. & Fr.: November – February.

Western Ghats: Evergreen forests of Maharashtra, Goa, Karnataka, Kerala and Tamil Nadu (Endemic).

IUCN: LC (IUCN, 2024).

Uses: Fruit yields yellow gum. Dried fruit pulp is used for substituting tamrind.

Garcinia travancorica Bedd., Fl. Sylv. S. India: t. 173. 1872. Vern.: *Malampongu* (Mal.).

Evergreen trees to 15 m tall; bark rough, brown; exudation milky, pale yellow, sticky; branchlets obtusely 4-angled. Leaves simple, opposite decussate, oblong, linear-oblong to subspathulate, 5-10 x 1.2-2.5 cm, base acute, apex obtuse or obtusely acute, margin entire, revolute, glabrous, coriaceous; lateral nerves many, parallel, almost right angle to the midrib, slender, intercostae reticulate, faint; petiole 6-18 mm, slender, glabrous, grooved above. Flowers polygamodioecious, white. Male flowers in trichotomous, short, few flowered, terminal or subterminal cymes; pedicels short, thickened. Sepals 4, orbicular, outer pairs smaller than the inner, concave, decussate. Petals 4, about twice as long as sepals, shortly clawed, round. Stamens numerous, in four multifid polyandrous masses, anthers bilocular, versatile; filaments short; pistillode columnar, stigma, circular, peltate. Female flowers in axillary, solitary or in pairs. Sepals and petals as in males; staminodes few, free, filaments complanate, linear, in 2 to 3-chotomous branches; anther loculi often divaricate, oblong.

Ovary superior, subglobose or pyriform, 4-locular. Berry oblong to subglobose, 3.7-5 x 1.2-2.5 cm, thick style with a broad imbricate stigma at the top. Seeds 1-2, brown, smooth, shiny. Fl. & Fr.: April – June.

Western Ghats: Evergreen forests of Kerala and Tamil Nadu (Endemic).

IUCN: CR-C2a(i). (IUCN, 2024).

Garcinia wightii T. Anderson, J.D.Hooker, Fl. Brit. India 1: 265. 1874. Vern.: *Attukaruka, Kolivala, Palimaranga, Puli maranga* (Mal.).

Trees to 10 m tall; bark thick, rough, brown, pustular; exudation orange yellow; branchlets 4 angled, glabrous. Leaves simple, opposite decussate, linear, linear-lanceolate or linear-oblong, 7.5-14 x 1.2-3 cm, base acute, decurrent into the petiole, apex acuminate, glabrous, coriaceous; lateral nerves 8-20 pairs, looped near the margin forming intramarginal nerve, intercostae reticulate, faint; petiole 6-8 mm long, slender, glabrous, grooved above. Flowers polygamodioecious, yellow. Male flowers in axillary, solitary or often 2-3, sessile. Sepals 4, equal, orbicular, concave, thinly coriaceous. Petals 4, yellow, 4.5-5 x 3.5 mm, obovate, distinctly concave. Stamens 18-20, often united in a column enclosing the tetragonal stylodium, filaments free above, anthers peltate; pistillode tetragonal. Female flowers in axillary, solitary, sessile. Sepals and petals as in male flowers. Ovary superior, almost globular, usually 4-locular, stigmas sessile, large. Berry subglobose, 11-13 x 9-11 mm, smooth, pale green with persistent stigma and sepals. Seeds 4, 9.5 x 4.5 mm. Fl. & Fr.: November – April.

Western Ghats: Rare in evergreen forests of Kerala and Tamil Nadu (Endemic).

IUCN: EN-D (IUCN, 2024).

Uses: Sap of this tree is bright yellow, it can be used for dying purposes.

Garcinia xanthochymus Hook.f. ex T. Anderson, J.D.Hooker, Fl. Brit. India 1: 269. 1874. *Garcinia pictorius* (Roxb.) D'Arcy, *Xanthochymu spictorius* Roxb., *Garcinia tinctoria* (DC.) Wight, *Xanthochymus tinctorius* DC. Vern.: *Ana vaya* (Mal.); *Periya Ilai kodampuli* (Tam.).

Evergreen trees to 20 m tall; bark thick, blackish or dark grey, exfoliating in small flakes; exudation white turning yellow; branchlets 4-angular. Leaves simple, opposite, linear oblong, oblong or oblong-lanceolate, 12-45 x 4-12 cm, base cuneate or acute, apex acute or acuminate. margin entire, glabrous, shining, coriaceous; lateral nerves 15-20 pairs, arched near the margin forming intramarginal nerves, intercostae reticulate, secondary laterals prominent; petiole 10-30 mm long, prominent ligulate projection at base, glabrous. Flowers polygamodioecious, white. Male flowers 1.5 cm across, 4-10 in fascicles, axillary or from the axils of the fallen leaves; pedicels thick, to 2.5 cm long. Sepals 4-5, orbicular-concave, unequal, apex ciliate. petals 4-5, 8 mm long, orbicular, incurved, spreading, greenish. Stamens in 5 broad bundles of 3-5 each, alternating with 5 fleshy glands; anthers bilocular. Female flowers solitary; staminodes few, complanate. Ovary ovoid, acuminate, 5 locular, ovules one in each cell; style short; stigmatic rays 5, oblong, spreading, entire. Berry yellow, when ripe, 4.5-5.5 x 4-4.5 cm, stigmatic lobes 5. Seeds 1-2, reniform, 2.5-3 x 2.5 cm, smooth. Fl. & Fr.: January – July.

Western Ghats: Evergreen forests of Maharashtra, Goa, Karnataka, Kerala and Tamil Nadu. World: Sri Lanka, Bangladesh, Myanmar, China, Thailand and Malaysia.

IUCN: LC (IUCN, 2024).

Uses: Fruit is edible. Sap is yellow and used for dying purposes.

Cultivated:

Garcinia affinis Wall. ex Pierre; Native: India, Bangladesh, Myanmar

Garcinia mangostana L.; Exotic: Native to Malaysia

Planted:

Clusia rosea Jacq.; Exotic: Native to Central America

Garcinia livingstonei T. Anderson; Exotic: Native to Tropical Africa

Rheedia floribunda (Miq.) Planch & Triana; Exotic: Native to Bolivia

COMBRETACEAE R.Br.

Terminalia L.

1a. Flowers in capitate heads 2

1b. Flowers in spikes or racemes 4

2a. Drupe broadly winged, margin glabrous, entire .... **T. coronata**

2b. Drupe minutely winged, margin ciliate, undulate 3

3a. Leaves suborbicular, mucronate at apex, greyish pilose above .... **T. phillyreifolia**

3b. Leaves elliptic-oblova, obtuse at apex, glabrous above . **T. anogeissiana**

4a. Fruit not winged, ovoid or subcompressed 5

4b. Fruit winged or angled 7

5a. Flowers in simple spike; leaves clustered at the ends of branchlets .... **T. bellirica**

5b. Flowers in terminal panicles; leaves scattered on the stem 6

6a. Leaves elliptic ovate or obovate, glabrous to dawny villous; fruit obovoid, faintly angled ....**T. chebula**

6b. Leaves ovate-lanceolate, pubescent when young; fruit ovoid, not angled ....**T. travancorensis**

7a. Leaves oblong-ellptic, usually crenate; fruits with short 5 equal wings.... ....**T. arjuna**

7b. Leaves obovate-oblong, usually margin entire; fruits with 3 or 5 braod wings..8

8a. Panicles with many slender spikes; flowers small with recurved bracteoles.. **T. paniculata**

8b. Panicles with a few branched spikes; flowers large with straight bracteols ....**T. elliptica**

Terminalia anogeissiana Gere & Boatwr., Bot. J. Linn. Soc. 184: 319. 2017. Synonyms: *Anogeissus latifolia* (Roxb. ex DC.) Wall. ex Guill. & Perr., *Conocarpus latifolia* Roxb. ex DC., *Anogeissus latifolia* var. *villosa* Clarke, *Anogeissus latifolia* var. *tomentosa* Haines, *Anogeissus latifolia* var. *glabra* Clarke. Vern.: *Maru kancharam* (Mal.); *Vellai Naaga, Vellai Namai, Namai, Vekkali* (Tam.).

Deciduous trees to 18 m tall; bark thick, surface grey or yellowish-grey, smooth, lenticellate; branchlets brown to reddish-brown, pubescent. Leaves simple, opposite to alternate, elliptic, elliptic-oblong, suborbicular or oblong-obovate, 2.5-18 x 1.7-9 cm, base obtuse, round, cuneate or acute, apex obtuse, emarginate or mucronate, margin entire, glabrous, subcoriaceous; lateral nerves 4-11 paired, arched towards the margin, intercostae scalariform, prominent; petiole 5-15 mm, slender, grooved above, glabrous. Flowers pale green or yellow, 5-6 mm across, in axillary aggregated globose heads; peduncle puberulous; pedicel reduced or absent. Sepals 5, tubular, tube 2 winged, 3-5.5 mm long, produced above the ovary, villous inside, teeth 5, ovate-triangular, to 1 mm long. Stamens 10, in 2 series, exserted, 1.5-3 mm long; anthers small. Ovary inferior, 1-celled, densely tomentose; ovules 2, pendulous; style 2-3 mm long, thickened at base, villous; stigma simple. Drupe greenish-yellow, 6-8 mm across, compressed, puberulous, circular; wings 2, margin entire or slightly undulate, beaked. Seed one, obovate. Fl. & Fr.: March – September.

Western Ghats: Deciduous forests of Maharashtra, Karnataka, Kerala and Tamil Nadu. World: Sri Lanka, Nepal and Pakistan.

IUCN: NE (IUCN, 2024).

Uses: The wood is highly valued and used for making agriculture implements, ship-building, furnitures, house cabinets and axe-handles. The wood is also produced excellent charcoal which is used as good fuel.

Terminalia arjuna (Roxb. ex DC.) Wight & Arn., Prodr. Fl. Ind. Orient. 1: 314. 1834. Synonyms: *Terminalia cuneata* Roth, *Pentaptera arjuna* Roxb. ex DC. *Terminalia arjuna* var. *angustifolia* (Roxb.) Clarke, *Pentaptera angustifolia* Roxb. *Terminalia berryi* Wight &Arn., *Terminalia glabra* (Roxb.) Wight &Arn., *Pentaptera glabra* Roxb., *Terminalia ovalifolia* Rottler ex C.B. Clarke. Vern.: *Kula maruthu, Marudhamaram, Neer maruthu, Vellai Maruthu* (Tam.)

Evergreen trees to 30 m tall; bark thick, surface pinkish-grey, smooth, flaking off in thin layers; branchlets drooping. Leaves simple, opposite to alternate, oblong, elliptic, oblong-ovate or oblong-obovate, 7.5-16 x 3.7-8.5 cm, base round, obtuse or cordate, apex obtuse or round, margin crenulate or entire, glabrous, coriaceous, 2 glands at the base on the lowerside; lateral nerves 10-17 pairs, intercostae reticulate, prominent; petiole 5-15 mm long, stout, grooved above, glabrous. Flowers dull yellow, 2-3 mm across, in short axillary spikes. Seplas tubular, tube 0.5-2 x 1-1.5 mm, constricted above the ovary, pubescent; lobes 5, cream, 1.5 mm, triangular, glabrous. Disc 5-lobed, pilose. Petals 0. Stamens 10, in 2 rows; filaments 3-4 mm, subulate, exserted; anthers small. Ovary ovoid, 1.5 mm, inferior, pubescent, 1-celled; ovules 2 or 3, pendulous; style subulate, small. Drupe oblong, 6 x 3 cm, 5-7 winged; wings equal, lines on the wings oblique and curving upwards, apex notched. Seed one. Fl. & Fr.: February – November.

Western Ghats: Low lying forests along the water course of Gujarat, Maharashtra, Goa, Karnataka, Kerala and Tamil Nadu. India: Andhra Pradesh and Odisha. World: Sri Lanka and Myanmar.

IUCN: NE (IUCN, 2024).

Uses; The wood is used for making carts, agriculture implements, boat-building, etc. The bark is yielding transparent gum, which is used for medicinal purposes.

Terminalia bellirica (Gaertn.) Roxb., Pl. Coromandel 2: t. 198. 1805. Synonyms: *Myrobalanus bellirica* Gaertn., *Terminalia bellirica* var. *laurinoides* Clarke, *Terminalia myrobalana* B.Heyne ex Roth, *Terminalia eglandulosa* Roxb. ex C.B.Clarke, *Terminalia moluccana* Roxb. Vern.: *Tani, Tandri* (Mal. & Tam.).

Trees deciduous to 35 m tall; bark gray, longitudinally ridged; branchlets with conspicuous, spirally ascending leaf scars. Leaves crowded into pseudowhorls at apices of branchlets, blade glossy, obovate, 18-26 × 6-12 cm, both surfaces glabrous, base obtuse-rounded or attenuate, apex obtuse or mucronate; lateral veins in 5-8 pairs; petiole 3-9 cm, glabrous but ferruginous tomentose when young, especially at base, with 2 glands above middle. Inflorescences axillary, simple spikes, 5-18 cm long, often grouped at branchlet apex and forming a panicle; axis densely ferruginous tomentose. Sepals 5, tube distally shallowly cupular, 4-5 mm, abaxially tomentose, adaxially long villous, lobes 5. Stamens 10, exserted, 4-5 mm long. Drupe stipitate, subglobose to broadly ellipsoid or ovoid, weakly to strongly 5-ridged, 2-3 × 1.8-2.5 cm, densely and finely velutinous or sericeous. Fl. & Fr.: March – November.

Western Ghats: Deciduous forests of Gujarat, Maharashtra, Goa, Karnataka, Kerala and Tamil Nadu. World: Sri Lanka, Nepal, Bangladesh, Myanmar, China, Laos, Cambodia, Vietnam, Thailand, Malaysia and Indonesia.

IUCN: LC (IUCN, 2024).

Uses: The wood is used for packaging cases and for house building.

Terminalia chebula Retz., Observ. Bot. 5: 31. 1788. Synonyms: *Terminalia chebula* var. *tomentella* (Kurz.) Clarke, *Terminalia tomentella* Kurz., *Terminalia argyrophylla* King & Prain, *Terminalia chebula* var. *gangetica* (Roxb.) C.B.Clarke, *Terminalia gangetica* Roxb., *Terminalia tomentella* Kurz Vern.: *Kadukkai* (Mal. & Tam.).

Trees to 30 m tall; bark grayish black to gray, coarsely split and thick; branchlets conspicuously white or yellowish long lenticellate, glabrous, or tomentose or appressed villous. Leaves alternate or subopposite, spaced along branchlets, blade elliptic, 7-18 × 4.5-10 cm, both surfaces glabrous, or appressed, villous at least when young, base obtuse-rounded or cuneate, oblique, apex mucronate; lateral veins in 6-12 pairs; petiole 1-3 cm, moderately stout, with 2 glands, 1-5 mm below apex. Inflorescences axillary or terminal, simple spikes, 5-10 cm long, numerous flowered, sometimes grouped at branchlet apex and forming a panicle; axis glabrous or sparsely hairy, with denser hairs near base of flowers. Flowers slightly fragrant, bisexual. Sepals 5, tube distally cupular,

2.5-3.5 mm long, abaxially glabrous, adaxially tawny tomentose, lobes 5, apex mucronate to aristate. Stamens 10, exserted, 3-4 mm. Drupe not stipitate, blackish brown when ripe, ovoid or broadly so, ellipsoid, or cylindric-ovoid, obtusely 5-ridged, 2-4.5 × 1.2-2.5 cm, rigid, becoming deeply wrinkled when dry, glabrous. Fl. & Fr.: March – December.

Western Ghats: Deciduous forests of Gujarat, Maharashtra, Goa, Karnataka, Kerala and Tamil Nadu. World: Sri Lanka, Nepal, Myanmar, Bangladesh, Thailand and Malaysia.

IUCN: LC (IUCN, 2024).

Uses: The wood is good valuable, for making furniture, carts, agriculture implements and house building. The fruit is black myrabalon of commercial importance and it is giving yellow dye.

Terminalia coronata (Stapf) Gere & Boatwr., Bot. J. Linn. Soc. 184: 319. 2017. Synonyms: *Anogeissus sericea* Brandis, *Anogeissus sericea* var. *nummularia* King ex Duthie, *Anogeissus rotundifolia* Blatt., *Anogeissus latifolia* var. *parviflora* Clarke, *Anogeissus coronata* Stapf

Large trees to 20 m tall; bark smooth; branchlets terete, with tomentose or sub tomentose. Leaves alternate, elliptic-sub-orbicular or orbicular, apex obtuse, emarginate or mucronate, silvery pubescent, nerves prominent beneath; lateral veins 4 pairs; petiole to 2 cm long, tomentose. Head globes, solitary, terminal or axillary, tomentose; peduncles to 1-2 cm long. Flowers yellow to yellowish brown. Sepals 5, tubular, tube compressed, 2-winged, attenuate dabove the ovary, limb long campanulate, pubescent, 5 dentate. Stamens exserted, 5. Drupe winged, margin glabrous, undulate- dentate, receptacle coronate, persistent. Seeds are very minute, wing ending with beaks. Fl. & Fr.: February – April.

Western Ghats: Semi-evergreen forests of Maharashtra (Endemic). World: Bangladesh.

IUCN: NE (IUCN, 2024).

Terminalia elliptica Willd., Sp. Pl., ed. 4. 4: 969. 1806. Synonyms: *Pentaptera crenulata* Roxb. *Terminalia alata* B.Heyne ex Roth, *Terminalia crenulata* Roth, *Terminalia coriacea* (Roxb.) Wight &Arn. *Pentaptera coriacea*Roxb. *Terminalia tomentosa* (Roxb. ex DC.) Wight &Arn., *Pentaptera tomentosa* Roxb. ex DC., *Terminalia tomentosa* Roxb. ex DC. *Terminalia tomentosa* var. *crenulata* (Roth) Clarke, *Terminalia tomentosa* var. *coriacea* (Roxb.) Clarke, *Terminalia alata* var.

tomentosa (Roxb. ex DC.) Parkins. *Terminalia tomentosa* var. *typica* Clarke, *Terminalia glabra* var. *tomentosa* (Roxb. ex DC.) Dalz. & A. Gibson, *Terminalia tomentosa* var. *nepalensis* Haines. Vern.: *Thembhavu, Karumaruthu* (Mal.); *Karumaruthu* (Tam.).

Deciduous trees to 30 m tall; bark thick, surface grey-black, very rough, deeply vertically fissured, horizontally cracked, forming thick flakes. Leaves simple, opposite to subopposite, oblong, oblong-ovate, elliptic-oblong or elliptic-ovate, 13-20 x 5-13 cm, base oblique, apex acute, round, or obtuse, margin entire or crenulate, glabrous, coriaceous, midrib with 2 stalked glands near the base beneath; lateral nerves 10-20 pairs, parallel, intercostae scalariform, prominent; petiole 10-20 mm long, stout, grooved above, glabrous. Flowers dull yellow, 2-3 mm across, in terminal and axillary paniculate spikes; peduncle tomentose; bracteoles 1-2 mm long, linear or subulate. Sepals 5, tubular, tube 1-2 x 1-1.8 mm, angled, base pubescent, constricted above the ovary, lobes 5, ovate, yellow, densely villous inside. Petals absent. Stamens 10 in 2 rows; filaments 4-5 mm. Disc 5-lobed, villous. Ovary 2.5 mm long, inferior, densely villous, 1-celled; ovules 2 or 3, pendulous; style 4 mm, subulate; stigma small. Drupe longitudinally 5-winged, 3.5 x 5.5 cm, glabrous; wings equal, thin, pubescent, lines on the wings horizontal, apex round, coriaceous, reddish-brown. Seed one. Fl. & Fr.: April – November.

Western Ghats: Deciduous forests of Maharashtra, Goa, Karnataka, Kerala and Tamil Nadu. World: Sri Lanka, Nepal, Myanmar, Laos, Cambodia, Vietnam, Thailand and Malaysia.

IUCN: NE (IUCN, 2024).

Uses: The wood logs has been ranked export quality for ship-building and for railway tracks.

Terminalia paniculata B.Heyne ex Roth., Nov. Pl. Sp.: 383. 1821. Synonyms: *Myrobalanus monaptera* (Roth) Kuntze, *Terminalia alata* D.Dietr., *Terminalia monaptera* Roth Vern.: *Maruthu, Pei-maruthu, Pillai-maruthu* (Mal.); *Vem-maruthu, Vellei-maruthu* (Tam.).

Deciduous trees to 30 m tall; bark thick, surface brown, rough, vertical fissures shallow, fibrous; branchlets reddish. Leaves simple, lower leaves subopposite, upper alternate, elliptic, oblong, ovate, elliptic-oblong or ovate-oblong, 10-24 x 5-10 cm, base round, obtuse, cordate or oblique, apex acute or acuminate, margin entire, rusty pubescent when young and glabrescent when mature, coriaceous, 2 sessile glands beneath the base of the lamina;

lateral nerve 5-15 pairs, arched towards the margin, prominent, intercostae reticulate, obscure; petiole 12-15 mm long, stout, slightly grooved above, glabrous. Flowers white, sessile, 6-12 mm across, in axillary and terminal panicles, tomentose; bracteoles 2-5 mm, linear-lanceolate, recurved. Sepals 5, tube 1-1.2× 0.8-1 mm, constricted above the ovary, pubescent, lobes 5, cream, triangular, villous within. Petals 0. Stamens 10 in 2 rows; filaments 3-4 mm. Disc 5-lobed, villous; ovary 1 mm, inferior, 1-celled; ovules 2 or 3, pendulous; style to 4 mm, subulate, stigma terminal. Fruit a drupe 13 x 12 mm, reddish-brown, rusty tomentose, with one large and 2 small wings. Seed one. Fl. & Fr.: October – February.

Western Ghats: Deciduous forests of Maharashtra, Goa, Karnataka, Kerala and Tamil Nadu (Endemic).

IUCN: NE (IUCN, 2024).

Uses: The wood is good bulding timber and for making furniture.

Terminalia phillyreifolia (Van Heurck & Müll.Arg.) Gere & Boatwr., Bot. J. Linn. Soc. 184: 321. 2017. Synonyms: *Anogeissus acuminata* (Roxb. ex DC.) Guill. & Perr. var. *acuminata*, *Conocarpus acuminata* Roxb. ex DC., *Anogeissus acuminata* var. *lanceolata* Clarke, *Anogeissus lanceolata* (Clarke) Prain, *Anogeissus pendula* Edgew., *Anogeissus acuminata* var. *phillyreifolia* (Van Heurck & Müll. Arg.) Kurz., *Anogeissus acuminata* (Roxb. ex DC.) Guill. & Perr. var. *fischeri* (M. Gangop. & Chakrab.) M. Gangop. & Chakrab. Vern.: *Namai*, *Numma* (Tam.).

Small trees to 10-20 m tall; bark grey, whitish or dark brown; branchlets blackish, brown or coppery, terete, grey-pubescent to glabrous. Leaves opposite to subopposite, elliptic to oblong-elliptic or ovate to suborbicular, 3-8 x 1.5-3 cm, acute, obtuse or mucronate or sometimes shortly acuminate at apex, base acute to obtuse or subcuneate and decurrent, thinly coriaceous, sparsely adpressed greyish-pilose to glabrous above; lateral nerves 4-8 pairs, obscure to faint above; intercostae inconspicuous above, scalariform to reticulate; petioles 1-8 mm long. Inflorescence in heads 1-2 cm across, axillary and terminal, solitary or occasionally in pairs; peduncles 0.5- 2 cm long; bracts on peduncles 1-2 pairs, subfoliaceous, ovate to lanceolate or oblanceolate, 3-10 mm long, subpersistent; bracteoles spathulate or ovate-elliptic, 1-2 mm long. Seplas 5, tube 2-4 mm long, tawny-or ferrugineous-tomentellous; expanded portion cupular or shortly campanulate, 1.5-2 x 2-2.5 mm, glabrous or pubescent only at base; teeth 5-6, triangular. Petals 0. Stamens 2-4 mm long; anthers oblong to suborbicular, 0.5 mm long. Ovary 1-celled, ovules 2, style 2-4 mm long. Drupe

2-winged, 3-7 x 5-8 mm, puberulous to glabrous, reddish-brown, often glossy; wings membranous, entire or slightly undulate along margins; beak 1.5-4 mm long. Seeds ovoid, 3-5 x 2.5-3 mm. Fl. & Fr.: February – July.

Western Ghats: Deciduous forests of Maharashtra, Kerala and Tamil Nadu. India: Andhra Pradesh and Odisha. World: Myanmar, Bangladesh, Thailand, Laos, Cambodia and Vietnam.

IUCN: NE (IUCN, 2024).

Terminalia travancorensis Wight & Arn., Prodr. Fl. Ind. Orient. 1: 314. 1834. Synonym: *Terminalia angustifolia* sensu Roxb., non Jacq.

Evergreen trees to 30 m tall; bark thick, surface yellowish-grey, smooth. Leaves simple, opposite to alternate, elliptic, ovate-lanceolate or oblong-lanceolate or elliptic-lanceolate, 8-16 x 2.5-5 cm, base unequally acute or round, apex acute or acuminate, margin entire, glabrous, coriaceous; lateral nerves 12-18 pairs, parallel, arched towards the margin, intercostae reticulate, obscure; petiole 12-30 mm long, grooved above, glabrous. Flowers white, 3-4 mm across, in axillary and terminal paniculate spikes; branches hispid. Sepals 5, tube constricted above the ovary; lobes 5, hispid outside, densely villous inside. Petals 0. Disc 5 lobed, villous. Stamens 10 in 2 rows; filaments 4 mm long. Ovary densely tomentose, inferior; style subulate; stigma terminal. Drupe ellipsoid, 3-3.5 x 1.3 cm, 5-angled, speckled with brown spots, glabrous, yellowish-brown. Seed one. Fl. & Fr.: May – January.

Western Ghats: Evergreen forests of Karnataka, Kerala and Tamil Nadu (Endemic).

IUCN: NE (IUCN, 2024).

Planted:

Terminalia catappa L.; Native: India, Sri Lanka, China, Tropical Asia, North Australia, Madagascar, Pacific Islands.

CONNARACEAE R.Br.

Ellipanthus Hook.f.

Ellipanthus tomentosus Kurz., J. Asiat. Soc. Bengal, Pt. 2, Nat. Hist. 41: 305. 1872. Synonyms: *Ellipanthus neglectus* Gamble, *E. thwaitesii* Hook.f., Vern.: *Padappan* (Mal.).

Small trees to 10 m tall; branchlets tomentose, at least when young. Leaves alternate, elliptic to lanceolate or obovate, 7-22 x 4-8 cm, apex pointed to obtusely acuminate, base acute to rounded or subpeltate, thinly coriaceous, subglabrous to tomentose below; lateral nerves 5-10 pairs; petiole 0.5-3 cm long. Flowers in axillary clusters on young twigs, white, densely hairy. Sepals ovate or acute, 1.5-2 mm long, inner surface tomentose to glabrous. Petals ovate-oblong to linear, 2-5 x 1-2 mm, obtuse, tomentose to pubescent outside, glabrous inside. Fertile stamens pubescent, staminodes glabrous. Ovary strigose; stigma bilobed. Fruits oblique-ovoid, flattened, 2-4 x 0.75-1.25 cm, apex shortly pointed, tomentose; stalk 0.5-3 cm long. Seeds 12-20 x 6-10 mm, often flattened, arillode obscurely lobed to rounded, covering one thrird portion at base. Fl. & Fr.: November – June.

Western Ghats: Evergreen forests of Kerala and Tamil Nadu. World: Sri Lanka, Thailand and Malaysia.

IUCN: NE (IUCN, 2024).

Note: In India, the tree species collected from the type locality of Colatoorpolay (Kerala) by Bourdillone in 1895. After that no further collection and not seen in any herbarium.

CORNACEAE Bercht. & J. Presl

Alangium Lam.

1a. Leaf apex acute; drupes less than 1 cm long .... **A. salvifolium**

1b. Leaf apex acuminate; drupes more than 1 cm long .... **A. hexapetalum**

Alangium hexapetalum Lam., Encycl. 1: 175. 1783. Synonyms: *Alangium salviifolium* var. *hexapetalum* (Lam.) Gamble, *Alangium sundanum* var. *insularum* Kurz Vern.: *Malai Azhinjil* (Tam.).

Small trees to 10 m tall; branchlets wiry, dense pubescent. Leaves obovate-oblong, 13-18 x 5-7 cm, apex acuminate, base round to subcordate; nerves 5 pairs, lowest pair from the base, rib-like; nervules reticulate, subparallel; petiole to 1 cm long. Flowers 2 cm long, in axillary fascicles; pedicels 8 mm long pubescent. Sepals cupular, 5 mm across, 7-teethed, densely tomentose. Petals 7, each 17 x 2 mm, linear-oblong, tomentose. Stamens 14; filaments 8 mm long, ciliate below; anthers 7 mm long, linear. Ovary inferior, 1-celled;

ovule solitary; style 1, elongate, stigma capitate. Berry ellipsoid, 2 x 1.3 cm, tomentose. Fl. & Fr.: March – August.

Western Ghats: Semi-evergreen forests of Kerala and Tamil Nadu. World: Sri Lanka and Indo-Malesia.

IUCN: NE (IUCN, 2024).

Alangium salvifolium (L.f.) Wang., H.G.A. Engler (ed.), P flanzenr., IV, 220b: 9. 1910. Synonyms: *Grewia salvifolia*L.f., *Alangium lamarckii* Thw., *Alangium salvifolium* ssp. *decapetalum* (Lam.) Wang., *Alangium decapetalum* Lam., *Alangium salvifolium* ssp. *hexapetalum* (Lam.) Wang. Vern.: *Alinjil, Valli-thondi* (Mal.); *Alinjilai, Achan chedi, Yerazhinjil* Tam.).

Small trees to 15 m tall; bark greyish-brown; branchlets tomentose. Leaves simple, alternate, oblong, elliptic or ovate, 7-15 x 3-4 cm, base oblique or obtuse, apex acute, margin entire, chartaceous, glabrous above, puberulous beneath; 3-5 nerved at base, lateral nerves 4-7 pairs, intercostae reticulate, prominent; petiole 7-10 mm long, slender, tomentose. Flowers white, fragrant, 1.5 cm across, in axillary cymes or fascicles; pedicel 4 mm, jointed, pubescent; bracts 4 mm, ovate. Sepals 5, 3 mm long, adnate to ovary, cupular, tomentose; lobes 10, triangular-ovate. Petals 10, 2.5 x 5 cm, linear-oblong, tomentose. Stamens to 20; filaments 1 cm long, base fleshy, villous, subconnate at base; anthers linear. Ovary infererior, 2 mm, turbinate, 1-celled, 1-ovuled, pendulous, ovary surmounted by a disc, enclosing the style; style 2 mm long, glabrous; stigma capitate. Berry subglobose, 2-3 x 1-1.5 cm, brownish-red, pubescent, crowned with calyx lobes. Seed one, ovoid. Fl. & Fr.: February – June.

Western Ghats: Dry deciduous forests of Gujarat, Maharashtra, Goa, Karnataka, Kerala and Tamil Nadu. India: Almost all the states of India. World: Sri Lanka, Myanmar, china, Thailand, Vietnam, Malaysia and Philippines.

IUCN: NE (IUCN, 2024).

Uses: Fruits are edible. The wood is used for making charcoal.

Cultivated

Cornus capitata Wall.; Native: India, Myanmar, Bhutan, Nepal, China

CORYNOCARPACEAE Engl.

Planted

Corynocarpus laevigata J. R. Forst. & G. Forst.; Exotic: Native to New Zealand

DAPHNIPHYLLACEAE Mull.-Arg.

Daphniphyllum Blume

Daphniphyllum neilgherrense (Wight) K.Rosenthal., H.G.A. Engler (ed.), Pflanzenr., IV, 147a: 7. 1919. Synonyms: *Daphniphyllum glaucescens* ssp. *neilgherrense* (Wight) T.C.Huang, *Goughia neilgherrensis* Wight, *Daphniphyllum glaucescens* var. *concolor* Müll.Arg.

Evergreen trees to 25 m tall; bark thick, brown, lenticellate; branchlets terete, glabrous. Leaves simple, alternate, elliptic-oblong or obovate, 7-12.5 x 3-7.5 cm, apex obtuse with apiculate tip, base obtuse, margin entire, coriaceous, glaucous beneath; secondary nerves 8 pairs, tertiary nerves reticulate; petiole up to 5 cm long. Flowers unisexual, yellowish-green, in axillary racemes. Male flowers: bracts 3 x 3 mm, ovate, obtuse, ciliate. Tepals 5, minute. Stamens 5-12, free; anthers oblong; connective apiculate. Female flowers: bracteate, perianth similar to male flowers. Ovary superior, imperfectly 2-celled; ovules 2 in each cell; stigmas thick, recurved, undivided. Drupe ovoid, 10-12 mm long, greenish-red. Fl. & Fr.: January – June.

Western Ghats: Evergreen forests of Karnataka, Kerala and Tamil Nadu. World: Sri Lanka.

IUCN: NE (IUCN, 2024).

Uses: Timber is used for fuel wood.

DICHAPETALACEAE Baill.

Dichapetalum Thouars

Dichapetalum gelonioides (Roxb.) Engl., H.G.A. Engler & K.A.E.Prantl, Nat. Pflanzenfam. 3(4): 348. 1896.Synonyms: *Moacurra gelionoides* Roxb., *Chailletia gelionoides* (Roxb.) Hook.f. Vern.: *Chrumaram* (Tam.).

Small trees to 10 m tall; young branches appressed pubescent, glabrescent, with sparsely white rounded lenticels. Leaves elliptic or oblong-elliptic, 6-16 x 2-6 cm, papery or subleathery, glabrous on both surfaces, or abaxially pubescent only along midrib and lateral veins, lateral veins 5 or 6 pairs, base cuneate or broadly cuneate, slightly oblique, margin entire, apex acuminate or obtusely acuminate; petiole 3-5 mm, glabrous or pilose. Flowers unisexual, dioecious, in few-flowered 0.75-1 cm cymes or single, axillary, minutely sparsely pilose. Sepal oblong, 2 mm, densely tomentose on exterior. Petals broadly spatulate, apex slightly lobed or subentire. Glands 0.25-0.75 mm, rounded to cordate, glabrous to minutely pubescent. Ovary of female flower 2-3-locular, densely yellow-brown pubescent. Pistillode of male flower densely white lanose; style 1, deeply lobed. Drupe 2 locules, obcordate, long and broad, 1.8 cm diam, densely yellow-brown pubescent when young, white pilose when mature, with distinct sutures on exterior. Fl. & Fr.: February – May.

Western Ghats: Evergreen forests of Maharashtra, Goa, Karnataka, Kerala and Tamil Nadu. India: Assam and Meghalaya. World: Sri Lanka, Indonesia, Malayasia, Myanmar, Philippines, Thailand, Vietnam, Bangladesh and Sumatra.

IUCN: LC (IUCN, 2024).

DILLENIACEAE Salisb.

Dillenia L.

1a. Flowers white 2

1b. Flwers yellow 3

2a. Flowers more than 12 cm across, leaf apex acute or acuminate .... **D. indica**

2b. Flowers less than 12 cm across, leaf apex retuse .... **D. retusa**

3a. Anther theca opening with longitudinal slit; flowers up to 5 cm across.. 4

3b. Anther theca opening with apical pore; flowers 10–12 cm across .... **D. bracteata**

4a. Flowers facicled on old branches .... **D. pentagyna**

4b. Flowers on terminal and lateral reacemes .... **D. suffruticosa**

Dillenia bracteata Wight., Icon. Pl. Ind. Orient. 2: t. 358. 1840. Synonyms: *Wormia bracteata* (Wight) Hook.f. &Thoms., *Dillenia repanda* Roxb. Vern.: *Koddapunna, Pattipunna* (Mal.); *Nai thekku, Kal Uva* (Tam.).

Trees to 20 m tall; bark greyish-brown, blotched with white; young shoots tomentose. leaf scars half the circumference of branches. Leaves simple, alternate, oblong-lanceolate, obovate to elliptic-obovate, 15-45 x 5-20 cm, base cuneate, acute, or obtuse, apex acute, acuminate or obtuse, margin serrate or rarely entire, coriaceous, glabrescent at maturity; lateral nerves 12-20 pairs, parallel, intercostae subscalariform, faint; petiole 10-50 mm long, pubescent, grooved above. Flowers yellow, 2.5-3 cm across, in racemes or in 2-7-flowered clusters on old branches; pedicels 4-10 mm long; bracts lanceolate or spathulate, hirsute, early cauducous. Sepals 5, ovate, 8-12 x 5-9 mm, coriaceous, reddish, silky pubescent. Petals 5, obovate, 15-20 x 5-10 mm, white turning to yellow. Stamens numerous, cohering slightly at base, in 2 series, purple blue; anthers linear, opening by small slits. Carpels 5, cohering at the axis, ovules many; styles 5, free, 9 mm. Pseudocarp orange, subglobose, enclosed within sepals, 2.5-3 cm across. Fruits 8 x 5 mm. Seeds 1-2, 5 x 4.5 mm, ovoid, dark reddish-brown to black, arillate, glabrous. Fl. & Fr.: November – June.

Western Ghats: Deciduous forests of Karnataka, Kerala and Tamil Nadu (Endemic). India: Andhra Pradesh and Eastern Ghats.

IUCN: NE (IUCN, 2024).

Uses: The wood is used for house posts and miscellaneous purposes.

Dillenia indica L., Sp. Pl.: 535. 1753. Synonyms: *Dillenia speciosa* Thunb., *Dillenia elongata* Miq., *Dillenia indica* f. *elongata* (Miq.) Miq. Vern.: *Chalita, Punna, Syalita* (Mal.); *Uga kai, Uvaa Thekku* (Tam.).

Evergreen trees to 30 m tall; bark reddish brown; young branchlets brown pubescent, glabrescent, leaf scars obvious. Leaves oblong or obovate-oblong, 15-40 x 7-14 cm, secondary veins 30-40 on either side, parallel, margin obviously serrate; petiole narrowly winged, stout, pubescent. Flowers solitary, 12-20 cm in diam, more than 5 cm in diam. in bud. Sepals 5, approximately rounded, 4-6 cm in diam, thickly fleshy. Petals white, obovate, 7-9 cm long. Stamens in 2 distinct groups, outer very numerous, slightly curved in bud, inner 25, apically reflexed outward in bud; anthers dehiscing with 2 pores. Carpels 16-20; stylodia spreading; ovules many per carpel. Aggregate fruit globose, 10-15 cm in diam, indehiscent with persistent sepals, fleshy, slightly swollen. Seeds 5 or more per carpel, exarillate. Fl. & Fr.: May – September.

Western Ghats: Deciduous to semi-evergreen forests of Maharashtra, Goa, Karnataka, Kerala and Tamil Nadu. India: Andhra Pradesh, Bihar, Madhya Pradesha and Meghalaya. World: Sri Lanka, Bhutan, Nepal, China, Laos, Vietnam, Thailand, Myanmar, Indonesia and Malaysia.

IUCN: LC (IUCN, 2024)

Dillenia pentagyna Roxb., Pl. Coromandel 1: 21. 1795. Synonyms: *Wormia pentagyna* var. *augusta* Kurz, *Dillenia augusta* Roxb., *Dillenia floribunda* Hook.f. & Thomson, Vern.: *Koddapunna, Pattipunna, Calapunna, Naithekku* (Mal.); *Nai thaekku, Kothukkai Uvaa* (Tam.).

Deciduous trees to 25 m tall; bark thick, surface grey, molted with white, smooth; branchlets stout, with prominent leaf scars. Leaves simple, alternate, clustered at the tip of branchlets, 15-obovate to oblong-lanceolate, 60 x 10-25 cm, scarious, base acute or attenuate, apex obtuse, margin serrate, rarely entire, glabrous above, puberulent beneath; lateral nerves many, parallel, prominent, intercostae scalariform, faint; petiole 15-60 mm long, sheathing, stout, glabrous, winged. Flowers fascicled on old branches, 2.5-3 cm across, yellow, fragrant; pedicels 2.5-6 cm long; bracts hairy. Sepals 5, 8-12 mm long, obovate or elliptic, glabrous, accrescent. Petals 5, 10-20 x 7-12 mm, obovate, obtuse, bright yellow. Stamens numerous, yellow in 2 series, outer series with 60-90 stamens, each 2.5-4 mm long, inner series with up to 10 stamens, each 6-9 mm long. Carpels 5-10, cohering at the axis, arranged on the narrow conical receptacle, unilocular, ovules many; styles free, 4 mm long. Fruit aggregate of berries, globose, 15 x 13 mm, indehiscent, fleshy, yellow, orange or red, subtended by persistent sepals. Seeds numerous, 5 x 3.5 mm, ovoid, black, glabrous, exarillate. Fl. & Fr.: March – June.

Western Ghats: Deciduous forests of Gujarat, Maharashtra, Goa, Karnataka, Kerala and Tamil Nadu. India: Andhra Pradesh, Assam, Bihar, Madhya Pradesh and Meghalaya. World: China, Nepal, Bhutan, Thailand, Myanmar, Indonesia, Vietnam and Malaysia.

IUCN: NE (IUCN, 2024).

Uses: The wood has little value, rarely used for hous-post.

Dillenia retusa Thunb., Trans. Linn. Soc. London 1: 200. 1791. Synonyms: *Wormia retusa* (Thunb.) Hook.f. & Thomson, *Wormia integra* (Thunb.) Hook.f. & Thomson Vern.: *Ceylon Naithekku* (Tam.).

Trees to 40 m tall; branchlets with prominent leaf scar. Leaves simple, large, alternate or spirally arranged, obovate-oblanceolate, 10-22 x 7-10 cm, serrate to undulate, apex obtuse or rounded, hirsute beneath; lateral veins 12-20, prominent, hirsute; petioles channeled above, hirsute when young, glabrous on maturity. Flowers solitary or clusters in terminal racemes, bracts, bracteoles caducous when present. Seplas 5, 5-toothed, spreading, enlarged and fleshy in fruit. Corolla 5, 5-lobed, white or yellow, usually larger than sepals, caducous.

Stamens free, numerous, filaments cylindric, anthers linear with parallel loculi, Carpels 5-20, borne on a conical receptacle, radiating styles, ovules 4-80. Fruits indehiscent pseudocarps, enclosed by enlarged thick sepals. Fruits subglobose, indehiscent, enclosed by thicked sepals, to 3 cm across, orange when rip. Seeds ovoid, 3 mm diam, exarillate. Fl. & Fr.: Apri – September.

Western Ghats: Deciduous to semi-evergreen forests of Karnataka and Tamil Nadu. World: Sri Lanka.

IUCN: NE (IUCN, 2024).

Dillenia suffruticosa (Griff. ex Hook.f. & Thomson) Martelli., Malesia 3: 163. 1887. Synonym.: *Wormia suffruticosa* Griff. Vern.: *Natchathrakai Uvaa, Seemai Uvaa* (Tam.).

Small trees to 6 m tall; bark smooth, greyinsh. Leaves broadly elliptic-oblong, 8-20 x 3-10 cm, base obtuse to acute, apex acuminate, margin serrate, shiny above, glaucous beneath; lateral nerves 15-20 pairs, parallel, glabrous; petiole to 5-7 cm long, stout, prominently winged, glabrous. Racemes large, terminal and lateral; pedicel to 3 cm long, pinkish purple, glabrous. Flowers bright yellow, large, 5 cm across. Sepals 5, subequal, inner petals slightly larger than the outer, orbicular, fleshy, margin entire. Petals 5, broadly orbicular, clawed at base, margin emarginate. Stamens to 25-40, white, clustered; anthers elongated, apext gracilis. Capsule star-shaped opened, 5-7-locular. Seeds with redish arillate. Fl. & Fr.: December–February

Western Ghats: Rare planted tree in Maharashtraa dn Kerala (Naturalised). World: Native to Malysia, Sumatra, Java, Borneo, Philippines

IUCN: NE (IUCN, 2024).

DIPTEROCARPACEAE Blume

Key to the genera

1a. Leaves folded in bud; stipules amplexicaulis; calyx tubular, shortly lobed .... **Dipterocarpus**

1b. Leaves not foled in bud; stipules not as above; calyx lobes free 2

2a. Fruiting calyx longer than fruit, erect; anthers with conspicuous awns 3

2b. Fruiting calyx shorter than fruit, spreading; anthers without awns 4

3a. Petals less than 6 mm long; stamens usually 10 .... **Hopea**

3b. Petals more than 9 mm long; stamens usually 15 .... **Shoera**

4a. Flowers in terminal panicles; sepals linear, obtuse; calyx scarcely enlarged in fruit .... **Vateria**

4b. Flowers in axillary racemes; sepals ovate-lanceolate, acuminate; calyx enlarged in fruit .... **Vatica**

Dipterocarpus C.F. Gaertn.

1a. Flowers in cymes or panicles, pedicle short ....**D. bourdillonii**

1b. Flowers in axillary racemes, pedicle long .... **D. indicus**

Dipterocarpus bourdillonii Brand., Hooker's Icon. Pl. 25: t. 2403. 1895. Vern.: *Charatta-anjili, Karangali, Kalpine* (Mal.).

Large trees to 30 m tall; bark grayish or orange brown or sometimes pinkish brown, flaked in patches, densely lenticellate and with aromatic white resin, branchlets with tubercular dots and subulate scales. Leaves simple, alternate, lanceolate-ovate to oblanceolate-elliptic, base obtuse to cuneate, margins entire usually sinuate towards the apex, apex obtuse to acute, leathery, coriaceous, midrib impressed above and prominent beneath, lateral veins straight, curved near margins; petiole stout, geniculate, stipules large encircling twigs near terminal buds, caducous, scars annular. Inflorescence axillary or terminal, irregular cymes or panicles. Flowers large, fragrant, pink or white. Sepals 5, 5-lobed, tubular with short valvate lobes, 2 longer, spathulate-oblong, 3 lobes short, forming a crown round apex. Petals 5, narrow oblong, large, white, with dark red or pink strip near centre. Stamens 15 or more, adnate persistent; filaments unequal base connate, broad tapering and filiform near yellow anthers. Ovary enclosed inside calyx tube, narrow ovoid-conical, puberulent, style filiform, glabrous, stigma dilated. Fruit large, nutlike, enclosed in calyx tube with 2 accrescent lobes like wings. Seeds adnate at the base, cotyledons large, unequal. Fl. & Fr.: January – June.

Western Ghats: Evergreen forests above 1300 m in Karnataka, Kerala and Tamil Nadu (Endemic).

IUCN: CR- C2a(i). (IUCN, 2024).

Uses: Wood is used for house construction and in plywood industry.

Dipterocarpus indicus Bedd., Conserv. For. Rep.: 17. 1864. Vern.: *Kalpayini, Vellaini, Vavangu* (Mal.); *Eunei, Vel ennei* (Tam.).

Lofty evergreen trees to 40 m tall; bark pale grey, smooth, exfoliations irregular; branchlets tomentose. Leaves simple, alternate, ovate, 7-25 x

4-17.5 cm, base acute, obtuse or cordate, apex acute or acuminate, margin crenate, undulate, glabrous, coriaceous, lateral nerves 10-15 pairs, prominent, intercostae scalariform, obscure; petiole 17-50 mm, slender, glabrous, swollen tipped. Flowers white, fragrant, 3-5 together in axillary racemes; pedicels 3 cm long. Sepals 5, cupular at base, lobes 5, unequal. Petals 5, tinged with pink. Stamens 30; filaments yellowish, often dilated at base; connective aristate. Ovary superior, slightly adherent to the calyx tube, 3-celled; ovules 2 in each cell. Fruit a nut, wings 2, oblong, reddish-brown, to 10 x 2 cm. Fl. & Fr.: February – June.

Western Ghats: Evergreen forests above 1200 m in Karnataka, Kerala and Tamil Nadu (Endemic).

IUCN: EN- C2a(i) (IUCN, 2024).

Uses: The wood is durable and ued for making furniture and building construction. A sweet-smelling gum exudate from stem bark which is used for trating rheumatism.

HOPEA Roxb.

1a. Raceme or panicle tomentose .... **H. parviflora**

1b. Raceme or panicle glabrous............ 2

2a. Bark peeling off in long strips upward becoming recurved and hanging on trunk; lateral nerves of leaf more than 5 pairs .... **H. racophloea**

2b. Bark not peeling off as above; lateral nerves of leaf more than 12 pairs 3

3a. Leaves linear-lanceolate; fruits covered gummy resins .... **H. utilis**

3b. Leaves elliptic-oblong or ovate; fruits not covered gummy resins 4

4a. Panicles several together in axils of leaves; petals 2-lobed............ **H. erosa**

4b. Panicles 1-3 together either terminal or leaf axile; petals not 2-lobed 5

5a. Petals glabrous, not ciliate at margin; anthers ovoid............ **H. ponga**

5b. Petals puberulous, ciliate at margin; anther orbicular 6

6a. Flowers less than 3 mm long; anthers forming a straight appendages .... **H. jacobi**

6b. Flower more than 3 mm long; anthers forming flagellate appendages...8

7a. Leaves elliptic-oblong, lateral nerves 6-8 pairs .... **H. glabra**

7b. Leaves ovate, lateral nerves 8-12 pairs .... **H. canarensis**

8a. Lateral nerves 4-5 pairs; petals white; fruiting pedicel present .... **H. modesta**

8b. Lateral nerves 5-9 pairs; petals yellow; fruiting pedicel absent .... **H. sasidharanii**

Hopea canarensis Hole., Indian Forester 44: 575. 1918.

Evergreen trees to 20 m tall; bark pale brown, smooth, at some length fissured or flaked, with white resin. Leaves simple, alternate, ovate, 10-18 x 3.5-8 cm, base obtuse to oblique, margins entire or very slightly undulate, apex acuminate, coriaceous, leathery, usually hairy domatia or gland-like swelling in the axils of the lateral veins, midrib impressed above and prominent beneath, lateral veins pinnate 6-10, tertiary veins parallel; petiole slender, stout, about 1-2 cm long, stipules linear, caducous. Inflorescence axillary, lax panicles or fascicled, glabrous. Flowers subsessile or about 2 mm long, bracts early caducous or ebracteate. Sepals 5, 5-lobed, tubular with very short, imbricate, glabrous, adnate to receptacle, 2 outer longer, ovate-deltoid, apex obtuse, thick, becoming much longer than others, 4-5 x 3 mm, 3 inner lobes short, suborbicular, thin near margins, apex usually mucronate, about 2-3 mm long. Petals 5, imbricate, falcate-oblong, fleshy, contorted to right or left, glabrous inside, densely stellate outside, 8-9 x 3 mm. Stamens 10-15, base connate, adnate to petals, usually caducous along with the petals; filaments subulate, usually inner ones shorter and sterils, outer ones longer; anthers medifixed, short, connective thin near apex. Ovary 3-locular, ovoid, glabrous or very sparsely puberulous; ovules in two rows; style filiform, short, cylindric, glabrous; stigma thick. Fruit large, nutlike, globose-ovoid, 1.2-1.5 cm, enclosed in calyx tube with 2 accrescent lobes like wings, about 5-8 cm long. Seed 1, adnate at the base, cotyledons unequal. Fl. & Fr.: March – July.

Western Ghats: Evergreen forests above 1200 m in Karnataka (Endemic).

IUCN: EN - B1ab (i,iii,v). (IUCN, 2024).

Hopea erosa (Bedd.) Stooten., Reinwardtia 3: 318. 1956. Synonyms: *Balanocarpus erosa* Bedd. Vern.: *Karakong* (Mal.); *Karumkongu* (Tam.).

Evergreen trees to 25 m tall; bark thick, pale brown, smooth. Leaves simple, alternate, oblong or oblong-lanceolate, 10-20 x 3-7 cm, apex acute or acuminate, base unequally cordate, margin entire, glabrous, coriaceous; lateral nerves 10-17 pairs, pinnate, arched, prominent, intercostae reticulate, prominent; petiole 17-50 mm, slender, glabrous, swollen tipped; stipule minute, lateral, deciduous. Flowers greyish-yellow, 1 cm across, subsessile, in unilateral, axillary racemed panicles. Sepals 5, tube short, adnate to the torus. 5-lobed, lobes suborbicular, subequal, margin ciliate, reddish, imbricate. Petals 5, bilobed, pilose outside, glabrous inside, pinkish-white. Stamens 15, rarely 10; filaments dilated at the base; anthers short, ovate, exceeded by the apical awn. Ovary superior, glabrous, 3-celled, ovules 2 in each cell; style short; stigma thick. Fruit a nut, ovoid or oblong, apiculate, enclosed at the base by the thickened and accrescent sepals which often attain 2.5 cm long and spread horizontally. Fl. & Fr.: April – January.

Western Ghats: Evergreen forests above 1000 m in Kerala and Tamil Nadu (Endemic).

IUCN: CR- C2a(i) (IUCN, 2024).

Hopea glabra Wight & Arn., Prodr. Fl. Ind. Orient. 1: 85. 1834. Synonyms: *Hopea wightiana* var. *glabra* (Wight & Arn.) Bedd. Vern.: *Illa ponga, Irumbakkam, Karukongu, Kongu* (Mal.); *Kara-pongu, Kongu* (Tam.).

Evergreen trees to 20 m tall; bark 5-6 mm, dark brown. Leaves elliptic, oblong-lanceolate, 6-10 x 2.5-4 cm, base acute, obtuse or round, apex acute or acuminate, margin entire, glabrous, coriaceous; lateral nerves 4-8 pairs, parallel, prominent, intercostae scalariform, prominent; petiole 8-20 mm, slender, glabrous. Flowers creamy yellow, 1-3 together in axillary racemose panicles. Sepals 5, glabrous. Petals 5, puberulous outside, margin ciliate. Stamens 15, slightly connate; anther orbicular, connective forms a subulate point. Ovary superior, pubescent, 3-celled, ovules 2 in each cell; stylopodium puberulous; style glabrous, short, subulate. Fruit a nut, ovoid; wings 2, reddish. Fl. & Fr.: January – November.

Western Ghats: Evergreen forests above 1300 m in Karnataka, Kerala and Tamil Nadu (Endemic).

IUCN: EN- C2a(i);D. (IUCN, 2024).

Hopea jacobi C.E.C. Fischer., Bull. Misc. Inform. Kew 1932: 245. 1932.

Trees to 15 m tall; bark thin, smooth, lenticellate; branchlets slender, dark brown, glabrous. Leaves altrenate, entire, rounded or obtuse at apex, 5.5-9 x 2.5-4 cm, chartaceous up to 7-flowered. Sepals 5, lobes suborbicular, 1.5 mm long, coriaceous, inner ones thin-margined. Petals 5, oblong, obtuse, 3-3.5 mm long, minutely ciliate, puberulous outside. Stamens 15; anthers orbicular, flat; appendages subulate, about 3 times longer than anthers forming a fine straight arista. Styles short, pointed. Fruits ovoid, to 1.7 cm long; larger 2 calyx lobes oblong to oblanceolate, obtuse at apex, 4.5 x 1.35 cm, 9-nerved. Fl. & Fr.: May – June.

Western Ghats: Rare in moist evergreen forests above 1000 m in Silent valley of Kerala, Madikeri hills of Karnataka and Tirunelveli hills of Tamil Nadu (Endemic).

IUCN: CR- B1ab(iii,v)+2ab(iii,v). (IUCN, 2024).

Hopea modesta (A.DC.) Kosterm., Ceylon J. Sci., Biol. Sci. 15: 47. 1982. Synonym: *Hopea jucunda* ssp. *modesta* A.DC. Vern.: *Kongu* (Tam.).

Trees to 20 m tall; bark thin, smooth, brow; branchlets green, terete, pendulous; latex white rsinous. Leaves simple, alternate, ovate-lanceolate, 5-9 x 2-4 cm, base rounded to obtuse, caudate acuminate at apex, margin entire, minutely tomentose on both sides, glabrous at maturity; lateral nerves 4-5 pairs, prominent on both sides, secondary nerves faint. Inflorescence branched panicle, terminal, tomentose. Flowers creamy white, small, fragrant. Sepals 5, toemntose. Petals 5, puberulous outside, margin ciliate. Stamens 15-20, slightly connate; anther orbicular, connective pointed above. Ovary superior, pubescent, 3-celled, ovules 2 in each cell; stylopodium puberulous; style glabrous, short, subulate. Fruit a nut, ovoid; wings 2, brownish. Fl. & Fr.: January – November.

Western Ghats: Evergreen forests above 1300 m in Tamil Nadu. World: Sri Lanka.

IUCN: CR- A1cd. (IUCN, 2024).

Hopea parviflora Bedd. Rep., Conservator Forests 1863-1864: 31. 1864. Vern.: *Thambagam, Kambagam, Irupu, Irumbagam* (Mal.); *Pongu, Kongu* (Tam.).

Evergreen trees to 35 m tall; bark 6-10 mm, light brown or grey, mottled with white, rough vertically fissured, fibrous; branchlets reddish-brown, slightly pubescent. Leaves simple, alternate, ovate, lanceolate or ovate-lanceolate, 8-11.5 x 3-5 cm, apex acute or glabrous, apiculate, base acute obtuse, cordate or subcordate, coriaceous; lateral nerves 8-12 pairs, intercostae scalariform, faint, domatia present; petiole 10-12 mm, slender, pubescent when young, glabrous when matured, grooved above; stipules small, lateral, deciduous. Flowers creamy yellow, 3-4 mm across, in unilateral terminal and upper axillary, tomentose racemose panicles. Sepals 5, 3 mm long, tomentose outside. Petals 5, 6 mm long, oblong, glabrous, fimbriate at apex. Stamens 5, rarely 10, slightly connate; connective of anthers produced into a subulate point. Ovary superior, glabrous, ovules 2 in each cell; style short, subulate. Fruit a nut, 5-6 mm long, terete, glabrous; wings 2, 6 x 1.8 cm, 8-10 nerved, glabrous. Fl. & Fr.: November – May.

Western Ghats: Evergreen forests above 1200 m in Karnataka, Kerala and Tamil Nadu (Endemic).

IUCN: LC (IUCN, 2024).

Uses: The wood is externsively used for house and bridge construction. It is also used for railway sleepters.

Hopea ponga (Dennst.) Mabber. var. **ponga.** Taxon 28: 587. 1979. Synonyms: *Artocarpus ponga* Dennst. Vern.: *Irumbagam, Ilampongu, Pongu* (Mal.)

Evergreen trees to 20 m tall; bark dark grey, smooth; branchlets pubescent. Leaves simple, alternate¸ovate-lanceolate or ovate-oblong or ovate, 7.5-24 x 3-7 cm, base rounded or obtuse, apex acute, glabrous, coriaceous; lateral nerves 5-10 pairs, intercostae scalariform, faint; petioles 5-15 mm long, stout, glabrous, grooved above; stipules small, lateral, deciduous. Flowers yellowish-pink, in axillary unilateral drooping racemose panicles. Sepals 5, shortly united at base, ovate, glabrous, two outer sepals obtuse, larger than the 3 acute inner ones. Petals 5, 5-8 mm long, ovate-lanceolate, pubescent. Stamens 10-15, alternate filaments with two anthers; appendages of anthers filiform, 4 times longer than anthers. Ovary superior, puberulous above, 3-celled; ovules 2 in each cell; stylopodium glabrous; style short, subulate. Fruit a nut, ovoid, 10-12 mm long, fruiting calyx lobes expanded to forms wings in fruit, dark red; 2 longer wings 5.5-10 x 1-1.5 cm, 8-nerved, 3 smaller ones unequal. Fl. & Fr.: March – September.

Western Ghats: Evergreen forests above 1200 m in Maharashtra, Goa, Karnataka, Kerala and Tamil Nadu (Endemic).

IUCN: VU- A2c. (IUCN, 2024).

Key to the varieties

1a. Leaves ovate-obong; fruit dark red .... var. **ponga**

1b. Leaves oblanceolate-oblong; fruit orange-brown .... var. **cauveriana**

Hopea ponga (Dennst.) Mabber. var. **cauveriana** Kesh. Murthy &Yogan. Curr. Sci. 56: 544. 1987.

Fl. & Fr.: March – April.

Western Ghats: Evergreen forests of Karnataka (Endemic).

IUCN: VU- A2c. (IUCN, 2024).

Hopea racophloea Dyer., J.D. Hooker, Fl. Brit. India 1: 310. 1874. Vern.: *Neduvali-pongu, Nai-kambagam* (Mal.); *Karum-kongu* (Tam.).

Lofty trees to 35 m tall; bark dark brown, peeling off in strips, with the lower ends of each strip detached from stem and the upper end still attached to it and curved upwards; branchlets black. Leaves simple, alternate, stipulate, ovate or oblong-lanceolate, 5-11 x 2.5-6 cm, base obtuse, round or acute, apex acute or obtusely acuminate, margin entire, glabrous, coriaceous, lateral nerves 4-6 pairs, intercostae scalariform, prominent, domatia present; petiole 5-10 mm, slender, glabrous. Flowers pinkish-yellow, 2-4 together, in axillary unilateral racemose panicles. Sepals 5, equal, spathulate, obtuse, glabrous. Petals 5, glabrous inside, hairy outside with a twisted, flattened appendage at apex. Stamens 15, slightly connate; connective produced into a subulate point. Ovary superior, obconical, 3-celled; ovules 2 in each cell; styles short, subulate; stylopodium nearly as long as ovary, cylindrical, slightly narrowed at base. Fruit a nut, glabrous, closely surrounded by the base of calyx; wings 2, 7.0-8.2 cm long, obtuse, with 7-11 longitudinal nerves, reddish. Fl. & Fr.: February – June.

Western Ghats: Evergreen forests above 1300 m in Karnataka and Kerala (Endemic).

IUCN: EN- C2a(i). (IUCN, 2024).

Hopea sasidharanii Robi & Sujanapal, Phytotaxa 429(2): 167. 2020.

Evergreen trees to 20 m tall; bark thick, pale brown, inner bark creamy or pale yellow; branchlets subterete, glabrous. Leaves simple, alternate, distichous, elliptic, elliptic-ovate, 7.5–21.5 × 3.5–8 cm, sub-rounded, obtuse or cuneate at base, margin entire, acute at apex; lateral nerves 5–9 pairs, basal 2–pairs opposite, secondary veins arcuate, narrowly depressed above, raised below, scalariform-reticulate, domatia absent; petiole 9–14 mm long, thick, glabrous. Inflorescence axillary to ramiflorous; panicles 7–13.5 cm long, lax, erect, sparsely puberulent; branches 5–9, slightly curved, bearing 4–8 secund flowers per branch; bracteole fugacious. Flowers 5–merous, actinomorphic, bisexual. Sepals 5, broadly ovate, subequal, 3–4 × 1.5–2 mm, minutely connate at base, hirsute, ciliolate along margin, obtuse at apex. Petals 5, pale yellow, twisted, falcate-oblong, 8.5 × 3 mm, partly hirsute dorsally, glabrous inside, connate at base, entire or ciliate, obtuse or undulate at apex, veins 7–10. Stamens 15 in 2 whorls, adnate to corolla base, 2 mm long; filaments to1 mm long, compressed at base, tapering, glabrous; anthers suborbicular, 0.3 mm long, dehiscence longitudinal; connective-appendages flagellate at apex, filiform, 0.5 mm long. Ovary ovoid, 1 mm long, hirsute; ovary and stylopodium hour-glass shaped, tapering into style; style thick, 0.6 mm long, glabrous. Fruits ovoid, 1.5 × 0.7 cm, obtuse at apex, enclosed by the persistent thick sepals, fruit pedicel absent. Fl. & Fr.: March – August.

Western Ghats: Rare in evergreen forests of Kerala (Endemic).

IUCN: NE (IUCN, 2024).

Hopea utilis (Bedd.) Bole, Kew Bull. 6: 146. 1951. Synonyms: *Balanocarpus utilis* Bedd., *Hopea longifolia* Dyer Vern.: *Kongu* (Tam.).

Evergreen trees to 25 m tall; bark surface dark brown, mottled with grey, smooth; young shoots pubescent. Leaves simple, alternate, stipulate, linear-lanceolate, ovate-lanceolate or oblong-ovate, 9-24 x 2.5-10 cm, base round or acute, apex acute, margin entire, glabrous, coriaceous; lateral nerves 10-12 pairs, intercostae scalariform, faint; petiole 10-12 mm long, stout, glabrous. Flowers yellowish-white, 6-8 mm long, in unilateral axillary, panicled racemes; pedicels short. Sepals 5, tube very short, adnate to the receptacle, lobes 5, slightly connate at base, hoary outside, 2 outer ones slightly longer, ovate, more or less obtuse, thickened, 3 inner suborbicular, often mucronate, thin along the margin. Petals 5, oblong, obtuse, crenulate, fleshy, pubescent, imbricate. Stamens 15; filaments dilated at base, subulate, anthers suborbicular, extended by the apicular awn. Ovary superior, pubescent, 3-celled; ovules 2 in each cell; style short, cylindric; stigmas thick. Fruit a nut 10-12 mm across, globose, pointed, shining, with tuberculate, subacute, accrescent calyx lobes. Fl. & Fr.: April – August.

Western Ghats: Evergreen forests of Kerala and Tamil Nadu (Endemic).

IUCN: EN- B1ab (ii,iii,v) + 2ab (ii,iii,v); C2a(i); D. (IUCN, 2024).

Shorea Roxb. ex C.F. Gaertn.

1a. Leaves ovate-laceolate; tip caudate or abrubtly acuminate 2

1b. Leaves elliptic oblong, tip obtuse or emarginate ....**S. roxburghii**

2a. Petals white, yellow with pinkish strips, glabrous inside, pubescent outside .... **S. robusta**

2b. Petals creamy white, not pinkish strips, glabrous inside, tomentose outside .... **S. tumbaggaia**

Shorea robusta C.F.Gaertn., Suppl. Carp.: 48. 1805. Synonyms: *Dryobalanops robusta* (C.F. Gaertn.) Oken, *Vatica robusta* (C.F. Gaertn.) Steud.

Large deciduous trees to 50 m tall; bark often fissured, flaky or smooth, reddish brown; branchlets tomentose, resin white. Leaves simple, alternate, lanceolate-ovate to oblong-elliptic, 8-40 x 4-24 cm, base obtuse to cordate, margins entire, apex acute with very short acuminate tip, reddish or pinkish when young becoming dark green when mature; lateral veins 11-13 pairs, subparallel, curved near margins, tertiary veins scalariform; petiole stout, glabrous, 2-2.5 cm long; stipules lanceolate-falcate, fugacious, caducous. Inflorescence axillary or terminal, panicles or unilateral racemes, to 25 cm long. Flowers large, fragrant, yellow or cream white; bracts minute, caducous. Sepals 5, tubular with short valvate lobes, 5 lobed, imbricate, ovate-triangular, adnate to the receptacle, outer 3 larger than the inner 2 lobes. Petals 5, white, pale yellow to pinkish strips, lanceolate, base slightly connate, apex acuminate, glabrous inside, buff pubescent outside, 10-13 nerved longitudinally. Stamens 25 or more, adnate persistent, short; filaments flattened or tapering and filiform; anthers subglobose, narrow oblong-ovate, connective subulate, minutely tri-fid towards beyond anther lobes. Ovary ovoid, pubescent, tri-locular; ovules 2 per locule; stylopodium distinct or not, style subulate; stigma tri-lobed or tri dentate. Capsule or nutlike, ovoid, apex acute, densely pubescent, 1.5 cm long, enclosed in calyx tube with 3 outer accrescent larger lobes, oblong-spathulate, apex obtuse, saccate, pubescent. Seeds 1, cotyledons large, fleshy. Fl. & Fr.: April-May

Western Ghats: Deciduous forests of Karnataka, Kerala and Tamil Nadu. World: Bangladesh, Himalaya, Nepal and Tibet.

IUCN: LC (IUCN, 2024).

Shorea roxburghii G. Don, Gen. Hist. 1: 813. 1831. Synonyms: *Shorea talura*Roxb., *Shorea talura* Roxb., *Shorea cochinchinensis* Pierre, *Shorea attopoensis* Pierre, *Shorea cochinchinensis* var. *saigonensis* Pierre Vern.: *Taluram* (Mal.); *Kungiliyam, Thalari, Thaloora* (Tam.).

Trees to 25 m tall; bark brownish to greyish-brown, longitudinally fissured; branchlets glabrous. Leaves simple, alternate, stipulate, elliptic, oblong, elliptic-ovate, elliptic-oblong or elliptic lanceolate, 6-18 x 3.5-9 cm, base round or subcordate, apex obtuse or emarginate, margin entire, glabrous, coriaceous; lateral nerves 12-16 pairs, parallel, intercostae scalariform, faint; petiole 10-25 mmlong, stout, glabrous. Flowers light pink, 10-13 mm long, in drooping on axillary and terminal panicles; pedicels to 5 mm. Sepals 5; 3 lobes much enlarged than the other 2, basally connate, ovate-lanceolate, glabrous, accrescent, imbricate. Petals 5, oblong, densely villous without. Stamens 15; filaments to 1 mm; anthers 1.5 mm, linear; connectives appendaged, to 1.5 mm. Ovary superior, globose, 3-celled, ovules 2 in each cell; style to 4 mm; stigma 3-lobed. Capsule enclosed within accrescent sepals; wings oblong, unequal, 4-7 cm, prominently nerved. seed solitary, obovoid 7 x 4 mm. Fl. & Fr.: December – May.

Western Ghats: Deciduous to semievergreen forests of Karnataka, Kerala and Tamil Nadu. India: Andhra Pradesh. World: Myanmar, Thailand, Laos, Cambodia, Vietnam and Malaysia.

IUCN: VU- A2cd. (IUCN, 2024).

Uses: Wood is used for making furniture, construction purposes, etc.

Shorea tumbuggaia Roxb., Fl. Ind., ed. 1832. 2: 617. 1832.

Deciduous trees to 12-18 m tall; bark often fissured longitudinally, thick, rough, dark brown with white resin. Leaves simple, alternate, broadly ovate to cordate, 6-21 x 3.5-12 cm, base truncate to shallow emarginate, margins entire, apex abruptly short acuminate tip, coriaceous, dark green, glabrous on both sides, midrib impressed above and prominent beneath; lateral veins 8-10 on either side of the midrib, subparallel, curved and branched near margins, tertiary veins scalariform; petiole stout, tomentose, about 2-5 cm long; stipules lanceolate-falcate, fugacious, caducous, covered with silvery peltate scales, about 8 mm long. Inflorescence axillary or terminal, panicles, up to 20 cm long, glabrous or hoary. Flowers cream white, pedicel subsessile, densely pubescent when buds. Sepals 5, 5- lobed, tubular with short valvate lobes, imbricate, outer 3 larger than the inner 2 lobes. Petals 5, white, lanceolate, base slightly connate, apex attenuate, glabrous inside, tomentose outside. Stamens 15 or more, adnate persistent, base slightly united, anthers subglobose, appendages hairy. Ovary enclosed inside calyx tube, narrow ovoid, pubescent, 3-locular, 2 ovules per locule, style subulate, stigma 3-lobed. Capsule or nutlike, ovoid, apex acute, densely pubescent, about 2 cm long, enclosed in calyx tube with 3 outer accrescent larger lobes and broader lobes. Seed 1, cotyledons large, fleshy. Fl. & Fr.: March-April

Western Ghats: Deciduous forests of Maharashtra, Karnataka and Tamil Nadu (Endemic). India: Andhra Pradesh and Telangana.

IUCN: EN- B1ab(i,ii,iii,v)+2ab(i,ii,iii,v). (IUCN, 2024).

Vateria L.

1a. Leaves pubescent beneath, lateral nerves 18-25; fruits 3-valved, indehiscent .... **V. copallifera**

1b. Leaves glabrous beneath, lateral nerves 12-20; fruits 3-valved dehiscent. 2

2a. Capsule ovoid oblong; seeds 1 per locule .... **V. indica**

2b. Capsule cylindric; seeds 2 per locule .... **V. macrocarpa**

Vateria copallifera (Retz.) Alston., H. Trimen, Handb. Fl. Ceylon 6(Suppl.): 26. 1931. Synonyms: *Elaeocarpus copallifera* Retz., *Vateria acuminata* Heyne Vern.: *Dhupa maram* (Tam.).

Evergreen trees to 30 m tall; bark smooth, grayish brown, thin, brittle, with little white resin. Leaves simple, alternate, lanceolate-oblong to elliptic, about 15-30 x 4.5-12 cm across, base rounded to subcordate, margins entire, apex obtuse with mucronate tip to shortly acuminate, thickly coriaceous, glabrous above and sparely pubescent on the midrib and veins beneath; lateral veins 18-25 like feathers on either side of the midrib; petiole stout, sparsely pubescent below; stipules narrow, linear-deltoid, apex attenuate, caducous. Inflorescence axillary or terminal, lax racemose panicles or corymbs, about 20 cm along. Flowers creamy white, pedicellate, bracts caducous. Sepals 5, 5-lobed, short imbricate lobes, base slightly connate, unequal, reflexed, ferruginous outside, persistent, scarcely enlarged in fruit. Petals 5, white, pale yellow or creamy white, orbicular-oblong, apex acute, spreading. Stamens 45-55, filaments sessile or very short; anthers linear oblong, yellow, puberulous, connective muticous. Ovary narrow ovoid, 3-locular, densely fulvous pubescent, 2-3 ovules per locule; style subulate; stigma minute entire or lobed. Capsule subglobose-ovoid, about 10 x 6-7 cm, apex pointed, densely fulvous pubescent, indehiscent, 3-valved, enclosed in scarcely accrescent calyx lobes. Seeds 1, cotyledons large, unequal, fleshy. Fl. & Fr.: June-August

Western Ghats: Evergreen forests above 1300 m in Tamil Nadu. World: Sri Lanka.

IUCN: EN- A1cd, C2a. (IUCN, 2024).

Vateria indica L., Sp. Pl.: 515. 1753. Synonym: *Vateria malabarica* Blume Vern.: *Payin, Payini, Perumpayini, Vella payini, Vella kutrukkam* (Mal.); *Vellei-kunthricum, Vellai kungiliyam* (Tam.).

Evergreen trees to 30 m tall; bark 10-12 mm, greyish, blotched with white and green, smooth, exudation resinous. Leaves simple, alternate, stipulate, oblong, 7-20 x 5-9 cm, base round, obtuse or cordate, apex acuminate or obtusely acute, margin entire, glabrous, coriaceous, lateral nerves 12-18 pairs, intercostae scalariform, prominent; petiole 25-40 mm, stout, stellate pubescent, swollen tipped. Flowers white, 2-3 cm across, fragrant, in terminal panicles, densely stellate puberulus. Sepals 5, free, lanceolate, covered with stellate hairs. Petals 5, obovate, spreading, shortly united at base. Stamens many, free; filaments hairy; anthers often slightly hairy at base; connective produced into a filiform appendage. Ovary superior, ovoid-oblong, tomentose, 3-celled, 2-ovules in each cell; style filiform, glabrous; stigma small. Capsule ovoid or oblong-ovoid, 11-15 x 5-6 cm, pale brown, lanceolate, tip acuminate. Seed one. Fl. & Fr.: March – May.

Western Ghats: Evergreen forests above 1300 m in Maharashtra, Karnataka, Kerala and Tamil Nadu (Endemic).

IUCN: VU- A2cd. (IUCN, 2024).

Uses: Wood is much demand in plywood industry and is sutiable wood for paper pulp. Gum resin extracted and in trade as Piney resin. Dried seed kernal yields the Malabar tallow or dhupa fat, which is used for making candle and soap.

Vateria macrocarpa Gupta., Indian Forester 55: 231. 1929. Vern.: *Perumpayin* (Mal.); *Vellei kunthricum* (Tam.).

Evergreen trees to 30 m tall; bark grey, mottled with white and green, smooth, exudation resinous, pale yellowish-white. Leaves simple, alternate, stipulate, oblong or oblong-lanceolate, 14-40 x 6-20 cm, base round, obtuse or subcordate, apex acute, obtuse or shortly acuminate, margin entire, glabrous, coriaceous; lateral nerves 12-20 pairs, intercostae scalariform, prominent; petiole 25-60 mm long, stout, minutely pubescent, swollen tipped. Flowers white, in panicles. Sepals 5, 3 mm long. Petals 5, 1.5 cm long. Anthers linear, glabrous. Ovary superior, densely tomentose, 3-celled, ovule 2 in each cell; styles slightly longer than stamens. Capsule cylindric, 15 x 5-6 cm, lanceolate, tip acuminate, curved, splitting from the top downwards into valves. Fl. & Fr.: September – June.

Western Ghats: Rare in evergreen forests above 1500 m Kerala and Tamil Nadu (Endemic)

IUCN: CR- B1ab(iii,v). (IUCN, 2024).

Uses: Wood is substituted for *Vateria indica*, due to unavailability of the wood not in trade.

Vatica L.

Vatica chinensis L., Mant. Pl. 2: 242. 1771. Synonyms: *Vatica roxbughiana* (Wight) Blume; *Vateria roxburghiana* Wight Vern.: *Adukka payni, Eheru piney, Vella payini* (Mal.); *Vellei Kunthriikam, Vellei pasa* (Tam.).

Evergreen trees to 25 m tall; bark pale green, smooth; exudation resinous. Leaves simple, alternate, stipulate, ovate or oblong, 9-25 x 3-11 cm, base obtuse or broadly cuneate, apex obtusely acute, margin entire, coriaceous, glabrous; lateral nerves 10-14 pairs, parallel, intercostae scalariform, prominent; petiole 20-50 mm long, stout, glabrous. Flowers white, in axillary spreading panicles; pedicels 5-ribbed; ribs alternating with sepals. Sepals 5, tube very short, adnate to the base of the ovary, lobes 5, ovoid-deltoid, acute, pubescent. Petals 5, white, oblong. Stamens 15 in 2 rows; filaments short, flattened at base; anthers oblong, shortly apiculate. Ovary superior, covered with large shallow pits, lepidote, 3-celled, ovules 2 in each cell; style about as long as ovary, ribbed; stigmas densely papillose, obscurely 3-lobed. Capsule lepidote, subglobose shortly pointed with 3 obscure, loculicidal furrows, puberulous; pericarp coriaceous; calyx persistent. Fl. & Fr.: February – May.

Western Ghats: Evergreen forests above 1400 m in Karnataka and Kerala. World: Sri Lanka.

IUCN: EN- A2c+3c. (IUCN, 2024).

Uses: Wood is durable and used for baot and ship construction. The transparent yellow resin obtained from the bark is used for making varnish.

EBENACEAE Gurke

Diospyros L.

1a. Calyx and corolla 3-lobed 2

1b. Calyx and corolla 4 or 5-lobed 6

2a. Corolla urceolate, glabrous on both sides 3

2b. Corolla tubular or salver-shaped, hairy on both sides 4

3a. Male flowers 3-5, in axillary clusters; stamens 12-14, staminode 6; fruit cylindric oblong, less than 1 cm in diam ....**D. sahayadryensis**

3b. Male flowers 7-15, in axillary cymes; stamens 13-22, staminodes 4; fruit globose, upto 2 cm diam ....**D. sylvatica**

4a. Calyx hairy or pubescent on both sides ....**D. neilgerrensis**

4b. Calyx glabrous within, pubesct out side only ....**D. oocarpa**

5a. Ovary 3-celled, cells 2-ovuled; staminodes absent; male flowers 1-3-flowered axillary cyme ....**D. ferrea**

5b. Ovary 6-celled, cells single ovuled; staminodes present; male flowers moer than 3-flowered axillary cymes or cluster 6

6a. Calyx lobes imbricate 7

6b. Calyx lobes valvate 18

7a. Corolla urceolate, campanulate or bell-shaped 8

7b. Corolla tubular or hypocrateriform 15

8a. Corolla completely glabrous both sides 9

8b. Corolla hairy or pubescent at least out side 12

9a. Stamens up to 20, staminodes up to 12; ovary 4-8-celled .... **D. ebenum**

9b. Stamens 24-64, staminode 5; ovary 6-celled 10

10a. Staemens seated on torus; pedicels of female flowers 1.5-2.5 cm long...... .... **D. foliolosa**

10b. Staemns inserted on corolla tube; pediecels of female flowers up to 8 mm long 11

11a. Female flowers 1-3-flowered; staminodes 4; .... **D. montana**

11b. Female flower solitary, staminodes 8; .... **D. lotus**

12a. Calyx and corolla 5-lobed; stamens 10-12; ovary 10-celled ....**D. sulcata**

12b. Calyx and corolla 4-lobed; stamens more than 12; ovary less than 10-celled...13

13a. Calyx lobes contorted; fertile anther apiculate or obtuse; ovary 4-celled . ... **D. Chloroxylon**

13b. Calyx lobes imbricate; fertile anther obtuse hairy; ovary 8-celled 14

14a. Female flowers 3-7-flowered clusters; fruit globose, 1-2 cm in diam......... ....**D. ovalifolia**

14b. Female flowersr solitary; fruit cylindric oblong, 4-8 mm in diam ....**D. buxifolia**

15a. Calyx and corolla 5-lobed 16
15b. Calyx and corolla 4-lobed..17
16a. Stamens 10-12, staminodes 5; ovary 10-celled; fruiting calyx not enclosing fruit ....**D. sulcata**
16b. Stamens 15-20, staminodes more than 5; ovary 4-6-celled; fruiting calyx enclosed the fruit...**D. paniculata**
17a. Calyx lobes imbricate in bud; leaves distinctly nerved; seeds oblong....... ....**D. pruriens**
17b. Calyx lobes contorted in bud; leaves faint nerved; seeds pear-shaped ....**D. saldanhae**
18a. Corolla campanulate; ovary 4-celled .... **D. candolleana**
18b. Corolla tubular; ovary 8-celled 19
19a. Corolla glabrous on both sides ....**D. undulata**
19b. Corolla pubescen or hairy both sides 20
20a. Stamens seated on torus, staminodes 4; ovary 4-celled ....**D. barberi**
20b. Stamens inserted on corolla tube, staminodes 8-16; ovary 6-8-celled....21
21a. Calyx lobes glabrous, margin ciliate ....**D. trichophylla**
21b. Calyx lobes hairy, margin ciliate or not 22
22a. Ovary glabrous; fertile anther obtuse or acute ....**D. ghatensis**
22b. Ovary hairy; fertile anther apiculate 23
23a. Stamens 4-20; female flowers 1-6-flowered clusters; ovary pubescent...... ....**D. insignis**
23b. Stamens 8-12; female flowers 5-7-flowered cymes; ovary glabrous 24
24a. Fruiting calyx flat, not enclosing fruit; endosperm ruminate ....**D. crumenata**
24b. Fruiting calyx cup-shpaed, enclosing fruit; endosperm not ruminate ..25
25a. Ovary 4-12-celled; stamens seated on torus 26
25b. Ovary 12-16-celled; stamens inserted on corolla tube 28
26a. Young branchlets glabrous; male cyme glabrate; calyx distinctly veined. ....**D. affinis**
26b. Young branchlets tomentose or pubescence; male cyme tomentose; calyx not veined 27
27a. Secondary veins reticulate reaised or obscure; endosperm smooth........... ....**D. melanoxylon**
27b. Secondary veins reticulate prominent; endosperm ruminate 28

28a. Stamens 8-16; ovary 4 or 8-10-celled 29

28b. Stamens 18-34; ovary 4-6-celled..**D. toposia**

29a. Female flowers subsessile; fruiting calyx woody, 3-3.5 cm diam ....**D. ridleyi**

29b. Female flowers pedicelled; fruiting calyx leafy, fruits more than 3 cm diam 30

30a. Female cyme on old wood; fruit globose ....**D. bourdilloni**

30b. Female cyme on leaf axile or old leaf scar; fruit subgloboe to cylindric.... 31

31a. Male and female inflorescence distinctly balck pubescent; stamens 20 or more ....**D. atrata**

31b. Male and female inflorescence sparse pubescent; stamens up to 20 32

32a. Leaves hairy on both sides; fertile stamens 5-8 .... **D. hirsuta**

32b. Leaves pellucid-punctate; fertile stamens 10-16 33

33a. Staminodes 5; ovary 8-celled; fruiting calyx foliaceous ..... **D. malabarica**

33b. Staminodes 8; ovary 4-8-celled; fruiting calyx bowl-shaped... **D. nilgrica**

Diospyros affinis Thw., Enum. Pl. Zeyl.: 179. 1860. Vern: *Kunchiganamara* (Kan.); *Illaccivicca, Kattuthuvara* (Mal.); *Somolpanachai, Chinnathuvarai* (Tam.).

Evergreen unarmed trees to 20 m tall; bark peeling off in thick flakes, fissured and ridged; branchlets glabrous or thinly puberulous. Leaves alternate, elliptic-lanceolate, 3–12 x 1–4 cm, narrowed at both ends, acuminate to obtuse at apex, glabrous on maturity, medifixed hairs when young, midrib canaliculate above, lateral nerves 8–10 pairs, not prominent. Male flowers 5–10, axillary, glabrate cymes. Sepals 4, 3–4 mm long, companulate-tubular, 4-toothed, lobes leafy, hairy both sides. Petals 4, 4-lobed, lobes 8 mm long, acute, glabrous inside, hairy outside. Stamens 12, mostly in pairs. Female flowers solitary, axillary, pedicelled. Sepals to 6–8 mm long, hairy both sides. Petals 8–10 mm long; staminodes 8. Ovary appressed hairy, 6-8 celled. Berry globose, 2.5–3 cm in diam. Seeds 4, ovate wedge shaped, 1.2–1.5 x 0.8 cm. Fl. & Fr.: February – October.

Western Ghats: Evergreen forests above 800 m in Karnataka, Kerala and Tamil Nadu. World: Sri Lanka.

IUCN: NE (IUCN, 2024).

Uses: Timber is used for bulding construction.

Diospyros atrata (Thw.) Alston, H. Trimen, Handb. Fl. Ceylon 6 (Suppl.): 180.

1931. Synonyms: *Diospyros embryopteris* var. *atrata* Thw. Vern.: *Gusvakendu, Thumbi* (Tam.).

Trees to 25 m tall; bark smooth, black, peeling off in plates; branchlets thinly pubescent. Leaves simple, alternate, elliptic-oblong to elliptic-lanceolate, 8.5-20 x 3-7 cm, apex acute to acuminate with blunt tip, base acute to attenuate, coriaceous, glabrous; midrib canaliculate above, stout beneath; secondary nerves 5-9 pairs, lower pairs closer than above ones; tertiary nerves strongly reticulate on both surfaces; petiole 1-1.5 cm long, grooved above, glabrous. Flowers dioecious. Male flowers pinkish-white, axillary, 3-8 in cymes. Female flowers solitary, axillary with short pedicels. Sepals 4, 4-5 x 5-6 mm, cup-shaped, black hairs both sides, lobes 4, thick, 2-3 mm long, ovate-triangular. Petals pinkish-white, tubular, 8-12 x 10-12 mm, tube short, fleshy, urceolate, glabrous both sides. Berry globose, up to 7 cm in diam, rusty brown scurfy tomentum, fruiting calyx foliaceous, densely black haired inside. Seeds 8, 2.5 x 1.5 cm. Fl. & Fr.: March – September.

Western Ghats: Evergreen forests of Kerala and Tamil Nadu. World: Sri Lanka.

IUCN: VU- B1+2c. (IUCN, 2024).

Uses: The leaves of this tree is cooked and eaten as vegetables by local people in Anamalai hills. Timber is occasionally used for mastss and yards. Fruits with plenty of gum and tannin.

Diopyros barberi Ramas., J. Proc. Asiat. Soc. Bengal 10: 47. 1914. Vern.: *Karunthuvarai, Karinthali* (Tam.).

Small trees to 2-5 m tall; bark rough, striated, lenticellate; branchlets and young leaves appressedly pubescent. Leaves alternate, simple, ovate-oblong or lanceolate-elliptic, 1.7-10 x 0.6-2.7 cm, glabrous, shining above, dull and punctate beneath, cuneate at base, entire, obtuse-acuminate, acumen up to 1.5 cm long, penninerved; midrib prominent and raised beneath, sunken or grooved above, hirsute beneath when young and glabrous when old, glabrous above; petiole up to 1 cm long, hirsute in young leaves, glabrescent-glabrous when old, grooved above. Male flowers in short cymes. Female flowers solitary with large leathery folded foliaceous calyx-lobes. Calyx tube pot-shaped up to 4 mm in buds, wrinkled, sparsely hairy without, glabrous within; lobes minute, margin ciliate, adnate to the ovary at base, lobes larger than male. Berry subglobose, 4 x 4 cm, pubescent; fruiting calyx cupular, accrescet, much enlarged in fruit. Fl. & Fr.: March – September.

Western Ghats: Evergreen forests above 1000 m in Kerala and Tamil Nadu. World: India including Andaman & Nicobar Islands.

IUCN: NE (IUCN, 2024), vulnerable (Narasimhan and Irwin, 2021).

Diospyros bourdillonii Brand., Indian Trees: 435. 1906. Vern.: *Karimaran, Karinthuvara, Kadal, Terika, Kari* (Mal.); *Karunthuvarai, Karuncattiyam, Karunduvarai, Karunchathi* (Tam.).

Evergreen dioecious trees to 25 m tall; bark thick, black, rough, flaking off in thin scales; young shoots strigose. Leaves simple, alternate, oblong, ovate-oblong, oblong-obovate or oblong-lanceolate, 7-17 x 2.5-8 cm, apex acuminate, base round or obtuse, margin entire, glabrous, coriaceous; lateral nerves 6-8 pairs, intercostae scalariform, prominent; petiole 5-10 mm long, stout, pubescent when young. Flowers unisexual, white. Male flowers in rusty tomentose axillary cymes, 10-12 mm long; pedicels 3 mm long. Female flowers in brown tomentose compound cymes from old wood; pedicels 5 mm long. Sepals 5, lobes 3 mm long, ovate, acute, brown tomentose. Petals 5, urceolate, hispid; lobes 5, white, orbicular, deflexed. Stamens 12-18, in 2 rows, equal; staminodes 10. Berry globose, 7.5 cm across, dark green; persistent calyx cup-like, woody with short pointed lobes. Fl. & Fr.: Throughout the year.

Western Ghats: Evergreen forests at 1200 m in Karnataka, Kerala and Tamil Nadu (Endemic).

IUCN: NE (IUCN, 2024).

Diospyros buxifolia (Bl.) Hiern., Trans. Cambridge Philos. Soc. 12: 218. 1873. Synonyms: *Leucoxylum buxifolium* Blume, *Diospyros microphylla* Bedd. Vern.: *Lunchiganamara* (Kan.); *Elicheviccha, Kattuthuvara, Thovarakarai, Corakkari* (Mal.); *Thokarakari, Anecola* (Mar.); *Chinnathuvarai, Chorakkari, Maligaithuvarai, Irumpalei* (Tam.).

Dioecious trees to 30 m tall; bark thick, blackish-grey, mottled with white; branchlets densely golden yellow hairy. Leaves simple, alternate, bifarious, elliptic-ovate, ovate, ovate-oblong or ovate-lanceolate, 1-4.5 x 0.6-2 cm, base acute or cuneate, apex acute, margin entire, coriaceous, glabrous above except midrib and pubescent beneath; lateral nerves 2-3 pairs, faint, intercostae obscure; petiole 2-3 mm, slender, hispid. Flowers unisexual, white. Male flowers: 2-3 mm, 1-4 together in subsessile small axillary cymes. Sepals 4, densely fulvous-hairy outside; lobes 4, broadly outside. Petals 4, campanulate, tube 0.1-0.15 cm long; lobes 4, orbicular, apiculate with a line of hairs on the back. Stamens 16, in pairs, united by their filaments at base; anthers ovate, dehiscence by apical pores; pistillodes minute, hairy. Female flowers: 7-8 mm, solitary, subsessile. Sepals and petals as in males; staminodes absent. Ovary superior, ovoid, pubescent, 4-celled, ovule 1 in each cell; style short 2 partite. Berry oblong, 15-18 x 10-12 mm, pointed, purple, glabrous. Seeds 1-2, endosperm equable. Fl. & Fr.: January – July.

Western Ghats: Deciduous forests of Maharashtra, Goa, Karnataka, Keralaand Tamil Nadu. World: Myanmar, Malaysia, Singapore, Indonesia, Laos, Cambodia, Vietnam, Maluku Islands, Philippines, Madagascar and New Guinea.

IUCN: LC (IUCN, 2024).

Uses: The wood is suitable for match boxes and splints. Fruits is said to be edible. This tree is having ornamental value.

Diopyros candolleana Wight., Icon. Pl. Ind. Orient. 4: t. 1221. 1848. Synonyms: *Diospyros canarica* Bedd., *Diospyros oligantha* Bedd. Vern.: *Karemara* (*Kan.*); *Kari, Karimaram* (*Mal.*); *Karamaran* (*Mar.*); *Nilavriksha* (*Sans.*); *Karimaram, Karikattai* (*Tam.*).

Dioecious trees to 15 m tall; bark thick, greenish-black, smooth; young shoots pubescent. Leaves simple, alternate, bifarious, oblong or elliptic-oblong, 7-15 x 2.5-5 cm, base acute, attenuate, cuneate or obtuse, apex obtusely acuminate, margin entire, coriaceous, glabrous when mature, lateral nerves 9-11 pairs, , faint, intercostae reticulate; petiole 8-14 mm long stout, grooved above, rugose. Flowers unisexual, pale yellow. Male flowers: 7-8 mm across, sessile, in dense tawny tomentose, axillary fascicles. Sepals 5, to 5 mm long, tomentose; lobes 5, ovate, acute or subobtuse. Petals 5, urceolate, more than twice as long as the calyx, fulvous-hairy outside; lobes 5, ovate, acute. Stamens 10, in unequal pairs; filaments glabrous; anthers lanceolate, acute. Female flowers: 10-12 mm across, sessile, 2-10 in tawny tomentose, axillary fascicles. Sepals hairy on both side; lobes 5, ovate, acute, margin reflexed; staminodes 4 or 5. Ovary superior, tomentose, 4-celled, ovules 1 in each cell; style 2-5, slender, tomentose; stigma hairy. Berry conical, 2.5 cm long, green, fruiting calyx flat or shallowly cupuliform, the lobes triangular with reflexed margins. Seeds 3-4, usually compressed. Fl. & Fr.: February – June.

Western Ghats: Semi-evergreen forests of Maharashtra, Goa, Karnataka, Kerala and Tamil Nadu. World: India, Nepal.

IUCN: NE (IUCN, 2024).

Diospyros chloroxylon Roxb., Pl. Coromandel 1: 38. 1795. Synonyms: *Diospyros capitulata* Wight, *Diospyros chloroxylon* var. *cupulosa* V.Singh, *Diospyros insculpta* Buch.-Ham., *Diospyros spinosa* Hiern, *Diospyros tomentosa* Poir. Vern.: *Karugariathumara* (Kan.); *Karuva-vakkanai, Kamuvakkanai, Periyulinci, Vakkanai* (Tam.).

Small trees to 10 m tall; bark scabrous, dark rust-colored. Leaves alternate, oval oblong to obovate-rhomboid, 1–8 x 1–4.5 cm, rounded to subcuneate at base, rounded, obtuse or mucronate at apex. Male flowers white, sessile or subsessile, 4-10 together, clustered on short, axillary, hairy oeduncles. Sepals 4, tube 2–3 mm long, campanulate, 4 lobed nearly at the base, lobes obovate densly hairy outside, glabrous inside. Petals 4, ureceolate, not much excerted from the calyx, lobes 4, ovate to lanceolate. Stamens 16, paired, glabrous, unequal in pairs. Female flowers sessile, axillary, solitary. Sepals campanulate, 2–3 mm long, densly hairy outside and glabrous inside. Petals 4, lobes ovate-lanceolate, contorted. Staminodes 7-8, rarely 9, inserted within the tube. Ovary flask-shaped, 3 mm long, glabrous 8 celled. Berry globose, glabrous, 7-10 mm in diam, pale olive colored. Seeds 2-3 usually. Fl. & Fr.: October - March.

Western Ghats: Deciduous to semi-evergreen forests of Gujarat, Maharashtra, Karnataka, Kerala, Tamil Nadu. India: Andhra Pradesh and Odisha. World: India, Sri Lanka, Bangladesh, Myanmar.

IUCN: NE (IUCN, 2024).

Diospyros crumenata Thw., Enum. Pl. Zeyl.: 179. 1860. Vern.: *Karumbudan* (Mal.); *Kantumari, Karitumari* (Kan.); *Chemelpaniche* (Tam.).

Evergreen trees to 25 m tall; bark black, thin scaly with small depression; branchlets glabrous. Leaves simple, alternate, oblong, elliptic, ovate-oblong or elliptic-oblong, 7.5-15 x 3.5-6.8 cm, base obtuse, obtuse or acute, apex bluntly acute or acuminate, margin entire, glabrous, glossy, coriaceous; lateral nerves 7-10 pairs, intercostae reticulate, prominent; petiole 7-13 mm, stout, grooved above, glabrous. Flowers unisexual, yellow. Male flowers; flowers 3-7, in shortly stalked cymes; peduncle 3-10 mm long, pilose; pedicels nodding, short, ultimate ones 1-4 mm long. Sepals large, cupular, tomentose on outside; lobes 4, obtuse. Petals 4, tubular, tube to 14 mm long, yellow; lobes 4, short, broad, rounded, recurved. Stamens to 12, 4-6 mm long, subequal; pistillode small or absent. Female flowers: solitary, axillary; pedicel short, stout. Sepals large, cupular, tomentose on both surfaces. Petals tubular, 15 mm long, densely tomentose, yellowish-white; lobes 4, broad acute, patent or reflexed;

staminodes 8-12, epipetalous, unequal. Ovary superior, pilose, 8-celled, ovule 1-in each cell; stigmas 4, fleshy, short. Berry subglobose, 3-5 cm across, smooth; fruiting calyx much enlarged, nearly flat, woody. Seeds 1-8, 2.2 x 1.2 cm, oblong, compressed, glossy black. Fl. & Fr.: February – June.

Western Ghats: Evergreen forests of Maharashtra, Goa, Karnataka and Kerala. World: India including Andaman & Nicobar Islands, Sri Lanka and Myanmar.

IUCN: CR- C2a(i). (IUCN, 2024).

Uses: Ripened fruits are edible. Wood is used for making agriculture implements and household articles.

Diospyros ebenum Koenig, Physiogr. Sälsk. Handl. 1: 176. 1780. Synonyms: *Diospyros assimils* Bedd., *Diospyros ebenaster* Retz., *D. glaberrima* L.f., *D. melanoxylon* Willd., *D. ebenum* var. *glaberrima* (L.f.) Bakh., *D. reticulata* Wall. *nom. nud.* Vern.: *Balemara, Dumbi, Kari, Karemara, Bale, Malli* (Kan.); *Karu, Karimaram, Mishatumbi, Varayi* (Mal.); *Abnus, Tai, Tendu* (Mar.); *Acha, Tai, Shengutam, Calam, Culli, Kakkayttali, Karai, Karunkali, Karuthali, Tumbai, Valaibattiram, Solakarimaram* (Tam.).

Dioecious trees to 20 m tall; bark brown, deeply fissured, ridges black, fissures lighter; young parts rusty-tomentose. Leaves simple, alternate or subopposite, elliptic, elliptic-oblong or ovate, 3.5-15 x 3-6.5 cm, base acute or obtuse, apex obtuse, round or subacute, often emarginate, margin entire, glabrescent above, thinly pubescent below, coriaceous; lateral nerves 5-9 pairs, intercostae subscalariform, slender, faint; petiole 4-10 mm long, slender, tomentose. Flowers unisexual, white. Male flowers in short densely tawny-tomentose drooping axillary cymes, sometimes at the base of branchlets; peduncle 7 mm, tomentose. Sepals yellowish-green, campanulate; lobes 5, ovate, tomentose, acute. Petal 5, tubular, thin-fleshy; tube 1 cm, appressed hairy without; lobes 5, ovate, 3 mm, acute, reflexed. Stamens 10-16, in pairs; anthers 2 mm, oblong, acute or apiculate; pistillode reduced, pubescent. Female flowers solitary, axillary; pedicel 4 mm long, stout, tomentose; bracteoles present. Sepals green to brown, deeply 5-lobed; lobes ovate, 7 mm, margin folded-crinkled, acute. Petals tubular, thin-fleshy, tube 1 cm, appressed hairy without; lobes 5, ovate, 3 mm, acute, reflexed; staminodes 8-10, 4 mm, lanceolate. Ovary superior, 4 mm across, globose, tomentose, 6-celled, ovules 1-in each cell; style 3 mm long, bifid. Berry globose, apiculate, rusty-tomentose when young, glabrous later, fruiting calyx slightly enlarged, fulvous-tomentose, margin reflexed. seeds 1-4, rarely 8, wedge-shaped, smooth, black. Fl. & Fr.: February – April.

Western Ghats: Common in deciduous forests of Maharashtra, Goa, Karnataka, Kerala, Tamil Nadu. World: Sri Lanka, Bangladesh, Myanmar, Japan, Malesia, Australia, Central America.

IUCN: VU- B1+2cde. (IUCN, 2024).

Uses: The wood is commonly called ebony wood that is known for making furniture, chop sticks, opium pipes, musical instruments, walking sticks, sport goods, brushes, carved stands, gymnastic apparatus, pedestals, etc. The fruit pulp is said to be edible.

Diospyros ferrea (Willd.) Bakh., Gard. Bull. Straits Settlem. 7: 162. 1933. Synonyms: *Diospyros vera* (Lour.) A.Chev., *Ebenoxylum verum* Lour., *Maba ebenus* Wight, *Ehretia ferrea* Willd., *Ehretia buxifoli*a (Rottb.) A. Juss., *Ferreola buxifolia* (Rottb.) Roxb., *Maba buxifolia* (Rottb.) A. Juss., *Ferreola guineensis* Schum. & Thonn., *Maba cumingiana* A.DC., *Maba vacciniaefolia* Benth., *Ebenus buxifolia* (Rottb.) O. Kuntze, *Diospyros ferrea* var. *buxifolia* (Rottb.) Bakh., *Diospyros ferrea* var. *guineensis* (Schum. & Thonn.) Bhak. Vern.: *Kargana, Piccane* (*Kan.*); *Irumbali, Karuvachi, Kirummal, Karianthovarai, Mariathovarai, Irumolampazhalm, Vakkanai* (Tam.).

Small trees to 4 m tall; bark dark grey, fissured, rough. Leaves alternate, obovate-oblong or spathulate, 1-2.5 x 1-1.8 cm, base cuneate to attenuate, apex obtuse to emarginate, coriaceous; petiole to 2 mm long. Flowers 1-3, in axillary fascicles, white. Male flower: sepals pale yellowish, 2 mm long, cup-shaped; lobes 3, 1-1.5 mm long, triangular-ovate, acute. Petlas 3-4, tubular, tube 5 -6 mm long, lobes 3, 1 -1.5 mm long, oblong-ovate, acute. Stamens 6, 2 mm long. Female flower: similar to male one. Ovary 2 mm long, globose, silky-pubescent; stigmas 3. Berry subglobose, 5-7 mm long, yellow to brownish red. Seeds 1-3, 4 mm long, wedge-shaped. Fl. & Fr.: February – September.

Western Ghats: Foot hills and deciduous forests of Karnataka, Kerala and Tamil Nadu. World: India, Sri Lanka, Myanmar, Malaysia, Singapore, Japan, N. Australia and South Africa

IUCN: EN- B1+2c. (IUCN, 2024).

Uses: Ripened fruits are edible among the tribes of Western Ghats. Wood is hard and durable, used for making baot anchor, cabinet work, handles and sheaths of weapons.

Diospyros foliolosa Wall. ex A. DC., A.P.De Candolle, Prodr. 3: 234. 1828. *D. calycina* Beddome. Vern.: *Kattupanachi* (Mal.); *Vellaithovarai, Palivellatithuvarai* (Tam.).

Glabrous dioecious trees to 10 m tall; bark dark brown, smooth; branchlets slender. Leaves simple, alternate, narrowly lanceolate, elliptic, oblong, 5-12 x 1-3.2 cm, base attenuate, apex acuminate, margin entire, coriaceous, glabrous, shining above; lateral nerves 4-5 pairs, intercostae reticulate, faint, pellucid; petiole about 6 mm long, slender, glabrous. Flowers unisexual, bright yellow. Male flowers: 5-9 in lax axillary cymes. Sepals 4, 0.2-0.3 cm long, 4-lobed to half way. Petals urceolate, very gibbous at base, tube 0.8-1 cm long; lobes 4, bright yellow. Stamens 12-16, in pairs, equal, apex of the anthers inflexed and long acuminate; filaments slightly hairy; pistillode rudimentary. Female flowers: solitary, axillary; pedicel 1.5-2 cm long. Sepals and petals as in male flowers; staminodes absent. Ovary superior, globose, hairy, 4-celled; ovule 1 in each cell; stigma 4-lobed. Berry globose, 2-2.5 cm across, covered with hair like scales, fruiting calyx enlarged, foliaceous, ovate-cordate. Seed 1 or 2, ellipsoid, laterally compressed, brownish-black. Fl. & Fr.: December – May.

Western Ghats: Evergreen forests of Karnataka, Kerala and Tamil Nadu (Endemic).

IUCN: NE (IUCN, 2024).

Uses: Wood is useful in making agriculture implements and building articles.

Diospyros ghatensis Ramesh & Franceschi., Blumea 38: 131. 1993.

Dioecious trees to 25 m tall; bark surface brownish-black, smooth; young branchlets hirsute and glabrous on maturity. Leaves simple, alternate, bifarious, elliptic or elliptic-lanceolate, 3-13.5 x 3-6 cm, base acute or attenuate, apex acuminate, margin entire, glabrous, coriaceous; lateral nerves 3-6 pairs, intercostae reticulate, faint; petiole 3-11 mm long, slender, glabrous. Flowers unisexual, creamy white. Male flowers sessile, in axillary cymes of 1.5 cm long, brown tomentose; bracts imbricately arranged, persistent, suborbicular, abaxial side tomentose, adaxial side glabrous, margin ciliate. Sepals 4, 3 mm long, tomentose outside, glabrous inside; lobes 4, 2 mm long, ovate, acute. Petals creamy white, sparsely hairy outside, glabrous inside; tube 6 mm long; lobes 4, obliquely ovate. Stamens 16, in opposite pairs, 2.5-4 mm long; connective slightly hairy; pistillode about 1 mm across, glabrous, with bifid apex. Female flowers unknown. Berry globose, to 2.5 cm across, solitary on a peduncle with several bract scars, glabrous, 4-celled; fruiting calyx not very accrescent, about 1 cm across. Seeds 1-4, flat, 1.3 cm long. Fl. & Fr.: April – November.

Western Ghats: Deciduous to semi-evergreen forests of Karnataka, Kerala and Tamil Nadu (Endemic).

IUCN: NE (IUCN, 2024).

Diospyros hirsuta L.f., Suppl. Pl.: 440. 1782.

Trees to 6 m tall; bark black, finely fissured; young parts densely pubescent. Leaves simple, alternate, ovate-lanceolate, elliptic or oblong, 7-17 x 3.5-7 cm, base round, truncate or acute, apex acuminate, caudate-acuminate or acute, margin entire, coriaceous, adpressed pubescent and glabrous above, and pilose beneath when mature; lateral nerves 5-8 pairs, secondary laterals present, intercostae reticulate, prominent; petiole 8-11 mm long, slender, grooved above, densely pubescent. Flowers unisexual, yellow. Male flowers in axillary, dense, sessile cymes; bracteoles small, pilose. Sepals campanulate, 2.5 mm long, pubescent; lobes 4-5, 1 mm long, acute. Petals 5 mm long, tubular, yellow, throat narrow; lobes 4-5, overlapping in bud, acute. Stamens 5, epipetalous, alternate with corolla lobes, glabrous, apiculate; filaments short; pistillode rudimentary. Female flowers 1-3, in sessile axillary clusters; bracteolate. Sepals to 7-9 mm long; lobes 4-5, silky brown, 3-4 mm long, triangular, reflexed. Petals upto 9 mm long, yellow; lobes 5, acute; staminodes 5, apiculate; filaments glabrous. Ovary superior, globose, densely pubescent, 8-10-celled; ovule 1-in each cell; styles 2, 2 mm long. Berry ellipsoid-ovoid to subglobose, 3 x 2.5 cm, densely, coppery-rusty haired, seated on enlarged reflexed calyx. Seeds 1-8, brownish-black. Fl. & Fr.: March – June.

Western Ghats: Evergreen forests of Maharashtra, Karnataka and Kerala. World: India, Sri Lanka.

IUCN: VU- A1c. (IUCN, 2024).

Diospyros insignis Thw., Enum. Pl. Zeyl.: 180. 1860. Synonyms: *Diospyros insignis* var. *parvifolia* Kosterm. Vern.: *Pottu thuvarai* (Tam.).

Small trees to 6 m tall; bark black, finely fissured, peeling off in flakes. Leaves usually alternate, rarely subopposite, oblong-elliptic, 7-35 x 2.5-15 cm, abruptly acuminate, subrounded or cuneate at base, glabrous except the midrib; lateral nerves 8-12 pairs, veinlets parallel or horizontal and conspicuous both sides; petioles 5-10 mm long. Male flowers and bisexual flowers in separate or same, axillary cymose clusters. Sepals campanulate, 3-5 mm long, glabrous within; lobes 4, divided half way down, ovate, valvate. Petals tubular, 8-10 mm long, tomentose-silky outside, glabrous within; lobes 4, ovate, acute, pilose outside, glabrous within, contorted. Stamens 14-20, unequal, free, attached at the base of corolla-tube; filaments glabrous or sparsely pilose; anthers glabrous, connectives produced, apiculate. Bisexual flowers like males, with well developed pistil. Female flowers axillary, subsessile, 1 to 3-together, rarely 6. Sepals as in males, pilose outside, glabrous within; lobes 4, 5 mm long. Petals as in males; lobes 4, somewhat recurved. Staminodes 4-5, epipetalous, with apiculate sterile anthers and glabrous filaments. Ovary globose, 4 or 8-celled, completely enclosed in the accrescent calyx; cells 1-ovuled; stigmas 4. Berry subglobose, 3-4 cm in diam, depressed at apex, smooth; fruiting-calyx cup-shaped, much enlarged, woody, thickened. Seeds 4 or 8, ovate-wedge-shaped, 2-3 x 0.75-1.5 cm, 7 mm thick, brown, smooth, obscurely transversely striated. Fl. & Fr.: March – September.

Western Ghats: Evergreen forests of Karnataka, Kerala and Tamil Nadu. World: India, Sri Lanka.

IUCN: NE (IUCN, 2024).

Diospyros lotus L., Sp. Pl.: 1057. 1753. Synonyms: *Diospyros lotus* var. *laevis* H. Ohba & S. Akiyama

Deciduous trees to 10 m tall; bark grayish black to grayish brown. Leaves elliptic to ovate-oblong, 5--13 x 2.5--6 cm, submembranous, abaxially drying green or glaucous and with dark veinlets, base obtuse, broadly cuneate, or subrounded, apex acuminate to acute; lateral veins 7-10 pairs, secondary veins prominent, flat, and dark; petiole to 1.5 cm long. Male flowers 1--3 together; pedicel to 6 mm. Sepals with lobes 4. Petals 4, reddish to pale yellow, urn-shaped, 4 mm long, lobes 4; stamens 16. Female flowers subsessile, pale green to reddish. Sepals with 4 lobes. Petals 4, tube urn-shaped, 6 mm long, lobes 4 or rarely 5; staminodes 8. Ovary 8-locular, glabrous except for apex; styles 4. Fruiting calyx lobes 4, ovate, apex obtuse. Berries pale yellow, becoming bluish black with a glaucous bloom, subglobose to ellipsoid, 1-2 cm in diam. Seeds brown, compressed, 10 x 6 mm. Fl. & Fr.: May – November.

Western Ghats: Deciduous forests of Maharashtra. World: India, Nepal, Pakistan, Temperate Asia, Europe, Mediterranean region.

IUCN: LC (IUCN, 2024).

Diospyros malabarica (Desr.) Kostel., Allg. Med.-Pharm. Fl. 3: 1099. 1834. Synonyms: *Garcinia malabarica* Desr., *Diospyros pergrina* (Gaertn.) Guerke,

Ebryopteri spergrina Gaertn., *Diospyros embryopteris* Pers., *Diospyros glutinosa* J.Koenig ex Roxb. Vern.: *Thumbika, Ettukottai pazham* (Tam.).

Dioecious trees to 10 m tall; bark black with greenish tinge, flaking off in thick scales; branchlets glabrous. Leaves simple, alternate, distichous, oblong or oblong-lanceolate, 13-25 x 4-7 cm, base round, obtuse or acute, apex acute or obtuse, margin entire, glabrous and shining, coriaceous; lateral nerves 6-8 pairs, intercostae reticulate, prominent, pellucid; petiole 8-12 mm long, stout, grooved above. Flowers unisexual, pale yellow. Male flowers: 2-7 in axillary, drooping cymes; peduncle short. Sepals 4, 6.35 mm long, lobes 4, broadly ovate, pubescent inside, margin ciliate. Petals 4, campanulate, 1 cm long, lobes 4, ovate, round at apex, with a pubescent patch on the back of each lobe, imbricate. Stamens usually 40, in pairs, inserted at the base of the corolla tube; filaments hairy; anthers nearly equal, linear with a line of hairs up the middle; pistillode fleshy. Female flowers: 2.5 cm across, usually solitary, axillary, drooping; pedicels very short. Sepals 4, campanulate, tube deeply divided; lobes 4, appressed-pubescent, ovate, acute, thick. Petasl campanulate, 12.5 mm long, deeply divided; lobes 4, ovate, obtuse, thick-fleshy, appressed hairy on the back, recurved; staminodes 4-12, hairy, inserted at the base of corolla tube. Ovary superior, globose, 8-celled; ovule 1-in each cell, subsessile, glandular-hairy without; style 4; stigma lobed, undulate. Berry subglobose, 5-6 cm across, yellow, covered with a rusty scruffiness; fruiting calyx much enlarged, green, woody, spreading reflexed. Seeds 4-8, flattened, reddish-brown. Fl. & Fr.: February – May.

Western Ghats: Deciduous to evergreen forests between 700 – 1100 m in Maharashtra, Goa, Karnataka, Kerala and Tamil Nadu. World: India, Sri Lanka, Nepal, Myanmar, Thailand, Malaysia, Sumatra, Java, Timor, Celebes.

IUCN: NE (IUCN, 2024).

Diospyros melanoxylon Roxb. var. **melanoxylon.** Pl. Coromandel 1: 36. 1795. Synonym: *Diospyros wightianan* Wall. ex Bedd. Vern.: *Balai, Thumbri, Tupra, Kendu, Makartendu, Toomrie, Temru* (Kan.); *Kari* (Mal.); *Karunthumbi, Karunthovarai, Tumballi, Tumbi, Karuppu, Karundumuki, Thumramaram, Oovamaram, Nallatumki* (Tam.).

Dioecious trees to 20 m tall; bark brown, deeply fissured, ridges black; young branchlets rusty-tomentose. Leaves simple, alternate or subopposite, elliptic, elliptic-oblong or ovate, 3.5-15 x 3-6.5 cm, base acute or obtuse, apex obtuse, round or subacute, often emarginate, margin entire, glabrescent above, thinly pubescent below, coriaceous, lateral nerves 5-9 pairs, intercostae subscalariform, slender, faint; petiole 4-10 mm long, slender, tomentose. Flowers unisexual, white. Male flowers in short densely tawny-tomentose drooping axillary cymes; peduncle 7 mm long, tomentose. Sepals yellowish-green, campanulate, lobes 5, ovate, tomentose, acute. Petals tubular, thin-fleshy; tube 1 cm, lobes 5, ovate, 3 mm, acute, reflexed. Stamens 10-16, in pairs; anthers 2 mm, oblong, acute or apiculate; pistillode reduced, pubescent. Female flowers solitary, axillary; pedicel 4 mm long, stout, tomentose; bracteoles present. Sepals green to brown, deeply 5-lobed; lobes ovate, 7 mm, margin folded-crinkled, acute. Petals tubular, thin-fleshy, tube 1 cm, lobes 5, ovate, 3 mm, acute; staminodes 8-10, 4 mm, lanceolate. Ovary superior, 4 mm across, globose, tomentose, 6-celled; ovules 1-in each cell; style 3 mm long, bifid. Berry globose, apiculate, rusty-tomentose when young, glabrous later; fruiting calyx slightly enlarged, fulvous-tomentose. Seeds 1-4, smooth. Fl. & Fr.: March – July.

Western Ghats: Deciduous forests of Gujarat, Maharashtra, Goa, Karnataka, Kerala and Tamil Nadu. World: Sri Lanka.

IUCN: NE (IUCN, 2024).

Key to the varieties

1a. Lateral nerves 5-9 pairs, pubescent beneath; staminodes 6-8 in pairs prominent in female flowers .... var. **mealanoxylon**

1b. Lateral nerves 8-12 pairs, tomentose beneath; staminodes 4-6 or absent in female flowers .... var. **tupru**

Diospyros melanoxylon Roxb. var. **tupru** (Buch.-Ham.) V. Singh., Monogr. Ind. Diospyros: 159. 2005. *Diospyros tupru* Buch.-Ham., *Diospyros exsculpta* Dalzell & A.Gibson., *Diospyros tomentosa*Roxb. Vern.: *Timburni, Tupru, Tindura, Bandadamara* (Kan.); *Thambbilpazham* (Mal.); *Tartar, Tumboorne* (Mar.); *Chilta-tumiki, Tumbil, Thumbai* (Tam.).

Fl. & Fr.: March – September.

Western Ghats: Deciduous forests of Maharashtra, Karnataka and Kerala. World: Nepal and Afghanistan.

IUCN: NE (IUCN. 2023).

Diospyros montana Roxb., Pl. Coromandel 1: 37. 1795. Synonyms: *Diospyros cordifolia* Roxb., *Diospyros montanta* var. *cordifolia* (Roxb.) Hiern., *Diospyros humilis* Bourd., *Diospyros kanjilali* Duthie, *Diospyros goindu* Dalz., *D.*

courtallumensis Bahadur & Gaur, *D. orixensis* Wight ex Hiern. Vern.: *Vakkanai, Vellei thuvarai* (Tam.).

Dioecious trees to 20 m tall; bark smooth, grey or yellowish-grey, exudation scanty, watery; branchlets slender, glabrous, divaricate spines on mature branches. Leaves simple, alternate, ovate, ovate-oblong, elliptic-oblong, 4-10 x 2-4 cm, base round, obtuse, subcordate or acute, apex obtuse, subacute or obtusely acuminate, margin entire, chartaceous, glabrous; lateral nerves 5-10 pairs, intercostae reticulate, faint; petiole 5-10 mm, slender, grooved above, glabrous. Flowers unisexual, white. Male flowers 2-6 in axillary umbels; peduncle to 5 mm. Sepals 4, 3 mm long; lobes 4, ovate, imbricate, thick, margin ciliate, obtuse. Petals 4, urceolate, 6-7 mm long, greenish-yellow, glabrous; lobes 4, ovate, imbricate, 2.5 mm long, subacute. Stamens 16, roughly in 8 pairs, unequal; anthers lanceolate, awned; pistillode conical, pointed. Female flowers solitary, axillary; pedicel to 4 mm. Sepals and petals as in male flowers; staminodes 4, 5 mm, linear-lanceolate. Ovary superior, 0.7 mm across, globose, glabrous, 8-celled; ovule one in each cell; styles 4, 1.5 mm long; stigma capitate. Berry ovoid to obconical, 1.5-2 cm across, yellow to orange, glaucous; fruting calyx foliaceous, slightly enlarged, reflexed. Seeds 3-6, rough, black. Fl. & Fr.: January – April.

Western Ghats: Deciduous forests of Gujarat, Maharashtra, Goa, Karnataka, Kerala and Tamil Nadu. World: Other states of India including Andaman & Nicobar Islands, Sri Lanka, Nepal and Java.

IUCN: NE (IUCN, 2024).

Diospyros neilgerrensis (Wight) Kostrem., Ceylon J. Sci., Biol. Sci. 12: 106. 1977. Synonyms: *Maba neilgerrensis* Wight., *Diospyros ferrea* var. *neilgerrensis* (Wight) Bakh., *Diospyros ferrea* var. *angustifolia* (Miq) Bakh., *Mababuxifolia* var. *angustifolia* (Miq.) Thw., *Maba buxifolia* sensu Clarke, *Maba angustifolia* Miq., *Diospyros angustifolia* (Miq) Kostrem., *Maba nigrescens* Dalz., *Diospyros nigrescens* (Dalz.) C.J. Saldanha, *Diospyros rheophytica* Kostrem. Vern.: *Karungali* (Tam.).

Small trees to 10 m tall; bark rough, black, lenticellate. Leaves simple, alternate, distichous, elliptic to elliptic-lanceolate, 4.5-8.5 x 1.5-3.5 cm, chartaceous, apex acute with blunt tip, base acute, glabrous; midrib canaliculate; secondary nerves 11 pairs, tertiary nerve closely reticulate; petiole up to 0.5 cm long, glabrous. Flowers unisexual, dioecious. Male flower in axillary cymes. Female flowers solitary, axillary, with cream and hairy petals. Berry with fruiting calyx. 1-3 seeded. Fl. & Fr.: November – February.

Western Ghats: Evergreen forests of Maharashtra, Goa, Karnataka, Kerala and Tamil Nadu. India: Andhra Pradesh. World: Sri Lanka.

IUCN: NE (IUCN, 2024).

Diospyros nilagirica Bedd., Icon. Pl. Ind. Or.: 27. 1871. Vern.: *Karunthuvarai* (Tam.).

Evergreen dioecious trees to 18 m tall; bark thick, black, smooth; branchlets densely rufous tomentose. Leaves simple, alternate, elliptic, oblong-lanceolate or ovate, 5-15 x 2-4 cm, apex acuminate, base attenuate, margin entire, reflexed, coriaceous, densely rufous tomentose when young, glabrous when mature; petiole 6-12 mm, lateral nerves 6-9 pairs, intercostae reticulate, faint. Flowers unisexual, yellowish-white. Male flowers: 3-6, in axillary cymes, 10-12 mm long. Sepals 4, campanulate, deeply 4 lobed; lobes broadly ovate or orbicular, obtuse, densely fulvous hairy, margin strongly reflexed. Petals cupular, salver shaped; lobes 4, ovate, obtuse, hairy without, twisted. Stamens 16, in pairs of unequal length; filaments of each pairs connate at base, villous; anthers lanceolate, apiculate; pistillode globose, densely hairy. Female flowers solitary, axillary, sessile, sepals and petals as in male flowers; staminodes 8, longer and shorter alternate with each other; filaments villous; anthers linear. Ovary superior, sessile, globose, 8-celled; ovule 1 in each cell, densely hairy; styles 4, much reflexed; stigma dilated. Berry globose, 2.5 x 2.5 cm, glabrous, fruiting calyx not enlarged. Seeds 12 x 7 mm, compressed. Fl. & Fr.: February – June.

Western Ghats: Evergreen forests of Karnataka, Kerala and Tamil Nadu (Endemic).

IUCN: NE (IUCN, 2024).

Diospyros oocarpa Thw., Enum. Pl. Zeyl.: 180. 1860. Synonym: *Diospyros arnottiana* Miq. ex Thw. Vern.: *Vellei karungali* (Tam.).

Trees to 20 m tall; bark yellowish to pale brown, peeling off in large irregular pieces with scalloped pits; branchlets softly hirsute. Leaves simple, alternate, ovate or elliptic-oblong, 4-9 x 2-5 cm, base round, subcordate or cuneate, apex obtusely acuminate or subacute, margin entire, fulvous hairy on both sides, chartaceous; lateral nerves 6-10 pairs, intercostae reticulate, obscure; petiole 5-9 mm long, slender, grooved above, densely hirsute. Flowers unisexual. Male flowers: 3-7 together, subsessile, in axillary fulvous-silky cymes; bracts round, concave, silky hairy. Sepals 4, barrel shaped, silky; lobes 4, indistinct. Petals tubular, yellow, pubescent outside, throat narrow; lobes 3-4, about equaling the tube, ovate-oblong, imbricate in bud. Stamens 9-14, unequal, single or in pairs; filaments glabrous; anthers pilose; pistillode none or a hairy disc present. Female flowers 1-3 together, similar to and slightly larger than the male. Sepals and petals as in male flowers; staminodes 3, epipetalous. Ovary superior, globose, hairy, 6-8-celled; ovule 1-in each cell; stigmas 1-3, short. Berry oblong-ovoid, wider at the top, apiculate, glabrous, seated on a slightly enlarged fruiting calyx. Seeds oblong with 2 longitudinal grooves. Fl. & Fr.: January – July.

Western Ghats: Evergreen forests of Maharashtra, Goa, Karnataka, Kerala and Tamil Nadu. World: Other states of India including Andaman & Nicobar Islands, Sri Lanka.

IUCN: NE (IUCN, 2024).

Diospyros ovalifolia Wight., Icon. Pl. Ind. Orient. 4: t. 1227. 1848. Vern.: *Karuvichi maram, Vedukkanari* (Tam.).

Dioecious trees to 20 m tall; bark dark brown with superficial ridges; branchlets slender, minutely adpressed-pilose apically. Leaves simple, alternate, elliptic, oblanceolate, obovate or obovate-oblong, 6.5-15 x 2-7 cm, base cuneate or acute, apex subacute, obtuse or round, margin entire, glabrous, coriaceous; lateral nerves 5-8 pairs, intercostae reticulate, prominent; petiole 7-15 mm long, stout, grooved above, glabrous. Flowers unisexual, yellowish-white, in short fascicles, usually in the axils of fallen leaves. Male flowers: sepals broadly campanulate, lobes 4-5, ovate, thick, appressed-hairy, margin ciliate, acute. Petals tubular-campanulate, thin fleshy; tube 4 mm long; lobes 4, ovate, acute, 1.5 mm long, twisted. Stamens 14-20, paired, unequal. Female flowers subsessile, 1 cm across. Sepals broadly campanulate; lobes 4-5, 4 mm long, ovate, appressed hairy without, acute, margin ciliate. Petals tubular-campanulate, 4 mm long, thin fleshy; lobes 5, 3 mm long, ovate, subacute, strongly reflexed; staminodes 4, spathulate, 4 mm long, sometimes lacking. Ovary superior, 3 mm long, globose, 4-celled; ovule 1-in each cell; stigma 4-lobed, subsessile. Berry globose, 1-2 cm across, pubescent when young, glabrous later; fruiting calyx reflexed, tawny-pubescent. Fl. & Fr.: March – August.

Western Ghats: Semi-evergreen to evergreen forests of Karnataka, Kerala and Tamil Nadu. India: Andaman & Nicobar Islands. World: Sri Lanka and Malaysia.

IUCN: NE (IUCN, 2024).

Diospyros paniculata Dalz., Hooker's J. Bot. Kew Gard. Misc. 4: 109. 1852. Vern.: *Karunthuvarai* (Tam.).

Dioecious trees to 15 m tall; bark smooth, black often with grey blotches; branchlets angular, pubescent. Leaves simple, alternate, oblong or oblong-lanceolate, 12-20 x 4-7 cm, base acute, round or cuneate, apex acute, acuminate or obtusely acuminate, margin entire, reflexed near the base, glabrous, coriaceous; lateral nerves 6-9 pairs, intercostae reticulate, surface finely reticulate; petiole 7-10 mm long, stout, puberulent when young. Flowers unisexual greenish-yellow. Male flowers 3-5 together in short panicled cymes with a pair of ovate-cordate bracts; bracts 1 x 0.8 cm, pubescent. Sepals 10 x 12 mm, globose, 5-angled, minutely hispid, deeply lobed; lobes-5, ovate-obtuse, margin reflexed, prominently veined, pubescent. Petals dull or creamy white, tube 12 mm long, pentagonal, narrowed above, hispid; lobes 5, 8 x 4 mm, oblong. Stamens 16-20 in pairs, equal; anthers linear, acute; pistillode absent. Female flowers solitary, axillary; pedicel to 2 cm long; bracts larger than those of the male flowers. Sepals glabrescent; staminodes 10. Ovary superior, 4-celled; ovule 1 in each cell, tomentose; stigma 4, unequal. Berry ovoid to obovoid, 3-4 x 2-3 cm, obtuse, obtusely 5-angled, yellowish-brown, puberulous, fruiting calyx 1.5 x 3 cm, saccate. Seeds few, 2 x 0.8 cm. Fl. & Fr.: November – April.

Western Ghats: Deciduous forests of Maharashtra, Goa, Karnataka, Kerala and Tamil Nadu. World: Malaysia.

IUCN: VU- A2cd. (IUCN, 2024).

Diospyros pruriens Dalz., Hooker's J. Bot. Kew Gard. Misc. 4: 110. 1852.

Dioecious trees to 10 m tall; bark black, scurfy scaly; branchlets densely hirsute-pubescent. Leaves simple, alternate, elliptic or elliptic-oblong or oblong-lanceolate, 6-16 x 2-5 cm, base round or subcordate, apex acute or obtusely acuminate, margin entire, fulvous-hirsute, membranous; lateral nerves 7-12 pairs, secondary laterals present, intercostae reticulate, prominent; petiole 3-5 mm long, stout, hirsute. Flowers unisexual, yellowish-white. Male flowers 1-2 flowered axillary cymes, close together in the upper axils; pedicels articulated at base; bracts small, ovate, cauducous, glabrous inside. Sepals 4, 6.35 mm long, divided nearly to the base, fulvous hairy; lobes 4, linear-oblong. Petals 4, 12.7-19.05 mm long, silky hairy outside; lobes 4, ovate-oblong, obtuse, about as long as the tube. Stamens 12-14, unequal, glabrous, connate at the base, shorter than the corolla tube; anthers as long as the filaments; pistillode densely hairy. Female flowers solitary, crowded in the upper axils. Sepals

6.35 mm long, 4 partite; lobes oblong, spreading, hairy. Petals 9.52-12.7 mm long, 4-fid, constricted about the middle; staminodes 4, small. Ovary superior, fulvous-hairy, 4-celled, ovule 1 in each cell; style 2, short, almost concealed by the hairs on the ovary, bifid at the apex. Berry ovoid-conical, 2.5 cm long, clothed with yellow stinging hairs, supported by the persistent, deflexed, slightly enlarged calyx lobes. Fl. & Fr.: February – August.

Western Gahts: Evergreen forests of Maharashtra, Goa, Karnataka, Kerala and Tamil Nadu. World: Sri Lanka.

IUCN: NE (IUCN, 2024).

Diospyros toposia Buch.-Ham., Trans. Linn. Soc. London 15: 115. 1827. Synonyms: *Diospyros racemosa* Roxb., *Diospyros toposioides* King & Gamble

Dioecious trees to 20 m tall; bark greyish-brown, rather smooth, hardly peeling off; branchlets drying pale brown, glabrous. Leaves simple, alternate, elliptic, ovate or oblong-lanceolate, 10-30 x 2-10 cm, base acute or round, apex acuminate, glabrous, coriaceous, surfaces areolate-reticulate; lateral nerves 8-9 pairs, intercostae reticulate, pellucid; petiole 5-12 mm long, slender, flat above, glabrous. Flowers unisexual, pale yellow. Male flowers 3-12 flowers in cymes, in the axils of cauducous bract or leaf; peduncle 8-12 mm long; pedicel upto 5 mm long, pilose; bract 2 mm long, hood-shaped; bracteole 1.5 mm, pilose. Sepals rough but not hairy, quite entire when young at length, lobes acute, unequal. Petlas yellow urceolate, with a very narrow throat; lobes 4, middle line pilose inside. Stamens 20-30, in pairs of unequal length; filaments glabrous; pistillode none. Female flowers solitary, axillary; pedicel 6-10 mm, slender, with minute adpressed hairs, slightly enlarged towards apex. Sepals 4 with tapered acute lobes, both sides seriaceous. Petals yellowish-white, urceolate to urceolate-tubular; lobes 4, triangular, acute; staminodes 12-16 in an epipetalous ring, short. Ovary densely pubescent, 4-6-celled, ovules 1-in each cell; stigma short. Berry ovoid-ellipsoid, top flattened, yellow covered with very short brown hairs, 2.5 x 2-4 cm; fruiting calyx 1.5-2.5 cm across, lobes triangular-ovate, 5-10 mm long, glabrescent. Fl. & Fr.: February – April.

Western Ghats: Evergreen forests of Kerala and Tamil Nadu. World: Sri Lanka, Bangladesh, Myanmar, Thailand, Philippines, Borneo and Timor.

IUCN: LC (IUCN, 2024).

Diospyros ridleyi Bakh., Gard. Bull. Straits Settlem. 7: 183. 1933. Synonyms: *Diospyros pyrrhocarpoides* Ramesh & Fransceschi, *Diospyros pyrrocarpa* var. *andamanica* Kurz.

Dioecious trees to 15 m tall; bark brown, mottled with green; branchlets slender, glabrous. Leaves simple, alternate, bifarious, oblong, elliptic-oblong or oblong-lanceolate, 2.5-7 x 3-13 cm, base round or truncate and slightly decurrent, apex obtusely acute, margin entire, glabrous, coriaceous; lateral nerves 8-11 pairs, intercostae reticulate, slender, prominent; petiole 6-20 mm long, slender, grooved above. Flowers unisexual, creamy white. Male flowers 3-4 in short subsessile cymes near the base of new shoots. Sepals 5.5 mm long, tomentose on either side; lobes 5, 2.5 mm long, triangular, acute. Petals tube 6 mm long, 3 mm long, tomentose outside, glabrous inside; lobes 5, broadly ovate, acute, tomentose on both side. Stamens about 20, 2.5 to 3 mm long; filaments 0-0.4 mm long; pistillode about 1 mm across, hairy. Female flowers solitary, sessile. Sepals 7 mm long, densely tomentose on either side; lobes 5, 4 mm long, ovate, acute. Petals tomentose except for the tube inside; tube 3 mm long; lobes ovate, bluntly acute at apex; staminodes 5 adnate to the base of corolla. Ovary superior, 3 mm across, orbicular, tomentose, 4-celled, ovule one in each cell; stigma subsessile, bilobed. Berry globose, 3 cm across, subsessile, densely brown tomentose; fruiting calyx accrescent, strongly reflexed with thin wavy margin. Fl. & Fr.: May – June.

Western Ghats: Evergreen forest of Karnataka and Kerala. World: Other states of India including Andaman & Nicobar Islands, Peninsular Malaysia.

IUCN: LC (IUCN, 2024).

Diospyros sahayadryensis P.Daniel &Vajr., J. Econ. Taxon. Bot. 3: 675. 1982. Synonyms: *Maba micrantha* (Dalz.) Hiern., *Holochilus micranthus* Dalz.

Dioecious trees to 12 m tall; branchlets glabrous. Leaves alternate, somewhat ovate with wider base or elliptic to oblong, 10-14.5 x 4.5-6 cm, obtusely acuminate at apex, subrounded or attenuated at base, coriaceous, glabrous; midrib flat above, glabrous; lateral veins more than 6 pairs, close; petioles 8-10 mm long, almost 4-gonous, glabrous. Male flowers sessile, in 3 to 5-flowered, sericeous, axillary clusters; epicalyx or scales, 3 mm long, sericeous. Sepals 5 mm long, campanulate, pubescent outside, glabrous within, minutely 3-toothed or almost truncate; teeth valvate. Petals 6-7 mm long, urceolate, 3-lobed; lobes contorted. Stamens 12-14, inserted at the base of corolla-tube. Female flowers subsessile, axillary, solitary; pedicels 1-2 mm long, sericeous. Sepals 3-4 mm long, tubular-campanulate, 3-toothed or almost truncate, pubescent outside, glabrous within; teeth valvate. Petals 1-1.2 cm long, tubular, 3-lobed, divided up to the middle; lobes ovate, obtuse, contorted.

Staminodes 6, inserted at the base of corolla-tube; filaments double the length of barren anthers. Ovary hemispheric, glabrous, 6-celled, each cell 1-ovuled; styles 3, rather thick, obtuse. Berry cylindric- oblong, 2.5 x 0.8 cm, glabrous, 6-seeded; fruiting pedicels upto 3 mm long, sericeous; fruiting-calyx funnel-shaped, truncate at apex. Fl. & Fr.: February – August.

Western Ghats: Evergreen forests of Maharashtra and Karnataka (Endemic).

IUCN: NE (IUCN, 2024).

Diospyros saldanhae Kostrem., J. Bombay Nat. Hist. Soc. 74: 326. 1977.

Trees to 8-10 m tall; branchlets very densely tomentose, interspaced with 2 mm long hairs. Leaves alternate, spirally arranged, subovate-elliptic, 3 x 6 - 6 x 14 cm, broadly acuminate with obtuse tip, base sub-cordate, both surfaces finely, prominulously reticulate, thinly coriaceous, glabrous above except on midrib; lateral nerves 8-10 pairs, slender, hardly prominulous; petiole very short, thick, 3 mm long, densely tomentellous. Fruit 2.5 cm diameter, solitary, subovoid-globose, densely tometellous; pedicel 2 mm long, pilose. Perianth lobes 4, oblong-ovate, obtuse, 5 mm long. Berry ovoid, 2 x 2.5 cm, bristly hairy, subsessile. Seeds 3, narrowly pear-shaped, slightly pointed, smooth, 15 mm long. Fl. & Fr.: March – November.

Western Ghats: Stream sides of evergreen forests of Maharashtra, Goa and Kerala (Endemic).

IUCN: NE (IUCN, 2024).

Diospyros sulcata Bourd., Forest Trees Travancore: 255. 1908.

Dioecious trees to 9 m tall; bark blackish-grey, smooth. Leaves simple, alternate, oblong or oblong-lanceolate, 12-19 15.2 x 5.5 cm, base cuneate, apex acuminate, margin entire, glabrous, pellucid, coriaceous; petiole mm long, glabrous. Flowers unisexual. Male flowers not known. Female flower solitary or 2-5 together, sessile axillary or on the old wood, 5-merous; staminodes 5, ovary superior, 10-celled, ovule 1 in each cell, covered with rusty tomentum; stigma 5. Berry truncate cone, 3 x 2.5 cm, apex at first covered with rusty tomentum, at length glabrous, marked with 8-10 deep longitudinal grooves. Fl. & Fr.: Throughout the year.

Western Ghats: Deciduous forests of Kerala (Endemic).

IUCN: NE (IUCN, 2024).

Diospyros sylvatica Roxb., Pl. Coromandel 1: 38. 1795. Vern.: *Karuppu thuvarai* (Tam.).

Dioecious trees to 18 m tall; bark dull grey to black, thin; branchlets puberulous. Leaves simple, alternate, lamina 5-15 x 2-5 cm, elliptic, ovate, elliptic-oblong or elliptic-obovate, base acute or cuneate, apex acute or acuminate, margin entire, glabrous, membranous; lateral nerves 5-8 pairs, intercostae reticulate, obscure; petiole 5-10 mm long, slender, glabrous. Flowers unisexual, white. Male flowers 6-10 in axillary cymes, 6 mm across; peduncle 5 mm long, puberulous. Sepals4, 3 mm long, campanulate; lobes 4, ovate, acute, pubescent outside. Petals campanulate, 6 mm long; lobes 4, triangular, obtuse. Stamens about 24; filaments in unequal pairs, ciliate at apex; female flowers; solitary or 3 together, 8-10 mm across. Sepals much longer than in male, lobes ovate-oblong, acute. Petals campanulate; lobes triangular, obtuse; staminodes 4, inserted on the corolla and alternate with its lobes. Ovary superior, globose, pubescent, 6-8-celled; 1-ovule in each cell; style 3-4, free, glabrous, bilobed with fringed margins. Berry globose, 16-18 mm across, dark green, fruiting calyx enlarged, flat, foliaceous, spreading. Seeds 2-8, oblong, rugose, shining. Fl. & Fr.: January – October.

Western Ghats: Evergreen forests of Maharashtra, Karnataka, Kerala and Tamil Nadu. World: Sri Lanka and Malesia.

IUCN: NE (IUCN, 2024).

Diospyros trichophylla Alston., H. Trimen, Handb. Fl. Ceylon 6(Suppl.): 181. 1931.

Dioecious trees to 8 m tall; bark black, rather smooth with narrow short longitudinal ridges; branchlets densely covered with long, stiff, 3-5 mm long hairs. Leaves alternate, thin, obtusely acuminate, base rounded to sub cordate, 1.5 x 5-4.5 x 10 cm, upper surface glabrous, venation lax, faintly pellucid; lateral nerves 9 pairs, erect-patent, obscure; petiole 2-5 mm, reddish-brown, with long, stiff hairs. Male inflorescence an axillary raceme of 3-4 pedicellate flowers; peduncle reddish-brown, 1 cm long, slender, widening at the flowers; pedicel short, basal bracts reddish-brown, orbicular, 3 mm long, caducous, very hairy outside. Flowers yellow, 13-25 mm long. Sepals green, lobes 4, chartaceous, linear oblong, narrow at base, tapering to acuminate apex, 8 mm long, hairy. Petals yellowish-white, tubular, 13-25 mm long, 2.5 mm diam, expanded top 2.5 cm diam, tube inside hairy; lobes 4, much longer than tube, rotate, acute, reddish-brown, 13 mm long, 6 mm wide at base. Stamens 12-14, connate into a central column. Pistil 0 or a rudimentary disc; anthers narrow, acute. Female flower solitary in upper leaf axils, yellow. Peduncle upto 16 mm long, with long stiff hairs. Sepals green, accrescent in flower; lobes 4-5, oblong, acute, 8 mm long, 3 mm wide at base, inside with long white hairs, outside subglabrous. Petals yellow, tubular, constricted towards middle, 10-28 mm long; lobes 4-5; staminodes 5 or more, epipetalous, linear, glabrous. Ovary depressed globose, 4-celled, 5 mm diam., densely long white pilose; styles 2, 4 mm long; stigmas 4, yellow. Berry globose, strongly apiculate, 2 cm diam, fruiting calyx lobes incurved. Seeds 1-4 per fruit, elliptic-wedge-shaped to globose-ovoid. Fl. & Fr.: October – December.

Western Ghats: Deciduous forest of Kerala. World: India including Andaman & Nicobar Islands, Sri Lanka.

IUCN: VU- A1c, B1+2c. (IUCN, 2024).

Diospyros undulata Wall. ex G.Don., Gen. Hist. 4: 40. 1837. *Diospyros walkeri* (Wight) Guerke, *Patonia walkeri* Wight, *Diospyros gardneri* Thw., *Diospyros undulata* var. *walkeri* (Wight) Bakh., *Diospyros cratericalyx* Craib, *Diospyros undulata* var. *cratericalyx* (Craib) Bakh.

Evergreen trees to 20 m tall; bark thin, longitudinally fissured. Leaves oblong to lanceolate, 12–18 by 3–7.5 cm, base acute, obtuse or rounded, apex acute or acuminate, subcoriaceous to coriaceous, glabrous on both surfaces; secondary nerves 10–15 pairs, impressed on upper surface, inconspicuous on lower surface, tertiary veins inconspicuous; petiole 1 cm long, glabrous. Male flowers cymose, 4–5-merous, pedicel 5–7 mm long, pubescent. Sepals campanulate, 3–5 mm long, divided to middle, tomentose outside, sericeous inside. Petals tubular, 8–13 mm long, divided to middle, sericeous outside, glabrous inside. Stamens 12–18, sericeous; rudimentary ovary pilose. Female flowers solitary, 4–5-merous, sessile or pedicellate. Sepals campanulate, divided to one third, woolly on both sides. Petals as in male flowers but larger. Ovary ovate, hirsute, 8-locular; style single, glabrous. Staminodes 4–8, glabrous. Berry sessile or distinctly stalked, globose, dry at maturity, tomentose, glabrescent, base rounded, apex rounded or truncate; fruiting calyx cup-shaped, woolly, lobes reflexed. Fl. & Fr.: February – June.

Western Ghats: Evergreen forests of Tamil Nadu. World: India, Sri Lanka.

IUCN: LC (IUCN, 2024).

Cultivated

Diospyros discolor Willd.; Exotic: Native to Philippines

Diospyros kaki Thunb.; Exotic: Native to China

Diospyros nigricans Wall. ex A. DC.; Native: India, Bangaldesh

ELAEOCARPACEAE Juss.

Elaeocarpus L.

1a. Leaves oblong to long-obovate; petal with inner surface densely hairy; outer anther tooth beaked; carpels 5-7 .... **E. angustifolius**

1b. Leaves elliptic-ovate or lanceolate; petals with inner surface glabrous; outer anther tooth not beaked; carples 4-5 2

2a. Outside of petal densely sericeous; outer anther tooth beaked; number of ovules per locule varies 9

2b. Outside of petal sparsely hairy; outer anther tooth not beaked; number of ovules 2-per locules 3

3a. Sepals to 4.5 mm long; petal divisions 13-21; stamens 17-21 .... **E. gaussenii**

3b. Sepals to 8.8 mm long; petal divisons 24-32; stamens 22-35 4

4a. Disc glands below the base of fruits, hairs very short; outer tooth of anthers with seatae .... **E. variabilis**

4b. Disc gland surrounding the basee of fruits, haris long; outer tooth of anthers without seatae 5

5a. Leaves with glabrous beneath 6

5b. Leaves with hairy beneath 8

6a. Drupe ellipsoid, style not persistent ....**E. balscoi**

6b. Drupe subglobose, style persistent 7

7a. Lateral nerves of leaf 4-5-pairs; petals pure white; drupe pointed at both ends.. **E. venustus**

7b. Lateral nerves of leaf 5-7-pairs; petals creamy white; drupe subglobe .... **E. serratus**

8a. Petals densely rugosus outside; ovary 2-celled .... **E. weibelii**

8b. Petals ferrrugineous tomentose outside; ovary 3-celled ....**E. rugosus**

9a. Leaves strongly concave .... **E. recurvatus**

9b. Leaves flat 10

10a. Leaves obovate to elliptic, 10-30 cm long; Seeds straight 11

10b. Leaves ovate to lanceolate, 5-12 cm long; seeds curved ....**E. munroi**

11a. Pyrenes pointed at both ends, prominently compressed, rugose .... **E. aristatus**

11b. Pyrenes obtuse at both ends, slightly compressed, coarsly rugose........12

12a. Ovules two series in a locule, many; fruit oblongoid ....**E. lanceifolius**

12b. Ouyles single in a locule; fruit subglobse .... **E. tuberculatus**

Elaeocarpus angustifolius Blume., Bijdr. Fl. Ned. Ind.: 120 (1825). Synonyms: *Elaeocarpus sphaericus* (Gaertn.) K. Schum., *Ganitrus sphaerica* Gaertn., *Elaeocarpus ganitrus* Roxb., *Elaeocarpus cyanocarpus* Maingay ex Mast. Vern.: *Ruthracham* (Mal.); *Utracham* (Tam.)

Trees to 30 m tall; young parts puberulous. Leaves alternate, oblong-lanceolate, oblanceolate or elliptic, 7-15 x 2.5-5 cm, cuneate or acute at base, acute or acuminate at apex, minutely crenate-serrate or subentire, thinly sericeous becoming glabrous, often with glands at nerve axils beneath, chartaceous; petioles 1-1.5 cm long, sharply margined, pubescent. Racemes 5-8 cm long, axillary, glabrous. Flowers white, 8-10 mm across; pedicels 8-9 mm long, hairy. Sepals 6-7 mm long, linear-lanceolate or oblong, acuminate, silky canescent outside, 1-ribbed inside. Petals white, 7-9 mm long, oblong or obtriangular laciniate, pubescent along margins near base. Stamens 30-40; filaments 1 mm long, puberulous; anthers 2.5-4 mm long, puberulous, acuminate at apex, locules unequal. Ovary globose, silky villous, 4-5 loculed, rarely 1-4 or 6-10 loculed; styles longer than stamens. Drupes globose, 1.5-2.5 cm across, deep blue or purple and succulent when ripe, usually 5-loculed; pyrenes globose, usually 5-locular, rarely 1-4 or 6-10 locular, 5-seeded. Fl. & Fr.: Novemebr – March.

Western Ghats: Evergreen forests of Maharashtra and Karnataka. World: Myanmar, Nepal, Bangladesh and Malaysia.

IUCN: LC (IUCN, 2024).

Elaeocarpus aristatus Roxb., Fl. Ind., ed. 1832. 2: 599. 1832. Synonym: *Monocera aristata* Jack

Large trees to 20-40 m tall. Leaves simple, alternate, spiral. obovate, elliptic or oblanceolate, 15-30 x 4-10 cm, cuneate or acute at base, rounded or subacute at apex, distantly serrulate or subentire, glabrous, thinly coriaceous; secondary nerves 12-18 pairs, obliquely arcuate; stipules subulate, caducous; petioles 1.5-5 cm long, swollen at both ends, geniculate, glabrous. Inflorescences axillary racemes 7-20 cm long, rusty villous. Flower-buds lanceolate, ribbed. Flowers white, 2 cm across, drooping; pedicels 1-2 cm long, rusty villous. Sepals 0-1.5 cm long, lanceolate, narrowly triangular or oblong, rusty tomentose without, glabrous within. Petals 12-16 mm long, dull white, silky, hairy outside, laciniate. Stamens 40-60, subulate; filaments to1 cm long, puberulous or glabrous; anthers 6 mm long, awned. Ovary ovoid, hairy, 2-loculed; styles subulate, longer than anthers, tapering from a conical base. Drupes ellipsoid, 3-3.5 cm long. Pyrenes pointed at both ends, compressed, oblong, rugose, 1-loculed. Fl. & Fr.: April – June.

Western Ghats: Evergreen forests of Maharashtra. World: Myanmar, Bangladesh, Bhutan.

IUCN: NE (IUCN, 2024).

Elaeocarpus blascoi Weibel., Candollea 27: 16. 1972. Vern.: *Kaattu Uthracham* (Tam.).

Evergreen trees to 25 m tall; bark rough, brown; branchlets with prominent leaf scar. Leaves simple, alternate, clustered at twig ends, elliptic or elliptic-ovate, 6-9.5 x 3-4.5 cm, apex acute to shortly acuminate with blunt tip, base rounded, margin shallowly serrate, glabrous; midrib slightly raised above; secondary nerves 7 pairs, with domatia in the axils beneath; tertiary nerves reticulo-percurrent; petioles 1.5-2.5 cm long, canaliculate, sparsely adpressed hairy; stipules caducous. Racemes terminal, to 7 cm long, 6-8-flowered. Flower white, 1.2 cm across; pedicel to1cm long. Sepals 5, lanceolate, 0.8 x 0.2 cm, rusty. Petals 5, cuneate,1.2 x 0.5 cm, apically laciniate. Anthers bearded, not ciliate. Drupe ellipsoid, 1.6 x 1 cm, 1-seeded. Fl. & Fr.: January – September.

Western Ghats: Evergreen forests of Tamil Nadu (Endemic).

IUCN: CR- B1ab(iii)+2ab(iii); D. (IUCN, 2024).

Elaeocarpus gaussenii Weibel, Candollea 27: 17. 1972. Vern.: *Moondrumuga ruthracham* (Tam.).

Trees to 15–20 m tall; bark deep red when cut; branchlets with closely adpressed hairs. Leaves ovate to obovate, 4.5–8 × 2.5–4 cm, apex broadly obtuse, base cuneate, ad-pressed hairs beneath, domantia regularly presentalong midvein, margin obviously serrate; petioles 6–9 mm long, glabrous. Racemes auxillary, behind the leaves, 5–9 cm long. Flowers 4 mm long; bracts early caducous, oblanceolate to ovate–elliptic, to1 mm long; pedicel 4–4.5 mm long. Sepals deep red col-oured, 4 mm long, verrucose. Petals white, 4–4.5 mm × 1.5–3 mm, surface glabrous. Disc with fused lobes, to 0.6–1 mm. Stamens 12–20, anthers 1.5–2 mm long; filament 1 mm long, arista absent. Ovary

3-4-carpellary, surrounded by the disc at base, to 0.9–1 mm long; style 2–2.5 mm long. Drupe broadly ellipsoid to ovoid, 2.5 cm long, apex rounded with or withou apical point, dark green when ripen. Seed ovoid to ellipsoid, 2–2.5 cm long, surface rugose to slightly sculp-tured, sutures obviously defined. Fl. & Fr.: April – October.

Western Ghats: Evergreen forests of Kerala and Tamil Nadu (Endemic).

IUCN: CR- B1ab(iii)+2ab(iii); D. (IUCN, 2024).

Elaeocarpus lanceifolius Roxb., Fl. Ind., ed. 1832. 2: 598. 1832. Synonyms: *Elaeocarpus serrulatus* Benth.

Trees to 10-15 m tall; bark greyish-brown, lenticellate; branchlets terete, silky. Leaves simple, alternate, elliptic, lanceolate, elliptic-lanceolate or oblanceolate, 6-15 x 3-5 cm, rounded at base, acute to acuminate at apex, distantly crenate-serrate, glabrous or sparsely pubescent beneath, thinly coriaceous; secondary nerves 6-8 pairs; petioles 0.5-1.5 cm long, puberulous or glabrous. Racemes axillary, 3-8.5 cm long, puberulous. Flowers white, 6-9 mm across; pedicels 5-12 mm long, slender, puberulous. Sepals lanceolate, 3-5 mm long, keeled, puberulous or glabrate without, silky puberulous within. Petals white, 4-6 mm long, slightly shorter than sepals, margin ciliate, laciniate. Stamens 15-30; filaments 3 mm long, pubescent; anthers 1.5-2 mm long, oblong with short bristles at apex. Disk 5-lobed, pubescent. Ovary pubescent, 3-loculed; styles villous at base. Drupe ovoid, 3-4 × 2-2.5 cm, conspicuously verrucose, unilocular. Seed 1, 2 cm long. Fl. & Fr.: April – June.

Western Gahts: Evergreen forests of Karnataka, Kerala and Tamil Nadu. World: Myanmar, Nepal, Bhutan, South China, Taiwan, hong Kong, Indonesia, Malaysia, Vietnam, Cambodia, Laos.

IUCN: NE (IUCN, 2024), vulnerable (Narasimhan and Irwin, 2021).

Elaeocarpus munroi (Wight) Mast., J.D. Hooker, Fl. Brit. India 1: 407. 1874. Synonym: *Monocera munroii* Wight Vern.: *Pungari* (Mal.); *Naribikki*, *Milachadhayan*, *Milacheriyan* (Tam.).

Trees to 20 m tall; bar dark grey, smooth; branchlets slender, glabrous. Leaves simple, alternate, clustered to the tip of branchlets, ovate or ovate-lanceolate, 4-10 x 1.7-5 cm, apex acuminate or caudate-acuminate, base round or obtuse, margin crenate or serrate, revolute, glabrous, chartaceous; lateral nerves 3-6 pairs, intercostae reticulate; petiole 25-50 mm long, very slender, glabrous. Flowers white, in short drooping axillary racemes. Sepals 5, 8 mm long, lanceolate, thinly tomentose. Petals 5, white with reddish lines at base, ovate-lanceolate, laciniate, densely silky on both sides, inserted round the base of glandular disc. Stamens numerous, inserted between the glands on the disc; anthers awned with long bristle. Ovary densely superior, silky hairy, placed on the torus, 2-celled, ovules 2 in each cell; style subulate, entire. Drupe oblong, blue, terete, glabrous, glaucous; stone 1 or 2 seeded. Fl. & Fr.: September – November.

Western Ghats: Evergreen forests of Maharashtra, Karnataka, Kerala and Tamil Nadu (Endemic).

IUCN: NT (Low risk/near threatened) (IUCN, 2024).

Elaeocarpus recurvatus Corner., Gard. Bull. Straits Settlem. 10: 325. 1939. Synonyms: *Elaeocarpus ferrugineus* (Wight) Bedd., *Monocera ferruginea* Wight. Vern.: *Naikotta* (Mal.); *Kaattu Uthracham* (Tam.).

Trees to 15 m tall. Leaves simple, alternate, clustered at the tip of branchlets, ovate or ovate-oblong, 7.5-15 x 4-8 cm, folded back, boat shaped, apex acute or caudate, base acute or cuneate, margin crenate or serrulate, glabrous above except downy midrib, softly pubescent beneath, coriaceous; lateral nerves 6-10 pairs, veins dividing dichotomously towards the margin, intercostae reticulate, obscure; petiole 15-50 mm, stout, tomentose. Flowers white, in short axillary racemes to 7 cm. Sepals 5, lanceolate, pubescent. Petals 5, 1.2 cm long, broadly ovate, sparsely within, inserted round the base of glandular disc. Stamens numerous, inserted between the glands on the disc; anthers ending in straight bristle. Ovary superior, subglobose, densely silky outside, 3-celled, ovules 2 in each cell; style subulate, entire. Drupe ovoid; stone single, compressed, coarsely tubercled. Seed one. Fl. & Fr.: February – May.

Western Ghats: Evergreen forests of Kerala and Tamil Nadu (Endemic).

IUCN: VU- B1+2c. (IUCN, 2024).

Uses: Wood logs can be used for building construction.

Elaeocarpus rugosus Roxb. ex G. Don., Gen. Hist. 1: 559. 1831. Synonyms: *Monocera rugosa* (Roxb. ex G. Don) Wight, *Terminalia sharmae* M. Gangop. & Chakrab., *Monocera roxburghii* Wight

Trees to 24 m tall; bark grey mottled with white, smooth; young stem, buds, petioles and peduncles densely brown villous. Leaves simple, alternate, clustered at the tip of branchlets, obovate, 8-24 x 4-15 cm, base cuneate or round, apex acute, obtuse or retuse, margin distantly serrate or crenate or subentire, glabrous above, ferruginous pubescent beneath; lateral nerves 8-12 pairs, intercostae scalariform, slender, domatia present; stipules free, lateral, brown-villous; petiole 15-30 mm, stout, pubescent. Flowers white, in axillary racemes to 12 cm long; pedicel 2 cm long, deflexed. Sepals 5, lanceolate, tomentose outside, valvate. Petals 5, white, fimbriate, fulvous tomentose inserted round the base of glandular disc. Stamens numerous, inserted between the glands on the disc; anthers thinly tomentose, terminating in long bristle. Ovary superior, subglobose, densely tomentose, placed on raised torus, 2-celled; ovules 2 in each cell; style subulate, tomentose, entire. Drupe oblong or ellipsoid, 3-3.5 × 2.5 cm, ferrugenous tomentose, stones single, compressed, coarsely tuberculate, 1-2-celled. Fl. & Fr.: March – April.

Westrn Ghats: Evergreen forests of Kerala and Tamil Nadu. India: Andaman & Nicobar Islands, Arunachal Pradesh, Assam and Meghalaya. World: Bangladesh, China, Mayanmar, Malaysia.

IUCN: VU- B1+2c. (IUCN, 2024).

Elaeocarpus serratus L., Sp. Pl.: 515. 1753. Synonyms: *Elaeocarpus oblongus* Gaertn., *Elaeocarpus cuneatus* Wight, *Elaeocarpus angustifolius* Wight, *Elaeocarpus monogynus* Murray. Vern.: *Nalla-kara, Valya-kara, Kara-mavu* (Mal.); *Olang-karai* (Tam.).

Trees to 20 m tall; bark 8-10 mm thick brown, mottled with grey; branchlets, petioles, peduncles and young leaves densely pubescent. Leaves simple, alternate, elliptic-obovate or elliptic ovate, 5-14 x 2-6.2 cm, apex acute, caudate-acuminate or obtusely acuminate, base acute or cuneate, margin crenate-serrate, glabrous, coriaceous; petiole 15-40 mm, slender, pubescent, swollen tipped, with 2 minute glands at tip; lateral nerves 3-8 pairs, intercostae reticulate, slender, domatia glandular. Flowers white, in axillary racemes to 8 cm long. Sepals 5, ovate, pubescent on outside, densely glandulose. Petals 5, laciniate, inserted round the base of glandular disc. Stamens many, inserted between the glands on the disc; anthers not awned, tipped with hairs. Ovary superior, densely tomentose, raised on torus, 3-celled; ovules 2 in each cell; style subulate, entire. Drupe subglobose, green, style persistent; stone tubercled. seed one. Fl. & Fr.: January – March.

Western Ghats: Evergreen forests of Maharashtra, Karnataka, Kerala and Tamil Nadu. World: Sri Lanka, Bhutan, Nepal, China, Myanmar, Java and Malaysia.

IUCN: NE (IUCN, 2024).

Elaeocarpus weibelii (Zmarzty) Shareef, S.P.Mathew & Shaju., Int. J. Advanced Res. 8(10): 1110. 2020. Synonyms: *Elaeocarpus serratus* var. *weibelii* Zmarzty

Large trees to 20 m tall; branchlets glabrous. Leaves obovate, 6-10 x 3-4 cm, apex acute, narrowed to the base, 3-5 ribbed, margin serrate, nerves with glandular domatia; petiole to 1.5 cm, glabrous. Raceme to 8 cm long; pedicel 5 mm long, slender. Flowers greenish white. Sepals 7 mm long, densely glandulose. Petals 7 mm long, glabrous. anthers 2.2 mm long, puberulus, bristles 3 or 4. Ovary densely tomentose, 3-celled. Drupe to 2.5 cm across, pyrene tubercled. Fl. & Fr.: September – March.

Western Ghats: Evergreen forests of Kerala and Karnataka (Endemic).

IUCN: NE (IUCN, 2024).

Elaeocarpus tuberculatus Roxb., Fl. Ind., ed. 1832. 2: 594. 1832. Synonyms: *Monocera tuberculata* (Roxb.) Wight & Arn., *Elaeocarpus bilocularis* Roxb. ex Wight & Arn., *Monocera subintegerrima* Miq. ex Mast. Vern.: *Pilahi, Naggara, Kodavasi, Ammakkaram* (Mal.); *Rutthracham* (Tam.).

Trees to 25 m tall; bark grey mottled with white, smooth; young stem, buds, petioles and peduncles densely brown villous. Leaves simple, alternate, clustered at the tip of branchlets, obovate, base cuneate or round, 8-24 x 4-15 cm, apex acute, obtuse or retuse, margin distantly serrate or crenate or subentire, glabrous above, ferruginous pubescent beneath; lateral nerves 8-12 pairs, parallel, intercostae scalariform, domatia present; petiole 15-30 mm, stout, pubescent; stipules free, lateral, brown-villous. Flowers white, in axillary racemes to 12 cm long; pedicel 2 cm long, deflexed. Sepals 5, lanceolate, tomentose outside. Petals 5, white, fimbriate, fulvous tomentose inserted round the base of glandular disc. Stamens numerous, inserted between the glands on the disc; anthers thinly tomentose, terminating in long bristle. Ovary superior, subglobose, densely tomentose, placed on raised torus, 2-celled, ovules 2 in each cell; style subulate, tomentose, entire. Drupe oblong or ellipsoid, 3-3.5 × 2.5 cm, ferrugenous tomentose, stones single, compressed, coarsely tuberculate, 1-2-celled. Fl. & Fr.: December – October.

Western Ghats: Stream side in evergreen forests of Maharashtra, Karnataka, Kerala and Tamil Nadu. World: Malaysia.

IUCN: NE (IUCN, 2024).

Elaeocarpus variabilis Zmartzy., Kew Bull. 56: 429. 2001. Syn.: *Elaeocarpus glandulosus* Wall. ex Merr.Vern.: Kara, *Kattakara, Kotlampazhamaram, Malamkara, Varali* (Mal.); *Kottampazham, Kotlampazham, Kaattukarai* (Tam.).

Trees to 25 m tall; branchlets, young leaves and buds densely pubescent. Leaves simple, alternate, elliptic-ovate, elliptic-obovate, elliptic-oblong or ovate, 5-12 x 2.3-6.5 cm, apex abruptly acuminate with emarginate tip, base cuneate or acute, margin distantly crenate or serrate, glabrous on both sides, coriaceous, punctate, domatia glandular; lateral nerves 4-8 pairs, intercostae reticulate, faint; petiole 15-35 mm, slender, swollen at base and tip, 2 minute glands at tip; stipules lateral, caudous. Flowers white, in axillary and terminal racemes; peduncle puberulus when young; pedicel 3 mm long. Sepals 5, 4-5 mm, ovate-lanceolate, puberulous outside, glandular inside. Petals 5, white, 6-8 mm, laciniate, glandular, inserted round the base of glandular disc. Stamens many, inserted between the glands on the disc; anthers tipped with hairs. Ovary superior, subglobose, placed on a raised torus, densely hairy outside, 3-celled, ovules 2 in each cell; style subulate, entire. Drupe oblong, 2.5 × 2.5 cm, stone one. Fl. & Fr.: Thorought the year.

Western Ghats: Evergreen forests of Maharashtra, Karnataka, Kerala and Tamil Nadu (Endemic)

IUCN: NE (IUCN, 2024).

Uses: Fruit is edible.

Elaeocarpus venustus Bedd., Fl. Sylv. S. India: t. 174. 1872. Vern.: *Kaattu Uthracham* (Tam.)

Trees to 15 m tall; branches glabrous. Leaves simple, alternate, elliptic-obovate or obovate, 6-12 x 3-5 cm, base attenuate, apex acute or obtusely acute, margin serrate, glabrous, coriaceous, lateral nerves 6-8 pairs, intercostae reticulate, prominent; stipules lateral, cauducous; petiole 10-20 mm long, stout, glabrous. Flowers white, in axillary racemes. Sepals 5, ovate-lanceolate, pubescent. Petals 5, laciniate, tomentose, inserted round the base of glandular disc. Stamens numerous, inserted between the glands the glands on the disc; anthers awned. Ovary superior, glabrous, 2-celled; ovules 2 in each cell; style subulate, entire. Drupe oblong-ellipsoid, 5 x 2.5 cm. Seed single. Fl. & Fr.: February – August.

Western Ghats: Evergreen forests of Kerala and Tamil Nadu (Endemic).

IUCN: CR- B1ab(iii,v). (IUCN, 2024).

Planted:

Muntingia calabura L.; Exotic: Native to South America

ERICACEAE Durande

Key to the genera

1a. Flowers in axillary or terminal long racemes, urceolate; berry globose, glabrous .... **Vaccinium**

1b. Flowers in terminal condensed racemes, campanulate; capsule cylindric, tomentose .... **Rhododendron**

Rhododendron L.

Rhododendron arboreum Sm., Exot. Bot. 1: 9. 1805. ssp. **arboreum** Vern.: *Kaattu-povarasu* (Mal.); *Alingi* (Tam.).

Evergreen trees to 20 m tall; bark gray-brownish, exfoliating into thin and small irregular flakes; young shoots stout, with distinct leaf traces, densely grayish tomentose, glabrescent. Leaves leathery, oblong-lanceolate or oblong-oblanceolate, 6–15 × 2–4.5 cm, base cuneate or ± rounded, margin revolute, apex acuminate or acute, abaxial surface with indumentum 1- or 2-layered, densely compacted, white to fawn tomentose, sometimes with loosely floccose brown upper layer, glabrescent, adaxial surface reticulate to bullate; lateral veins 15–26-paired; petiole 10–25 mm long with dense indumentum intermixed with glands. Inflorescence dense 20-flowered, rachis 10–15 mm long, tomentose; pedicel 0.9 cm long, pilose, glandular. Sepals 5, lobes 5, 1–2 mm long, small, triangular, sparsely glandular and hairy. Petals 5, tubular-campanulate, pink to deeply crimson, 3.5–4 cm, with 5 black-blotched basal nectar pouches and dark flecks, lobes 5, 1.2–1.7 cm, apex emarginate. Stamens 10, unequal, 1.7–2.7 cm, filaments glabrous. Ovary conoid, 4–6 mm, white-tomentose, sometimes also glandular; style to 3.3 cm, glabrous. Capsule cylindric, 30.6 mm. Fl. & Fr.: March-May.

Western Gahts: Evergreen forests above 1200 m in Karnataka, Kerala and Tamil Nadu. World: Sri Lanka, Nepal, Mayanmar, Bhutan, China, Vietnam and Malaysia.

IUCN: LC (IUCN, 2024).

Key to the subspecies

1a. Flwoers deeply pink crimson, nectar guide usually black spots on inner face of petals 2

1b. Flowers white, small blood red or dark brown spots on the inner base of the petals ....ssp. **cinnamomeum**

2a. Leaves greyish tomentose beneath, lateral nerves 15-26 pairs; cymes 20-flwoered ....ssp. **arboreum**

2b. Leaves ferrugineous tomentose beneath, lateral nerves 12-16 pairs; cymes 10-flowered ....ssp. **nilagiricum**

Rhododendron arboreum ssp. **cinnamomeum** (Wall. ex G. Don) Tagg. J. B. Stevenson, Spec. Rhododendron: 15. 1930. Synonyms: *Rhododendron cinnamomeum* Wall. ex G. Don, *Rhododendron arboreum* ssp. *campbelliae* (Hook.f.) Tagg, *Rhododendron campbelliae* Hook.f.

Rhododendron arboreum Sm. ssp. **nilagiricum** (Zneker) Tagg., J.B.Stevenson, Spec. Rhododendron: 15. 1930. Synonyms: *Rhododnedron nilagiricum* Zenker, *Rhododendron arboreum* var. *nilagirica* (Zenker) Clarke, *Rhododendron arboretum* sensu Bourd. Vern.: *Alinji* (Tam.).

Fl. & Fr.: January – May.

Western Ghats: Evergreen forests above 1200 in Kerala and Tamil Nadu (Endemic).

World: NE (IUCN, 2024).

Vaccinium L.

1a. Leaves lanceolate-elliptic, to 10 cm long; bracts persistent ....**V. nilgherrensis**

1b. Leaves ovate-elliptic, to 5 cm long; bracts caducous ....**V. symplocifolium**

Vaccinium symplocifolium (D.Don ex G.Don) Alston., H. Trimen, Handb. Fl. Ceylon 6(Suppl.): 176. 1931. Synonyms: *Vaccinium leschenaultii* Wight var. *leschenaultia. Agapetes symplocifolia* D.Don ex G.Don, *Epigynium leschenaultii* (Wight) Klotzsch, *Vaccinium leschenaultii* Wight var. *rotundifolia* (Wight) Clarke, *Vaccinium rotundifolium* Wight, *Vaccinium leschenaultii* Wight var. *zeylanica* Clarke Vern.: *Kalavu* (Tam.).

Small trees to 7 m tall. Leaves simple, alternate, ovate, elliptic or elliptic-obovate, 4.5-6.5 x 2.5-3 cm, sometimes suborbicular, apex acute or acuminate, base attenuate to acute or obtuse, margin serrate, glabrous, coriaceous, glands on the midrib beneath; midrib slightly raised or flat above; secondary nerves 4-6 pairs, tertiary nerves broadly reticulate; petiole 0.5-1 cm long, planoconvex in cross section, pubescent when young. Racemes axillary and terminal. Flowers urceolate, pink or rarely whitish. Petals glabrous. Stamens with awns; filaments to 0.3 cm long. Berry smooth, globose. Seeds many. Fl. & Fr.: November – May.

Western Gahts: Evergreen forests above 1000 m in Kerala and Tamil Nadu. World: Sri Lanka.

IUCN: NE (IUCN, 2024).

Vaccinium neilgherrense Wight., Calcutta J. Nat. Hist. 8: 173. 1847. Synonyms: *Epigynium neilgherense* (Wight) Klotzsch Vern.: *Kalavu* (Tam.).

Small trees to 10 m tall; branchlets drooping, lenticellate. Leaves simple, alternate, lanceolate to elliptic-lanceolate, 4.5-10 x 1-2.2 cm, apex acuminate, base attenuate to cuneate, margin crenate-serrate, glabrous, coriaceous, glands on the midrib beneath; secondary nerves 4-6 pairs, tertiary nerves broadly reticulate, not prominent; petiole to 0.3 cm long, planoconvex in cross section, minutely pubescent when young. Racemes axillary and terminal, lax. Flowers white to rose colored; pedicel up to 0.7 cm long. Petals outside pubescent. Berry smooth, globose. Seeds many. Fl. & Fr.: March – June.

Western Ghats: Evergreen forests above 1200 in Karnataka, Kerala and Tamil Nadu (Endemic).

IUCN: NE (IUCN, 2024).

Planted:

Andromeda leschenaultii J. Broughton

Arbutus unedo L.; Exotic: Native to Europe, Northern Africa, Siberia

Clethra arborea Aiton; Exotic: Native to Europe, Northern Africa, Siberia

Rhododendron ponticum L.; Exotic: Native to Spain

ERYTHROXYLACEAE Kunth

Erythroxylum P. Browne

1a. Stipules persistent .... **E. monogynum**

1b. Stipules caducous 2

2a. Leaf apex obtuse or rounded 3
2b. Leaf apex acute or acuminate ....**E. lanceolatum**
3a. Drupe elliptic-oblong, slightly curved .... **E. moonii**
3b. Drupe linear-oblong, straight ....**E. obtusifolium**

Erythroxylum lanceolatum (Wight) Walp., Repert. Bot. Syst. 1: 407. 1842. Synonyms: *Sethia lanceolata* Wight, *Sethia erythroxyloides* Wight, *Erythroxylum wightii* Voigt

Trees to 15 m tall; bark reddish-brown, smooth. Leaves simple, alternate, distichous, elliptic, elliptic-lanceolate or elliptic-obovate, 3-11.5 x 1.5-3 cm, base narrowly attenuate or acute, apex acute or caudate-acuminate, margin entire, glabrous, chartaceous; lateral nerves 4-12 pairs, intercostae reticulate, obscure; stipules intrapetiolar, on the arrested shoots imbricating; petiole 3-7 mm long, grooved above, glabrous. Flowers white, axillary, solitary or paired. Sepals 5-6, campanulate, lobes 5-6, ovate, connate at base. Petals 5-6, white, with an erect double ligule on the inner side, ovate-lanceolate. Stamens 10-12, monodelphous. Ovary superior, oblong, 3-4-celled; ovules 1-2 in each cell; style 3-4, united into a style with capitate stigma. Drupe oblong, 1.3-1.5 x 0.2-0.3 cm. Seeds 1, oblong. Fl. & Fr.: March – June.

Western Ghats: Evergreen forests of Kerala and Tamil Nadu. World: Sri Lanka.

IUCN: NE (IUCN, 2024).

Erythroxylum monogynum Roxb., Pl. Coromandel 1: 61. 1798. Synonyms: *Sethia indica* DC., *Erythroxylum sideroxyloides* Roxb., *Erythroxylum indicum* (DC.) Elliot Vern.: *Sempulinchan, Devadhaaru* (Tam.).

Small trees to 7 m tall; bark dark-brown, very rough, lenticellate; branchlets compressed. Leaves simple, alternate, rarely distichous, obovate or oblanceolate, 2-6.5 x 1-3 cm, base attenuate or cuneate, apex obtuse, margin entire, lateral nerves 6-14 pairs, intercostae reticulate, faint; stipules intrapetiolar, on the arrested shoots imbricating; petioles 3-8 mm long, slender, grooved above, glabrous. Flowers white, axillary, solitary or 1-4 in axillary fascicles. Sepals 5-6, ovate, acute, glabrous. Petals 5-6, white, oblong, with an erect double ligule on inner side, imbricate. Stamens 10-12, Monadelphous. Ovary 3-4-celled; styles 3-4, united into a style with capitate stigmas. Ovary superior, 3-4-celled; ovules 1-2 in each cell. Drupe ellipsoid, some what 3-angled, apiculate, glabrous, reddish. Seed one. Fl. & Fr.: Throughout the year.

Western Ghats: Deciduous to semi-evergreen forests of Maharashtra, Karnataka, Kerala and Tamil Nadu. India: Andhra Pradesh and Telangana. World: Sri Lanka.

IUCN: NE (IUCN, 2024).

Erythroxylum moonii Hochr., Bull. Inst. Bot. Buitenzorg 22: 54. 1905. Synonyms: *Erythroxylum acuminatum* (Arn.) Walp., *Sethia acuminata* Arn., *Erythroxylum lucidum* Moon ex Trimen

Small trees 3-6 m tall. Leaves lanceolate or elliptic, cuneate to obtuse at base, caudate-acuminate at apex, 4-15 x 1.5-4 cm; lateral nerves 7-13 pairs, faint on both surfaces; stipules lanceolate, acuminate at apex, 5 mm long; petioles 2-5 mm long. Flower axillary, solitary, pendulous, white or greenish to yellowish white; pedicel 5 mm long. Sepals 5, lobes ovate, acute at apex, 1.5 mm long. Petals 5, oblong, obtuse at apex, 3.5 x 1.5 mm. Stamens equal; staminal tube 2 mm long. Ovary 2 x 1 mm; styles 2.5 mm long, free; stigma depressed-capitate. Drupes oblong, often slightly curved, 1.5 x 0.3 cm, longitudinally ridged, pink when mature. Fl. & Fr.: April - August.

Western Ghats: Evergreen forests of Kerala and Tamil Nadu. India: Andhra Pradesh. World: Sri Lanka.

IUCN: NE (IUCN, 2024).

Erythroxylum obtusifolium (Wight) Hook.f., Fl. Brit. India 1: 415. 1874. Synonyms: *Sethia lanceolata* var. *obtusifolia* Wight, *Erythroxylum lanceolatum* var. *obtusifolium* (Wight) Walp., *Sethia obtusifolia* Thwaites

Small tree to 7 m tall. Leaves elliptic to obovate, cuneate to obtuse at base, acute at apex with an obtuse tip, 4-9 x 2-5 cm, reddish brown when dry, lateral nerves 8-11 pairs, distinct on both surfaces; petioles 4-7 mm long; stipules lanceolate, 5-7 mm long, longer than or as long as petiole. Flowers axillary, solitary or in pairs, yellowish white; pedicels 3 mm long. Sepala 5, lobes ovate, acute at apex, 1.5 mm long. Petals oblong, rounded at apex, 4.5 x 1.5 mm. Stamens equal; staminal tube 1.5 mm long. Ovary obovoid, 2 mm long; styles 1 mm long, free; stigma depressed-capitate. Drupe oblong, 1.3 x 0.4 cm, longitudinally ridged, red when mature, free parts of filaments also persistent along with staminal tube. Fl. & Fr.: February - August.

Western Ghats: Evergreen forests of Karnataka, Kerala and Tamil Nadu. World: Sri Lanka.

IUCN: VU- A1c, B1+2c. (IUCN, 2024).

Planted:

Erythroxylum coca Lam.; Exotic: Native to South America

EUPHORBIACEAE Juss.

Key to the genera

1a. Leaves reduced, if developed caducous; thorny or thornless trees; latex milky; inflorescence cyathia; tepals absent or minute; staemns 1 .... **Euphorbia**

1b. Leaves well developed; usually thornless trees; latex yellow or colourless; inflorescence other than cyathium; tepals conspicuous; stamens 2-10 2

2a. Cells of the ovary 1-ovuled 3

2b. Cells of the ovary 2-ovuled 17

3a. Filaments inflexed in bud 4

3b. Flowers erect in bud 5

4a. Stamens 3-5, connate .... **Trigonostemon**

4b. Stamens more than 5, not connate .... **Croton**

5a. Inner tepals present in male flowers; staemens not branched ... **Aleurites**

5b. Inner tepals absent in male flowers; stamens branched 6

6a. Leaves deeply-palmately lobed; staminal bundles indefinite....... **Ricinus**

6b. Leaves entire, rarely slightly-lobed 7

7a. Anther 4-celled 8

7b. Anther 2-celled 10

8a. Indumentum simple, anther connective enlarged .... **Agrostistachys**

8b. Indumentum malpighiaceous; anter connective not enlarged 9

9a. Styles bifid; anthers globose .... **Celidion**

9b. Styles simple; anthers cross-shaped .... **Macaranga**

10a. Stamens numerous ....12.

10b. Stamnes 2-9 11

11a. Inflorescence spicate; styles connate into column, shallowly bifid .... **Epiprinus**

11b. Infloresecence terminal heads; styles free, distally multifid . **Koilodepas**

12a. Outer tepals valvate in male flowers .... **Mallotus**

12b. Outer tepals imbricate in male flowers 13
13a. Inner tepals present 14
13a. Inner tepals absent 15
14a. Flowers glomerate, spicate or racemose .... **Tristaxis**
14b. Flowers paniculate, not spicate ....**Givotia**
15a. Flowers in terminal reacems or spike or umbellate.................................16
16a. Inflorescence axillary raceme; tepals absent in male flwoers ..**Falconeria**
16b. Inflorescence umbellate; tepals present in male flowers .... **Blachia**
17b. Stamens more than 3, filaments to 5 mm long 18
17a. Stamens 3, filaments 2 mm long ....**Suregada**
18a. Branchlets and inflorescence with yellow hairs; lateral nerves less than 7 ....**Shirakiopsis**
18b. Branchlets and inflrescence glabrous; lateral nerves more than 8..........19
19a. Fruits deeply 3-lobed; glabrous ....**Excoecaria**
19b. Fruits not lobed; cottony-hairy ....**Paracroton**

Agrostistachys Dalz.

1a. Spike short to 7 cm long; bracts closely imbricate, male bract single flowered; stipules to 10 mm long .... **A. indica**

1b. Spike elongate to 15 cm long; bracts distant, male bracts many-flowered; stipules to 2 mm long ....**A. borneensis**

Agrostistachys borneensis Becc., Nelle Forest. Borneo: 331. 1904. Synonyms: *Agrostistachys longifolia* (Wight) Trimen, non Kurz., *Sarcoclinum longifolium* Wight, *Agrostistachys latifolia* (Hook.f.) Pax & K. Hoffm., *Agrostistachys longifolia* var. *latifolia* Hook.f., *Agrostistachys meeboldii* Pax & K. Hoffm. Vern.: *Mulimpala* (Mal.); *Mancherai* (Tam.).

Dioecious trees to 15 m; bark blackish-grey mottled with white, smooth; branchlets stout, marked with the scales of petioles and stipules, exudation milky. Leaves simple, alternate, clustered at the tips of branchlets, oblanceolate or obovate-spathulate, 10-25 x 5-7.5 cm, base long attenuate, apex acuminate, margin entire, recurved, glabrous, coriaceous; lateral nerves 10-18 pairs, intercostae very obscure; stipules lateral, short, broad, acute, cauducous; petiole 5-10 mm long, glabrous. Flowers unisexual, yellow. Male flowers in axillary lax spiciform racemes to 15 cm long, much shorter than the leaves, few flowers per bract; bracts broadly ovate, ciliate. Tepals 8-10, biseriate, outer tepals connate, cupular, splitting into 2-5 valvate segments; inner tepals petalloid, alternating with outer, ovate, obtuse fleshy, fimbriate, shorter than outer petals. Disc glands 5, large, alternating with inner tepals. Stamens 10, exserted; outer 5 alternating with disc; inner 5 shortly connate, around pistillode; anthers versatile; pistillode 2-3 clefted. Female flowers in axillary lax spiciform racemes much shorter than the leaves, solitary in each bract. Tepals 8-10, biseriate, outer tepals ovate, stiff, spreading; inner tepals petalloid, longer than the outer tepals, alternating with outer, cauducous; entire. Disc glands 5, partly connate. Ovary superior, pubescent, 3-celled, ovules 1 in each cell, laterally pendulous; styles thick, short, spreading. Capsule globose, 3-lobed cocci, 7-15 mm across, reddish, crustaceous, smooth. Seeds globose, smooth, shining. Fl. & Fr.: August - October.

Western Ghats: Evergreen forests above 1000 m in Karnataka, Kerala and Tamil Nadu. World: Sri Lanka, Malaysia, Thailand, Sumatra, Borneo, Philippines and New Guinea.

IUCN: LC (IUCN, 2024).

Agrostistachys indica Dalz., Hooker's J. Bot. Kew Gard. Misc. 2: 41. 1850. Synonyms: *Agrostistachys indica* var. *genuina* Müll.Arg., *Agrostistachys indica* var. *longifolia* Müll.Arg., *Agrostistachys longifolia* (Müll.Arg.) Kurz Vern.: *Vepaala maram* (Tam.).

Small tree to 5 m tall. Leaves elliptic-oblong, 21 x 9 cm, acute, coriaceous, base attenuate, spineus serrate; lateral nerves 13 pairs, prominent, nervules parallel; petiole 1 cm long. Male flowers in short axillary racemes, to 7 cm long; bracts broadly ovate, densely tomentose. Sepals white, 4 x 2.5 mm, obovate, obtuse. Petals smaller than sepals, white. Stamens 12, filaments united at base. Female flowers solitary or 2-3 together, axillary; bracts many, 3 x 3 mm, imbricating, ciliate; pedicel 7 mm long, stout. Sepal oblong, densely hairy. Capsule 15 mm across, cocci minutely hispid. Seeds globose, glabrous. Fl. & Fr.: September - April.

Western Ghats: Evergreen forests of Maharashtra, Goa, Karnataka, Kerala and Tamil Nadu. World: Sri Lanka, Laos, Cambodia, Vietnam, Malayasia, Thailand, Borneo, Philippines and Papua New Guinea.

IUCN: LC (IUCN, 2024).

Aleurites J.R. Forst. & G. Forst.

Aleurites moluccanus (L.) Willd., Sp. Pl., ed. 4. 4: 590. 1805. Synonyms: *Jatropha moluccana* L., *Juglans camirum* Lour., *Aleurites angustifolius* Vieill., *Aleurites* *moluccanus* var. *floccosus* Airy Shaw, *Ricinus dicoccus* Roxb. Vern.: *Katakrote, Naatu Akrootu* (Tam.).

Evergreen trees to 20 m tall; bark dark gray, shallowly smooth; branchlets densely gray-brown stellate-puberulent, subglabrous at maturity. Leaves ovate to elliptic-lanceolate, entire or up to 5-lobed, 14- 20 × 7-17 cm, both surfaces stellate-puberulent when young, stellate-puberulent or glabrescent at maturity, base broadly cuneate or obtuse, rarely shallowly cordate, with 2 sessile discoid adaxial glands, apex acute to acuminate; basal veins 3-5; petiole 6-12 cm long, densely stellate-puberulent. Panicles terminal, to 20 cm long. Male flowers: pedicels 7-10 mm. Sepals 5, lobes ovate, 3-3.5 × 2-3 mm, apex acute, abaxially stellate-tomentose, adaxially glabrous. Petals narrowly oblong-spatulate, 6-8 × 2 mm, obtuse, glabrous, yellowish white, disk glands shallowly 3-lobed. Stamens 15-20; inner filaments connate into hairy column, 4 mm; free filaments 1 mm long, puberulent. Female flowers: pedicels 3-4.5 mm. Sepals and petals as in male but petals to1.5 mm wide. Disc 2 mm across, flat. Ovary densely stellate-puberulent, 2-3-locular; styles 2, short, bipartite. Drupes subglobose or slightly obliquely globose, 4-5 × 4-6 cm, shallowly 2-lobed, 1- or 2-seeded. Seeds broadly ovoid, 2.5 × 2.75 × 2.25 cm, lateral margin flattened, tuberculate-angular. Fl. & Fr.: December - May.

Western Ghats: Evergreen forests of Maharashtra, Karnataka, Kerala and Tamil Nadu. India: Odisha and West Bengal. World: Sri Lanka, Myanmar, China, Indonesia, Malaysia, Philippines, Thailand and Vietnam.

IUCN: LC (IUCN, 2024).

Uses: Kernal of the seed is said to be edible. The oil is stracting from the seeds which is used for making candles, in varnish and paint industry. The oil-cake is used as fertilizers. The wood is very soft and used for making packaging cases and pulpwood.

Blachia Baill.

1a. Female sepals deeiduous; leaves often weakly to clearly trinerved at the base; first pair of lateral nerves distinct, not weaker than subsequent pairs .... **B. denudata**

1b. Female sepals persistent or accrescent; leaves penninerved; first pair of lateral nerves not distinct, weaker than subsequent pairs 2

2a. Female sepals large, 5-15 x 2.5-9 mm; fruits 4-5-coccous .... **B. calycina**

2b. Female sepals small, 2-5 x 1-3 mm; fruits tricoccous .... **B. umbellata**

Blachia denudata Benth., J. Linn. Soc., Bot. 17: 226. 1878. Synonym: *Blachia andamanica* ssp. *denudata* (Benth.) N. Balakr. & Chakrab.

Small trees to 5 m tall; bark smooth, black, blotched with green, yellow inside; branchlets glabrous. Leaves elliptic or oblong, 10 x 5 cm, acute or acuminate, base truncate or obtuse, membranous; lateral nerves 5-7 pairs; petiole 0.5-1 cm. Male flowers in racemes, pedicels slender, gracilis, smooth. Perianth lobes ovate, concave, yellow, 2 mm long. Female flowers in fascicles of 2 or 3; pedicels thick. Capsule subglobose, 3-lobed, 1 cm across, galbrous or appressed strigose. Seeds oblong-ellipsoid to subglobose, 5-10 mm long. Fl. & Fr.: Septemebr - March.

Western Ghats: Semi-evergreen forests of Maharashtra, Goa, Karnataka and Kerala (Endemic).

IUCN: NE (IUCN, 2024).

Blachia calycina Benth., J. Linn. Soc., Bot. 17: 226. 1878. Synonym: *Tetracera heyneana* Wall.

Small trees to 6 m tall; branches glabrous. Leaves lanceate or oblanceate, elliptic-oblong to obovate, 4.5 -9.5 x 1.5 - 4 cm, subequal sided, entire, glabrous, bluntly acuminate at apex, narrowed at base; lateral nerves 5-16 pairs, secondary nerves obscure; petioles up to 2-10 mm long. Flowers greenish yellow, monoecious. Male flowers in lax, terminal racemes. Female flowers solitary, axillary. Sepals 4-6, elliptic, oblong to ovate or lanceolate, accrescent. Ovary subglobose, 3.5 mm across, 4-5-ovuled; style 4 or 5, 8 mm long, bifid. Capsules 1 cm long, with accrescent calyx about 1.5 cm long, strongly 3-5-lobed. Fl. & Fr.: November – June.

Western Ghats: Common in semi-evergreen foersts along the water course of Karnataka, Kerala and Tamil Nadu (Endemic).

IUCN: NE (IUCN, 2024).

Blachia umbellata (Willd.) Baill., Etude Euphorb.: 387. 1858. Synonyms: *Croton umbellatus* Willd., *Blachia reflexa* Benth., *Ehretia indica* (Dennst. ex Kostel.) M.R. Almeida & S.M. Almeida

Small trees to 8 m tall; branchlets slender, glabrous. Leaves elliptic, oblong to oblong-lanceolate, 15 x 5 cm, acuminate at apex, base acute, membranous, coriaceous; lateral nerves 5-15 pairs, irregular, looping; petiole 1 cm long. Male flowers in umbellate, 5 mm across, 5-10 together in a terminal umbel; pedicel 1.5 cm long. Sepals 4, 2.5 x 2 mm, ovate to orbicular, acute, glabrous. Stamens

15-20; anthers orbicular, leass than 1 mm across. Female umbels 1-8-flowered; flowers 8 mm across; 3-5, on stout peduncle. Sepals 4-6, 4 x 2 mm, oblong, acute, glabrous. Ovary subglobose, glabrous; styles 3, 8 mm long, bifid. Capsule 18 mm across, 3-lobed, glabrous, brown when dry. Seeds broadly oblong-ellipsoid, 7 mm long, smooth Fl. & Fr.: January - August.

Western Ghats: Evergreen forests along the steram banks of Karnataka, Kerala and Tamil Nadu. World: Sri Lanka.

IUCN: NE (IUCN, 2024).

Cleidion Blume

1a. Male spike 5-20 cm long; female peduncle 5-7 cm long; seeds 10-15 mm in diam... **C. javanicum**

1b. Male spike 10-35 cm long; female peducle 4-5 cm long; seeds 6-8 mm diam .... **C. nitidum**

Cleidion javanicum Blume., Bijdr. Fl. Ned. Ind.: 613. 1826. Synonyms: *Acalypha spiciflora* Burm.f. (*nom.confusum*), *Cleidion spiciflorum* (auct.non Burm.f.) Merr., *C. javanicum* var. *monoliflorum* (Airy Shaw) Chakrab. & M. Gangop., *Cleidion moniliflorum* Airy Shaw, *C. spiciflorum* var. *moniliflorum* (Airy Shaw) Radcl.

Evergreen trees to 20 m tall; bark greenish-white, smooth; branchlets glabrous. Leaves simple, alternate, elliptic, elliptic-oblong or elliptic-obovate, 7.5-20 x 2.5-7.5 cm, base acute, apex acuminate, margin serrate, glabrous, coriaceous, 2 glands at the base beneath; lateral nerves 4-6 pairs, intercostae reticulate, prominent; stipules lateral, ovate, acute, entire; petiole 15-70 mm long, slender, grooved above, glabrous. Flowers unisexual, pale green. Male flowers in long slender axillary racemes. Tepals 2-4 lobed, valvate. Stamens many, in a globose mass on a conical receptacle; filaments free; anther cells transversely didymous on a broad connective. Female flowers single or in pairs on long axillary peduncle; pedicel 4-5 cm long, slender. Tepals 3-5 lobed; lobes imbricate. Ovary superior, 2-celled; ovules 1 in each cell; styles 2, connate at base, arms 2-3, filiform. Capsule 2-lobed, 10-12 x 12-25 mm, broader than long, globose. Seeds 1 in each cell, 10-12 mm across, round, grey-black. Fl. & Fr.: September – April.

Western Ghats: Deciduous to evergreen forests of Maharashtra, Goa, Karnataka, Kerala and Tamil Nadu. India: Andaman & Nicobar Islands, Assam, Arunachal Pradesh, Bihar, Manipur, Meghalaya, Mizoram, Nagaland,

Sikkim and Tripura. World: Sri Lanka, China, Myanmar, Malesia and South East Asia.

IUCN: LC (IUCN, 2024).

Cleidion nitidum (Müll.Arg.) Thwaites ex Kurz., J. Asiat. Soc. Bengal, Pt. 2, Nat. Hist. 42: 245. 1873. Synonyms: *Mallotus nitidus* Müll.Arg.

Small tree to 7 m tall; bark silvery greyoutside, reddish inside; young branches lenticellate. Leaves simple, alternate, broadly to elliptically lanceolate, rarely they turn obovate, 10 - 15 × 4 - 6.5 cm, acute to attenuate at base, crenate to serrate above the middle, acute toacuminate at apex, thicker and rarely glandular in the margins, coriaceous, glabrous; secondary nerves 5 - 9 pairs, raised beneath; petioles to 1cm long, often bent and faintly bulged at both ends, with 3-5 glands abaxially at the junction of lamina; stipules early caducous. Pistillate and staminate flowers on separate branches in monoecious habit. Staminate flowers leaf-opposed, confining to terminal branches, single, red granulated, up to 20 cm long, pale yellow, 3-5 flowers per node; pedicels minutely bristly, up to 1mm long. Sepals ovate, 1–2 mm long, thick, tufted hairy at tip outside, reflexed in open flowers. Stamens 50, packed in a cluster on a raised thalamus; filaments to 1mm long, attached dorsally; anthers monothecous, with a tuft of bristles at tip. Pistillate inflorescences axillary, 3 - 5cm long, with 1 - 5 pistillate flowers, older flowers above, devoid of abscission mark; bracts ovate, 1mm long, persistent. Sepals 3, 1mm long, ovate, acute at tip, glabrous, caducous in fruit. Ovary 3-loculed, styles 3, connate, 0.5mm long, bifurcated above; stigmas 6, 1.5 – 2 cm long, purplish and hairy on adaxial face, glabrous on dorsalface. Capsules 3-lobed, 4 - 6 × 6 -10 mm, brownish when dry, glabrous. Seeds 3, with elliptic hilum. Fl. & Fr.: January - November

Western Ghats: Semi-evergreen forests of Tamil Nadu including Andaman and Nicobar Islands. World: Sri Lanka

IUCN: LC (IUCN, 2024).

Croton L.

1a. Foliar glands stipitate; style brances twice or thrice bifid . **C. zeylanicus**

1b. Foliar glands sessile; style branches simply bifid 2

2a. Fruits obovoid, to 25 mm long, shortly stipitate ....**C. malabaricus**

2b. Fruit subglobose, smaller to 15 mm long, without basal stipe 3

3a. Indumentum of young shoots lepidote 4

3b. Indumentum of young shoots pubescent or lacking 5

4a. Leaves very coarsely dentate-serrate-creante; fruits prominently lobed .. .. **C. persimilis**

4b. Leaves scarcely dentate; fruits globose, not lobed .... **C. gibsonianus**

5a. Fruit strongly muricate, hairs present ....**C. aromaticus**

5b. Fruit not muricate, galbrous .... **C. tiglium**

Croton aromaticus L., Sp. Pl.: 1005. 1753. Synonyms: *Croton laccifer* L. (as *'lacciferum'*), *Cascarilla aromatica* (L.) Raf., *Aleurites laccifer* (L.) Willd. Vern.: *Teppadai, Vid-Pune* (Tam.).

Small trees to 15 m tall; branchlets stellate, brown tomentose or scabrid. Leaves simple, ovate, 12 x 6 cm, acuminate, cordate at base, 3-ribbed from base, glands 2 pairs; upper leaves with acute base; lateral nerves 2-9 pairs, prominent; petiole 2-5 cm long, glabrous; stipules linear to filiform, 15 mm long. Inflorescence to15 cm long, densely stellate hairy, white; bracts petaloid. Male flower 3-5mm long, above, unisexual, bracteate. Perianth biseriate, tepels 10, outer layer 1-2 mm, 5 fused, obovate, stellately hairy; inner layer spathulate, 0.6 -1mm, 5 fused, white, stellately hairy. Stamens 10, 0.8-1mm long, basifixed, white, polyandrous; anther globose. Female flower 7 mm across. Sepals ovate, acute. Petals obovate, ciliate. Styles 3, bifid, glabrous. Capsule ovoid to subglobose, 8-10 mm long, trilocular, trilobedhirsute. Seeds 4- 6mm, 3, black and cream patches, glabrous. Fl. & Fr.: April – September.

Western Ghats: Fairly common in evergreen and semi-evergreen forests of Maharashtra, Karnataka, Kerala and Tamil Nadu. World: Bangladesh and Sri Lanka.

IUCN: NE (IUCN, 2024).

Croton gibsonianus Nimmo, J. Graham, Cat. Pl. Bomaby: 251. 1839. Synonym: *Oxydectes gibsoniana* (Nimmo) Kuntze.

Small trees to 6 m tall; branchlets terete, covered with stellate hairs, with translucent exudate. Leaves simple, alternate, distichous, ovate, 4.5 -10 × 2.5 - 5 cm, apex acuminate to caudate, margin entire, ciliate, base usually cuneate to rounded, with two lateral glands, characteristically. tri-nerved at the base, adaxial surface glabrous, abaxial surface densely stellate pubescent; petiole 4 – 8 mm long, densely stellate hairy. Racemes terminal and axillary, 5 cm long, sparsely stellate hairy; bracts lanceolate, 3 mm long, often caducous. Male flowers: pedicel 8 mm long, stellate-pubescent, bracteole inconspicuous. Sepals

5, equal, ovate, 3 × 2 - 2.5 mm, acute, abaxial surface stellate hairy, glabrous adaxially, faintly 1 - 2-veined. Petals 5, membranous, cream to yellowish, obovate, 4 × 3 mm, reflexed, glabrous outside, villous inside. Stamens inflexed in buds, 15 - 21 in two whorls, outer whorl antepetalous; filament glabrous, 2 mm long; anthers basifixed, oblong, 1 × 0.6 mm; glands 5, conspicuous. Female flowers: pedicel 2 - 5 mm long, stellate hairy; bracteole as in male flower. Sepals 5, equal, ovate-elliptic, 4 - 10 × 3 - 5 mm, apex acute, abaxial surface sparsely stellate pubescent, adaxial surface glabrous, margin ciliate, faintly reticulate. Petals rudimentary; gland oblate, 1 × 1.8 mm, otherwise similar to the male flowers. Ovary ovoid, 3 × 3 mm, sparsely stellate hairy; styles 3, linear, 5 mm long, bifid twice; stigma simple. Capsule greenish yellow, tricoccus, ovoid to subglobose, 14 - 18 × 15 -17 mm, glabrous to sparsely stellate-pubescent. Seeds 3, 1 per locule, ellipsoid, 8 - 13 × 5 - 7 mm, mature seed stramineous, mottled with vinaceous-purple. Fl. & Fr.: December - February

Western Ghats: Rare in moist deciduous forests amd outskirts of forest undergrowth at 600 m in Kerala, Karnataka and Maharasthra (Endemic).

IUCN: NE (IUCN, 2024).

Croton malabaricus Bedd, Fl. Sylv. S. India: cciv. 1873. Synonyms: *Oxydectes malabarica* (Bedd.) Kuntze Vern.: *Anakkuru, Kolavacchi, Koluvasi, Pimbaram, Thenadal* (Mal.); *Anakkutu, Malamgouri, Thavattu-polavu* (Tam.).

Trees to 20 m tall; bark greyish-white, smooth; branchlets stellate-hairy. Leaves simple, alternate, rhombic-ovate, broadly ovate or elliptic, 7-24 x 3-12 cm, apex acuminate, base cuneate, obtuse or round, margin entire, glabrous or with silvery stellate hairs and reddish glands beneath, prominently, 3-4-ribbed from base; lateral nerves 4-6 pairs, intercostae scalariform, obscure; stipules 12-18 mm long, lateral, filiform, scarious; petiole 10-125 mm long, slender, stellate scales present. Flowers unisexual, 5-10 mm across, white, in erect terminal racemes of 5-12.7 cm long; bracts small. Perianth 2-seriate; 5-partite, glandular at base. Sepals ovate, obtuse, silky tomentose. Petals 5, smaller than sepals, obovate. Ovary stellate hairy, 3-celled, ovules one in each cell; styles long, slender, pistillode absent in males. Stamens 10-15, filaments bearded. Capsule obovoid, 2.5 x 2 cm, depressed above, brown tomentose. Seeds 13 x 8 mm, oblong, mottled with brown. Fl. & Fr.: March – December.

Western Ghats: Evergreen forests at 1000 m in Karnataka, Kerala and Tamil Nadu (Endemic).

IUCN: NE (IUCN, 2024).

Croton persimilis Muell.-Arg, Linnaea 34: 116. 1865. Synonyms: *Croton persimilis* var. *genuinus* Müll.Arg., *Croton roxburghii* N.P. Balak., *Croton oblongifolius* Roxb., *Croton oblongifolius* Roxb., *Croton roxburghii* N.P.Balakr., *Croton virbalae* M.R.Almeida. Vern.: *Kote, Putol, Pongalam* (Mal.); *Ganasura, Gunsur, Mikir, Marthu-arong* (Mar.); *Millakumari, Milagu-naari* (Tam.).

Medium trees to 20 m tall; branchlets and leaves lepidote. Leaves oblong, elliptic to ovate-oblong or obovate, 10-45 x 5-15 cm, rounded or shortly cordate at base, coarsely dentate-serrate or crenate along margins, acute to short-acuminate at apex, glabrous; lateral nerves 8 - 22 pairs; basal glands sessile; petioles 1- 7 cm long, scattered lepidote, glabrescent at maturity; stipules linear, 4-7 mm long. Inflorescences 12-55 cm long, sometimes unisexual. Male flowers: pedicels 2-5 mm long. Sepals ovate, elliptic to triangular, 2-3.5 x 1 - 2.5 mm. Petals spathulate or narrowly oblong, 2-3.5 x 0.5-1.5 mm. Stamens 10-15,3.5-4.5 mm long. Female flowers: pedicels 2-5 mm long. Sepals ovate to elliptic. 2-3.5 x 1-2.5 mm. Petals spathulate to filiform, 0.3-1.5 mm long. Ovary subglobose, 2-4 mm in diam.; styles 4-6 mm long. Fl. & Fr.: November – April.

Western Ghats: Evergreen forests of Maharashtra, Goa, Karnataka, Kerala and Tamil Nadu. India: Assam, Andhra Pradesh, Madhya Pradesh, Meghalaya, Nagaland, Odisha, Sillim, Uttar Pradesh and West Bengal. World: Sri Lanka, Nepal, Bhutan, Bangladesh, Myanmar, China, Thailand and Vietnam.

IUCN: NE (IUCN, 2024).

Uses: The stem and root bark is locally used for medicinal purposes to treate hepatitis, liver complaints, bruises and rheumatic swellings. Oil is extracted from the seeds which is used to substitute the croton oil, often used for insecticide.

Croton tiglium L, Sp. Pl.: 1004. 1753. Synonyms: *Croton officinalis* (Klotzsch) Alston, *Kurkas tiglium* (L.) Raf., *Croton birmanicus* Müll.Arg., *Croton tiglium* var. *globosus* J.J.Sm. Vern.: *Danti, Japalabeeja, Neepala, Nervalam* (Kan.); *Dantibijam, Neervalam, Valam* (Mal.); *Jamalgota, Japal* (Mar.); *Kaattukattai, Naganam, Naeravalam, Nigumbam, Sambari, Sevalangottai, Siduram, Tendi* (Tam.).

Small trees to 7 m tall; branches sparsely stellate-hairy, glabrous at maturity. Leaves ovate, ovate-elliptic or ovate-lanceolate, 5-15 x 2-7 cm, glabrous or glabrescent, yellowish to brownish when dry, base cuneate or broadly rounded, rarely slightly cordate, with discoid glands, margins serrulate or subentire, apex acute or acuminate; basal veins 3-5, lateral veins 3 or 4; petiole 2.5-6 cm, subglabrous. Racemes terminal, 8-20 cm long; bracts subulate, 2 mm.

Male flowers: bud subglobose, sparsely stellate-hairy or glabrescent. Female flowers: sepals oblong-lanceolate, 2.5 mm, glabrescent. Ovary densely stellate-hairy; styles bipartite. Capsules ellipsoidal, oblong-ovoid, or subglobose, 1-2 x 1-2 cm, 3-4-lobed, sparsely stellate-hairy or glabrescent. Seeds elliptic or oblong-ovate, 8-12 x 6-7 mm, gray-brown. Fl. & Fr.: Throughout the year.

Western Ghats: Cultivated in deciduous to evergreen forests areas of Maharashtra, Kerala and Tamil Nadu. India: Andhra Pradesh, Assam, Madhya Pradesh, Meghalaya, Nagaland, Odisha, Sikkim, Uttar Pradesh and West Bengal. World: Sri Lanka, Nepal, Bhutan, Myanmar, Bnagladesh, China, Japan, Cambodia, Indonesia, Malaysia, Philippines, Thailand, Vietnam.

IUCN: LC (IUCN, 2024).

Uses: The popular Croton oil extracted from the seeds, which is very drastic purgative and useful in rheumatism and termatitis.

Croton zeylanicus Muell.-Arg, Linnaea 34: 107. 1865. Synonyms: *Oxydectes zeylanica* (Müll.Arg.) Kuntze, *Croton reticulates* Heyne ex Muell.-Arg., *Croton hypoleucs* Dalz. Vern.: *Panduray, Pondhari, Pandharisala* (Mar.).

Small tree to 6 m tall: branchlets densely covered with brown, fimbriate scales. Leaves ovate-lanceolate, 20 x 6 cm, acuminate, rounded at base, white-scaly below; lateral nerves 6-10 pairs, glands 1 pairs, sessile; petiole 1-3 cm long. Racemes to 6 cm long. Female flowers pedicelled, 4 mm across. Sepals ovate, obtuse. Petals obovate, smaller, tomentose. Ovary brown stellate hairy; styles 3, each 3 fid and then 2 fid, glabrous. Male flowers smaller. Stamens many, filaments glabrous. Capsule ovoid, 1.3 x 1 cm. Seeds 8 x7 mm, oblong, mottled with brown. Fl. & Fr.: Throughout the year.

Western Ghats: Evergreen forests of Maharashtra, Karnataka, Kerala and Tamil Nadu. World: Sri Lanka.

IUCN: NE (IUCN, 2024).

Epiprinus Griff.

Epiprinus mallotiformis (Muell.- Arg.) Croizat, J. Arnold Arbor. 23: 53. 1942. Synonyms: *Symphyllia mallotiformis* Muell.-Arg., *Adenochlaena indica* (Bedd.) Bedd. ex Hook.f., *Cephalocroton indicum* Bedd., *Adenochlaena mallotiformis* (Müll.Arg.) M.R. Almeida Vern.: *Karinjikkada* (Mal. & Tam.).

Trees to 12 m tall; branchlets stellate-tomentose. Leaves simple, alternate, elliptic, elliptic-ovate or elliptic-obovate, 7-20 x 1-8 cm, base round, obtuse or acute, apex acute or acuminate, margin entire, glabrous, coriaceous; lateral nerves 6-10 pairs, intercostae reticulate, prominent; petiole 4-20 mm long, stout, stellate-tomentose when young. Flowers unisexual, pale green, in long axillary spikes. Male flowers densely clustered along the upper part of the raceme. Tepals 4-6, ovate, acute, thinly tomentose. Stamens 4-6, free, inflexed in bud; pistillode columnar. Female flowers tepals 5-8, unequal, 2 mm, shortly connate, lanceolate, puberulous without. Ovary superior, 2 mm long, globose, pubescent, 3-locular; ovules one in each cell; styles 3, stout, 3 mm, twice forked, papillose. Capsule 3-lobed, 2.5-3 cm across, lobes globular, densely stellate hairy. Seeds 3, globose, slightly scabrous rugose. Fl. & Fr.: December – March.

Western Ghats: Evergreen forests of Maharashtra, Karnataka, Kerala and Tamil Nadu (Endemic). India: Andhra Pradesh.

IUCN: NE (IUCN, 2024).

Euphorbia L.

1a. Plants non-spiny .... **E. tirucalli**

1b. Plants spiny 2

2a. Branches terete; spine-shields rounded .... **E. nivulia**

2b. Branches winged; spine-shields conical 3

3a. Branches crowded at apex, space in between the wings flat .... **E. susan-holmsiae**

3b. Branches not crowded at apex; space in between the wings not flat 4

4a. Branches 6-10-winged; cyathia always solitary; bracts pale pink .... **E. santapaui**

4b. Branches 2-5-winged; cyathia usually in diads or triads; bracts pale yellow 5

5a. Branches twisted; capsules acutely keeled 6

5b. Branches not twisted; capsules obtusely keeled 7

6a. Branches 3-winged; leaves less than 1 cm long; primary peduncles in fasicles .... **E. tortilis**

6b. Branches 5-winged; leaves more than 5 cm long; primary peduncles 1-2 .... **E. neriifolia**

7a. Branches 2-winged in juvenile stage, later 4-5-winged; wings less than 1 cm thick; peduncles up to 1 cm long; involucre ca 5 by 5 mm .... **E. vajravelui**

7b. Branches constantly 3-winged; wings more than 1 cm thick; peduncles more than 1 cm long; involucre 8 mm diam 8

8a. Leaves well developed, large obovate to 7 cm long; cyathia usually tetrad .... **E. lacei**

8b. Leaves absent; cyathia usually triad 9

9a. Wings of stem 3-5- ribbed; cyanthia pink or redish; capsule tri-lobed, depressed above .... **E. antiquorum**

9b. Wings of stem 2-3-ribbed; cyathis greenish; capsule globose, conical above .... **E. trigona**

Euphorbia antiquorum L., Sp. Pl. 450. 1753. Synonyms: *Euphorbia mayuranathanii* Croizar, *Tithymalus antiquorus* (L.) Moench, *Euphorbia antiquorum* var. *polygona* Wight, *Euphorbia arborescens* Roxb. Vern.: *Mulaje-mudu* (Kan.); *Chathura-kalli* (Mal.); *Narsej* (Mar.); *Mukkona Paal Kalli, Paal Kalli, Sadrua-kalli* (Tam.).

Succulent trees to 8 m tall; bark whitish, peeling off in mature stem. Stems 5-7 cm thick, green, 3-4-ribbed, branching from upper parts; ribs prominent, winglike, up to 1-2 cm wide, 3-5 mm thick, prominently triangular toothed. Leaves caducous, spine shields small; stipules prickle-like; spines paired, 2-5 mm; petiole very short. Cyathia subterminal, single, peduncle 2-3 mm; cyathophylls as long as involucre, membranous; involucre broadly campanulate, 3 × 5 mm, lobes 5, rounded, denticulate; glands 5, entire. Male flowers many; bracts linear. Female flower: ovary pedicellate, exserted from involucre, smooth and glabrous; styles free; stigma slightly 2-lobed. Capsule 3-lobed, compressed, 3.4-4 × 4-5 mm, separating into 3-bivalved 1-seeded cocci. Seeds subglobose, 2 × 2 mm, brown-yellow, smooth. Fl. & Fr.: November – March.

Western Gahts: Dry scrub forests of Maharashtra, Goa, Karnataka, Kerala and Tamil Nadu. India: Andaman & Nicobar Islands, Andhra Pradesh, Arunachal Pradesh, Assam, Bihar, Gujarat, Madhya Pradesh, Meghalaya and Odisha. World: Sri Lanka, Myanmar, Bangladesh, Pakistan, Mayasia and Indonesia.

IUCN: LC (IUCN, 2024).

Uses: The latex is used for bone-setting and for preparing plaster to joint the broken bone. The stem and root bark used to cure external rheumatism and as purgative.

Euphorbia lacei Craib, Bull. Misc. Infrom. Kew 1911: 456. 1911. Synonyms: *Euphorbia barnhartii* Croizat, *E. trigona* Roxb. *nom.illeg*. Vern.: *Thirugu Kalli* (Tam.).

Evergreen succulent trees to 5 m tall; branches 3-winged, long, flexible, looping downwards. Leaves obovate to oblong-obovate, 3-3 x 2-3 cm, fleshy, narrowed at base, entire at margin, blunt to somewhat pointed tip; leaf bases projected, 1.5-2 cm long; spines 2, spreading, 2-3 mm long, straight, black; petiole minute or absent. Cyathia solitary or in traids; middle one stalked, lateral sessile, one gland stalked. Involucre 5, top-shaped, transversely oblong, 4-5 mm long. Capsule ovoid, 4-5 mm diam. Fl. & Fr.: December – February.

Wesern Ghats: Wet tropical forests of Tamil Nadu and Karnataka. India: Andaman & Nicobar Islands, Andhra Pradesh, Bihar, Madhya Pradesh, Meghalaya, Nagaland, Uttar Pradesh and West Bengal. World: Borneo, Jawa, Laos, Lesser Sunda Is., Myanmar, Philippines and Thailand.

IUCN: NE (IUCN, 2024).

Euphorbia neriifolia L., Sp. Pl.: 451. 1753. Synonym: *Elaeophorbia neriifolia* (L.) A. Chev., *Euphorbia ligularia* Roxb. *nom.illeg*. Vern.; *Ilaikalli* (Tam.).

Small flessy trees to 5 m tall; branches crowded, jointed, cylindrical or obscurely 5-angled, twisted. Stem slightly confluent in 5 irregular rows, projected up to 8 mm length, 2-3 cm apart; spines 1 or 2, divaricate, 4-5 mm long, sharp, black, persistent. Leaves alternate, sessile or subsessile, crowded at ends of branches, oblong-obovate, oblanceolate or subspathulate, 6-25 x 2-4 cm, cuneate at base, entire, rounded, obtuse or acute at apex, fleshy, shiny dark green above, pale beneath, deciduous; petioles 1 - 5 mm long. Cyathia solitary, in diads or triads, the central one male, sessile, lateral ones bisexual, pedunculate; peduncles to 5 mm long. Involucral cup subtended by 2 ovate 2.5 x 2 mm bracts; lobes 5, suborbicular; glands 5, transversely oblong, 2 x 4 mm. Male florets in 4 or 5 fascicles of 3 or 4 flowers each; pedicels to1 mm long; anthers oblong. Bracteoles wedge-shaped, fimbriate at apex, 1 mm long, covering the male florets when young. Female floret gynophore 3 mm long. Ovary obscurely 4 or 5 angled; styles connate up to middle; stigma papillose. Capsule deeply trilobed, 5-8 mm in diam, smooth, glabrous. Seeds globose, 3 mm in diam, smooth; caruncle small, white. Fl. & Fr.: December - June

Western Gahts: Scrub jungles and dry deciduous forests of Maharashtra, Goa, Karnataka, Kerala and Tamil Nadu. India: Gujarat, Madhya Pradesh, Punjab, Rajasthan, Tripura, Uttar Pradesh and West Bengal. World: West Tropical Asia, South & South East Asia, Myanmar to W. Indonesia.

IUCN: LC (IUCN, 2024).

Euphorbia nivulia Buch.-Ham., Trans. Linn. Soc. London 14: 286. 1824. Synonyms: *Euphorbia nivulia* var. *helicothele* (Lem.) Boiss. Vern.: *Yelli-kalli* (Kan.); *Ela-kalli, Kalli-pala* (Mal.); *Kaudenivdung, Newrang, Nivudung* (Mar.); *Ila-kalli, Manchenvi, Naga-kalli, Nananglli* (Tam.).

Fleshy deciduous trees to 8 m tall; branches terete; spines straight, 3-6 mm long, paired on flat corky base. Leaves alternate, sessile, oblanceolate-oblong or obovate, 7-18 x 3-6.5 cm, base cuneate, apex obtuse, fleshy. Cyathia usually paired, in subterminal cymes. Involucre broadly cupular, yellow; lobes fimbriate; glands 5, transversely oblong, fleshy, appendage absent. Male flowers with sterile florets, bracteolate. Female florets laterally pendulous. Style stout, branched above; stigmatic lobes flattened and slightly expanded. Capsule 5 mm across, 3-lobed. Seeds 4-angular, smooth. Fl. & Fr.: December - May

Western Gahts: Dry deciduous to semi-evergreen forests of Maharashtra, Karnataka, Kerala and Tamil Nadu. India: Andhra Pradesh, Bihar, Delhi, Gujarat, Haryana, Madhya Pradesh, Odisha, Punjab, Rajasthan and Uttar Pradesh. World: Myanmar, Java and Madagascar

IUCN: NE (IUCN, 2024).

Uses: The juice of leaves is useful in treating purgative and diuretic. The latex is mixed with jaggery and applied the plaster for bone setting.

Euphorbia santapaui A.N. Henry., Bull. Bot. Surv. India 6: 329. 1965. Vern.: *Kalli* (Tam. & Mal.); *Paal Kalli* (Tam.).

Succulent trees to 10 m high; branchelets condensed, glabrous. Leaves minute, 0.1-0.3 cm diam, ovate-orbicular, convex on the upper side, early deciduous. Cyathia single, near the apex of the branches, 0.5-1 x 0.5-0.6 cm. Male flowers 5-8; pedicel to 2 mm long; stamens single, filament 0.1-0.15 cm long, attached and joined with pedicel; anthers subglobose. Ovary 0.2-0.3 cm across; style 3, connate, 0.08-0.12 cm long; stigma 2-fid. Capsule 0.6-0.8 cm across, glabrous or slightly pubescent without, separating into 3-bivalved cocci. Seeds ovoid, 3 mm diam, smooth, brown, caruncle yellowish brown. Fl. & Fr.; May – September.

Western Ghats: Evergreen forest above 1200 m in Kerala and Tamil Nadu (Endemic).

IUCN: EN- B1+2cd. (IUCN, 2024).

Uses: Kani tribes of southern Western Ghats apply the milky latex on skin diseases and scabies.

Euphorbia susan-holmsiae Binojk. & R. Gopalan., Cact. Succ. J. (Los Angeles) 65: 277, f. l. 1993. Vern.: *Kalli* (Tam.).

Small trees to 3 m high; stem terete, branches crowded at apex, slender, usually 3-4-winged, with flat intervening spaces between wings. Spines 2, divaricate, 2-4 mm long. Leaves present only on young branches, sessile, orbicular, entire, obtuse, 2-5 x 24 mm, fleshy, early caducous. Cyathia above the spines, either in triads or solitary, subsessile; primary peduncles 1 - 3, above the spines, 3-5 mm long; secondary peduncles 2-4 mm long; involucre cupular, 3-3.5 x 3-4 mm; bracts 2, triangular, fimbriate at apex, 2 x 1.2 mm; lobes 5, suborbicular, 1 x 1.2 mm; glands 5, yellow. Male florets in 5 fascicles of 3-6 florets each, covered by 1.5 mm long laciniate bracteoles; pedicels to 1 mm long; anthers subglobose. Female floret gynophore 2-4 mm long. Ovary globose, 2 mm in diam., glabrous; styles 1 mm long, connate at base, each bifid at apex; stigma plumose. Capsule subglobose, 3-winged, 3-4 mm across; cocci 3, acutely keeled. bivalved, glabrous, slightly pinkish. Seeds subglobose, 2.5 x 2 mm. Fl. & Fr.: July – April

Western Ghats: Deciduous forests of Tamil Nadu (only in Tirunelveli district). World: Sri Lanka.

IUCN: NE (IUCN, 2024).

Euphorbia tirucalli L., Sp. Pl.: 452. 1753. Synonyms: *Arthrothamnus tirucalli* (L.) Klotzsch & Garcke, *Tirucalia indica* Raf., *Euphorbia geayi* Costantin & Gallaud, *Euphorbia media* N.E.Br., *Euphorbia media* var. *bagshawei* N.E.Br. Vern.: *Bontakalli* (Kan.); *Kol-kalli, Tiri-calli, Tir-calli* (Mal.); *Nevil, Niwala, Niwarang* (Mar.); *Kombu-kalli, Pachan-kilai, Tirukalli* (Tam.)

Trees to 12 m tall; branchlets terete, succulent, articulated. Leaves deciduous, 5-10 mm long, linear-oblong, base cuneate, apex obtuse to subacute. Cyathia clustered in the forks of the branchlets, shortly pedicelled, mostly female. Involucre campanulate; glands 3-5, transversely oval, peltate, lobes short, hairy; appendage 0. Male florets bracteolate, bracteoles laciniate at tip. Styles short, recurved, 2-lobed. Capsule globose, 5 mm diam, cocci compressed, velvety. Seeds ovoid, smooth. Fl. & Fr.: June – Septemebr.

Western Ghats: Scrub forests to deciduous forests of Maharashtra, Karnataka, Kerala and Tamil Nadu. India: Andhra Pradesh, Bihar, Gujarat, Odisha and Uttar Pradesh. World: Native to Africa and Madagascar.

IUCN: LC (IUCN, 2024).

Uses: The milky latex of this palnt is purgative, carminative and useful in gonorrhoea, whooping cough, asthma, dropsy, leprosy, dyspepsia, jaundice, colic, tumours and baldder stones.

Euphorbia tortilis Rottler ex Ainslie., Materia indica 2: 424. 1826. Vern.: *Murukalli* (Mal.); *Tirukukalli, Valankalli* (Tam.).

Flessy trees to 8 m tall; branchlets spirally twisted, 3 angled, spines to 1 cm. Leaves small, deciduous. Cyathia axillary, in clusters, triads or solitary; peduncles 0.5 cm long; involucre cupular, 3 x 5 mm; glands 5, thick. Male flowers in groups of 5; stalk 1.5 mm long, glabrous; anthers subglobose; bracteoles spathaceous, deeply laciniate, fimbriate at apex. Female flowers-laterally pendulous, gynophote to 1 mm long, glabrous. Ovary globose, 3 x 7 mm, sharply 3-angled; style connate; stigma rapand. Capsule 8 mm across, sharply keeled. Seeds subglobose, tetragonous, 1.5 mm diam, pale black. Fl. & Fr.: May – August

Western Ghats: Foot hills and scrub forests of Kerala and Tamil Nadu. India: Andhra Pradesh. World: Sri Lanka.

IUCN: NE (IUCN, 2024).

Euphorbia trigona Mill., Gard. Dict., ed. 8.: n.° 3. 1768. Synonym: *Euphorbia hermentiana* Lem. Vern.: *Mandagalli* (Kan.); *Kallimullu* (Mal.); *Kattomantu, Palkalli* (Tam.).

Succulent trees to 8 m tall; branchlets 3-angled, spines 3 mm, divergent, cauducous. Leaves to 2.5 mm, ovate, cauducous. Cyathia solitary or in groups, bracts paired to 3.5 x 4 mm, ovate, obtuse, dentate, glands 5 to 2.2 mm; bracteoles 2.5 mm, obovate, apex fringed. Stamens numerous, filaments jointed, 2.6 mm. Female flowers erect, ovary 3.7 mm across, 3-celled, style 3, ovule 1. Capsule obovoid, 7 x 3.5 mm, 3-lobed. Fl. & Fr.: Throughout the year.

Western Gahts: Semi-evergreen foersts and deciduous forests of Kerala and Tamil Nadu. India: Andhra Pradesh. World: Sri Lanka and Malesia.

IUCN: DD (IUCN, 2024).

Euphorbia vajravelui Binojk. & N.P. Balak., Cact. Succ. J. (Los Angeles) 63: 229. 1991.

Stout succulent trees to 10 m tall; stems thick, flattened or 3-angled, 3-4 cm broad, glabrous; angles armed with paired black spines, 1-10 mm long. Leaves minute, 5 mm long, ovate. Cyathea solitary or 2-5 together, 5-8 mm across, sessile above the spines along the angles of the stem, involucral lobes obtuse, glabrous, 4-glanded along the angles. Male florets fascicled in 4 or 5 bundles with 4-6 in each bundle. Ovary glabrous; styles connate to the middle. Capsules globose, stalked to 1.5 cm long, glabrous. Seeds ovoid, black. Fl. & Fr.: January – April.

Western Ghats: Evergreen foersts of Kerala and Tamil Nadu (Endemic).

IUCN: EN- B1ab(iii)+2ab(iii); D. (IUCN, 2024).

Key to the varieties

1a. Branchlets erect, usually trigonous; capsule globose, not agular ....var. **vajravelui**

1b. Branchlets bendulous, usually flattened; capsule traiangular, depressed above ....var. **theniensis**

Euphorbia vajravelui var. **theniensis** B.De Jong & R.W.Stewart., Euphorbia World 15(2): 8. 2019.

Fl. & Fr.: January – April.

Western Ghats: Deciduous to semi-evergrren forests of Kerala and Tamil Nadu (Endemic).

IUCN: EN- B1ab(iii)+2ab(iii); D. (IUCN, 2024).

Excoecaria L.

1a. Leaves alternate, margin entire or rarely sinuate-crenate; filaments long .... **E. agallocha**

1b. Leaves opposite, margin crenate or rarely entire; filaments short .... **E. oppositifolia**

Excoecaria agallocha L., Syst. Nat., ed. 10. 2: 1288. 1759. Synonyms: *Excoecaria agallocha* var. *genuina* Mull.Arg., *Commia cochinchinensis* Lour., *Excoecaria agallocha* var. *camettia* (Willd.) Mull.Arg., *Excoecaria agallocha* var. *orthostichalis* Mull.Arg. Vern.: *Harogida, Thilla* (Kan.); *Komatti* (Mal.); *Gewa, Phungali, Surund* (Mar.); *Kampetti, Thillai* (Tam.).

Trees to 10 m tall; bark grey, smooth; branchlets 2-5 mm thick, brownish or greyish. Leaves simple, alternate, elliptic, oblong-lanceolate or ovate-lanceolate,

3-13 x 1.5-5 cm, base acute, obtuse or round, apex acuminate or acute, margin crenulate-serrulate or entire, glabrous, glossy above, coriaceous; lateral nerves 5-10 pairs, intercostae reticulate, prominent; petiole 10-30 mm long, slender, glabrous. Flowers unisexual, pale green. Mmale flowers in axillary spikes of 3.5-12.5 cm long, 2-3 together in an axil with a series of spirally arranged bracts; each bract subtending one male flower, bracteoles lacinate, large glands on either side, sessile. Tepals 3, 1.5 × 1 mm, lacinate. Stamens 3, yellow; filaments free, to 2 mm long; anthers basifixed. Female flowers in axillary racemes of 4-8 cm long, 1-2 in each axil; pedicel 3 mm long; bracteoles as in male flower. Tepals 3, 1.5 × 1 mm, lacinate, closely adpressed to the ovary. Ovary superior, 1-1.5 mm across, tricarpellary, ovules one in each locule, pendulous; styles 3, simple, stout, spreading, spreading, recurved; stigma 3, glandular. Capsule depressed globose, 4-5 x 8-10 mm, crustaceous, 3-celled, black, pedicels 3-5 mm long. Seeds 3, globose. Fl. & Fr.: February – September.

Western Ghats: Along the seashore and tidal forests of Maharashtra, Goa, Karnataka, Kerala and Tamil Nadu. India: Andaman & Nicobar Islands, Andhra Pradesh, Odisha and West Bengal. World: Sri Lanka, Myanmar, Bangladesh, Japan, Taiwan, Malesia, Indonesia, Philippines, Pacific Islands and Australia.

IUCN: LC (IUCN, 2024).

Uses: Wood is useful in carpentry, packaging cases, tous, legs of bedsteads and floats for fishing nets. Wood is also useful in making charcoal.

Excoecaria oppositifolia var. **crenulata** (Wight) Chakrab. & Gangop., J. Econ. Taxon. Bot. 18: 208. 1994. Synonyms: *Excoecaria crenulata* Wight, *Excoecaria robusta* Hook.f. Vern.: *Paalamaram, Malai Thillai* (Tam.).

Trees upto 8 m tall; branchlets 3-7 mm thick, flattened or angled when young, terete in age, brown. Leaves simple, opposite, oblong-lanceolate, oblanceolate, elliptic or obovate, 8-20 x 2-4.5 cm, base acute or obtuse, apex acuminate, apiculate or subacute, margins shallowly crenate-serrate, dentate or entire, coriaceous, glabrous; lateral nerves, 10-22 pairs, slender, intercostae reticulate, prominent; petiole 5-30 mm long, stout, grooved above, glabrous. Flowers unisexual, yellowish-green; bracteoles with large glands on either side. Male flowers in terminal and axillary 3.5-18 cm long spikes; bracts 1.5-2 x 0.8-2 mm, broadly ovate, deltoid, triangular, denticulate to entire. Tepals 3, broadly ovate, ovate-oblong or suborbicular, 1-1.5 x 0.8-1.3 mm, dentate or erose or occasionally subentire. Stamens 2-3, 1.5-2 mm long; anthers 0.5-1 mm across. Female flowers: in axillary or terminal 1-3 flowered clusters; bracts

1-2 x 1-2.5 mm, broadly triangular to deltoid or suborbicular, entire to erose; pedicel 1-5 mm long. Tepals 3, 2-3.5 x 1.5-3 mm, broadly ovate to triangular, lanceolate to entire. Ovary superior, 1.5-2 mm across, subglobose, 3-celled; ovules 1 in each cell; styles 3, 3-7 mm long, simple or occasionally shortly bifid above. Capsule somewhat depressed, 1-3 cm across, shallowly or sometimes deeply 3 lobed. Seeds globose. Fl. & Fr.: January – May.

Western Ghats: Evergreen forests of Karnataka, Kerala and Tamil Nadu. World: Sri Lanka

IUCN: NE (IUCN, 2024).

Cultivated:

Excoecaria cochinchinensis Lour.

Falconeria Royle

Falconeria insignis Royle., Ill. Bot. Himal. Mts.: 354. 1839. Synonyms: *Sapium insigne* (Royle) Kurz., *Excoecaria insignis* (Royle) Muell.-Arg., *Falconeria malabarica* Wight, *Sapium insigne* var. *malabaricum* (Wight) Hook.f. Vern.: *Kurda* (Kan.); *Dudla* (Mar.); *Sakkarakalli* (Tam.).

Deciduous trees to 15 m tall; branchlets stout, terete; exudation milky. Leaves simple, alternate, elliptic-oblong, elliptic-ovate or obovate, 7-20 x 2.5-7 cm, base acute, apex acuminate, margin crenate, membranous, glabrous; lateral nerves 10-18 pairs, intercostae reticulate, prominent; petiole 10-50 mm, slender glabrous, with 2 conspicuous glands at the apex. Flowers unisexual, reddish, in terminal spikes. Male flowers above, female ones below or in separate spikes; bracts 2, orbicular. Male flowers many in each bract; bracteoles linear. Tepals 2, orbicular, concave, stamens 2, free, exserted. Female flowers solitary in each bract; bracteoles concave imbricate, basally connate, mucronate. Ttepals 2, ovate, acuminate. Ovary superior, ovoid, glabrous, 3-locular; ovules 1 in each cell, pendulous; styles 2-3; stigmas 3, subsessile, recurved. Capsule, loculicidal, fleshy, subglobose. Seeds 3, globose, smooth. Fl. & Fr.: December – May.

Western Ghats: Semi-evergreen to evergreen forests of Gujarat, Maharashtra, Goa, Karnataka, Kerala and Tamil Nadu. India: Andhra Pradesh, Assam, Bihar, Himachal Pradesh, Meghalaya, Odisha, Sikkim and Uttar Pradesh. World: Sri Lanka, Nepal, Bhutan, Bangladesh, Myanmar, South West China, Laos, Cambodia, Vietnam, Thailand and Malaysia.

IUCN: NE (IUCN, 2024).

Givotia Griff.

Givotia moluccana (L.) Sreem., Taxon 24: 696. 1975. Synonyms: *Croton moluccanum* L., *Givotia rottleriformis* Griff. Vern.: *Pubeer* (Kan.); *Polki* (Mar.); *Kottaithanakku, Thalamaram, Vendalai, Vellai-poothalai* (Tam.).

Dioecious trees to 20 m tall; bark brown, smooth, peeling off in thick scales; branchlets densely stellate-tomentose. Leaves simple, alternate, 6-18 x 5-20 cm, broadly ovate or orbicular, base cordate, apex acute or shortly acuminate, margin coarsely repand-dentate, glabrous above except nerves and white stellate tomentose beneath, 5-ribbed from base, palmate, intercostae scalariform; petiole 8-12 cm long, swollen at base, stellate-tomentose, 2 prominent glands near the tip. Flowers unisexual, in axillary or subterminal racemose or cymose panicles; pedicels jointed; bracts linear, conspicuous. Tepals 10, in 2 series; outer 5 5 x 3 mm, incurved, subacute, outside, stellate tomentose; inner petalloid 7 x 4 mm, shortly connate at base, obtuse. Stamens 8-20, exserted; filaments free, unequal to 6 mm, villous at base; anthers oblong. Disc 4, free. Ovary superior, globose, stellate tomentose, 2-3-celled; ovules in each cell; styles short, 2-fid, spreading. Drupe subglobose or ellipsoid, 2.5 cm across, densely stellate tomentose. Seeds 3, oblong; testa bony. Fl. & Fr.: May – October.

Western Ghats: Dry deciduous forest to evergreen forests of Mahrashtra, Karnataka, Keala and Tamil Nadu. India: Andhra Pradesh and Telangana. World: Sri Lanka.

IUCN: NE (IUCN, 2024).

Uses: Wood is useful in making toys, decorative articles. Lubricant oil is extracted from seeds.

Koilodepas Hassk.

Koilodepas calycinum Bedd., Fl. Sylv. S. India: ccvii. 1873. Synonyms: *Adenochlaena calycina* Bedd. Vern.: *Kattupira* (Tam.).

Small trees to 8 m tall; young parts furfuraceous. Leaves lanceolate-oblong, rounded at base, dentate- serrate along margins, acuminate at apex, 4 - 18 x 1.5-6 cm, glabrous or sparsely stellate furfuraceous on major nerves beneath; lateral nerves 8 - 12 pairs; petioles 5-10 mm long. Inflorescences axillary or terminal, 7.5 - 14 cm long, usually branched. Male flowers: bracts subulate, 1 mm long; sepals 4 -5, deltoid-ovate, 2 x 3 mm. Stamens 4 -5; filaments to 2 mm long; anther-thecae oblong. Female flowers; bracts subulate, 2 mm long;

pedicels 5-8 m long. Sepals 5-8, orbicular-ovate, 4 x 3 mm. Ovary subglobose, 2 mm across, hairy; styles 3, 5 mm long. Capsule subglobose, slightly depressed, 15 mm in diam, surrounded by membranous, broadly and irregularly lobed greatly enlarged calyx. Fl. & Fr.: October – December

Western Ghats: Rare in evergreen forests of Kerala and Tamil Nadu (Endemic).

IUCN: EN- B1+2c. (IUCN, 2024).

Macaranga Thours

1a. Bracts sparsely dentate, teeth often glandular inside, rarely eglandular; stamens 2-5 .... **M. peltata**

1b. Bracts entire, thick, glandular outside, eglandular inside; stamens 6-8 ... . **M. indica**

Macaranga indica Wight., Icon. Pl. Ind. Orient. 5: t. 1883. 1852. Vern.: *Bhoota-thamara* (Mal.); *Vatta thamarai* (Tam.).

Trees to 25 m tall; branchlets yellowish brown pubescent when young. Leaves ovate-orbicular, 14-25 × 13-23 cm, thinly leathery, abaxially pubescent and glandular-scaly, adaxially glabrous or pilose along veins, base rounded and broadly peltate, with glands, margin serrulate, apex acuminate; veins 9, palmate; stipules broadly ovate, acuminate, 1.5-2 × 1 cm, pubescent, deciduous; petiole 11-14 cm long. Male inflorescences branched, 10-15 cm, branchlets zigzag; bracts linear-spatulate, 4-7 mm, with 1-3 elliptic glands. Male flowers many per bract; bracts linear spathulate, 5-7 mm long; pedicel 1 mm, pilose. Sepals 3, lobes ovate, 1 mm long, glabrous. Stamens 5-7. Female inflorescences branched, 5-6 cm long panicles; bracts triangular, 1 mm long, pubescent. Female flower solitary. Sepals 4, 1.5 mm, pilose, persistent. Ovary 1-locular; style 1, subulate, 1.5 mm long. Capsule globose, 4-8 mm in diam, sparsely glandular-scaly. Seeds smooth, black. Fl. & Fr.: October – December.

Western Ghats: Semi-evergreen forests above 700 m in Karnataka, Kerala and Tamil Nadu. India: Andaman & Nicobar Islands, Andhra Pradesh, Arunachal Pradesh, Assam, Bihar, Meghalaya, Odisha, Sikkim, Uttar Pradesh and West Bengal. World: Sri Lanka, Bangladesh, Bhutan, South West China, Thailand, Laos and Malaysia.

IUCN: LC (IUCN, 2024).

Macaranga peltata (Roxb.) Muell.-Arg., A.P.de Candolle, Prodr. 15(2): 1010. 1866. Synonyms: *Osyris peltata* Roxb., *Macaranga roxburghii* wight, *Macaranga tomentosa* Wight, *Macaranga flexuosa* Wight Vern.: *Chandakanni, Chandkal,*

Chenthakani, Chinthakanni (Kan.); *Vatta* (Mal.); *Chanda, Chandora, Chandwar* (Mar.); *Vatta, Vatakanni* (Tam.).

Dioecious trees to 15 m tall; bark greyish-brown mottled with white, smooth, lenticellate; exudation red, gummy; branchlets terete, glaucous. Leaves simple, alternate, ovate-orbicular, deltoid-ovate, ovate or orbicular, 11-25 x 9-20 cm, base peltate, apex acute or acuminate, margin entire, glabrous or glabrescent, pubescent with reddish glands beneath, coriaceous; 8-10 ribs from base, prominent beneath; lateral nerves 7-8 pairs, parallel, intercostae scalariform, much prominent; petiole 12-35 mm long, stout, glabrous. Flowers unisexual, greenish-yellow. Male flowers in axillary, much branched, dense, tomentose, panicles, concealed in large bracts; bracteoles concave. Tepals 3, minute, obovate, cuneate. Stamens 2-8, free, shortly connate below, exserted. Female flowers in panicles simpler than in males, branches racemes with larger bracts. Tepals 4, basally connate at base. Ovary superior, 2-6-celled, densely glandular; ovule one in each cell; style lateral; stigma sessile, often embracing one side of the ovary. Capsule globose, depressed abive, 5-6 mm across, hairy, yellow resinous glandular, black. Seed one, black. Fl. & Fr.: December – July.

Western Ghats: Evergreen forests of Maharashtra, Goa, Karnataka, Kerala and Tamil Nadu. India: Andhra Pradesh, Bihar, Madhya Pradesh, Odisha, Sikkim and West Bengal. World: Sri Lanka, Myanmar and Thailand.

IUCN: NE (IUCN, 2024).

Uses: The wood is suitable for making matches and paper pulp. A reddish resinous gum obtained from the bark is called kino which is used for adhesive and as medicine for veneral sore.

Mallotus Lour.

1a. Leaves penninerved, not palmately nerved at base 2

1b. Leaves palmately 3 - 9-nerved at base, penninerved above 3

2a. Leaves strictly opposite.. **M. resinosus**

2b. Leaves alternate or opposite in young shoots .... **M. atrovirens**

3a. Leaves opposite to subopposite, epeltate.. 4

3b. Leaves alternate, usually peltate, rarely epeltate 6

4a. Fruits 1 cm across, densely fulvous stellate tomentose, glandular ....**M.distans**

4b. Fruits more than 1 cm across, glabrous, sub glabrous or rugose..............5

5a. Fruits 2 - 3.5 cm across, with very thick pericarp................. **M. nudiflora**

5b. Fruits 1 - 1.5 cm across, with very thin pericarp .... **M. polycarpus**

6a. Fruits smooth, not echinate or tubercled..7

6b. Fruits echinate or tubercled 8

7a. Leaves opposite and alternate, subglabrous beneath; male receptacle central; fruits minutely pale brown stellate tomentose; seeds brown .... **M. rhamnifolius**

7b. Leaves alternate, tomentose beneath; male receptacle not central; fruits densely fulvous tomentose; seeds black...9

8a. Leaves narrowly oblong-ovate, mostly twice as long as broad, with red granules beneath; stamens 15 - 30; styles 2; fruits 3 or 4-loculed, densely clothed with crimson-red resinous powder......................**M. philippensis**

8b. Leaves broadly ovate-deltoid, mostly as long as broad, yellow tomentose and glandular beneath; stamens 40 - 80; styles 2; fruits 2-loculed, densely fulvous tomentose, not with red resinous powdery glands .... **M. repandus**

9a. Leaves shortly 3-lobed; basal nerves 9 or 11; spine-layer on fruits continuous ....**M. barbatus**

9b. Leaves not 3-lobed; basal nerves 3 or 5 (rarely 9); spine-layer on fruits not continuous ....**M. tetracoccus**

Mallotus atrovirens Wall. ex Muell.-Arg., Linnaea 34: 195. 1865.

Small trees to 7 m tall; bark ash-grey; branchlets puberulous. Leaves simple, alternate or opposite, elliptic-obovate or obovate, 5.5-17 x 2-6 cm, base acute or cuneate, apex acute or shortly acuminate, margin entire, glabrous above, yellow glandular beneath, chartaceous; lateral nerves 5-7 pairs, intercostae scalariform, prominent; petiole 5-20 mm long, glabrous, grooved above. Flowers unisexual, yellow, in axillary and terminal racemes; pedicels yellow glandular. Perianth globose, glabrescent, densely glandular, separating into 3-5 lobes. Ovary 2-4-celled; ovules one in each cell; Styles 2, to 2.5 mm long; stigma sessile, 2, plumose. Stamens many, exserted, free. Capsule smooth, 5 x 7 mm, 2-lobed, yellow glandular. Seeds ovoid-globse, 2 per capsule, plano-convex or hemispherical, glandular on outer surface. Fl. & Fr.: December – July.

Western Gahts: Evergreen foresrts and dense hill slopes of Maharashtra, Karnataka, Kerala and Tamil Nadu (Endemic).

IUCN: VU- B1+2c. (IUCN, 2024).

Mallotus barbatus Muell.-Arg., Linnaea 34: 184. 1865. Synonym: *Rottlera barbata* Wall.

Small trees to 6 m tall; branchlets, petioles, and inflorescences densely brownish, floccose-tomentose. Leaves suborbicular or broadly ovate, often tricuspidate or shortly and acutely 1- or 2-lobed, 8-35 × 6-28 cm, adaxially stellate-villous along nerves, abaxially stellate-tomentose, base rounded, broadly peltate, with up to 4 basal glands, margin repand-denticulate, apex acuminate; basal veins 5-7 pairs; petiole to 20 cm long. Male inflorescences terminal, branched or unbranched, 15-34 cm; bracts subulate, 5-7 mm. Male flowers 3-5-fascicled; pedicel 3-4 mm. Sepal 4-5, lobes 4 or 5, oblong, 3-3.5 mm, stellate-pubescent. Stamens 40-60. Female inflorescences branched or unbranched; peduncle stout, erect, 10-20 cm long; bracts subulate, 4-5 mm. Female flowers pedicel 2-3 mm. Sepals 4-5, 5-lobed, 4-5 mm, tomentose. Ovary 3-5-locular, densely setose, tomentose; style 3-4 mm long, plumose. Fruiting pedicel to 30 mm lomg. Capsule subglobose, 3-lobed, 12-20 mm in diam, densely stellate-tomentose and softly spiny. Seeds ovoid, 5 mm, black, verruculose. Fl. & Fr.: February – November.

Western Ghats: Dense evergreen forests of Maharashtra. World: Myanmar, China, Thailand and Malesia.

IUCN: LC (IUCN, 2024).

Uses: Oil extracted from the seeds is often used for lighting the lamps and for making candles.

Mallotus distans Muell.-Arg., Linnaea 34: 194. 1865. Synonym: *Croton distance* B. Heyne ex Wall.

Small trees to 9 m tall; branchlets stellate pubescent. Leaves opposite and alternate, 5-20 x 3.5-15 cm, ovate-lanceolate or oblong, acute or acuminate, base rounded, sessile glands at base prominent, pubescent or glabrous above, fulvous tomentose below; lateral nerves 3-5 pairs; petiole to 9 cm long. Male flowers on axillary reacemes, to 15 cm long. Sepals 3-5, ovate to orbicular. Stamens many; filaments free, to 2.5 mm long; anther oblong-linear, 0.5 mm long. Female flowers on axillary racemes, to 14 cm long; pedicles to 6 mm long. Sepals spathaceous, 4-5-lobed, lobes lanceolate. Ovary 3-loculed, 3-4 mm across, densely yellowish tomentose, glandular; style 3, to 5 mm long. Capsule 3-lobed, 1 cm across, brown stellate-tomentose. Fl. & Fr.: July- March.

Western Ghats: Evergreen forests and hill slopes of Kerala and Tamil Nadu. World: Sri Lanka.

IUCN: NE (IUCN, 2024).

Mallotus tetracoccus (Roxb.) Kurz., J. Asiat. Soc. Bengal, Pt. 2, Nat. Hist. 42: 245. 1873. Synonyms: *Mallotus ferrugineus* (Roxb.) Muell.-Arg., *Rottlera ferruginea* Roxb., *Rottlera tetracocca* Roxb., *Mallotus albus* var, *occidentalis* Hook.f. Vern.: *Adukanni, Vattakambil* (Mal.); *Mullu-polavu* (Tam.).

Large trees to 25 m tall; bark thick, surface grey, smooth, fibrous; branchlets grey tawny tomentose. Leaves simple, alternate, orbicular or deltoid-ovate, 9-21 x 7.5-20 cm, apex acuminate, base truncate, peltate, subcordate or round, margin entire or glandular dentate, glabrous above and tomentose below when matured, coriaceous; 5-ribbed from base, lateral nerves 6-8 pairs, intercostae scalariform, prominent; petiole 4.5-10 cm long, stout, swollen at base, with 2 large glands at the top, stellate tomentose. Flowers unisexual, yellow, in terminal branching tomentose spicate panicles. Male flowers 7 mm across. Tepals 4, 2 mm, ovate. Stamens many, crowed on a fleshy receptacle; filaments to 3 mm; anther cells unequal. Female flowers 5 mm across. Tepals 4, 2 mm, connate at base. Ovary superior, 0.8 x 0.7 cm, 3-4 locular, densely stellate hairy; ovule one in each cell; style 3 mm, 3, plumose; stigma acute. Capsule globose, 7-8 mm across, loculicidal 3-4 valved, echinate with soft, woolly spines, white-tomentose. Seeds ovoid, angular, black, minutely tubercled. Fl. & Fr.: September – December.

Western Ghats: Evergreen forests of Maharashtra, Goa, Karnataka, Kerala and Tamil Nadu. India: Arunachal Pradesh, Assam, Meghalaya and West Bengal. World: Sri Lanka, Nepal, Bangladesh, Myanmar and China.

IUCN: NE (IUCN, 2024).

Mallotus nudiflorus (L.) Kulju &Welzen., Blumea 52: 124. 2007. Synonyms: *Trevia nudiflora* L., *Trevia macrophylla* Roth, *Trevia macrostachya* Klotzsch, *Trevia nudiflora* var. *toemntosa* Susila & N.P. Balak. Vern.: *Kaadukayi, Katkumbala* (Kan.); *Kaatu-kumi, Mala-kumi, Pambara-kumbil* (Mal.); *Petari* (Mar.); *Annathuvarei, Attarasu, Attu-puvarasu, Raypkunul, Atthuvanji, Kanchi* (Tam.).

Deciduous trees to 20 m tall; bark greyish-brown, smooth; young parts tomentose. Leaves simple, opposite decussate, broadly ovate or deltoid base subcuneate, 6-20 x 9-15 cm, subtruncate, truncate, subcordate or cordate at base, apex acuminate to cuspidate or cordate-acuminate, margin entire, glabrous above, glaucous beneath; 3-ribbed from base, prominent, lateral nerves 3-6 pairs, intercostae scalariform, faint; stipules inter petiolar, 2-3 x 1 mm, linear lanceolate; petiole 3-12 cm long, slender, glabrous. Flowers unisexual, pale green. Male flowers 3-5 mm across, numerous on 10-20 cm long peduncle with abundant floral remains over the peduncle; bracts 3 mm across, ovate-lanceolate, pubescent; pedicels 4-5 mm long. Tepals globose, ovate, splitting into 3-4 broad, concave. Stamens many, clustered on a convex receptacle; filaments 1.5-3 mm long, free, slender; anthers oblong. Female flowers; 5-9 mm across, 4-5 on axillary racemes; pedicels 5-10 mm long, finely hairy. Tepals 3-5, 4-5 mm long, broadly ovate, densely tomentose beneath, cauducous. Ovary superior, 3-4 mm across, 2-4-loculed; ovule one in each cell; styles 3, 1.5-3 cm long, spreading, connate below, oblong. Capsule woody, 2-3 cm across, dark brown, broadly rounded, 3-4 loculed. Seeds globose, ovoid. Fl. & Fr.: December – March.

Western Ghats: Stream sides and river banks of low-lying forest above 500 m in Gujarat, Maharashtra, Karnataka, Kerala and Tamil Nadu. World: Sri Lanka, Nepal, Bangladesh, Myanmar, South China, Thailand. Laos, Cambodia, Vietnam and Malesia.

IUCN: LC (IUCN, 2024).

Uses: Wood is soft which is used for carving, planking, in match industry, packing case and plywood. The fruit pulp is said to be edible.

Mallotus philippensis (Lamk.) Muell.-Arg., Linnaea 34: 196. 1865. Synonyms: *Croton philippnesis* Lam., *Croton montanus* Willd., *Rottlera tinctoria* Roxb., *Rottlera tinctoria* var. *monstruosa* Ham. ex Dillwyn, *Mallotus philippensis* var. *tomentosa* Gamble. Vern.: *Kumkumadamara* (Kan.); *Chenkolli, Kapila, Kuramadakku, Manchana, Mannana, Ponnagam* (Mal.); *Kapila, Roem, Shendri* (Mar.); *Avam, Curangicari, Kabilam, Kaisalppodi, Kapli, Kungumam* (Tam.).

Dioecious trees to 12 m tall; bark pale brown; branchlets, young leaves and inflorescence tawny or rusty pubescent. Leaves simple, alternate, ovate or ovate-lanceolate, 5-20 x 2-8 cm, apex acuminate or acute, base acute or round, margin entire or sparsely serrate, glabrous above, greyish pubescent to fulvous tomentose with minute red glands beneath; 3-ribbed from base, lateral nerves upto 6 pairs, intercostae scalariform, prominent; stipules lateral, caudučous; petiole 7-60 mm long, swollen at base, fulvous-pubescent with 2 small sessile glands on each side. Flowers unisexual, brick red, in rusty puberulous, terminal spicate panicles. Male flowers 4.5 mm across. Tepals 4, lanceolate, stamens many. Female flowers 4 mm across. Tepals 3 or 4, thicker than in males, ovate-lanceolate. Ovary with red glands, superior, 3-celled, ovules one in each cell; styles 3, to 3.5 mm long, papillose. Capsule globose, 7-8 mm across, 3-lobed, loculicidally 3 valved, densely red-glandular, pubescent. Seeds 1-4, globose, glabrous, black. Fl. & Fr.: October – February.

Western Ghats: common in deciduous to semi-evergreen forests of Gujarat, Maharashtra, Goa, Karnataka, Kerala and Tamil Nadu. India: Almost all the states. World: Sri Lanka, Myanmar, Bhutan, Bangladesh, China, Japan, Taiwan, Malaysia, Philippines and Australia.

IUCN: LC (IUCN, 2024).

Uses: Kamala dye obtained from the fruits is used for dying the silk and wool. Oil is extracted from the fruits is used for fixer in hair dye and oinments.

Mallotus polycarpus (Benth.) Kuliju & Welzen., Blumea 52: 130. 2007. Synonyms: *Trevia polycarpa* Benth. *Trevia nudiflora* var. *polycarpa* (Benth.) Susila & N.P. Balak.

Small trees to 8 m tall; branches ribbed when dry. Leaves opposite to sub-opposite, ovate, 12.4–17.9 x 7.3–14.5 cm, base cordate to truncate, margin subentire, apex acuminate, extrafloral nectaries, lower surface sparsely hairy, domatia absent; 3-5-nerved, nerves 4 or 5 per side, looped and closed near the margin; petiole 3.5–7.4 cm long, often with a basal constriction when dry, subglabrous. Staminate inflorescences and flowers unknown. Pistillate inflorescences axillary racemes, behind the leaves, 3 –5.5 cm long, basally 1.5 mm thick, axis subglabrous to sparsely hairy; flowers 6–17. Pistillate flowers; pedicel 3 –6 mm long, subglabrous to sparsely hairy. Ovary ovoid, laterally flattened, with loculicidal grooves, densely hairy, glabrescent, locules 2; style 1–2 mm long, hairy; stigmas 8–14 mm long, papillose to somewhat plumose inside. Capsules indehiscent, ovoid to ellipsoid, 12–14 x 10.5–13, surface rugose, brown. Seeds broadly ellipsoid with the side against the septum flattened, the two seeds of a fruit still attached to each other after dehiscence, surface brown. Fl. & Fr.: January – June.

Western Ghats: Evergreen forests of Goa (Endemic). India: Madhya Pradesh.

IUCN: NE (IUCN, 2024).

Mallotus repandus (Rottler ex Willd.) Muell.-Arg., Linnaea 34: 197. 1865. Synonyms: *Croton repandus* Rottler ex Willd., *Croton rhombifolius* Willd., *Rottlera dicocca* Roxb., *Rottlera viscida* Blume, *Trevia nudiflora* var. *dentata* Susila & N.P.Balakr. Vern.: *Kumati* (Kan.); *Thavatti* (Mal.).

Scandent trees to 10 m tall; branchlets, dull yellowish-brownish stellate-tomentulose, thorns prominet. Leaves triangular-ovate, oblong-ovate, or ovate, 3.5-10 x 2.5-7 cm, adaxially glabrous, abaxially stellate-pubescent, scatteredly yellowish granular-glandular, base broadly cuneate, with 4 basal glands, margin entire or repand, apex acute or acuminate; basal veins 3; petiole 1.5-6 cm long. Male inflorescences terminal, rarely axillary, branches few or absent, 5-15 cm; bracts subulate, 1.5 mm long. Male flowers 2-5-fascicled; pedicel 2-4 mm. Sepal lobes 3 or 4, oblong, 3 mm, tomentulose. Stamens 40-75. Female inflorescences 5-8 cm long; bracts lanceolate, 2 mm. Female flowers: pedicel 2-3 mm. Sepals 4 or 5, lanceolate, 2-3 mm, tomentulose. Ovary 2-3-locular, dull yellow tomentulose; style 3-5 mm, almost free, plumose. Capsule 2-locular, 10 mm in diam, yellowish-brownish tomentulose, scattered glandular-scaly. Seeds subglobose, 4-5 mm in diam, black. Fl. & Fr.: September – November.

Wesern Ghats: Evergreen forests of Maharashtra, Goa, Karnataka, Kerala and Tamil Nadu. India: Andaman & Nicobar Islands, Andhra Pradesh, Assam, Bihar, Odisha, Sikkim, Tripura, Uttar Pradesh and West Bengal. World: Sri Lanka, Bhutan, China, Bangladesh, Nepal, Myanmar, Laos, Cambodia, Vietnam, Thailand, Indonesia, Malaysia, Philippines, Australia and New Caledonia.

IUCN: LC (IUCN, 2024).

Mallotus resinosus (Blanco) Merr., Sp. Blancoan.: 222. 1918. Synonyms: *Adelia resinosus* Blanco, *Claoxylon muricatum* Wight, *Mallotus andamanicus* Hook.f., *Mallotus aureo-punctatus* (Dalzell) Müll.Arg., *Mallotus beddomei* Hook.f., *Mallotus intermedius* (Baill.) N.P.Balakr., *Mallotus resinosus* var. *muricatus* (Wight) N.P.Balakr. & Chakr., *Mallotus resinosus* var. *stenanthus* (Müll.Arg.) Susila & N.P.Balakr., *Mallotus resinosus* var. *subramanyamii* (J.L.Ellis) Chakrab., *Mallotus subramanyamii* J.L.Ellis, *Mallotus walkerae* Hook.f., *Mallotus walkerae* var. *laxiflorus* Hook.f. Vern.: *Karuvalichai* (Tam.).

Small trees to 10 m tall; branchlets rusty tomentose to puberulous. Leaves unequal, oblanceolate, 13 x 4.5 cm, acuminate at apex, acute at base, distantly serrate or repand dentate, glandular; nerves 9-14 pairs, prominent, nervules closely parallel; petioles unequal, 0.3-1.8 cm long. Male racemes 1-4 cm long, axillary or terminal, rusty tomentose; bracts ovate-triangular, 2 mm long. Flowers 3 mm across, closely packed, tomentose. Sepals ovate, 3-4, ovate. Staemns 20-40. Female flowers 5-10 mm long, solitary in axile of bracts, raceme to 15 cm long. Sepals 3-6, connate at base, linear-lanceolate, 1 mm long. Ovary 3-loculed, globose, 4 mm long, covered with conical tubercles. Capsule 15 mm across, covered with curved spines and yellow glands. Seeds 5 mm across, globose, dull brown. Fl. & Fr.: October – June.

Western Ghats: Evergreen forests above 1000 m in Maharashtra, Goa, Karnataka, Kerala and Tamil Nadu. India: Andhra Pradesh and Telangana. World: Sri Lanka, Myanmar, Laos, Cambodia, Vietnam, Thailand and Malaysia.

IUCN: LC (IUCN, 2024).

Mallotus rhamnifolius (Willd.) Muell.-Arg., Linnaea 34: 196. 1865. Synonyms: *Croton rhamnifolius* Willd., *Croton reticulatus* Willd., *Mallotus rhamnifolius* var. *ovatifolius* Hook.f.

Small trees to 10 m tall; young parts densely brown tomentose. Leaves opposite or alternate, ovate-lanceolate or elliptic-lanceolate, 7-16 x 3.5-6.5 cm, base cuneate or subrounded, apex acuminate, glabrous above, finely pubescent and yellow-glandular below, basally 3-ribbed; petiole to 10 cm long. Spikes terminal, unbranched, 8-15 cm long. Male flowers distantly fascicled, 1-2 mm across. Tepals 1-seriate, splitting to 3-5 valves. Stamens many. Female flowers 1.5-2 mm across. Tepals 5, lanceolate; ovary 3-celled; styles 3, spreading. Capsule 3-4 mm across, 3-valvular, brown stellate-tomentose. Seeds 2 mm across, globose. Fl. & Fr.: October – August.

Western Ghats: Evergreen and semi-evergreen forests of Kerala and Tamil Nadu. India: Andhra Pradesh. World: Sri Lanka.

IUCN: NE (IUCN, 2024).

Paracroton Miq.

1a. Leaves entire; male inflorescence to 30 cm long; stamens 10-20 .... **P. integrifolius**

1b. Leaves dentate or serrate; male infloresence to 65 cm long; stamens 14-32 .... **P. pendulus** ssp. **zeylanicus**

Paracroton integrifolius (Airy Shaw) N.P. Balak. & Chakrab., Kew Bull. 48: 718. 1993. Synonyms: *Fahrenheitia integrifolia* Airy-Shaw., *Ostodes integrifolia* Airy Shaw, *Ostodes zeyanicus* var. *minor* sensu Gamble

Trees to 20 m tall; bark surface brown, smooth, brittle or semifibrous; branchlets pubescent. Leaves simple, alternate, elliptic or elliptic-obovate, 10-16 x 4-7 cm, apex shortly acuminate, base cuneate, margin entire or distantly indistinctly serrate, glabrous, coriaceous; lateral nerves 10-13 pairs, intercostae scalariform, prominent; stipules lateral, caducous; petiole 10-25 mm long, stout, 4 glands at the apex of the petiole. Flowers unisexual, white, in terminal and axillary racemes. Tepals 10 in 2 series; outer 5 lobed, unequal; inner 5 petalloid, free. Stamens 10-20, free, connective broad. Ovary superior, glabrous, 3-celled, 4 mm across, strigose; ovules 1 in each cell; styles 3, short, bifid. Disc with 5-10 glands. Capsule trilobed, depressed globose, 2 cm across, 6-ribbed, subglobose, tomentose, breaking into 3, 2-valved cocci. Seeds mottled. Fl. & Fr.: June – December.

Western Ghats: Evergreen forests above 900 m in Kerala and Tamil Nadu (Endemic).

IUCN: NE (IUCN, 2024).

Paracroton pendulous ssp. **zeylanicus** (Thw.) N.P. Balak. & Chakrab., Kew Bull. 48: 719. 1993. Synonyms: *Desmostemon zeylanicus* Thw., *Ostodes zeyalnicus* (Thw.) Muell.-Arg., *Fahrenheitia zeylanica* (Thw.) Airy-Shaw. Vern.: *Bale, Sotega* (Kan.).

Tree to 20 m tall; bark smpoth. Leaves oblong to obovate, oblong-oblanceolate, 16-50 × 5-15 cm, glabrous above, sparsely hairy beneath, margin with glandular serrations, apex acuminate; lateral nerves 8-10 pairs; petiole stout, 3-12 cm long. Male inflorescences terminal and axillary, 40-65 cm long. Petals 5, 6-9 mm long. Stamens many. Female inflorescences terminal, 18-50 cm long. Sepals 5, 8 mm across, lobes 5, orbicular, ovate-triangular, dorasally appendiculate. Petals 5, oblong, triangular-ovate, 8 x 4 mm, caducous. Disc glands connate to form a ring, lobed. Ovary densely pale brownish-yellow and stellate. Capsule spherical, 2.5-3.5 cm in diameter, scarcely 3-lobed, tomentellous on surface. Fl. & Fr.: March – November.

Western Gahts: Evergreen coastal forests of Karnataka, Kerala and Tamil Nadu. World: Sri Lanka.

IUCN: NE (IUCN, 2024).

Ricinus L.

Ricinus communis L., Sp. Pl.: 1007. 1753. Synonyms: *Ricinus communis* var. *genuinus* Müll.Arg., *Croton spinosus* L.

Glaucous branched small tree to 4 m tall. Leaves alternate, palmately 6-8-lobed, peltate, to 20 x 24 cm; lobes 9-15 x 3-6 cm, lanceolate, margin coarsely serrate, apex acuminate; petiole to 18 cm long. Flowers in terminal paniculate racemes, pale yellow. Male flowers below, female ones above. Male flowers perianth cupular, 3-5-lobed, 4 mm long, lanceolate. Stamens many, filaments connate, repeatedly branched. Female flowers tepals 5, subequal, 5 mm long, lanceolate. Ovary globose, 3-locular, echinate; ovule 1-per locule; styles 3, papillose. Capsule 3-lobed, 1.6-2 cm across, prickly. Seeds oblong, smooth, marbled, carunculate. Fl. & Fr.: Throughout the year.

Western Ghats: Cultivated and naturalized in all forested areas of Gujarat, Goa, Maharashtra, Karnataka, Kerala and Tamil Nadu. World: Native to Tropical Africa, naturalized everywhere.

IUCN: NE (IUCN, 2024).

Shirakiopsis Esser

Shirakiopsis indica (Willd.) Esser., Blumea 44: 185. 1999. Synonyms: *Sapium indicum* Willd., *Excoecaria indica* (Willd.) Muell.-Arg., *Stillingia diversifolia* Miq., *Excoecaria diversifolia* (Miq.) Meull.-Arg.

Evergreen trees to 8 m tall; bark greyish or pale brown, smooth, lenticellate; exudation milky acrid; branchlets greyish, terete. Leaves simple, alternate, elliptic-lanceolate, lanceolate, oblong or oblong-elliptic, 4-15 x 2-4 cm, base acute or cuneate, apex caudate, acuminate or acute, margin crenate, coriaceous, glabrous; lateral nerves 5-22 pairs, intercostae reticulate, prominent; petiole 5-15 mm long, grooved above, glabrous. Flowers unisexual, yellowish-green, in terminal or lateral racemes, to 10 cm long with a series of spirally arranged bracts; bracts subtending 5-7 male flowers and large glands on either side. Female flowers: 1-2 at base; pedicels upto 4 cm long, slender. Tepals 3, connate, 1-1.5 × 1.4-2 mm. Ovary superior, 1.5-2 mm across, 3-celled; ovules one in each cell; styles 3, simple, spreading, recurved, shortly connate at base; stigma linear, glandular. Capsule globose, 2.5-3 cm across, unlobed or scarcely lobed, 3-locular, dehiscent. Seeds 12-12 x 5-7 mm, dark brown, oblong-ovoid. Fl. & Fr.: November – June.

Western Ghats: Coastal forests of Maharashtra and Kerala. World: Sri Lanka, Thailand, Caroline Islands, Solomon Islands, Malaysia, Sumatra, Borneo, Celebes, Maluku Islands and New Guinea.

IUCN: LC (IUCN, 2024).

Uses: Leaves can be obtained greenish yellow dye.

Suregada Roxb. ex Rottler

1a. Leaves obovate-oblanceolate, sharply acuminate at apex, to 15 cm long; flowers small in fascicles; fruits smooth, deeply lobed ....**S. lanceolata**

1b. Leaves elliptic-oblong, rounded or obtuse at apex, to 25 cm long; flowers in cymes, rather large; fruits rough, obscurely lobed ....**S. multiflora**

Suregada lanceolata (Wild.) Kuntze., Revis. Gen. Pl. 2: 619. 1891. Synonyms: *Gelonium angustifolium* Muell.-Arg., *Gelonium lanceolatum* auct. non Willd., *Suregada angustifolia* (Müll.Arg.) Airy Shaw Vern.: *Kakaipalai* (Tam.).

Small deciduous trees to 8 m tall; branchlets glabrous. Leaves simple, alternate, obovate or oblanceolate, 5-15 x 3-5 cm, apex obtuse, base cuneate, midrib flat above; secondary nerves 6 pairs, not prominent, tertiary nerves obscure; petiole 0.3-0.5 cm long, glabrous; stipules caducous. Flowers unisexual, dioecious. Male flowers small, in axillary fascicles; pedicel to 5 mm long. Sepals 5-6, lobes concave, 3 mm long, scaly. Stamens 15-25, mixed with rugose glands; filaments 1.5 mm long. Female flowers solitary, axillary. Sepals 5-6, 3 mm across, ovate, finely ciliate. Capsule ovoid, smooth, deeply 3-lobed, keeled on the lobes, 0.6 cm in diam. Seeds 1 per lobe. Fl. & Fr.: September – May.

Western Ghats: Deciduous forest to 750 m in Karnataka, Kerala and Tamil Nadu. India: Andhra Pradesh. World: Sri Lanka.

IUCN: NE (IUCN, 2024).

Suregada multiflora (A. Juss.) Baill., Etude Euphorb.: 396. 1858. Synonyms: *Gelonium multiflorum* A.Juss., *Gelonium fascuculatum* Roxb., *Suregada glabra* Roxb., *Suregada multiflora* var. *lamellata* Airy Shaw

Small trees to 12 m tall; bark grey, yellowish-white, smooth. Leaves alternate, elliptic or elliptic-oblong, 8-17 x 3-6 cm, acute, with few serrations towards the apex, coriaceous. Flowers large, many in cymes or clusters. Sepals in male of 5, concave orbicular imbricate lobes, in female of 5-6 narrower lobes. Petals 0. Disc in male 0, in female copular, the glands large, peripheric. Stamens 10-60, free, central on a convex receptacle, often mixed with rugose glands; filaments filiform; anthers oblong, dorsifixed, the cells parallel. Ovary 2-4 celled; ovule 1 in each cell; style minute, reniform semilunate or bifid, depressed. Capsule globose-subglobose, rough, obscurely lobed, 1.2-1.8 cm in diameter. Seeds subglobose, arillate. Fl. & Fr.: September – May.

Western Ghats: Evergreen forests of Kerala and Tamil Nadu. World: Myanmar, China, Thailand and Malesia.

IUCN: NE (IUCN, 2024).

Trigonostemon Blume

Trigonostemon nemoralis Thw., Enum. Pl. Zeyl.: 277. 1861.

Small trees to 10 m tall; young parts silkily hairy. Leaves alternate, elliptic-oblong or lanceolate, 15 x 5 cm, sinuate-lobed; nerves 10-12 pairs, slender, base narrowed into the short stout channelled petiole, 2-glands at the top. Racemes with a stout rachis, bracts subulate, lower 2 mm long and more. Flowers monoecious, red, males clustered shortly pedicelled, female solitary longer pedicelled. Seplas 5, with imbricate lobes. Petals 5, sometimes 0 in female. Disc of 5 glands, often united in a lobed cup. Stamens 3 or 5, connate in a short column with free spreading tips; anthers erect or horizontal. Ovary 3-celled, hispid; ovule 1 in each cell; styles very short and broad, bifid or twice bifid, rarely entire. Capsule of 3-2 valved cocci. Seeds ovoid or globose, mottled. Fl.& Fr.: September – May.

Western Ghats: Evergreen forests above 900 m in Kerala and Tamil Nadu. World: Sri Lanka.

IUCN: NE (IUCN, 2024).

Tritaxis Baill

1a. Inflorescences dichasial; stamens 8; ovary glabrous; styles 2-3 mm long, free; fruits glabrous .... **T. beddomei**

1b. Inflorescences not dichasial; stamens 10 - 16; ovary pubescent; styles 3 - 7 mm long, connate below into a column; fruits evanescently pubescent... .... **T. glabellus**

Tritaxis beddomei Benth., J. Linn. Soc., Bot. 17: 221. 1878. Synonyms: *Dimorphocalyx beddomei* (Benth.) Airy Shaw, *Trigonostemon beddomei* (Benth.) N.P. Balak.

Small monoecious trees to 6 m tall; branchlets terete, glabrous. Leaves alternate, entire or slightly toothed, 7-16 x 3-5 cm, ovate-lanceolate, acuminate at apex, attenuate at base. Flowers in terminal cymes; central flower of each cyme female, surrounded by male flowers. Perianth 2-seriate. Sepals 5, 0.5-0.6 cm long. Petals 5, white, 1-1.5 x 0.4-0.5 cm, ovate- lanceolate. Disc of 5 glands. Stamens 8, in 2 series. Ovary 3-celled; ovule 1 in each cell; styles 3, each 2-fid. Capsule globose, 1-1.5 cm diam., of 3, two valved cocci. Seeds globose to trigonous-ovoid. Fl. & Fr.: May – October.

Western Ghats: Evergreen forests of Kerala and Tamil Nadu above 1000 m altitude (Endemic).

IUCN: EN- B1ab(i,iii). (IUCN, 2024).

Tritaxis glabellus (Thwaites) R.Y.Yu & Welzen., Taxon 68: 932. 2019. Synonym: *Dimorphocalyx glabellus* Thwaites Vern: Malamkuluki, Malakulukki (Tam.).

Small trees to 8 m tall; branchlets nearly white. Leaves simple, alternate, 12 x 5 cm, elliptic or lanceolate, obtuse or acute, reddish brown when dry and rather thin; petiole 0.5-1 cm long. Male flowers shortly pedicelled; female long pedicelled; pedicels with few small obtuse bracts. Petals of the male longer than the sepals, of the female shorter. Stamens about 10. Ovary strigose. Capsule slightly hispid-villous, shortly lobed; seeds oblong, mottled. Fruiting calyx obovate, obtuse, with a glandular pit on back near the tip. Fl. & Fr.: Throughout the year.

Western Ghats: Evergreen forests above 1000 in Maharashtra, Karnataka, Kerala and Tamil Nadu. India: Andhra Pradesh. World: Sri Lanka.

IUCN: NE (IUCN, 2024).

Key to the varieties

1a. Stamens 10-11; fruiting sepls subequal, relatively smaller to 10 mm long .... var. **glabellus**

1b. Stamens 11-16; fruiting sepals unequal, relatively larger to 18 mm long.. .... var. **praetervisa**

Tritaxis glabella *var.* **praetervisa** Chakrab. & G. Krishna., Pl. Sci. Today 7: 303. 2020. Synonyms: *Dimorphocalyx glabellus* **var.** *lawianus* (Hook.f.) Chakrab. & N.P. Balakr., *Dimorphocalyx lawianus* Hook.f.

Trees to 8 m tall; bark smooth, yellowish-grey; branchlets slender, greyish-white. Leaves simple, alternate, oblong, elliptic-lanceolate, elliptic-oblong or ovate-lanceolate, 5-17.5 x 2-7 cm, base acute, apex acute, obtusely acuminate or acuminate, margin entire or distantly dentate or crenate, glabrous, dark green above, paler beneath, coriaceous; lateral nerves 7-13 pairs, parallel, intercostae reticulate, faint; petiole 5-20 mm long, grooved above, glabrous. Flowers unisexual, white. Male flowers 8-10 mm across, in axillary cymes. Sepals cupular, unequally 5-lobed, 4-5 mm long. Petals 5, oblong, 10 x 5 mm, obtuse, imbricate. Disc glands adnate with petals. Stamens 10-20, form a short column in 2 whorls; filaments thick; anthers dorsifixed. Female flowers solitary or 2-3 together, terminal or leaf-opposed. Sepals 5, 2.5 cm long, lobes oblong, erect, imbricate. Petals as in male flower. Ddisc annular. Ovary superior, ovoid, 3-celled, densely hairy, ovule one in each cell; styles 3, connate at base, bifid, spreading. Capsule obovoid, 2 x 2.2 cm, 3-lobed, surrounded by persistent calyx lobes. Seeds yellowish mottled with brown, ellipsoid. Fl. & Fr.: October – June.

Western Ghats: Evergreen forests of Karnataka and Kerala (Endemic).

IUCN: NE (IUCN, 2024).

Cultivated:

Codiaenum variegatum (L.) A. Juss.; Exotic: Native to Maluku Islands, Malay Archipelago.

Euphorbia lactea Haw.; Exotic: Native to Sri Lanka, Pakistan, Bangladesh, Africa, Maluku Islands, Europe.

Euphorbia leucocephala Lotsy; Exotic: Native to South America

Hevea brasiliensis (Willd. ex A.Juss.) Mull.-Arg.; Exotic: Native to South America.

Hura crepitans L.; Exotic: Native to South America

Manihot carthagenensis ssp. **glaziovii** (Müll.Arg.) Allem; Exotic: Native to Brazil

Vernicia cordata (Thunb.) Airy Shaw; Exotic: Native to Japan, Korea, Chinna

Vernicia montana Lour.; Exotic: Native to Japan, South China, Vietanam, Myanmar, Thailand

FABACEAE Lindl.

Key to the genera

1a. Calyx divided nearly to the base; petals free, slightly unequal; stamens free....................2

1b. Calyx gamosepalous; petals more or less connate or lobes free; staemns monadelphous or diadelphous or free....................13

2a. Leaves simple, bilobed .... **Bauhinia**

2b. Leaves pinnate or bipinnate, leaflets 2-many 3

3a. Calyx campanulate; petals oblanceolate .... **Acrocarpus**

3b. Calyx with short or long tube; petals ovate with clawed base or absent 4

4a. Petals 5, yellow or red coloured 5

4b. Petals absent 7

5a. Anthers dehiscing by apical pores 6

5b. Anthers dehiscing by longitudinal split..9

6a. Pod cylindrical; leaflets 8-15 pairs .... **Cassia**

6b. Pod flat; leaflets more than 15 pairs .... **Senna**

7a. Stamens 2; leaves with 7-9 foliolate .... **Dialium**

7b. Stamens 10; leaves with 2-7 foliolate..8

8a. Leaflets 1-pair; stigma large, peltate; pod samaroid below . . .**Hardwickia**

8b. Leaflets 4-7; stigma minute; pods thickened below.......................... **Prioria**

9a. Calyx with short tube; pod one seeded 10

9b. Calyx with long tube; pod many-seeded 11

10a. Leaflets many, small, caducous; flowers bright yellow ....**Parkinsonia**

10b. Leaflets 2-6, large, persistent; flwoers white or purple ....**Cynometra**

11a. Stamens 2, monadelphous ....**Tamarindus**

11b. Stamens 3-9, free 12

12a. Petals absent; calyx long tubular .... **Saraca**

12b. Petals present; calyx tube capanulate ....**Humboldtia**

13a. Petals free, imbricate, posterior one outside; stamens usually diadelphous..14

13b. Petals connate, valvate, posterior one inside; staemns free or Monadelphous 22

14a. Leaves trifoliolate 15

14b. Leaves odd-pinnate 17

15a. Petals unequal; pod 2-10 seeded 16

15b. Petals equal; pod 1-seeded .... **Butea**

16a. Flwoers bright red coloured; standard petal longer than others ....**Erythrina**

16b. Flowers creamy or brown coloured; standard petal equal to lateral .... **Ougeinia**

17a. Pod dehiscent 18

17b. Pod indehiscent 19

18a. Seeds 1-2 per pod, arillated, aril bright red coloured .... **Ormosia**
18b. Seeds 8-10 per pod, not arillated .... **Mundulea**
19a. Leaflets opposite 20
19b. Leaflets alternate 21
20a. Branchlets pubescent; racemes terminal; flowers purplish .... **Milletia**
20b. Branchlets glabrous; racemes axillary; flowers white ....**Pongamia**
21a. Flowers white or purple; pod oblong ....**Dalbergia**
21b. Flowers yellow; pod round .... **Pterocarpus**
22a. Stamens 10; anthers with deciduous apical glands 23
22b. Stamens many; anthers without glands 24
23a. Flowers in racemes; calyx campanulate; pod narrow .... **Adenanthera**
23b. Flwoers in heads; calyx tubular; pod woody flat .... **Xylia**
24a. Armed trees; stamens free 25
24b. Unarmed trees; stamens monadelphous 29
25a. Leaflets less than 0.5 long; stamens two colours in same inflorescence; pod curved .... **Dichrostachys**
25b. Leafletes more than 1 cm long; stamens uniform coloured; pod usually straight...26
26a. Pod constricted between the seeds or moniliform .... **Vachellia**
26b. Pod not constricted, margin entire 27
26a. Pod indehiscent; usually yellow when ripe 28
27b. Pod dehiscent; usually papery or dry when ripe .... **Senegalia**
28a. Stipular thorns to 3 cm long; pinna to 3-5 pairs ....**Neltuma**
28b. Stipular thorns to 1.5 cm long; pinna to 5-10 paris .... **Prosopis**
29a. Leaves bifoliolate 30
29b. Leaves many-foliolate 31
30a. Flowers in 5-8 per head; staemns more than 2 cm long; pod not curved.. ....**Sanjappa**
30b. Flowers more than 15 per head; stamens less than 1 cm long; pod curved .... **Pithecellobium**
31a. Leaflets below 1 cm long; petals usually caducous ....**Albizia**
31b. Leaflets more than 2 cm long; petals not caducous 32
32a. Pod curved, constricted; seeds black ....**Archedendron**
32b. Pod straight, not constricted; seeds brown ....**Samanea**

Acrocarpus Wight & Arn.

Acrocarpus fraxinifolius Wight & Arn., Mag. Zool. Bot. 2: 547. 1839. Synonyms: *Acrocarpus combretiflorus* Teijsm. & Binn., *Acrocarpus grandis* (Miq.) Miq. Vern.: *Malai kondrai, Nelrai* (Tam.).

Trees to 35 m tall; bark light-grey or yellowish-grey. Leaves bipinnate, alternate, pinnae 3-5 pairs, paripinnate, 20-30 cm long, slender glabrous; leaflets 10-16 per pinnae, opposite, estipellate; petiolules 4 mm long, glabrous; pinnules 7.5-12 x 2.5-4 cm, elliptic, elliptic-lanceolate or elliptic-oblong, base oblique, acute or obtuse, apex acuminate, margin entire, glabrous; lateral nerves 7-15 pairs, intercostae reticulate, faint; rachis 13-15 cm long, stout, glabrous, pulvinate. Flowers bisexual, in dense, erect, axillary racemes; pedicels short; bracts small, cauducous. Sepals campanulate, disc-lined, lobes 5, lanceolate, equal. Petals 5, narrow, subequal, greenish. Stamens 5, exserted, crimson coloured; anthers versatile. Ovary half inferior, stipitate; ovules many; style short, incurved; stigma minute. Pod flat ligulate, 12 x 2 cm, dehiscent, the upper suture winged. Seeds 5-10, obovate, oblique, compressed. Fl. & Fr.: February – April.

Western Ghats: Evergreen forests of Maharashtra, Karnataka, Kerala and Tamil Nadu. India: Arunachal Pradesh, Assam, Madhya Pradesh, Meghalaya, Mizoram, Punjab, Sikkim and West Bengal. World: Sri Lanka, Bangladesh, Nepal, Bhutan, Myanmar, China, Laos, Sumatra and Java.

IUCN: NE (IUCN, 2024).

Adenanthera L.

Adenanthera pavonina L., Sp. Pl.: 384. 1753. Vern.: *Perumkundumani* (Tam.).

Deciduous treess to 20 m tall; bark grey to reddish-brown, smooth; branchlets sparsely to densely minutely tomentellous to glabrous. Leaves bipinnate, alternate, pinnae 2-3 pairs, 2-42 cm, opposite or rarely subopposite, even pinnate, grooved above, pulvinate, puberulent; leaflets 8-20, alternate; petiolule up to 3 mm long; pinnules 1.3-8×. 8-4.2 cm, oblong, oblong-elliptic, base slightly oblique and truncate, apex round, emarginate or obtuse and mucronate, margin entire, puberulent above and glaucous beneath; lateral nerves 9-10 pairs, parallel, intercostae reticulate, faint; stipules minute, free, lateral, cauducous; rachis 14.5-62.5 cm long, stout, pulvinate, grooved above, glabrous, with a gland at the tip. Flowers bisexual, pale yellow, 6 mm across, clustered in axillary spiciform racemes; bracts to 5 mm, linear, bracteole to 1 mm, pedicel to 3 mm. Sepals campanulate, 1 mm, 5 toothed, pubescent. Petals 5, connate below, linear-lanceolate, upto 4.5 mm long, glabrous. Stamens 10, free; filaments filiform, alternately long and short; anthers oblong, ending in a stipitate gland. Ovary subsessile, flat, 3 mm; style 2 mm, filiform; stigma simple. Pod straight to falcate to slightly twisted, 10-25 x 0.7-1.5 cm, spirally coiled after dehiscence, thin, septate within, tapered at base, apex acute, glabrous. Seeds 6-15, 8-10 x 7-9 mm, elliptic-lenticular, red. Fl. & Fr.: March - August.

Western Ghats: Cultivated in Maharashtra, Goa, Karnataka, Kerala and Tamil Nadu. World: Sri Lanka, Myanmar, Thailand, Timor, China and Malesia.

IUCN: LC (IUCN, 2024).

Albizia Durazz.

1a. Pinnae not more than 6 pairs (2-6) 2
1b. Pinnae more than 6 pairs 6
2a. Leaf rachis glabrous 3
2b. Leaf rachis hairy 4
3a. Flowers sessile, greenish white .... **A. procera**
3b. Flowers pedicelled, whitish yellow .... **A. lebbeck**
4a. Discoid gland below middle of the petiole 5
4b. Discoid gland above middle of the petiole .... **A. thompsoni**
5a. Leaflets 6-15 pairs, puberulous or glabrescent, margin entire .... **A. odoratissima**
5b. Leaflets 10-20 pairs, tomentellous, margin entire, ciliate .... **A. lathamii**
6a. Seeds 6–8, pods strongly nerved, greyish brown .... **A. amara**
6b. Seeds 8–12, pods obscurely nerved, pale brown .... **A. chinensis**

Albizia amara (Roxb.) Boiv. P.-S. Laurentie & al., Encycl. Dix-Neuvième Siécle 2: 34. 1837. Synonyms: *Mimosa amara* Roxb., *Acacia amara* (Roxb.) Willd., *Feuilleea amara* (Roxb.) Kuntze Vern.: *Usilai, Unjal, Thurinji, Arappu* (Tam.).

Multibranched trees to 13 m tall; bark thin, surface grey, rough, scaly; branchlets brown, pubescent. Leaves bipinnate, alternate, pinnae 3-16 pairs, opposite, even pinnate, 2.5-7 cm, slender, rachis with a bristle at the terminal end, with a gland in between terminal pairs on the upper side or between all pairs; leaflets 20-60, opposite, even pinnate, estipellate, pinnule 3-8 x.1-2 mm, sessile, linear, base oblique, apex subacute or obtuse, margin entire, ciliate or glabrous; lateral nerves obscure; stipules minute, free, lateral, lanceolate, caducous; rachis 3-15 cm, slender, pulvinate. Flowers bisexual, white, heads solitary or 2-3 in axillary fascicles, 6-8 mm long; peduncle to 5 cm, slender, densely pubescent; bracts to 5 mm, caducous; pedicel to 1 mm, puberulous. Calyx tubular, funnel-shaped, 1-2×1-1.5 mm, puberulous, teeth 5, short. Petals 3-4 mm long, glabrous except the bearded apex, lobes 5, oblong-lanceolate. Stamens many, 1-1.5 cm long, monadelphous at base, tube as long as or slightly shorter than the corolla tube; filaments long exserted. Ovary stipitate, pubescent, style filiform, stigma minute. Pod flat, 10-24 x 2.5-4 cm, greyish-brown, straight or wavy along margin, indehiscent. Seeds 6-13, compressed, ovate-orbicular. Fl. & Fr.: April - August.

Western Ghats: Foot hills and scrub forests slopes to 600 m in Maharashtra, Goa, Karnataka, Kerala and Tamil Nadu. India: Almost all the states. World: Sri Lanka and Tropical Africa.

IUCN: LC (IUCN, 2024).

Albizia chinensis (Osbeck) Merr., Amer. J. Bot. 3: 575. 1916. Synonyms: *Mimosa chinensis* Osbeck, *Albizia stipulacea* (Roxb.) Boivin, *Mimosa stipulacea* Roxb., *Albizia marginata* (Lam.) Merr., *Mimosa marginata* Lam., *Acacia stipulacea* DC., *Albizia chinensis* var. *smithiana* (Roxb.) K.C. Sahni *et al.* Vern.: *Pilla vagai* (Tam.).

Deciduous trees to 25 m tall; bark surface brownish-grey, smooth, horizontally fissured; branchlets minutely golden tomentous. Leaves bipinnate, alternate, pinnae 4-20 pairs, opposite, even pinnate, 8-12.5 cm long, slender, sparsely tomentellous; with a gland in between each pairs or on distal 2-3 pairs; leaflets 50-90, opposite, even pinnate, sessile, 3-6 x 15-18 mm, falcate-lanceolate or falcate, apex acute, base oblique, margin entire, ciliate, glabrous above; stipules 2.5-4 x 1 cm, free, lateral, crisped, orange-pink, caducous; rachis 18-35 cm, slender, grooved above, pulvinate, with a gland near its base on upper side. Flowers white, in globose heads, 2-6 in axillary fascicles or arranged in terminal and axillary panicles; peduncle slender, pubescent; pedicel 0.5-1 mm, pubescent; bracts stipular, persistent. Sepals cubular, 2.5-5 × 1-2 mm, funnel shaped, pubescent, teeth short. Petals connate, 6-10 mm long, funnel shaped, pubescent, lobes 5, ovate-lanceolate. Stamens many, yellowish, monadelphous at the base, tube slightly shorter to or longer than the corolla tube. Ovary 1.5 mm long, sessile, glabrous; style upto 3 cm long. Pod strap shaped, 15-20 x 2.5 cm, reddish-brown, straight or wavy along the margin, slightly bullate over the seeds. Seeds 8-20, 6-10 x 5-8 mm, ovate, flat, dull brown. Fl. & Fr.: April - December.

Western Ghats: Deciduous forests of Maharashtra, Goa, Karnataka, Kerala and Tamil Nadu. India: Most of the states. World: Sri Lanka, Bhutan, South China, Myanmar, Thailand, Laos, Cambodia, Vietnam and Malesia.

IUCN: NE (IUCN, 2024).

Albizia lathamii Hole., Indian Forest Rec. 4: 153. 1913. Vern.: *Kaattu vagai* (Tam.).

Trees to 7 m tall; bark greyish-brown, rough, lenticellate; branchlets brownish or blackish, terete, white lenticellate. Leaves bipinnate, alternate, stipulate; stipule free, lateral, caducous;rachis 25-50 mm, slender, pulvinate, brownish or greyish pilose, with a gland near the base; pinnae 2-4 pairs, 1.5-5 cm long, opposite, evenly pinnate, slender, pilose, with a gland in between terminal 2 pinnae; leaflets 10-20, opposite, even pinnate, estipellate, sessile; pinnule 1-1.4 x 0.3-0.8 cm, oblong-obovate, base unequal, apex round or retuse, margin entire, ciliate, glabrous above on maturity and sparsely pilose beneath, midrib more or less central; lateral nerves 3-4 pairs, intercostae reticulate. Flowers white, in globose heads arranged in axillary corymbs. Sepals campanulate, 4 x 2.5 mm, puberulous, teeth triangular. Petals 5, funnel shaped, lobes 5, ovate-triangular. Stamens many, monadelphous at the base, tube shorter than the corolla tube. Ovary sessile, oblong, pubescent; style filiform; stigma minute. Pod strap shaped, 7.5-12 x 1.5-2 cm, flattened, often slightly undulate along margins, puberulous, obtuse to rounded and mucronulate at apex, slightly bullate along seeds, blackish-brown when dry, very short. Seeds ellipsoid. Fl. & Fr.: February - June.

Western Ghats: Deciduous forests of Kerala and, Tamil Nadu (Endemic). India: Madhya Pradesh.

IUCN: NE (IUCN, 2024).

Albizia lebbeck (L.) Benth., London J. Bot. 3: 87. 1844. Synonyms: *Mimosa lebbeck* L., *Acacia lebbeck* (L.) Willd., *Albizia lebbeck* var. *rostrata* Haines, *Albizia lebbeck* var. *pubescens* Haines, *Mimosa flexuosa* Rottler ex Wight & Arn., *Mimosa sirissa* Roxb. Vern.: *Vagai* (Tam.).

Deciduous trees to 25 m tell; bark surface yellowish-brown, deeply fissured. Leaves bipinnate, alternate, pinnae 2-4 pairs, 5.5-12 cm long, slender, even pinnate; leaflets 8-20, opposite, petiolule to 1 mm long, a gland in between each leaflets; leaflets 2.7-5 x 1-2.5 cm, obliquely oblong, base of one half cuneate, margin entire, coriaceous; midrib towards distal margin, lateral nerves 6-8 pairs, intercostae reticulate; stipules small, free, lateral, cauducous; rachis 7-9 cm long, slender, grooved above, with a gland at its base, glabrous, pulvinate. Flowers bisexual, greenish-white, in subglobose heads, 3.5-3.7 cm long, solitary or 2-4 together in axillary corymbose racemes; peduncle 6-8 cm; bracts linear, cauducous; pedicels to 3 mm. Sepals 5, tubular, tube funnel-shaped pubescent, 3-4.5×1.5-3 mm, teeth triangular. Petals infundibuliform, 7-10 mm long, greenish, lobes 5, lanceolate, pubescent without. Stamens many, filament tube shorter than corolla tube, filaments long exserted, green or pink; anthers very small. Ovary subsessile, glabrous, style 2.5 cm, filiform; stigma minute. Pod oblong, 20-30 x 4-5 cm, compressed, straw coloured, base and apex obtuse, turgid above the seeded region. Seeds 8-12, 6-10 x 5-8 mm, ovate, dull dark brown, flattened. Fl. & Fr.: March – December.

Western Ghats: Common cultivated avenue trees in Gujarat, Maharashtra, Goa, Karnataka, Kerala and Tamil Nadu. India: Most of the states. World: Sri Lanka, Malay Archipelago, China, North Australia and Tropical Africa.

IUCN: LC (IUCN, 2024).

Uses: Wood is used for making furnitures, veneering, general constructions, carving, firewood and charcoal.

Albizia odoratissima (L.f.) Benth., London J. Bot. 3: 88. 1844. Synonyms: *Mimosa odoratissima* L.f., *Albizia odoratissima* var. *mollis* Benth. ex Baker, *Albizia orissensis* K.C. Sahni & Bennet, *Acacia odoratissima* (L.f.) Willd. Vern.: *Karuvagai* (Tam.).

Deciduous trees to 30 m tall; bark surface greyish-brown to dark brown, rough, irregularly cracked; branchlets blackish to brown, terete, glabrous. Leaves bipinnate, alternate, pinnae 2-8 pairs, opposite, even pinnate, 5-13 cm long, slender, puberulent, glands between the junctions of 1-2 distal pairs of pinnae, leaflets 14-40, opposite, even pinnate, sessile, 1.8-2.5 x 0.5-1.2 cm, oblong, apex obtuse and apiculate, margin entire, glabrous; stipule free, lateral, caducous; rachis 20-30 cm long, stout, grooved above, spulvinate, with a gland at the base; lateral nerves 3-6 pairs, pinnate, intercostae reticulate, prominent. Flowers white, in globose heads forming terminal panicles; peduncle, slender, puberulous. Sepals cupular, tube 1-1.5 x 1-1.3 mm, teeth minute, deltoid, pubescent. Petals connate, broadly funnel shaped, lobes 5, ovate-lanceolate, pubescent. Stamems many, 1.2-2 cm long, monadelphous at base, tube as long as or shorter than the corolla tube. Ovary stipitate, glabrous or pilose; style filiform; stigma terminal. Pod strap-shaped,15-20 x 2.5-3.7 cm, seed portion constricted, rounded to rostrate at apex, glabrous, reddish-brown to dark brown, finely reticulate. Seeds oblong, 6-12, orbicular, compressed. Fl. & Fr.: May – December.

Western Ghats: Deciduous forests of Maharashtra, Goa, Karnataka, Kerala and Tamil Nadu. India: Most of the states. World: Sri Lanka, Bhutan, Nepal, Pakistan, China and South East Asia.

IUCN: LC (IUCN, 2024).

Uses: Wood is used for making household implements and garden poles.

Albizia procera (Roxb.) Benth., London J. Bot. 3: 89. 1844. Synonyms: *Mimosa procera* Roxb., *Albizia procera* var. *elata* (Roxb.) Baker, *Inga gracilis* Jungh. ex Miq., *Inga kihian* Blume ex Miq., *Mimosa elata* Roxb. Vern.: *Koda vagai, Nalla vagai* (Tam.).

Deciduous trees to 30 m tall; bark greenish-yellow, smooth, exfoliation small, irregular; branchlets brownish to blackish or yellowish, terete, silky pubescent. Leaves bipinnate, alternate, pinnae 2-5 pairs, 9.7-27 cm long, slender; leaflets 4-20, opposite, even pinnate, petiolule 2-3 mm long; pinnule 2.5-5 x 1.8-3 cm, obliquely oblong, base unequal, apex round or subacute, margin entire, glabrous above, with a gland between terminal pairs of leaflets; lateral nerves 6-10 pairs, intercostae reticulate, obscure; stipule free, lateral; rachis 13-30 cm long, slender, pulvinate, puberulous. Flowers yellowish-white, in globose heads, forming axillary or terminal panicles. Calyx tubular, tube narrowly obconic, 2.5-3 x 1.5-2 mm, glabrous, teeth triangular. Petals 5-7 mm long, 5-lobed; lobes oblong-elliptic, puberulous towards apex. Stamens many, 8-12 mm long, monadelphous at base, tube longer than the corolla tube. Ovary oblong, stipitate, glabrous, style filiform; stigma terminal. Pod strap-shaped, 10-20 x 1.8-2.5 cm, glabrous, often glossy, faintly reticulately veined, black, brown or yellow when dry. Seeds 5-12, ovate-elliptic, compressed, greenish-brown. Fl. & Fr.: May – December.

Western Ghats: Deciduous forests above 500 m in Maharashtra, Karnataka, Kerala and Tamil Nadu. India: Most of the states. World: China, Malay Archipelago and North East Australia.

IUCN: LC (IUCN, 2024).

Albizia thompsonii Brandis., Indian Forester 25: 284. 1899. Synonym: *Albizia thompsonii* var. *galbana* Haines Vern.: *Menmaeni vagai* (Tam.).

Large trees to 13 m tall; bark black, rough; branchlets pubescent. Leaves bipinnate, 16-19 cm long, pinnae 4-6 pairs, 8-16 cm long, leaflets 15-19 pairs, overlapping, oblong, slightly falcate, 1-1.5 x 0.4-0.5 cm, cordate at base, margin entire, apex obtuse, pubescent when young; petiole short, a gland above middle of the petiole, rachis tomentose with a gland between the upper most pair of pinnae. Peduncle 2-4 cm long, flowers in globose heads, pedicel short to 3 mm long. Seplas 5, campanulate, 5-lobed, to 3 mm long, pubescent. Petals 5, 5-toothed, to 3.5 mm long, pubescent without. Stamens many; filaments 5 mm long, Monadelphous at the base. Ovary sessile or shortly stalked, 1.5 mm long, filiform. Pods moderately thick, 10-12 x 1-1.5 cm, shining, dark brownish. Seeds 6-8, brown, glabrous. Fl. &Fr.: April – October.

Western Ghats: Tamil Nadu (Endemic). It is distributed in Andhra Pradesh, Bihar, Odisha and Telangana states.

IUCN: NT (Low risk/near threatened) (IUCN, 2024).

Archidendron F. Muell.

1a. Branchlets glabrous; leaflets 2-4 pairs ....**A. bigeminum**

1b. Branchlets rusty tomentose; leaflets 3-12 pairs ....**A. clypearia**

Archidendron bigeminum (L.) I.C. Nielsen., Opera Bot. 76: 73. 1984 publ. 1985. Synonyms: *Abarema bigemina* (L.) Kosterm., *Pithecellobium bigeminum* (L.) Mart., *Mimosa monadelpha* Roxb., *Pithecellobium gracile* Bedd., *Pithecellobium nicobaricum* Prain Vern.: *Kalpaaku, Poopathumaram* (Tam.).

Evergreen trees to 12 m tall; bark reddish-brown, prominently lenticellate; branchlets glabrous. Leaves bipinnate, alternate, pinnae 2, opposite, 11-13 cm long, slender, puberulent, a gland in between 2 pinnae; leaflets 4-8, opposite, with a gland in between every pair of leaflets, 6.5-12 x 2.5-5 cm, obovate, oblong or elliptic-acuminate, apex acuminate, base unequally acute, margin entire; stipules inconspicuous; rachis 10-11 cm long, slender, pulvinate, pubescent, with a gland near its base; lateral nerves 8-14 pairs, pinnate, prominent, secondary lateral also prominent, intercostae reticulate. Flowers white, in globose heads in terminal and axillary panicles; bracteoles lanceolate, persistent. Sepals campanulate, to 1.5 mm long, lobes 4, minute, acute. Petals 4, basally connate, 4 mm long, lobes ovate, acute. Stamens 4, monadelphous, much exserted. Ovary sessile; style filiform; stigma minute; ovules many. Pod twisted, 10 x 1.5-2 cm, strap-shaped, dehiscent, red within. Seeds 3-8, ovoid, smooth, without aril. Fl. & Fr.: January – March.

Western Ghats: Evergreen forests above 1200 m in Maharashtra, Goa, Karnataka, Kerala and Tamil Nadu. India: Andaman & Nicobar Islands, Arunachal Pradesh, Assam, Meghalaya, Mizoram, Sikkim, Tripura and West Bengal. World: Sri Lanka, Bhutan, Nepal, Bangladesh, Malay Archipelago and Philippines.

IUCN: VU-A1c. (IUCN, 2024).

Archidendron clypearia (Jack) I.C. Neilsen ssp. **clypearia.** Adansonia, n.s., 19: 15. 1979. Synonyms: *Inga clypearia* Jack, *Pithecellobium clypearia* (Jack) Benth., *Abarema clypearia* (Jack) Kostrem. Vern.: *Kaattu poopathumaram* (Tam.).

Unarmed trees to 12 m tall; bark smooth, greyish; young branchlets rusty tomentose. Leaves bipinnate, alternate pinnae 3-12 pairs, opposite, 2.3-6.5 cm, slender, pubescent, gland one half way between each pair of pinnae; leaflets 14-50, sessile, opposite, stipels absent; glands in between each pair of leaflets; pinnules 0.6-1 x 0.1 x 0.3-0.4 cm, oblong or oblong-trapezoid, base oblique, apex obtusely acute, mucronate; margin entire, recurved, glabrous above, villous beneath; lateral nerves obscure; rachis 5-16 cm long, slender, tomentose, pulvinate; glands at the base on upperside. Flowers bisexual, greenish-white, arranged in 6-flowered globular spike, in terminal panicles 2.5 cm across. Sepals campanulate, 1.2 ×1-1.5 mm, golden pubescent, 5 toothed. Petals funnel shaped, 4-7 mm, golden pubescent; lobes ovate to oblong. Stamens many, monadelphous at base, 7-10 mm long, exserted. Ovary stipitate, pubescent, 1-celled, ovules many; style filiform; stigma minute. Pod to 7-11 cm long, hispid, curved in to circle, orange within, dorsal suture coriaceous. Seeds 4-12, 5 mm across, black, compressed. Fl. & Fr.: April - March.

Western Ghats: Semi-evergreen to evergreen forests of Kerala and Tamil Nadu. India: Andaman & Nicobar Islands, Arunachal Pradesh, Assam, Meghalaya, Nagaland, Sikkim, Tripura and West Bengal. World: Nepal, Bangladesh, Mayanmar, China, Thailand, Philippines, and Cambodia.

IUCN: LC (IUCN, 2024).

Key to the subspecies

1a. Pinna 4-5 pairs; leaflets glabrous; pod grabrous; seeds with basal white aril .... ssp. **glypearia**

1b. Pinna 8-12 pairs; leaflets scabrous; pod pubescent; seeds with basal redish aril .... ssp. **subcoriaceus**

Archidendron clypearia (Jack) I.C. Neilsen ssp. **subcoriaceum** (Thw.) I.C. Nielsen., Opera Bot. 76: 57. 1984 publ. 1985. Synonyms: *Pithecellobium* *subcoriaceum* Thw., *Archidendron clypearia* var. *subcoriaceum* (Thw.) M. Gangop. & Chrakrab., *Abarema subcoriacea* (Thw.) Ksotrem., *Pithecellobium anamallayanum* Bedd.

Fl. & Fr.: October - December.

Western Ghats: Evergreen forests of Kerala and Tamil Nadu (Endemic).

IUCN: NE (IUCN, 2024).

Bauhinia L.

1a. Fertile stamens 10..2

1b. Fertile stamens 3–5..3

2a. Flowers white in colour, leaflets connate for two-third of their length .... **B. racemosa**

2b. Flowers yellow, leaflets connate for about half their length....................... .... **B. tomentosa**

3a. Wood without patches and streaks, flowers rose in color ..... .**B. purpurea**

3b. Wood with patches and streaks, flowers pink in color ....**B. variegata**

Bauhinia purpurea L., Sp. Pl.: 375. 1753. Synonyms: *Phanera purpurea* (L.) Benth., *Bauhinia coromandeliana* DC., *Bauhinia kurzii* Prain, *Bauhinia triandra* Roxb., *Phanera kurzii* (Prain) Thoth. Vern.: *Mantharai* (Tam.).

Trees to 10 m tall; bark ashy brown or dark brown, smooth; young branchlets pubescent. Leaves petiolate, rounded, deeply bilobed, 7.0-18 cm long, rather longer than broad, 9-11 nerved, cleft about halfway down into 2 acute or rounded lobes, minutely pubescent below when young; petiole 2.5-5 cm long. Inflorescence few flowered panicles at the ends of the branches. Flowers pedicellate; pedicel 5-13 mm long; tomentose; bract 3 mm long; bracteole 2 mm long. Hypanthium 7-10 mm long. Sepals lanceolate, lobes 2.5-3.0 cm long, usually splitting into two reflexed segments, one emarginate the other 3 toothed. Petals to 5 cm long, oblanceolate, long clawed, spreading, veined. Stamens usually 3, fertile, others reduced to antherless filaments. Ovary downy, long stalked; style long, stigma oblique. Pod 15-25 cm long, 1.5-2 cm broad. Seeds 12-15, almost round, 1.2-1.3 cm in diam, brown, smooth. Fl. & Fr.: October – January.

Western Ghats: Cultivated avenue trees in Maharashtra, Goa, Karnataka, Kerala and Tamil Nadu. India: Most of the states. World: Sri Lanka and South East Asia.

IUCN: LC (IUCN, 2024).

Bauhinia racemosa Lamk., Encycl. 1: 390. 1785. Synonyms: *Bauhinia parviflora* Vahl, *Pilostigma racemosum* (Lam.) Benth. *Pauletia racemosa* (Lam.) A. Schmitz., *Bauhinia timorana* Baker Vern.: *Aathi, Kalar Aathi, Kokkumandharai* (Tam.).

Deciduous trees to 10 m tall; bark grey to black, rough, thinly scaly, with numerous vertical cracks. Leaves simple, bilobed, alternate, broader than long, ovate-orbicular, 2-5 x 3.5-11 cm, base cordate, apex obtuse, bilobed and mucronate at cleft, margin entire, glabrous above, pubescent beneath; 7-9 nerves from the base, palmate, intercostae reticulate, prominent; petiole 10-33 mm long, pubescent. Flowers yellowish-white, 10-12 mm across, in terminal and leaf opposed few flowered racemes; bracts and bracteoles 2-3 toothed, minute, pubescent; pedicel 1-2 mm, slender. Sepals spathaceous, reflexed, 5-toothed at apex, pubersulous. Petals 5, 1.5 x 0.2 cm, linear-oblong. Stamens 10, all perfect; filaments equal, filiform, pilose at base; anthers versatile. Ovary half-inferior, stalked, pubescent, ovules many; style filiform; stigma peltate. Pod oblong, 15-22 x 1.5-2 cm, blackish-brown, apex horned indehiscent. Seeds 10-20, ovoid. Fl. & Fr.: March –December.

Western Ghats: Dry deciduous forests in Gujarat, Maharashtra, Goa, Karnataka, Kerala and Tamil Nadu. India: Most of the states. World: Sri Lanka, Bangladesh, Myanmar, China and Malay Archipelago.

IUCN: NE (IUCN, 2024).

Bauhinia tomentosa L., Sp. Pl.: 375. 1753. Synonyms: *Bauhinia pubescens* DC., *Alvesia tomentosa* (L.) Britton & Rose, *Pauletia tomentosa* (L.) A.Schmitz, *Bauhinia volkensii* Taub. Vern.: *Iruvatchi, Saerndhadum paavu* (Tam.).

Small tree to 3 m tall. Leaves petiolate, 2.5-5 cm long, 4.5-6.5 cm broad, bilobed, lobes rounded at apex; 7-9 nerved at the base, glabrous above, hairy below; petiole 2.5 cm long. Flowers axillary in pairs, sometimes 1-3, each on a pedicel bearing a pair of subulate persistent bracteoles; bracteole 3 mm long. Sepals 1.3-1.4 cm long, with 2 teeth at the apex in bud, splitting down on one side when the flower opens. Petals 5, 1.8-5 cm long, obovate, not clawed, yellow, conspicuously veined, one with a red blotch on the inner surface. Stamens 10. Ovary stipitate; style 1.8-1.9 cm long. Pods dehiscent, 6-10 seeded. Fl. & Fr.: September – January.

Western Ghats: Cultivated in gardens and avenues of Maharashtra, Goa, Karnataka, Kerala and Tamil Nadu. India: Most of the states. World: Sri Lanka, Bangladesh and Africa.

IUCN: LC (IUCN, 2024).

Bauhinia variegata L., Sp. Pl. 375. 1753. Synonyms: *Perlebia variegata* (L.) A.Schmitz, *Bauhinia variegata* var. *purpurascens* Voigt Vern.: *Sivappu mantharai* (Tam.).

Trees to 10 m tall; bark dark brown, smooth; young shoots pubescent. Leaves 4.5-15 cm long, as broad as or rather broader than long, bilobed, lobes obtuse, the base is deeply heart shaped; 9-15 nerved, pubescent beneath when young; petiole 2-3.8 cm lon. Inflorescence few flowered pubescent raceme; pedicel short or absent, bracteole minute. Hypanthium slender, 1.2-2.5 cm long. Sepals 2.0-2.7 cm long, tomentose, 5 toothed at the apex. Petals 5-6.3 cm long, obovate, with long rather broad claw, all white or 4 petals pale purple and fifth darker with purple veins. Stamens 5, fertile, no staminodes. Ovary hairy, stipe 10-17 mm long; style long, stigma capitate. Pods 15.0-30 cm long, 1-2.5 cm broad, hard, flat, dehiscent 10-15 seeded. Fl. & Fr.:

Western Ghats: All the states cultivated for its showy flowers. World: Bangladesh, China, Nepal and Pakistan.

IUCN: LC (IUCN, 2024).

Butea J. Koenig ex Roxb

Butea monosperma (Lamk.) Kuntze., Revis. Gen. Pl. 1: 202. 1891. Synonyms: *Erythrina monosperma* Lam., *Butea frondosa* J. Koenig ex Roxb., *Butea monosperma* var. *lutea* (Witt) Maheshw., *Butea frondosa* var. *lutea* Witt, *Butea monosperma* var. *rubra* Kuntze, *Butea monosperma* var. *flava* Kuntze. Vern.: *Muthuga* (Kan.); *Palas* (Mar.); *Brahmaviruksham, Chamatha, Mukkapuyam* (Mal.); *Porasu* (Tam.).

Deciduous trees to10 m tall; bark grey to greyish-brown, exudation red; branchlets densely tomentose. Leaves trifoliate, alternate; lateral leaflets 8.8-13.7 x 5.5-11 cm, broadly oblong-ovate or suborbicular, base oblique, apex obtuse, terminal leaflet 11-15 x 12.5-15 cm, widely rhomboid, base obtuse, apex emarginate, silky pubescent on both sides when young, glabrous above, silky pubescent beneath when mature, margin entire; lateral nerves 4-8 pairs, intercostae scalariform, prominent; stipules small, lateral, cauducous; rachis 12-20 cm long, stout, pubescent, pulvinate; petiolule 5-10 mm long, stout, pubescent. Flowers bright red, 5 cm long, in terminal or axillary, densely fascicled, racemes. Sepals broadly campanulate, teeth 5, deltoid, short, upper 2 connate, velvety. Petals 5, standard petal 5 x 2.5 cm, lanceolate, clawed, wings falcate 4.5 x 1.5 cm adnate to keel, keel united 4.5 x 3 cm, curved. Stamens 9

+ 1; vexillary stamens free; anthers uniform. Ovary 2.5 cm, inferior, 1-celled, ovules2; style long, incurved, beardless; stigma small. Pod oblong, 12.5-28 cm long, base flat, wing-like and indehiscent, the tip splitting round the apical seed. Seed obovate, compressed. Fl. & Fr.: December – June.

Western Ghats: Deciduous forests of Gujarat, Maharashtra, Goa, Karnataka, Kerala and Tamil Nadu. World: Nepal, Pakisttan, Myanmar, Laos, Cambodia, Vietnam, Thailand and Malesia.

IUCN: LC (IUCN, 2024).

Cassia L.

1a. Flowers yellow, leaflets 4–8 pairs .... **C. fistula**

1b. Flowers reddish to flame coloured, leaflets 10–15 pairs 2

2a. Racemes paniculate .... **C. javanica**

2b. Racemes corymbose .... **C. roxburghii**

Cassia fistula L., Sp. Pl.: 377. 1753. Synonyms: *Cassia rhombifolia* Roxb., *Cathartocarpus fistuloides* (Collad.) G. Don, *Cathartocarpus rhombifolius* (Roxb.) G. Don Vern.: *Kondrai, Manjakondrai, Sarakkondrai* (Tam.).

Trees to 20 m tall; bark smooth, ash coloured becoming rough and dark brown in old. Leaves compound 22.5-40 cm long, with 3-8 pairs of opposite leaflets, 6-10 cm long, smooth above, hairy below; Stipules deltoid, acute, 1-2 mm long, caducous; rachis 12-25 cm long, terete, glabrous. Flowers arranged in drooping racemes, each raceme 10-45 cm long; peduncle 2-10 cm long; pedicel 3.7-5.8 cm long, slender, slightly hairy or quite smooth, bracts 8-10 mm long, ovate, acute, hairy. Sepals 5, green, hairy, ovate, 9 mm long. Petals 5, obovate, distinctly veined. Stamens 10, 3 longest stamens much curled and bear large oblong anthers, 4 smaller medium ones are quite straight, 3 remaining stamens are quite short, erect and sterile. Ovary slender, thinly appressed hairy, style sturdy; stigma punctiform. Pods terete, glabrous, indehiscent, 40-60 cm long, 1.5-2 cm broad, black glossy brown, 40-100 seeded. Fl. & Fr.: March – November.

Western Ghats: Cultivated and escaped in deciduous forests of Maharashtra, Goa, Karnataka, Kerala and Tamil Nadu. India: Most of the states. World: Sri Lanka, Myanmar, Nepal, Bangladesh and Malesia.

IUCN: LC (IUCN, 2024).

Cassia javanica L., Sp. Pl.: 379. 1753. Synonyms: *Bactyrilobium javanica* (L.) Hornem., *Cathartocarpus javanicus* (L.) Pers. Vern.: *Seemai kondrai* (Tam.).

Trees to 15 m tall; bark rough, dark brown, longitudinally fissure. Leaves 15-40 cm; leaflets 5-12 pairs, 2-8 × 1.2-3.3 cm, both surfaces pilose or abaxially pilose, adaxially sparsely puberulent, base slightly asymmetric, apex acute, obtuse, or shortly acuminate. Inflorescences lateral on short side branches or terminal on young leafy shoots, paniculate. Sepals ovate or broadly lanceolate. Petals deep yellow or pink, 2.5-4.5 × 1-2 cm. Stamens 10, among them 3 abaxial antesepalous with filaments longer than other 7. Ovary pubescent. Pod terete, 30-50 × 0.2-1.5 cm, with annular nodes. Fl. & Fr.: April – June.

Western Ghats: Cultivated trees in Maharashtra, Karnataka, Kerala and Tamil Nadu. India: Most of the states. World: Native to Indonesia, Malesia and Australia.

IUCN: LC (IUCN, 2024).

Cassia roxburghii DC., Prodr. 2: 489. 1825. Synonyms: *Cassia marginata* Roxb., *Cathartocarpus marginatus* G.Don, *Cathartocarpus roxburghii* (DC.) G.Don Vern.: *Sengkondrai* (Tam.).

Trees to 15 m tall; bark brown, rough; branchlets velvety pubescent. Leaves paripinnate, alternate, leaflets 16-30, subopposite, pinnule 1.5-4.5 x 1-2 cm, elliptic-oblong or ovate, base oblique or subacute, apex obtuse, retuse or mucronate, margin entire, glabrous above, puberulent beneath; lateral nerves10-16 pairs, slender, faint, intercostae reticulate; rachis 14-25 cm long, slender, pulvinate, pubescent; stipules ovate; petiolule to 2 mm, puberulent. Flowers flame coloured, 2.5 cm across, in axillary corymbose racemes to 8 cm; pedicels to 1 cm. Sepals 5, tube campanulate, lobes 5, 4-7 mm long, ovate, obtuse. Petals 5, 1.5-2 cm long, pink to rose, ovate- oblong, concave, pubescent without, clawed. Stamens 10, the lower 3 filaments 2.5-3 cm long, curved; others shorter; upper 3 short with reduced filaments and anthers; 4-medium with erect filaments. Ovary stipitate, 2 cm long, pubescent, ovules many; style to 6 mm. Pod oblong, 20-30 x 2 cm, terete, torulose, black, indehiscent. Seeds many, transverse, imbedded in pith. Fl. & Fr.: July – December.

Western Ghats: Planted trees in Maharashtra, Karnataka, Kerala and Tamil Nadu. India: Most of the states. World: Sri Lanka.

IUCN: NE (IUCN, 2024).

Cynometra L.

1a. Leaves uniformly 2-foliolate, sometimes with 1 or 2 stipule-like much reduced leaflets below the normal pair of leaflets .... **C. travancorica**

1b. Leaves more than 2-foliolate, if 2-foliolate then intermixedwith 4–6-foliolate leaves.. 2

2a. Petals 2–5 mm long, if up to 5 mm, then leaves never more than 4-foliolate; filaments usually less than 7 mm long, if more, then style 0.5–3.5 mm long 3

2b. Petals 6–8 mm long; filaments more than 8 mm long; style4–8 mm long. .... **C. beddomei**

3a. Sepals reflexed with tips curled up; distal part of pods with beak......... 4

3b. Sepals not as above; distal part of pods not as above....... **C. bourdillonii**

4a. Petals less than 2 mm long; ovary wall hairy inside; podsobovoid, laterally compressed and beaked, deeply rugose, scurfy, rusty brown..... .... **C. iripa**

4b. Petals up to 5 mm long; ovary wall glabrous inside; podsboat-shaped, laterally compressed, obscurely beaked at the tip .... **C. sampathkumaraniana**

Cynometra beddomei Prain., J. Asiat. Soc. Bengal, Pt. 2, Nat. Hist. 66: 478. 1897.

Evergreen trees to 20 m tall; bark blackish-green. Leaves paripinnate, alternate, leaflets 4-6, opposite, 2.5-5 x 1-1.8 cm, obliquely obovate-oblong or obovate, base oblique, acute or cuneate, apex obtusely acuminate, emarginate, margin entire, glabrous; lateral nerves 6-8 pairs, faint, intercostae reticulate; stipules free, lateral, cauducous; rachis 30-40 mm, slender, pulvinate, grooved above, pubescent. Flowers creamy-white, in axillary clusters from conspicuous imbricate bracts; peduncle 12-20 mm, densely-patent-hairy; bracts ovate, appressed, hairy, ciliate, striate, imbricate; pedicel 5 mm, pubescent; receptacle 1-2 mm deep, campanulate, circum scissile under the ripening fruit. Sepals 3 mm long, hairy, ciliate. Petals 5, 3.5 x 1 mm, free, oblanceolate, subequal, glabrous. Disc 0; stamens 10, alternately 5 and 7 mm; filaments filiform, connective cleft at base, apex apiculate; anthers versatile. Ovary half inferior, densely long brown hairy; style 2.5 mm, slender, hairy upto half way; stigma capitate. Pod reniform-globose, grooved near sutures, indehiscent. Seed one. Fl. & Fr.: April – Ausugt.

Western Ghats: Evergreen forests of Kerala (Endemic).

IUCN: EN-D (IUCN, 2024).

Cynometra bourdillonii Gamble., Bull. Misc. Inform. Kew 1908: 446. 1908.

Evergreen trees to 20 m tall; bark blackish-green. Leaves paripinnate, alternate, leaflets 4-6, opposite, pinnules 2.5-5 x 1-1.8 cm, obliquely obovate-oblong or obovate, base oblique, acute or cuneate, apex obtusely acuminate, emarginate, margin entire, glabrous; lateral nerves 6-8 pairs, secondary laterals present, faint, intercostae reticulate; stipules free, lateral, cauducous; petiolule 1-2 mm; rachis 30-40 mm, slender, pulvinate, grooved above, pubescent. Flowers creamy-white, in axillary clusters from conspicuous imbricate bracts; peduncle 12-20 mm, densely-patent-hairy; bracts ovate, appressed, hairy; pedicel 5 mm long, pubescent; receptacle 1-2 mm deep, campanulate. Sepals 3 mm long, hairy, ciliate. Petals 5, 3.5 x 1 mm, free, oblanceolate, subequal, glabrous. Disc absent. stamens 10, alternately 5 and 7 mm; filaments filiform, connective cleft at base, apex apiculate; anthers versatile. Ovary half inferior, densely long brown hairy; stipe 0.8 mm long, exocentric; style 2.5 mm, slender, hairy upto half way; stigma capitate. Pod reniform-globose, grooved near sutures, indehiscent. Seed one. Fl. & Fr.: November – January.

Western Ghats: Evergreen forestrs above 1300 m in Karnataka and Kerala (Endemic).

IUCN: EN- B1+2c. (IUCN, 2024).

Cynometra iripa Kostel., Allg. Med.-Pharm. Fl. 4: 1341. 1835. Synonyms: *Cynometra mimosoides* (Baker) Wall. ex Prain, *Cynometra ramiflora* var. *mimosoides* Baker

Small trees to 6 m tall; bark finely fissured, lenticelate, pale. Leaves bi-jugate, leaflets discolorous, asymmetrical, obliquely-elliptical, obliquelyoblong, obliquely-obovate or obliquely-ovate; basal pair 8.5–50 x 5–30 mm; terminal pair 23–85 x 11–50 mm, coriaceous, glabrous, base oblique, cuneate, attenuate or obtuse, apex obtuse and emarginate, margin entire; secondary veins 6–10 pairs, tertiary venation reticulate; petioles 1–4 mm long, minutely pubescent; rachis 9–30 mm long, channelled on the upperside, puberulent; petiolules up to 3 mm long, wrinkled, mostly enclosed by leaflet base, puberulent. Inflorescences axillary, terminal or ramiflorous, solitary racemes or a fascicle of racemes on a swelling, racemes 3–9-flowered; bracteoles 1 or 2 towards pedicel base, filiform, 1.5–3 mm long, sparsely hairy, caducous. Flowers fragrant, erect, 6.5–8.5 x 5.5–8 mm wide; receptacle inverted cone-shaped, 1.5 x 1.5 mm. Sepals 4, unequal in width, lanceolate or oblong-ovate, becoming reflexed and often incurved at apex, white or very pale pink, margin at apex ciliate, glabrous or with few sparse hairs on abaxial surface. Petals 5, white or very pale pink, lanceolate, 2–3 x 0.5–0.7 mm, glabrous, caducous. Stamens 10; filaments terete, 4–7 mm long, straight or curved, glabrous; anthers 1 x 1 mm, brown. Ovary inserted excentrically on a short stipe, 2.2 x 1.2 mm, pink, white pubescent externally, appressed-pubescent internally; ovules 1; style slender, excentric to varying degrees, 2–4 mm long, sparsely hairy from median section to base; stigma capitate. Pod to 6–10 mm long, tip with a distinct beak at apex of dorsal suture and partway along dorsal side, suborbicular, laterally compressed, deeply rugose, glabrescent, brown. Seeds 1-2, 25–29 x 17–20 mm. Fl. & Fr.: January – May.

Western Ghats: Rare in evergreen forests of Maharashtra, Goa, Karnataka and Kerala. World: Sri Lanka, Bangladesh, Myanmar, Indo-China, Malesia and Australia.

IUCN: LC (IUCN, 2024).

Cynometra sampathkumaraniana Sanjappa, Sringesw. & Dalavi., Nelumbo. 64: 2. 2022

Evergreen trees 10 m tall; bark smooth, brownish-grey, on young twigs papery, shining brown, cracked and flaking; branchlets slender, glabrous, lenticellate. Leaves alternate, 4-foliolate with 2 pairs of opposite leaflets, the terminal pair of leaflets much larger sometimes mixed with 2-foliolate and 3-foliolate; leaflets light green or reddish-green when young, coriaceous, glabrous, elliptic, elliptic-lanceolate to somewhat obovate or obovate-oblong, obtuse at apex, entire, obscurely reticulate-veined; lateral nerves 7–9 pairs, irregularly connected in loops near the margin; petiole dark brown, transversely wrinkled like pulvinus; stipules filiform, 5–8 mm long, early caducous. Racemes axillary and cauliflorous on stems and old branches, often 3–5-fascicled; racemes 1–1.5 cm, long; rachis 7–10 mm long; bracts broadly ovate, striate, brownish, ciliate at margin, caducous after anthesis; pedicels slender, 3–4 mm long, minutely pubescent. Hypanthium 1–1.5 mm long, campanulate, ircumscissile under developing pod. Sepals creamy white, 4, free, thin, reflexed in open flower, sometimes tips curved upwards, 3.5–5 x 1.5–2.5 mm, elliptic-oblong with one of them ovate-oblong, abruptly obscurely apiculate, glabrous. Petals white, 5, free, erect, linear-oblong, 2.5–5 x 1–1.5 mm, thin, abruptly short acuminate, caducous. Stamens 10, filaments slender, glabrous, 7–9 mm long; anthers medifixed, dark brownish, 1–1.5 mm long. Ovary 2.5–3 x 1.5–2 mm, villous, 1-ovuled; stipe 1–1.5 mm long, pubescent; style 4.5–5 mm long, slender, faintly curved, glabrous; stigma capitate. Pod smooth, boat-shaped, 2 x 1 cm, laterally compressed, green becoming brown, finely pubescent, obscurely wrinkled, obliquely rounded at base, abruptly narrowed into a blunt beak. Seeds not seen. Fl. & Fr.: May – December.

Western Ghats: Rare in evergreen forests of Karnataka (Endemic).

IUCN: NE (IUCN, 2024).

Cynometra travancorica Bedd., Fl. Sylv. S. India: t. 316. 1873.

Evergreen trees to 20 m tall; bark reddish-brown, smooth, lenticellate. Leaves alternate, 2-foliate, leaflet 8-12 x 2-3 cm, often falcate, ovate-oblong, oblong or lanceolate, base oblique, apex acute or acuminate, margin entire, glabrous, coriaceous; lateral nerves 6-8 pairs, intercostae reticulate, prominent; rachis 3-8 mm, stout, pulvinate, glabrous; petiolule 2-3 mm, opposite. Flowers rose-white, 6-8 mm across, in short dense flowered axillary racemes; peduncle puberulous; bracts ovate, 1-3 mm long, puberulous and ciliate; bracteole ovate, 2 mm long, very narrow, with only few hairs at the tip; pedicel 2-4.5 mm, slender, pubescent. Sepals 4, 3 mm long, oblong, reflexed, glabrous. Petals 5, 4 mm long, subequal, clawed. Stamens 10, alternate long and short, connective cleft at base, very slightly apiculate; filaments filiform; anthers small, versatile. Ovary half inferior, semi-lunar, glabrous; stipe 0.8 mm; style 1 mm; stigma capitate. Pod oblanceolate, 2.5-3 cm long, smooth, rugose along the margin, semicircular, indehiscent. Seed one. Fl. & Fr.: September – May.

Western Ghats: Evergreen forests of Karnataka, Kerala and Tamil Nadu (Endemic).

IUCN: EN- B1+2c. (IUCN, 2024).

Dalbergia L.f.

1a. Stamens monadelphous, usually 9, united basally a sheath 2
1b. Stamens diadelphous, usally 9 +1, clefted deeply 5
2a. Leaflets 3-7, dark green when young; flower axullary .... **D. latifolia**
2b. Leaflets 7-10, pale green when young; flowers terminal 3
3a. Leaflets orbiculate with apiculate tip .... **D. sissoo**
3b. Leaflets elliptic-lanceolate or obovate with acuminate tip 4
4a. Branchlets with thorns; flowers pure white .... **D. melanoxylon**
4b. Branchlets without thorns; flweors with purplish streaks ... **D. sissoides**
5a. Leaflets 10-17, glabrous; flowers axillary .... **D. lanceolaria**
5b. Leaflets 17-21, pubescent; flowers terminal .... **D. sericea**

Dalbergia lanceolaria L.f., Suppl. Pl.: 316. 1782 ssp. **lanceolaria.** Synonyms: *Dalbergia frondosa* Roxb., *Dalbergia bhutanica* Thoth., *Dalbergia zeylanica* Roxb. Vern.:*Erigei, Nalvellangu* (Tam.).

Large trees to 20 m tall; bark smooth, gryish white. Leaves 15-25 cm long; leaflets 13 or 15, to 4 x 2 cm, oblong-elliptic, apex emarginate, slightly coriaceous; petiole to 2 cm long; petiolule 3-5 mm long; stipules small, deciduous. Panicles terminal and axillary, 5-15 cm long. Pod oblong-lanceolate, 8 x 2 cm, glabrous, stipitate, coriaceous. acute at both ends. seeds 1-3. Fl. & Fr.: March – June.

Western Ghats: Deciduous forests above 800 m in Gujarat, Maharashtra, Goa, Karnataka, Kerala and Tamil Nadu. India: Most of the states. World: Sri Lanka, Bangladesh, Pakistan, Nepal, Bhutan, Myanmar, Thailand, Vietnam, Cambodia and Indonesia.

IUCN: LC (IUCN, 2024).

Key to the subspecies

1a. Inflorescence much branched panicle; pod with 1 or 2 seeded ....ssp. **paniculata**

1b. Inflorescence effuse panicle; pod to 3 seeded .... ssp. **lanceolaria**

Dalbergia laceolaria L.f., ssp. **paniculata** (Roxb.) Thoth. Bull. Bot. Surv. India 25: 171. 1985. Synonyms: *Dalbergia paniculata* Roxb., *Dalbergia hemsleyi* Prain, *Da lbergia lanceolaria* var. *hemsleyi* (Prain) Thoth. Vern.: *Iravu, Naaivellaangu, Pandrivaragi, Velluluvai* (Tam.).

Small trees to 15 m tall; branchlets pubescent. Leaflets alternate, 5 or 6 pairs, 3-4 x 2.5 mm, oblong-obovate, apex obtuse, base cuneate; petiole 4 mm long. Panicles terminal or axillary on tender shoot, pubescent. Flowers to 1 cm across. Sepals tubular, tube 4 mm long, rusty tomentose. Corolla white, standard 1 x 0.5 mm, ovate, oblong, wings and keels to 8 x 3 mm. Stamens 10, diadelphous. Ovary 3 mm diam. Pod to 7 x 11.5 cm, lanceolate, acute. Seeds 1 or 2. Fl. & Fr.: February – May.

Western Ghats: Deciduous forests of Gujarat, Maharashtra, Goa, Karnataka, Kerala and Tamil Nadu. India: Most of the states. World: Nepal, Myanmar, Laos, Cambodia and Vietnam.

IUCN: LC (IUCN, 2024).

Dalbergia latifolia Roxb., Pl. Coromandel 2: 7. 1799. Synonyms: *Dalbergia emarginata* Roxb., *Dalbergia javanica* Miq. Vern.: *Illuvadi, Eeti maram, Thodhakathi* (Tam.).

Deciduous trees to 25 m tall; bark grey, smooth with short irregular cracks, exfoliating in thin fibrous longitudinal flakes. Leaves imparipinnate, rarely 1-foliate, alternate, leaflets 3-9, petioluleupto 10 mm, stout, grooved above, glabrous; 2.5-10 x 2.5-10 cm, orbicular, base acute or very rarely truncate, apex obtuse or emarginate, margin entire, glabrous, glaucous beneath, coriaceous; lateral nerves 5-10 pairs, intercostae reticulate, faint; rachis 7.5-20.5 cm, slender, pulvinate, glabrous. Flowers white, 5-6 mm long, in short corymbose panicles from the axils of fallen leaves of previous years shoot; bracteoles small, deciduous. Sepals tubular, tube campanulate; lobes 5, subequal, upper 2 connate, lateral obtuse, lower acute. Petals 5, standard petal obovate, cuneate at base, reflexed, wings 6 x 2.5 mm, clawed, base auricled, keel 6 mm long, hastate, united, clawed. Stamens 9, monadelphous; staminal tube 4 mm, split open one side. Ovary stipitate, 5-7 mm, glabrous, 1-celled; ovules 1-few; style slender, incurved, glabrous; stigma capitate. Pod oblong-lanceolate, 5-7.5 x 1.5-1.8 cm, tip obtuse, distinctly marked in the place of seeds, but not wrinkled, indehiscent. Seeds 1-4, reniform, brown. Fl. & Fr.: August – December.

Western Ghats: Deciduous forests to semi-evergreen forests of Maharashtra, Goa, Karnataka, Kerala and Tamil Nadu. India: Most of the states except North-east. World: Nepal, Pakistan, Indonesia and Malaysia.

IUCN: VU- A1cd. (IUCN, 2024).

Dalbergia melanoxylon Guill & Perr., Fl. Seneg. Tent.: 227. 1832. Synonyms: *Amerimnon melanoxylon* (Guill. & Perr.) Kuntze, *Amerimnon stocksii* (Benth.) Kuntze, *Dalbergia stocksii* Benth.

Small trees to 15 m tall; bark grey with spiny shoots. Leaves alternate, 6–22 cm long, pinnately compound, with 8-13 alternately arranged leaflets; leaflets 1.2-3.6 x 1.2-1.8 cm, obovate, broadly notched at the tip, or almost flat, more or less rounded at the base; stalks slender, 2 mm long. Flowers white, in dense clusters. Stamens 9, united or variously divided. Pod linear-oblong, 3-7 cm long, containing one to two seeds. Fl. & Fr.: April – June.

Western Ghats: Deciduous to semi-evergreen foersts of Maharashtra and Tamil Nadu. India: Uttar Pradesh and West Bengal. World: Native to Tropical Africa.

IUCN: NT (Low risk/near threatened) (IUCN, 2024).

Dalbergia sericea G.Don., Gen. Hist. 2: 375 (1832), *nom. cons.* Synonyms: *Dalbergia stenocarpa* Kurz, *Dalbergia hircina* Buch.-Ham. ex Benth.

Trees to 5-6 m tall; young branchlets silky puberulent, gradually glabrous. Leaves alternate, 15-25 cm; leaflets 17-21, ovate or oblong, 2.5-5 x 2-2.5 cm, base rounded or cuneate, apex slightly obtuse stipules caducous, subfalcate-lanceolate, silky puberulent. Panicles axillary, 2.5-5 x 1.2-2 cm; peduncles branches; pedicels densely silky puberulent. Sepals campanulate, densely puberulent, 5-toothed; teeth obtuse, shorter than tube except lowest lanceolate. Petals white, rather long clawed; standard suborbicular, emarginate. Stamens 10, diadelphous. Ovary shortly stipitate, pubescent, 4-5-ovuled; style slender; stigma small. Pod narrowly ligulate, 2.5-5 x 0.6-1 cm, reticulate opposite 1-3 seeds, base attenuate to short stipe, apex acute. Seeds reniform, compressed, 5 x 2.5 mm. Fl. & Fr.: February – September.

Western Ghats: Deciduous forests of Kerala and Tamil Nadu. World: Nepal, Bhutan and China.

IUCN: LC (IUCN, 2024).

Dalbergia sissoides Graham ex Wight & Arn., Prodr. Fl. Ind. Orient. 1: 265. 1834. Synonym: *Dalbergia latifolia* var. *sissoides* (Graham ex Wight &Arn.) Baker Vern.: *Thodhagathi, Eeti, Vel-eeti, Iruvadi* (Tam.).

Deciduous trees to 30 m tall; bark grey to pale brown, smooth, exfoliations small, irregular. Leaves imparipinnate, alternate, leaflets 5-11, 3-7.5 x 2-6 cm, elliptic-obovate, orbicular or obovate, apex acute, base cuneate or acute, obtuse or retuse, margin entire, membranous; stipules small, lateral, cauducous; lateral nerves 4-12 pairs, intercostae reticulate, prominent; rachis 5-15 cm, pulvinate, glabrous. Flowers white, 7-8 mm long, in loose terminal panicles. Sepals campanulate, glabrous; lobes 5, subequal. Petals 5; standard cuneate at base, reflexed; keel petals hastate. Stamens monadelphous; filaments alternately longer and shorter. Ovary stipitate, inferior, 1-celled; ovule 1-few; style slender; stigma capitate. Pod oblong, 5-7.5 x 1-1.2 cm, tapering to both ends, stipitate. Seeds 2-4, reniform, pale brown. Fl. &Fr.: February – November.

Western Ghats: Cultivated avenue trees in Karnataka, Kerala and Tamil Nadu. India: Andhra Pradesh and Telangana. World: Indonesia.

IUCN: VU- A4bcd. (IUCN, 2024).

Dalbergia sissoo Roxb. ex DC., Prodr. 2: 416. 1825. Synonyms: *Amerimnon sissoo* (Roxb.) Kuntze, *Pterocarpus sissoo* (Roxb. ex DC.) Wight & Arn., *Endespermum diversifolium* Blume ex Miq. Vern.: *Eeti, Sisu maram* (Tam.).

Trees to 10 m tall; bark greyish or pale brown, thick. Leaves imparipinnate; leaflets 3-5, 2.5-6 x 2-5.5 cm, ovate, obovate or orbicular, base rounded to cuneate, apex cuspidate or acuminate, coriaceous, puberulous to pubescent when young, glabrous when mature; petiolules 3-5 mm long. Panicles axillary or terminal, 5-7 cm long, densely pubescent. Flowers cream coloured, 6-8 mm long, subsessile. Calyx 4-6 mm long, campanulate, puberulous without; lobes oblong. Petals exserted; standard to 8 x 5 mm, obovate, long-clawed. Stamens 9, monadelphous. Ovary long-stipitate, pubescent. Pod linear-oblong, 2.5-7 x 0.8-1 cm, base cuneate, apex acute. Seeds 1-3. Fl. & Fr.: February – October.

Western Ghats: commonly cultivated in low levels and plains of Maharashtra, Goa, Karnataka, Kerala and Tamil Nadu. India: Most of the states. World: Bangladesh, Bhutan, Myanmar, Nepal, Afghanistan, Pakistan, Iran and Iraq.

IUCN: LC (IUCN, 2024).

Dialium L.

Dialium travancoricum Bedd., Indian Forester 30: 243. 1904. Vern.: *Malam puli* (Mal.).

Evergreen trees to 30 m tall; bark pale brown, smooth, exudation red, scanty. Leaves imparipinnate, alternate; leaflets 5-11, subopposite or alternate, 5-7.6 x 1.8-3.2 cm, ovate, ovate-lanceolate or elliptic-lanceolate, base round or acute, apex caudate-acuminate, margin entire, glabrous; lateral nerves 11-14 pairs, intercostae reticulate; stipules small, lateral; rachis 10-15 cm, slender, glabrous, pulvinate. Flowers 2-3 mm long, brown, in axillary or terminal fulvous-pubescent panicles; bracts and bracteoles small, cauducous. Sepals tubular, tube short, divided to base; lobes 5, petalloid, imbricate. Petals absent. Stamens 2, free; filaments short; anthers erect, oblong; ovary half inferior, sessile, oblique, hairy; ovules 2, style subulate; stigma small, terminal. Pod ovoid, 2-2.2 cm broad, globose, dark velvety-brown, indehiscent, red. Seed one, more or less compressed, pale brown, thin. Fl. & Fr.: September – October.

Western Ghats: Evergreen forests of above 1300 m in Gujarat, Maharashtra, Goa, Karnataka, Kerala and Tamil Nadu (Endemic).

IUCN: CR- B1ab(i,ii,iii)+2ab(i,ii,iii); D. (IUCN, 2024).

Dichrostachys (DC.) Wight & Arn.

Dichrostachys cinerea (L.) Wight & Arn., Prodr. Fl. Ind. Orient. 1: 271. 1834. var. **cinerea** Synonyms: *Mimosa cinerea* L., *Acacia cinerea* (L.) Spreng., *Cailliea cinerea* (L.) J.F.Macbr., *Desmanthus cinereus* (L.) Willd., *Dichrostachys cinerea* ssp. *cinerea, Dichrostachys santapaui* Sebastine & Ramam. Vern.: *Vidathalan, Vedathala* (Tam.).

Small shrubby trees to 6 m tall; bark with vertical fissure, peeling off, whtish; branches usually densely pubescent when young. Leaves bipinnate, glands present on the rachis, sessile or stipitate; pinnae 1-2 cm long, leaflets 7-19 pairs, 2-5 mm, sometimes up to 9 mm long, obtuse, subacute or acute at the apex, margins hairy. Inflorescence fascicled spike; peduncles hairy. Flowers polygamous, 2 mm across, in solitary or axillary paired spikes; upper flowers bisexual, yellow; lower ones neutar, pink or purple. Sepals campanulate, lobes 5. Petals 5, lanceolate connate below, pubescent, valvate. Stamens 10, free, exserted, anthers ending in stalked glands. Ovary 0.5 mm long, subssessile, pubescent; style 3 mm; stigma terminal, truncate. Pod linear, flat, coiled, pubescent, indehiscent or opening from apex. Seeds 4-6, ovoid, compressed. Fl. & Fr.: August - January.

Western Ghats: Common in foot hills and scrub forests of Maharashtra, Goa, Karnataka, Kerala and Tamil Nadu. India: Andhra Pradesh, Delhi, Madhya Pradesh, Odisha, Punjab, Rajasthan and Uttar Pradesh. World: Sri Lanka, Malay Archipelago, Northern Australia and Tropical Africa.

IUCN: LC (IUCN, 2024).

Key to the varieties

1a. Leaves and branchlets densely pubescent; pods pubescent .... var. **cinerea**

1b. Leaves and branchlets glabrous or glabrescent; pods glabrous .... var. **indica**

Dichrostachys cinerea (L.) Wight &Arn var. **indica** Brenan &Brummitt. Bol. Soc. Brot., sér. 2, 39: 111. 1965.

Fl. & Fr.: August – December.

Western Ghats: Maharashtra (Endemic).

IUCN: NE (IUCN, 2024).

Uses: Bark and leaves are used for medicinal purposes.

Erythrina R. Geesink

1a. Calyx spathaceous, split on one side 2

1b. Calyx not spathaceous, split into two lips 4

2a. Calyx spilt up to the base; petals sub-equal; pod 6-8-seeded .... **E. variegata**

2b. Calyx split half way down, petals equal; pod 2-5-seeded 3

3a. Leaflets broadly deltoid ovate, membranous .... **E. stricta**

3b. Leaflets broadly ovate, coriaceous .... **E. subumbrans**

4a. Brickles on branches yellowish-white; leaflets rhomboid ovate, densely tomentose beneath .... **E. suberosa**

4b. Brickles on branches brown to black or absent; leaflets broadly ovate-deltoid, glabrous 5

5a. Pod sub-cylindric, 8-10 seeded; branchlets without brickles ....**E. fusca**

5b. Pod laterally compressed, 5-8-seeded; branchlets densely brickled .... **E. arborescens**

Erythrina arborescens Roxb., Pl. Coromandel 3: 14. 1811.

Trees to 15 m tall; bark rough, brownish, with prickles. Leaves compound, trifoliolate; terminal leaflet reniform, 8-20 x 4-10 cm, both surfaces glabrous, base truncate or almost cordate, margin entire, apex acute; lateral leaflets obliquely broadly cordate; petiole longer than leaflets, without or with a few prickles. Raceme solitary, in axil of terminal leaf, erect, longer than leaves. Flowers in clusters of 3, red, pendulous; bract 1, ovate. Sepals turbinate, truncate or unequally 2-lobed. Petals red; standard subovate, boat-shaped, 3.2 cm; wings obliquely ovate, shorter than keels; keels half as long as stamens. Filament slightly shorter than standard, connate near base. Ovary long stipitate; style as long as stamens. Pod curved, 12-19 × 2-3 cm, with obvious beak and stipe. Seeds 5-10, white or brown, reniform. Fl. & Fr.: March.

Western Ghats: Deciduous forests of Maharashtra. World: Myanmar, Nepal, Bhutan, Bangladesh, china and Thailand.

IUCN: NE (IUCN, 2024).

Erythrina fusca Lour., Fl. Cochinch.: 427. 1790. Synonyms: *Erythrina glauca* Willd., *Erythrina ovalifolia* Roxb., *Erythrina argentea* Blume ex Miq., *Erythrina ovalifolia* Roxb., *Erythrina patens* Moc. & Sessé ex DC.

Medium branched deciduous trees to 10 m tall; bark pale brown, thick and fluted, with scattered brikles. Leaves trifoliate, alternate; leaflets 8-15 x 4.5-8 cm, elliptic or ovate, base rounded, apex obtuse or rounded, subcoriaceous, lateral nerves 6-8 pairs; stipels 1 mm long, persistent. Flowers crimson, mostly 3-nate on the ferruginous hairy rachis of the racemes; bracts and bracteoles

2.5 x 2 mm, ovate; pedicels 8-15 mm long. Sepals campanulate, 1.4-1.7 cm long, ferruginous pubescent, often irregularly splitting. Standard 3.5-4.5 long, rounded-rhomboid, the claw 1.5 cm long; wings 2 x 1.3 cm, obliquely obovate; keels 3 x 1.5 cm, ovate-falcate. Ovary long-stipitate; ovules many. Pods linear, 8-15 x 1.2-1.8 cm, torulose, beaked, adpressed hairy, 5-8-seeded. Seeds 1-1.5 x 0.5-0.8 cm, oblong-ellipsoid, blackish. Fl. & Fr.: January – May.

Western Ghats: Cultivated trees in tea estates of Maharashtra and Kerala. World: Sri Lanka, Myanmar, Bangladesh, Malesia, Australia, Afica and America.

IUCN: LC (IUCN, 2024).

Erythrina stricta Roxb., Fl. Ind., ed. 1832. 3: 251. 1832. Synonyms: *Corallodendron strictum* (Roxb.) Kuntze, *Erythrina stricta* var. *yunnanensis* (H.T.Tsai & T.T.Yu ex S.K.Lee) R.Sha Vern.: *Muruku* (Tam.).

Trees to 15 m tall; bark yellowish, vertically fissured, armed with bossed prickles; branchlets apically stellate-pubescent, basally glabrescent, densely prickled. Leaves trifoliate, alternate, leaflets 5-12.5 x 3.7-10 cm, deltoid, truncate or obtuse, apex subacute or acuminate, margin entire, glabrous, membranous; nerves 3-7 from base, palmate, lateral nerves 5-7, pinnate, prominent, intercostae reticulate, prominent; stipules small, lateral; rachis 10-13 cm long, slender, glabrous, pulvinate, prickled or not. Flowers red, 5 x 1.8 cm, clustered on the brown-tomentose rachis, in terminal racemes; bracts ovate; bracteole 3 mm; pedicel 3 in a cluster, to 8 mm. Sepals spathaceous, 1-1.5 cm, split half way down, glabrous, erect. Petals 5, deep red; standard 5 x 2.5 cm, oblong-lanceolate; wings 5.5 x 3 mm, obovate; keel 2 x 0.7 cm, ovate. Stamens10, monadelphous, vexillary filament free; staminal sheath 2.5 cm; filaments 1 and 1.5 cm long; anthers uniform. Ovary inferior, stipitate, 2 cm, pubescent, 1-celled; ovules many; style 1.5 cm, suberect; stigma capitate. Pod lanceolate, 5-10 cm long. Seeds 3-6, black. Fl. & Fr.: January – June.

Western Ghats: Deciduous forests of Maharashtra, Goa, Karnataka, Kerala and Tamil Nadu. India: Most of the states. World: Bhutan, Nepal, Bangladesh, China, Tibet, Myanmar, Laos, Cambodia, Vietnam and Thailand.

IUCN: NE (IUCN, 2024).

Erythrina suberosa Roxb., Fl. Ind., ed. 1832. 3: 253. 1832. Synonyms: *Erythrina stricta* var. *suberosa* (Roxb.) Niyomdham, *Erythrina alba* Roxb. ex Wight & Arn.,

Erythrina glabrescens (Prain) R.Parker, *Erythrina maxima* Roxb. ex Wight & Arn., *Erythrina suberosa* var. *glabrescens* Prain, *Erythrina sublobata* Roxb., *Erythrina suberosa* var. *sublobata* Roxb. ex Haines Vern.: *Mullumurukku* (Tam.).

Trees to 17 m tal; branchlets armed with prickles. Leaf sometimes armed, trifoliolate, petiole 10-20 cm long; leaflets 7.5-15 cm long and broad, the terminal rhomboid ovate, the lateral obliquely deltoid, glabrous above, densely pubescent below; stipels glanduliform; stipules linear-lanceolate, caducous. Racemes to 14 cm long, near the tip of branches; peduncle short. Flowers in fascicles of 1-3; pedicel 12-13 mm long; bracts caducous, bracteoles minute. Sepals 6-10 mm long, pubescent, becoming bilabiate. Petals scarlet vexillum 3.7-5.0 cm long, 12-16 mm broad. Pod stipitate, 12.5-15.0 cm long, torulose, 2-5-seeded. Fl. & Fr.: April – October.

Western Ghats: Cultivated trees in tea and coffee estates of Gujarat, Maharashtra, Goa, Karnataka, Kerala and Tamil Nadu. India: Most of the states. World: Sri Lanka, Nepal, Bhutan, Myanmar, China, Thailand, Vietnam, Malaysia and Cambodia.

IUCN: NE (IUCN, 2024).

Erythrina subumbrans (Hassk.) Merr., Philipp. J. Sci., C 5: 113. 1910. Synonyms: *Hypophorus subumbrans* Hassk., *Erythrina lithosperma* Blume ex Miq., *Erythrina mysorensis* Gamble., *Erythrina lithosperma* var. *armata* Miq., *Erythrina lithosperma* var. *inermis* Miq. Vern.: *Seemai Kalyanamurungai, Mullu-murungai* (Tam.).

Trees to 15 m tall; branchlets terete, glabrous. Leaves trifoliate, alternate, leaflets 8-16 x 3.8-10.5, terminal leaflets broadly elliptic-obovate or rhomboid, lateral leaflets elliptic-ovate or ovate, base obtuse, apex acuminate, margin entire, glabrous; 3-ribbed from the base, lateral nerves 5-9 pairs, intercostae scalariform; rachis 10.5-17.5 cm long, slender, glabrous, pulvinate; petiolule to 10 mm long, opposite, slender, glabrous. Flowers orange-red, 3 cm long, in terminal pyramidal racemes to 10 cm long; bracts and bracteoles lanceolate. Sepals erect, split halfway down; tube 4 mm, 2-lipped; lobes 4 mm. Standard petals 2.5-3.5 cm long, oblanceolate; wings and keel about 1.5 cm long, keel petals free. Stamens 9 +1, diadelphous, sheath 1.5 cm, alternately long and short; anthers uniform. Ovary half inferior, stipitate, ovules many; style curved, subulate; stigma capitate. Pod narrow-oblong, 8-12 x 1.5-2 cm, turgid, proximate half seedless, distal half with 1-5 seeds. Fl. & Fr.: August – May.

Western Ghats: commonly cultivated shad tree in coffee estates in Karnataka, Kerala and Tamil Nadu. India: Andhra Pradesh, Rajasthan and Tripura. World: Myanmar, China, Thailand, Laos, Vietnam, Malesiaa and Fiji.

IUCN: NE (IUCN, 2024).

Erythrina variegata L., Herb. Amb.: 122. 1754. *Erythrina variegata* var. *orientalis* (L.) Merr., *Erythrina corallodendron* var. *orientalis* L., *Erythrina indica* Lam., *Erythrina indica* var. *alba* Blatt. & Mill., *Erythrina variegata* var. *alba* (Blatt. & Millard) M.R.Almeida, *Erythrina variegata* f. *alba* (Blatt. & Millard) Maheshw., *Erythrina variegata* f. *marmorata* Maheshw., *Erythrina variegata* f. *mysorensis* Maheshw. Vern.: *Kalyana murungai, Kakkapoo, Kavir, Mullumurungai, Soodu kottai* (Tam.).

Deciduous trees to 10 m tall; bark yellowish or greenish-grey, smooth, peeling off in papery flakes, prickles black on branches. Leaves trifoliate, alternate, leaflet 10-13.5 x 8.5-12.5 cm, widely ovate, rhomboid or deltoid-ovate, apex acuminate, base truncate or obtuse, margin entire, glabrous, membranous; 3 ribbed from base; lateral nerves 4-7 pairs, pinnate, prominent, ascending, intercostae reticulate, faint; stipules small, lateral; rachis 17-21 cm, slender, pulvinate, glabrous; stipels gland-like; petiolule 7-10 mm, stout, glabrous Flowers bright red, in dense racemes. Sepals spathaceous, oblique, recurved, split to the base on one side, 5-toothed at the tip. Petals 5, sessile; standard 6.5x2.5 cm, oblong-elliptic, apex obtuse; wings 1.5 x 1 cm, obovate; keel 1. 5 x 1.7 cm, oblong-falcate, free. Stamens 10, monadelphous, alternately longer and shorter, vexillary filament free; anthers uniform. Ovary stipitate, inferior, pubescent,1-celled, ovules many; style curved, subulate at the apex, not bearded; stigma capitate. Pod constricted between seeds, 15-30.5 cm long, dehiscent. Seeds 6-8. Fl. & Fr.: January – June.

Western Ghats: Cultivated for its edible leaves in Maharashtra, Goa, Karnataka, Kerala and Tamil Nadu. India: Most of the states. World: Sri Lanka, China, Thailand, Myanmar, Laos, Cambodia, Vietnam, Malesia, Africa, Madagascar and Australia.

IUCN: LC (IUCN, 2024).

Hardwickia Roxb.

Hardwickia binata Roxb., Pl. Coromandel 3: 6. 1819. Synonyms: *Hardwickia trapeziformis* Graham Vern.: *Anjan* (Tam.).

Large deciduous trees to 20 m tall; bark dark brown, rough, with irregular cracks; branchlets slender, drooping. Leaves alternate, 2-foliate; leaflets 2-6 x0.9-3.5 cm, sessile, with a minute bristle in between, obliquely ovate or obliquely oblong-ovate, base obliquely acute, apex obtuse, margin entire, glabrous; nerves 4-5 from base, arcuate, intercostae reticulate; stipules small, lateral, cauducous; rachis 8-75 mm long, slender, swollen at base and at tip, glabrous. Flowers greenish-yellow, in lax racemose panicles; bracts minute; bracteoles scale like. Sepals tubular, tube scarcely any; lobes usually 5, orbicular, petalloid, round the small basal disc; Petals 0. Stamens 10, alternately longer and shorter; filaments filiform; anthers versatile. Ovary half inferior, sessile, free; ovules 2; style filiform, stigma large, peltate. Pod oblong-lanceolate, 50-80 x 10-12 mm, seed at the apex where the pod dehisces. Seed pendulous, obovate, compressed, 2-furrowed. Fl. & Fr.: August – January.

Western Ghats: Deciduous forests of Gujarat, Maharashtra, Goa, Karnataka, Kerala and Tamil Nadu (Endemic).

IUCN: LC (IUCN, 2024).

Humboldtia Vahl

1a. Stipules appendiculate at base 2

1b. Stipules not appendiculate at base 7

2a. Leaflets 4 .... **H.brunonis**

2b. Leaflets 6–12 3

3a. Rachis of leaves terete; plagiotropous branches swollen and hollow (often in habitedby ants) .... **H. laurifolia**

3b. Rachis of leaves flattened and winged; internodes of branches neither swollen nor hollow. 4

4a. Stipules 5–8 cm long 5

4b. Stipules 2–3 cm long 6

5a. Flowers with 1–1.2 cm long pedicels .... **H. decurrense**

5b. Flowers sessile or rarely up to 0.2 cm long pedicels .... **H.ponmudiana**

6a. Inflorescence racemose; pods straight, tomentose .... **H.vahliana**

6b. Inflorescence corymbose; pods falcate, velvety .... **H. bourdillonii**

7a. Flowers crimson; staminal filaments glabrous .... **H. unijuga**

7b. Flowers white; staminal filaments pilose at base .... **H. sanjappae**

Humboldtia bourdillonii Prain., J. Asiat. Soc. Bengal, Pt. 2, Nat. Hist. 73: 200. 1904.

Trees to 15 m tall; bark reddish-brown, smooth; branchlets solid, glabrous. Leaves paripinnate, alternate; leaflets 6-12, opposite; petiolule 3-5 mm, stout, glabrous; leaflets 7.5-25 x 1.5-6 cm, oblong-lanceolate or linear-lanceolate, base unequal or obtuse, margin entire, glabrous, chartaceous; lateral nerves 7-10 pairs, intercostae reticulate, prominent; stipules lateral, ovate, acute, prominently veined, glabrous; rachis 10-12 cm long, pulvinate, obcordately winged between leaflets, glabrous. Flowers pinkish, 7 mm long, in corymbs on tubercles on stem and old branches, tawny velvety; pedicels 1.4-1.7 cm long, brown velvety; bracts ovate, acute, densely brown tomentose on both surface; bracteoles 2, connate almost throughout when young, ovate-oblong, gland present towards apex without; Sepals tubular, tube 1.4 cm long, obconical, brown tomentose, lined by the disc, persistent; lobes 4, subequal, crimson, imbricate. Petals 5, white with pink veins, 3 longer and 2 shorter, clawed, obovate, obtuse to rounded at apex, sparsely pilose within, glabrous with out. Stamens 5, free, alternating with 5 minute staminodes; filaments broad at base, pilose; anthers versatile, shortly apiculate at one end. Ovary half inferior, stipitate, obliquely oblong, densely pilose; ovules 4-6; style filiform, long, pilose; stigma capitate. Pod oblong, 8-11 x 2-2.5 cm, velvety brown pubescent, bright red or crimson, dehiscent. Seeds 3-5. Fl. & Fr.: December – July.

Western Ghats: Evergreen forests above 1300 m in Kerala and Tamil Nadu (Endemic).

IUCN: EN- B1+2c. (IUCN, 2024).

Humboldtia brunonis Wall., Pl. Asiat. Rar. 3: 17. 1831 var. **brunonis.** Synonyms: *Batschia brunonis* (Wall.) Kuntze

Trees to 10 m tall; bark dark brown to black; branchlets solid, pubescent when young, glabrous at maturity, lenticellate. Leaves paripinnate, alternate; leaflets 2-4, opposite, 7-20 x 2.2-8 cm, obovate oblong, oblong or elliptic-oblong, apex obtusely acuminate, base unequal, margin entire, glabrous, coriaceous; lateral nerves 7-13 pairs, intercostae reticulate, prominent; stipules 1-2 x 0.25-0.5 cm, lateral, linear, prominently parallel veined, glabrous; rachis 20-40 mm long, obscurely winged, shallowly grooved above, tomentose, pulvinate; petiolules 3-5 mm long, stout. Flowers white, 2-3 cm long, in axillary pendulous, brown pubescent, racemes; pedicels 4-5 mm long, pilose; bracts ovate, acute, pubescent without; bracteoles 2, connate below, obovate or obovate-oblong, obtuse at apex, pubescent on both surfaces. Sepals tubular, tube 8-10 mm long, lined by the disc, pubescent without; lobes 4, concave, acute or obtuse at tip, densely pubescent without and thinly within. Petals 3, white, pink or orange, ovate, shortly clawed, veined, glabrous. Stamens 5, perfect, free, alternating with 5 minute staminodes, deep pinkish; filaments 2 cm long, glabrous; anthers 2-3 mm long, versatile. Ovary 5 mm long, half inferior, obliquely linear; ovules 3-5, densely pubescent; style filiform, 10 mm long, pilose at base; stigma capitate. Pod dolabriform, 5-10 x 2-4 cm, brown pubescent when young. Seeds 3-4. Fl. & Fr.: November – June,

Western Ghats: Evergreen forests of Karnataka, Kerala and Tamil Nadu (Endemic).

IUCN: NE (IUCN, 2024).

Key to the varieties

1a. Leaflets 4-foliolate; flowers up to 2 cm long .... var. **raktapushpa**

1b. Leaflets 4-8-foliolate; flower more than 2 cm long .... var. **brunonis**

Humboldtia brunonis Wall. var. **raktapushpa** Udayan *et al.*, J. Bot. Res. Inst. Texas 1: 121. 2007.

Fl. & Fr.: January – April.

Western Ghats: Evergreen forests of Kerala (Endemic).

IUCN: NE (IUCN, 2024).

Humboldtia decurrens Bedd. ex Oliv., Hooker's Icon. Pl. 24: t. 2368. 1895.

Trees to 15 m tall; bark brown with prominent lenticels, rough; branches dark brown, pubescent when young, glabrous at maturity. Leaves paripinnate alternate; leaflets 8-12, opposite, estipellate; petiolule 3-5 mm long, stout, 15-38 x 4-12 cm, oblong-lanceolate, base obtuse or round, apex long acuminate, margin entire, glabrous; lateral nerves 10-16 pairs, parallel, intercostae reticulate, prominent; stipules lateral, ovate to narrowly ovate, acuminate to cuspidate at apex; rachis 15-36 cm long, pulvinate, obcordately or decurrently winged. Flowers pink, 17-18 mm long, in pendulous axillary or cauliflorus racemes, tawny villous; pedicels 1-1.2 cm long, villous; bracts broadly ovate, obtuse to acute at apex; bracteoles 2, connate at base, ovate, obtuse, brown villous on both surfaces. Sepals tubular, tube upto 1.2 cm long, obconical, lined by the disc, villous; lobes 4, oblong, rounded at apex. Petals 5, pink or white, 2-2.5 x 0.5 cm, obovate or oblanceolate, clawed, obtuse to round at apex, pilose within at base. Stamens 5, 3.5-4 cm long, perfect, alternating with 5 minute staminodes; filaments filiform, pilose at base; anthers versatile. Ovary half inferior, stipitate, obliquely linear, villous; ovules 3-6; style filiform, glabrous; stigma capitate. Pod oblong, 7-12 x 2.7-4 cm, brown, veined, rugulose and dark brown tomentose when young. Seeds 2-3, thick, glabrous. Fl. & Fr.: January – June.

Western Ghats: Evergreen forests of Kerala and Tamil Nadu (Endemic).

IUCN: NT (Low risk/near threatened) (IUCN, 2024).

Humboldtia laurifolia Vahl., Symb. Bot. 3: 106. 1794.

Trees to about 10 m tall; young stems glabrous, often swollen and inhabited by ants. Leaves pinnate, alternate, leaflets 3-5 pairs, ovate to ovate-oblong or lanceolate-oblong, 4-15 cm long, 2-5.5 cm wide, acuminate at the apex, obliquely rounded to cuneate at the base, glabrous; stipules leaflike above, 2-3.5 x 1-2 cm, acute, the lower portion falcate; rachis 1-20 cm long, glabrous or nearly so. Inflorescences axillary, racemose or paniculate, 8-12 cm long; bracts deltoid, 4 mm long; bracteoles oblong-spathulate, 6-7 mm long; bracts deltoid, to 4 mm long; bracteoles oblong-spathulate, 6-7 mm long. Sepals reddish or orangish with tube 3 mm long, segments 8-10 mm long, subequal. Petals 5, white, 10-12 mm long. Stamens exserted to 12-20 mm long. Pod puberulent or subglabrous when young, glabrous at maturity, ellipsoid, compressed, 5-10 cm long. Seeds 3-4. Fl. & Fr.: January – June.

Western Ghats: Evergreen forests of Kerala. World: Sri Lanka.

IUCN: VU- A1c, B1+2cd. (IUCN, 2024).

Humboldtia ponmudiana E.S.S. Kumar, Shareef & Raj Vikr., Phytotaxa 552: 116. 2022.

Trees to 12 m tall; bark black, warty; young shoots densely brown tomentose, glabrescent at maturity. Leaves up to 45 cm long, subsessile, 8–12-foliolate; rachis obcordately or decurrently winged, wings reticulately veined, densely brown tomentose on both surfaces; petiolules 2–4 mm long, not covered by rachis wings, densely brown tomentose, depressed glandular; leaflets 10.5–37 × 3.0–7.0 cm, lanceolate or ovate-lanceolate, subcordate or rounded at base, acuminate to caudate-acuminate at apex, slightly undulate at margins, subcoriaceous, depressed glandular beneath; lateral nerves 10–19 pairs, prominently raised beneath, margin forming a prominent intramarginal vein; stipules 5.5–6.0 × 1.5–2.0 cm, ovate lanceolate, slightly falcate, acuminate to cuspidate at apex, prominently parallel veined, depressed glandular, densely brown tomentose when young and glabrescent at maturity. Racemes

5.0–9.0 cm long, pendulous, axillary or cauliflorous, subsessile; rachis terete, brown tomentose, many flowered. Flowers 3.5 cm across at anthesis, white, sessile or subsessile; pedicels absent or rarely up to 0.2 cm long, brown villous; bracts 1.0–1.2 × 0.7 cm, broadly ovate, greenish-white, acuminate at apex, brown villous and eglandular without, glabrous within, fugacious; bracteoles 2, 1.0–1.3 × 0.7 cm, ovate, obtuse at apex, greenish-white, connate to middle in cauliflorus flowers or connate to middle on one side and free up to the base on the other side in axillary flowers, glabrous within. Calyx obconic tube, silky villous; lobes 4, 1.2–1.8 × 0.4–0.7 cm, subequal, one larger than the rest, elliptic-oblong or linear oblong, creamy-white, rounded at apex. Petals obovate to oblanceolate, 5, 1.6–2.3 × 0.6–1.0 cm, white, clawed at base, acute or slightly acuminate at apex, glabrous or minutely silky pilose along the midrib abaxially. Stamens 5, 3.5–4.0 cm long; filaments filiform, pink, broad and silky pilose at base; anthers 3.5–4.0 mm long, pink, obtuse at both ends. Ovary 0.8–1.0 cm long, stipe 1.0–1.5mm long, obliquely linear, silky villous, 1–4 ovuled; style filiform, 2.7–3.0 cm long, rarely coiled, glabrous, pink; stigma capitate. Pods 13–15 × 3–4 cm, oblong, silky tomentose; beak 2–2.5 cm long. Seeds 2–3, thick, flat, glabrous. Fl. & Fr.: December – February.

Western Ghats: Ponmudi hills of Tiruvandrum district, Kerala (Endemic).

IUCN: NE (IUCN, 2024).

Humboldtia sanjappae Sasidh. & Sujanapal., Rheedea 17: 21. 2007.

Trees to 20 m tall; bark greenish brown, smooth; branchlets stout, terete, glabrous. Leaves pinnate, to 25 cm long, leaflets 4, rarely 2; petiolule 3-5 mm long, glabrous, 5-19 x 3-7.5 cm, elliptic to elliptic-obovate, acute, obtuse or rounded at base, acute or obtusely acuminate at apex, glabrous; lateral nerves 3-5 pairs, slender, intercostae reticulate, faint; stipule 1 x 0.5 cm, obliquely ovate, acute, 5 or 6 nerved, glabrous; rachis to 10 cm long, terete glabrous. Flowers in racemes, axillary or from leaf scars, solitary or groups of 2 or 3, to 13 cm long, grayish- white tomentose; bracteoles 2, 4 x 4 mm, lobes oblong, rounded at apex, villous, united at base for 1 mm long; pedicels 3 mm long, villous. Sepals tubular, tube 2.5-3 mm long, lobes 4, subequal, larger one 5 x 2.5 mm, oblong, obtuse, villous, others 5 x 2 mm. Petals 5, subequal, white, larger one obovate, acute, 7 x 5 mm, glabrous. Stamens 5, filaments 15-17 mm long, sparsely pilose to the middle, anthers oblong, 2 mm long, versatile. Ovary 1.5 mm long, pilose; ovules 3, style 8-10 mm long, glabrous, stigma capitate. Pod oblong, 15-17 x 4 cm, oblique, widening towards apex, beaked, compressed, glabrous. Seeds 1-2, orbicular, c. 3 cm across, brown, glabrous. Fl. & Fr.: December – April.

Western Ghats: Rare in evergreen forests above 1200 m in Kerala (Endemic).

IUCN: NE (IUCN, 2024).

Humboldtia unijuga Bedd. var. **unijuga** Pl. Ind. Or.: 22. 1870. Synonym: *Batschia unijuga* (Bedd.) Kuntze

Trees to 8 m tall; branches slender, solid, terete, strigose when young, glabrous at mature. Leaves bifoliate, alternate, leaflets subsessile, 4-15 x 2-5 cm, elliptic-lanceolate or lanceolate or narrowly ovate, base unequal, apex obtusely acuminate, margin entire, glabrous; lateral nerves 5-15, looped at the margin forming intramarginal nerve, prominent; intercostae reticulate; stipule lateral, 5-7 x 2.5 mm narrowly ovate, acute, prominently parallel veined, glabrous; rachis 5-8 mm long, glabrous, gland at the top. Flowers crimson, 1.5-2.5 cm long, in axillary, rarely cauliflorus racemes; sterile bracts present at the base of racemes; peduncle pubescent; pedicel 5-10 mm long, finely brown pubescent; bracteoles 2, 4-5 x 1.5-2 mm, ovate, rounded at apex. Sepals in tubular, tube 8 mm long, obconical lined by the disc; lobes 4, pubescent, obovate, rounded at apex. Petals 5, 10-15 mm long, obovate, narrowed at base, glabrous. Stamens 5, free, filaments 10 mm long, filiform; anthers versatile; staminodes absent. Ovary 5-6 mm long, half inferior, stipitate, obliquely oblong, densely brown pubescent; style 8-10 mm long, glabrous; stigma capitate. Pod oblong, 5-8 x 2-3 cm, compressed, deep red, veined. Seeds 2-3. Fl. & Fr.: December – March.

Western Ghats: Evergreen forests above 1300 m in Kerala and Tamil Nadu (Endemic).

IUCN: NE (IUCN, 2024), endangered (Narasimhan and Irwin, 2021).

Key to the varieties

1a. Leaflets bifoliate, elliptic lanceolate to narrowly lanceolate; flowers crimson red; staminodes absent; pod 8-10 cm long ....var. **unijuga**

1b. Leaflets 6-foliate, oblanceolate to elliptic-oblong; flowers pink; staminodes 5; pod 6-8 cm long ....var. **trijuga**

Humboldtia unijuga Bedd. var. **trijuga** Joseph & Chandras., J. Bombay Nat. Hist. Soc. 81: 729. 1985. Synonyms: *Humboldtia trijuga* (J. Joseph & V. Chandras.) M.Mohanan

Fl. & Fr.: December – March.

Western Ghats: Evergreen forests of Kerala (Endemic).

IUCN: NE (IUCN, 2024).

Humboldtia vahliana Wight., Icon. Pl. Ind. Orient. 4: t. 1607. 1850. Synonym: *Batschia vahliana* (Wight) Kuntze

Trees to 18 m tall; bark dark brown mottled with white; branchlets solid, terete, glabrous, sometimes zig-zag. Leaves paripinnate, alternate; leaflets 4-8, opposite, 10-25 x 3-6.4 cm, oblong-lanceolate, base obtuse, apex acuminate, glabrous; lateral nerves 8-14 pairs, intercostae reticulate, prominent; stipules 1.5-3 x 2.2 cm, persistent, ovate-cordate, parallel veined, glandular, glabrous; rachis 10-16.5 cm long, obscurely winged, pulvinate, tomentose without. Flowers white, 27-30 mm long, in many flowered, axillary brown velvety racemes; pedicel 5-6 mm long; bracts lanceolate, acute, villous on both surfaces, persistent; bracteoles 2, connate by base, obovate. Calyx tubular, tube 8-10 mm long, lined by the disc, obconical, brown villous; lobes 4, obovate-oblong, rounded at apex. Petals 5, 10-12 x 4 mm, obovate, shortly clawed, broadly rounded at apex, glabrous. Stamens 5, perfect, alternating with 5 minute staminodes; filaments 10 mm long, red, pilose upto the middle; anthers versatile. Ovary 4-5 mm long, half inferior, obliquely linear; ovules 3-4; style 10-15 mm long, filiform, pilose at base, reddish; stigma capitate. Pod elliptic, 15-20 x 3.7-6 cm, compressed, brown villous. Seeds 3-4, almost orbicular, glabrous. Fl. & Fr.: December – July.

Western Ghats: Evergreen forests of Kerala and Tamil Nadu (Endemic).

IUCN: EN- B2ab(i,ii,iii,v). (IUCN, 2024).

Millettia Wight &Arn.

1a. Stipule strap shape; Flowers in pendulous, brown pseudoracemes .... **M. dura**

1b. Stipule ovate; Flowers in axillary simple or branched racemes .... **M. rubiginosa**

Millettia dura Dunn, in J. Bot. 49: 221. 1911. Synonyms: *Millettia pseudoracemosa* Thoth. & Ravik., *M. cyanantha* Dunn, *Millettia pulchra* var. *munnarensis* Balan & Pradeep

Trees to 13 m tall; young branchlets brownish pubescent. Leaflets 15-19, oblong, 8-9 x 2-3 cm, glabrous, base asymmetric, apex acuminate; nerves and margin hairy; stipules strap shaped, 8 mm long; rachis pubescent, up to 20 cm long; petiole 3-5 cm long; petiolules 3-4 mm long. Flowers in pendulous, brown pseudoracemes, up to 15 cm long, peduncle 5 cm, floral bracts oval 1-2 mm long, pedicels up to 13 mm. Sepals in tubular, brown pubescent; tube 6 mm. Petals mauve; standard white, silky outside, 25 mm long. Ovary 18 mm long, pubescent 8-11 ovulate. Pod flat, dehiscent, 14-20 cm long, up to 21 mm wide, glabrescent. Seeds ellipsoid, oblique, funicle dilated proximally. Fl. & Fr.: May – August.

Western Ghats: Evergreen forests of Kerala and Tamil Nadu. World: Sri Lanka, Angloa, Brundi, Kenya, Tanzania, Uganda and Zaire.

IUCN: NE (IUCN, 2024).

Note: *Millettia pseudoracemosa* Thothathri & Ravikumar was described on the basis of a single collection (S. Ravikumar 334), which was collected on 21 April 1995 from Anamalai Hills, Tamil Nadu, South India. The new species was considered to be similar to *M. racemosa* (Roxb.) Benth. in habit, shape, number and hairiness of leaflets, floral color, and pod shape. The latter species was first described based on specimens also from India, but it has been treated as a monospecific genus *Endosamara* Geesink belonging to the tribe Wisterieae because of its true panicles with ebracteole flowers and very peculiar fruits with a lomented endocarp around each seed (Compton *et al.* 2019). Thothathri & Ravikumar (1997) also pointed out that their new species has a unique character in the genus, possessed the large sized (2.5–3.0 cm long) and violet flowers.

Millettia rubiginosa Wight & Arn., Prodr. Fl. Ind. Orient. 1: 263. 1834. Synonym: *Phaseoloides rubiginosum* Kuntze

Large climbing trees to 20 m tall; bark yellow pubescent. Leaves pinnately 3-5-foliolate; leaflets to 20 x 8 cm, ovate-oblong, acuminate at apex; minutely golden brown hairy, petiole to 10 cm long; stipule ovate; stipel filiform. Flowers in axillary simple or branched racemes; bracts ovate. Sepals 5, campanulate, 3 mm long, minutely lobed. Standard petals 13 x 10 mm, obovate, emarginate, densely fulvous hairy on back; wings 10 mm long, oblong, with two patches of golden hairs; keel similar to wings. Stamens 9 + 1; anthers uniform. Ovary sessile, few-ovuled. Pods obovate, 8 x 4.5 cm, compressed, woody, densely hairy. Seeds 1-3, compressed, ovoid. Fl. & Fr.: September – November.

Western Ghats: Semi-evergreen forests of Karala and Tamil Nadu (Endemic).

IUCN: NE (IUCN, 2024).

Mundulea (DC.) Benth.

Mundulea sericea (Willd.) A. Cheval., Compt. Rend. Hebd. Séances Acad. Sci. 180: 1521. 1925. Synonyms: *Cystisus sericeus* Willd., *Dalbergia sericea* (Willd.) Bojer, *Tephrosia sericea* (Willd.) DC.

Small trees to 6 m tall; bark longitudinally fissured, greyinsh; branchlets whitish pubescent. Leaves imparipinnate; leaflets opposite and subopposite, to 4.5 x 1.5 cm, elliptic, apex emarginate, base cuneate, silvery pubescent below; petiolule 2 mm, stipule 2 mm. Flowers fascicled, in pseudo-racemes, violet; pedicels 2 cm. Sepals tubular, campanulate, tube 5 mm, pubescent, lobes 2 mm, triangular, upper connate. Standard petal 2.5 x 1.5 cm, obovate. Staminal tube 2 cm long, filaments 3 cm long, alternate ones dilated at the base. Ovary 2 cm, pubescent. Pod linear, pubescent, acute at apex. Fl. & Fr.: September – October.

Western Ghats: Scrub to deciduous forests of Maharashtra, Goa, Karnataka, Kerala and Tamil Nadu. World: Sri Lanka, Africa and Madagascar.

IUCN: LC (IUCN, 2024).

Neltuma Raf.

1a. Leaflets linear, 12-20 pairs; pod coiled .... **.N. chilensis**

1b. Leaflets linear-elliptic, 20-36 pairs; pod straight .... **.N. juliflora**

Neltuma chilensis (Molina) C.E.Hughes & G.P.Lewis., PhytoKeys 205: 174. 2022. Synonyms: *Ceratonia chilensis* Molina, *Prosopis chilensis* (Molina) Stuntz, *Ceratonia chilensis* Molina

Deciduous trees to 14 m tall; bark pale brown, thick, with cracks and vertical fissures. Leaves pinnate, 20 cm long, leaflets 12-20 pairs, ovoid, margin entire, yellowish green upper surface; stipular thorns strong in pair, curved, borne at each whorl of leaves. Flowers in dense axillary spikes, yellow. Sepals 5. Stamens 10. Pods coiled, to 15 cm long. Seeds many, brown coloured. Fl. & Fr.: October – April.

Western Ghats: Scrub to deciduous forests of Tamil Nadu. World: Native to South America, Naturalized here.

IUCN: NE (IUCN, 2024).

Neltuma juliflora (Sw.) Raf., Sylva Tellur.: 119. 1838. *Mimosa juliflora* Sw., *Acacia juliflora* (Sw.) Willd., *Entada juliflora* (Sw.) Roberty, *Mimosa juliflora* Sw. Vern.: *Velikkaathan, Seemai mullu, Vaelikarauvai, Seemai karuvai* (Tam.).

Trees to 10 m tall; bark deeply fissured, yellowish-brown; branchlets zig-zag, glabrous. Leaves bi-pinnate, alternate; pinnae 2-4, even pinnate, 3-11 cm long, ending in a short bristle, a circular gland near the base of lower pinnae; leaflets 22-36, opposite, 6-23 x 1.5-5 mm, oblong or linear-oblong, base obtuse, apex obtuse, emarginate or mucronulate, margin entire, glabrous; lateral nerves prominent beneath; stipular spines straight, 5 mm long in young, solitary or paired, divergent, upto 5 cm long on older branches. Flowers greenish-yellow, 1.5 mm across, subsessile, in axillary spikes, to 12 cm long; bracteoles minute. Sepals campanulate, 1.5 mm long, glabrous, 5-toothed. Petals 5, ligulate, sub coherent at base. Stamens 10, free, slightly exserted; anthers with an apical gland. Oovary superior, stipitate, pubescent; ovules many; style filiform; stigma minute. Pod pale yellow, 20-30 x 1.5 cm, smooth, with straight parallel sutures or irregularly sub moniliform. Seeds ovoid, brown, embedded in pulpy mesocarp. Fl. & Fr.: Febrary - May.

Western Ghats: Common in foot hills of Maharashtra, Karnataka, Kerala and Tamil Nadu. World: Native to America, now naturalized everywhere in Tropics.

IUCN: LC (IUCN, 2024).

Ormosia G. Jackson

Ormosia travancorica Bedd., Fl. Sylv. S. India: t. 45. 1870. Synonyms: *Placolobium travancoricum* (Bedd.) Yakovlev Vern.: *Malaimanchedi* (Tam.).

Evergreen trees to 25 m tall; bark brown, mottled with grey, smooth. Leaves imparipinnate, alternate, clustered at the tips of branchlets; leaflets 9-13, opposite, 3-15 x 2-4.3 cm, elliptic-oblanceolate, elliptic-oblong, or obovate, base acute, obtuse, truncate or oblique, apex acuminate, margin entire, glabrous above, downy beneath; lateral nerves 4-9 pairs, intercostae reticulate, obscure; stipules small, lateral, pubescent; rachis 4.5-20 cm long, slender, pulvinate, pubescent. Flowers purple, striped with green, in axillary and terminal racemes; bracts small; bracteoles minute. Sepals campanulate, deeply 5-cleft, the upper teeth the shortest. Petals 5, claws short, the standard orbicular, wings oblong, often more or less sagittate at base, keel petals free, somewhat incurved. Stamens 10, free, alternately long and short; anthers versatile. Ovary subsessile, inferior, glabrous, 1-celled; ovule 2-many; style subulate, involute; stigma oblique. Pod oblong, 5-7.5 cm long, red, dehiscent. Seed one, rarely 2, obovate or suborbicular, bright red. Fl. & Fr.: February – April.

Western Ghats: Evergreen forests of Karnataka, Kerala and Tamil Nadu (Endemic including Andaman & Nicobar Islands).

IUCN: NE (IUCN, 2024), vulnerable (Narasimhan and Irwin, 2021).

Ougeinia Benth.

Ougeinia oojeinensis (Roxb.) Hochr., Annuaire Conserv. Jard. Bot. Genève 13-14: 51. 1909. *Dalbergia oojeinensis* Roxb., *Desmodium oojeinense* (Roxb.) H.Ohashi, *Ougeinia dalbergioides* Benth. Vern.: *Malavenga, Nemi, Thodukara* (Mal.).

A deciduous tree to 12 m tall; bark grey or dark brown, deeply cracked. Leaves alternate, trifoliate, to 30 cm long; leaflets broadly ovate, 7.5-15 cm long and 3.8-10 cm wide, coriaceous, glabrous above, sparsely pubescent beneath, base cordate, margins shallowly crenate, main nerves 4-8 pairs; petiolules 3 mm long; petioles 3.8-5 cm long; stipules 6 mm long, lanceolate, acute. Flowers white or pink, somewhat fragrant, borne in short-fascicled racemes arising from the nodes of old branches; pedicels 1.2-2 cm long, filiform; bracts 1.3 mm long, ovate, acuminate; bracteole 1 beneath the calyx, minute. Sepals 4-6 mm long, pubescent, teeth short, triangular. Petals1-1.3 cm long. Pod linear-oblong, 5-7.5 cm long, light brown, flat, joints 2-3 times as long as broad, reticulately veined. Seeds 2-5. Fl. & Fr.: February – March.

Western Ghats: Deciduous forests of Kerala, Karnataka, Maharashtra and Tamil Nadu. India: Western Himalaya. World: Nepal.

IUCN: NE (IUCN,2024)

Parkinsonia L.

Parkinsonia aculeata L., Sp. Pl.: 375. 1753. Synonyms: *Parkinsonia inermis* Spreng., *Parkinsonia thornberi* M.E.Jones

Small armed trees to 8 me tall; branchlets tomentose, glabrescent when old, spines erect, broad based, woody. Leaves bipinnate, 35 cm long, pinnae 2-4 pairs; leaflets 2, 2-7 x 1-3 mm, oblanceolate, rachis flattered. Racemes axillary, to 15 cm long. Flowers yellow; pedicels 1 to 1.5 cm; bracts causucous. Sepals tubular, tube short, lobes 6 x 2 mm, unequal. Petals 10 x 10 mm, obovate, clawed, villous. stamens 10, filaments unequal, villous. Pod linear lanceolate, 10 x 0.6 cm, tomentose, striated; 1-5 seeded. Fl. & Fr.: November – March.

Western Ghats: Foot hills of Maharashtra, Kerala and Tamil Nadu. World: Native to Tropical America, naturalized everywhere.

IUCN: LC (IUCN, 2024).

Pithecellobium Matt.

Pithecellobium dulce (Roxb.) Benth., London J. Bot. 3: 199. 1844. Synonyms: *Mimosa dulcis* Roxb., *Inga javana* DC., *Inga pungens* Humb. & Bonpl. ex Willd., *Mimosa pungens* Poir., *Pithecellobium bertolonii* Benth. Vern.: *Kodukkaipuli, Konapuli* (Tam.).

Trees to 20 m tall; branchlets armed with spinescent stipules. Leaves bifoliolate, glands at junction of pinnae and leaflets; leaflets sessile, 1 pair per pinna, elliptic or obovate-elliptic, 2-5 × 0.2-2.5 cm, both surfaces glabrous, reticulate veins raised abaxially, base slightly oblique, apex obtuse or emarginate. Inflorescence pedunculate heads, aggregated in terminal panicles. Sepals funnel-shaped, 1-1.5 mm, tomentose. Petals to 6 mm long. Stamens numerous, connate into a tube at base. Pod blackish or pinkish brown, curved into a circle, flat, 5-7 cm in diam. Seeds dark brown, shiny, ovoid-ellipsoid, 1.5 cm, hard, covered with fleshy white aril. Fl. & Fr.: January – July.

Western Ghats: Occasional in foot hills of Maharashtra, Goa, Karnataka, Kerala and Tamil Nadu. India: Most of the states. World: Native to Tropical America and naturalized everywhere.

IUCN: LC (IUCN, 2024).

Pongamia Vent.

Pongamia pinnata (L.) Pierre., Fl. Forest. Cochinch. 4: t. 385. 1898. *Cytisu spinnatus* L., *Pongamia glabra* Vent., *Millettia pinnata* (L.) Panigrahi, *Pongamia glabra* var. *xerocarpa* (Hassk.) Prain, *Pogamia xerocarpa* Hassk. Vern.: *Pungamaram* (Tam.).

Evergreen trees to18 m tall; bark grey, smooth, speckled with brown, lenticellate. Leaves imparipinnate, alternate, leaflets 5-7, opposite, leaflet 4.5-12 x 2-7 cm, elliptic-acuminate, elliptic-lanceolate, ovate or ovate-oblong, apex acuminate, margin entire, glabrous; lateral nerves 5-8 pairs, faint; intercostae reticulate, obscure; stipules lateral, small, oblong, cauducous; rachis 10-15 cm long, slender, pulvinate, pubescent. Flowers purplish-white, 15-18 mm long, in lax axillary racemes, axis pubescent; bracts small, cauducous. Sepals campanulate, minutely 5 toothed, apex truncate. Petals 5, clawed; standard suborbicular with curved folds above the claw; wings obliquely oblong, slightly adnate above the claws to the obtuse keel petals which are joined near the tip. Stamens 10, monadelphous, the vexillary stamen free below and above; anthers uniform. Ovary subsessile, inferior, 1-celled; ovules 2; style in curved, beardless; stigma capitate. Pod obliquely oblong, 4-5 x 2-2.5 cm, flat, thick, pointed at both ends, indehiscent, slightly falcate. seed one, reniform. Fl. & Fr.: December – June.

Western Ghats: Common cultivated avenue trees in Gujarat, Maharashtra, Goa, Karnataka, Kerala and Tamil Nadu. India: Most of the states. World: Sri Lanka, Bangladesh, Pakistan, Myanmar, China, Malesia, Vietnam, Thailand, Australia and Polynesia.

IUCN: LC (IUCN, 2024).

Prioria Griseb.

Prioria pinnata (Roxb. ex DC.) Breteler., Wageningen Agric. Univ. Pap. 99(3): 52. 1999. Synonyms: *Kingiodendron pinnatum* (Roxb. ex DC.) Harms., *Hardwickia pinnata* Roxb. ex DC. Vern.: *Kulavu* (Mal.); *Kolavu, Madayan Sambrani* (Tam.).

Evergreen trees to 30 m tall; bark greyish-brown with green blotches, rough, exuding a reddish sticky resin. Leaves imparipinnate, alternate; leaflets 5-9, 4.5-10.5 x 2-4.5 cm, ovate-lanceolate or oblong, apex acuminate, falcate or oblique, margin entire, glabrous; lateral nerves 8-13, secondary laterals present, intercostae reticulate, prominent; stipules minute, lateral, cauducous; rachis 10-15.2 cm, slender, pulvinate, glabrous; petiolule 5-10 mm, grooved above, glabrous. Flowers white, 2-3 mm across, in axillary and terminal panicled racemes. Sepals in tubular, tube almost wanting; lobes 5, broadly ovate, imbricate. Petals 0. Disc very small. Stamens 10, equal, filaments filiform, villous at base; anthers versatile. Ovary half inferior, sessile, villous at base; ovules 2; style subulate; stigma minute, oblique. Pod ovate-ellipsoid, 4-5 x 2-2.5 cm, turgid, obtusely beaked, dark brown, indehiscent. Seed one, pendulous. Fl. & Fr.: April – August.

Western Ghats: Rare in evergreen forests of Karnataka, Kerala and Tamil Nadu (Endemic).

IUCN: VU- A1acd. (IUCN, 2024), Endangered (Narasimhan and Irwin, 2021).

Prosopis L.

Prosopis cineraria (L.) Druce., Rep. Bot. Exch. Club Soc. Brit. Isles 3: 422. 1913 publ. 1914. Synonyms: *Mimosa cineraria* L. *Prosopis spigicera* L. *Prosopis spicata* Burm. Vern.: *Vannimaram* (Tam.).

Medium trees to 10 m tall; branches prickly, prickles curved, compressed. Leaves alternate, bipinnate, pinnae 1-2 pairs, 2.5-8.0 cm long, leaflets 7-12 pairs, more or less sessile, 3-5 x 2-4 mm, oblong, oblique, apex usually mucronate, base rounded, 3-nerved; rachis 1.2-5 cm long, glabrous. Flowers creamy white in pedunculate spikes, nearly 5-12.5 cm long; peduncle 1.0-2.5 cm long. Sepals cup-shaped, 1-1.5 mm long, truncate or obscurely 5 toothed. Petals oblong, 3-4 mm long, tips recurved. Stamens 10, free, shortly exserted; anthers tipped with deciduous glands. Pods slender, 12.5-25 cm long, 5-8 mm broad, pendulous, cylindric, turgid. Seeds 10-15, oblong, compressed. Fl. & Fr.: November - April.

Western Ghats: Common in foot hills to mixed deciduous scrub forests of Maharashtra and Tamil Nadu. India: Almost all the states except North-east. World: Pakistan, Afghanistan, Iran, Arabia.

IUCN: NE (IUCN, 2024).

Pterocarpus Jacq.

Pterocarpus marsupium Roxb. ssp. **marsupium.** Pl. Coromandel 2: 9. 1799. Synonyms: *Lingoum marsupium* (Roxb.) Kuntze, *Pterocarpus marsupium* var. *marsupium* f. *bilobus* (Roxb. ex G.Don) Prain, *Pterocarpus bilobus* Roxb. ex G.Don, *Pterocarpus marsupium* f. *acuta* Prain Vern.: *Vaengai* (Tam.).

Deciduous trees to 30 m tall; bark grey or greyish-black, rough, deeply vertically cracked, exudation blood-red. Leaves imparipinnate, alternate, leaflets 5-7, 3.5-12.5 x 2-7 cm, elliptic-oblong, oblong-ovate or oblong, apex obtuse and emarginate, base obtuse or acute, margin entire, glabrous; lateral nerves 9-20 pairs, secondary laterals prominent, intercostae reticulate, prominent; stipules small, lateral, cauducous; rachis 6.5-11.1 cm long, pulvinate, glabrous; petiolule 6-10 mm, glabrous. Flowers yellow, in terminal and axillary panicles; 10-12 mm long; bracts small, dioecious; bracteoles 2, cauducous. Sepals campanulate, lobes short, the upper 2 often connate. Petals 5, all long-clawed, crisped along the margins; standard orbicular, wings oblique, obovate, auricled; keel petals oblique, small, slightly connate. Stamens 10, monadelphous; filaments subequal; anthers uniform. Ovary shortly stalked, inferior, tomentose, 1-celled; ovules 2; style filiform, in curved, beardless; stigma capitate. Pod orbicular-reniform, 2.5-5 cm across, broadly winged. Seed one, subreniform. Fl. & Fr.: December – July.

Western Ghats: Deciduous to semi-evergreen forests of Gujarat, Maharashtra, Goa, Karnataka, Kerala and Tamil Nadu. India: Most of the states except North-east. World: Sri Lanka, Nepal, Taiwan and Madagascar.

IUCN: NT (Low risk/near threatened) (IUCN, 2024).

Key to the subspecies

1a. Leaves elliptic oblong, laeral nerves 8-10 pairs, tip acuminate .... ssp. **acuminata**

1b. Leaves ovate or orbicular-oblong, lateral nerves 7-8 pairs, tip emarginate .... ssp. **marsupium**

Pterocarpus marsupium Roxb. ssp. **acuminatus** (Prain) Thoth., Taxon. Revis. Tribe Dalbergieae in Indian Subcont.: 227. 1987. Synonyms: *Pterocarpus marsupium* f. *acuminata* (Prain) Prain, *Pterocarpus marsupium* var. *acuminata* Prain

Fl. & Fr.: October – January.

Western Ghats: Semi-evergreen forests of Maharashtra, Karnataka and Kerala (Endemic).

IUCN: NE (IUCN, 2024).

Samanea Merr.

Samanea saman (Jacq.) Merr., J. Washington Acad. Sci. 6: 47. 1916. Synonyms: *Mimosa saman* Jacq., *Enterolobium saman* (Jacq.) Prain., *Pithecellobium saman* (Jacq.) Benth., *Albizia saman* (Jacq.) Merr., *Albizia saman* (Jacq.) F. Muell. Vern.: *Thoongumunji maram* (Tam.).

Trees to 25 m tall; branchlets puberulent to tomentose. Leaves bipairypinnate, pinnae 3-5 pairs, to 15 cm long, glands at junctions of pinnae and leaflets; leaflets 3-8 pairs per pinna, oblong, 2-4 x 1-1.8 cm, abaxially pubescent, adaxially shiny, apex rounded or obtuse, often emarginate and mucronulate; petiole 15-40 cm long. Heads 1-5, together axillary, 5-6 cm in diam; peduncles 5-6 cm long. Marginal flowers pedicellate. Sepals funnel-shaped, 6 mm long. Petals red or yellowish red, 1.2 cm long. Stamens white at base, purple toward apex, 3.5 cm long; staminal tube shorter than corolla tube. Central flowers sessile. Petals to 1.2 cm. Staminal tube longer than corolla. Pod black, oblong, compressed, 10-20 x 1.2-2.5 cm, septate between seeds. Seeds brown, ellipsoidal, 8 × 5 mm. Fl. & Fr.: May - September.

Western Ghats: Planted avenue trees of Maharashtra, Karnataka, Kerala and Tamil Nadu. India: Andhra Pradesh, Bihar, Gujarat, Madhya Pradesh, Meghalaya, Sikkim, Tripura, Uttar Pradesh and West Bengal. World: Native to Central and South America.

IUCN: LC (IUCN, 2024).

Sanjappa E.R.Souza & Krishnaraj

Sanjappa cynometroides (Bedd.) E.R.Souza & Krishnaraj., Rheedea 26: 6. 2016. Synonyms: *Calliandra cynometroides* Bedd., *Inga cynometroides* (Bedd.) Baker, *Feuilleea cynometroides* (Bedd.) Kuntze

Small trees to 6 m tall; bark greyish-brown, spiny. Leaves alternate, bifoliate, leaflets 2, 4.0-12 x 1.5-5.5 cm, elliptic lanceolate or elliptic-oblong, base obliquely attenuate, apex acute or acuminate, margin entire, glabrous; lateral nerves 4-8 pairs, intercostae reticulate; stipule spiny; rachis 18-35 mm long, pulvinate, glabrous. Flowers bisexual or unisexual, in axillary solitary globose heads, 16-18 mm long. Sepals 3, 3-lobed. Petals 3, greenish-yellow, connate, valvate. Stamens many, long-exserted, monadelphous; anthers minute. Ovary sessile; style filiform; stigma minute. Pod linear, 8-12 x 1.5-1.8 cm, pale brown, subsessile, falcate, with a hook at the tip, glabrous. Seeds 5-6, orbicular or obovate, compressed. Fl. & Fr.: Throughout the year.

Western Ghats: Evergreen forests of Kerala and Tamil Nadu (Endemic).

IUCN: NE (IUCN, 2024).

Saraca L.

Saraca asoca (Roxb.) Wilde., Blumea 15: 393. 1968. Synonyms: *Jonesia asoca* Roxb.

Trees to 10 m tall; bark brown or brownish-black, lenticellate. Leaves paripinnate, alternate; leaflets 6-12, opposite, 7-28.5 x 2-8.5 cm, narrowly oblong, oblong-ovate or ovate-lanceolate, round, cuneate or acute, apex acute or acuminate, base obtuse, margin entire, glabrous; lateral nerves 10-15 pairs, intercostae reticulate, faint; stipules 7-20 mm long, intra petiolar, scarious, ovate, connate; rachis 4-25 cm long, slender, pulvinate, glabrous; petiolule 2-10 mm long, stout, glabrous. Flowers yellow-orange or red, in dense sessile paniculate corymbs, axillary to leaves or leaf scars; bracts ovate, small deciduous; bracteoles 4 mm long oblong-spathulate, ciliolate. Sepals 4, 4 cm long, petalloid, cylindric, enclosing a lobed disc; lobes 4, ovate-oblong, unequal, spreading, imbricate. Petals 0. Stamens 7 or 8, much exserted, free; filaments long, filiform, coloured, glabrous; anthers versatile. Ovary half inferior, stipitate, the stipe adnate below to one side of the disc, pubescent; style incurved, glabrous, filiform; stigma small, capitate; ovules many. Pod oblong, 10-5 x 2-5 cm, flat, coriaceous or almost woody, tapering at both ends. Seeds 2-8, 3.8 cm long, ovoid, slightly compressed. Fl. & Fr.: December – July.

Western Ghats: Moist deciduous to evergreen forests above 900 m in Maharashtra, Goa, Karnataka, Kerala and Tamil Nadu. World: Sri Lanka, Bangladesh, Bhutan, Nepal, Myanmar and Malaysia.

IUCN: VU- B1+2c. (IUCN, 2024).

Senegalia Raf.

1a. Flower white; pod constricted .... **S. senegal**
1b. Flowers yellow or pale yellow 2
2a. Pinna 2-6 pairs; seeds 4-10 in a pod 3
2b. Pinna 8-20 pairs; seeds more than 10 per pod 4
3a. Pod glabrous, tip caudate apiculate, seeds 4-5 .... **S. ferruginea**
3b. Pod tomentose, tip rounded, seeds 6-10 ....**S. modesta**
4a. Pod brown, linear-elliptic, tip mucronate, seeds 10 .... **S. catechu**
4b. Pod whitish, broadly elliptic, tip rounded, seeds to 5 ..... **S. hohenackeri**

Senegalia catechu (L.f.) P.J.H.Hurter & Mabb., D.J.Mabberley, Plant-book, ed. 3: 1021. 2008. Synonyms: *Acacia catechu* (L.f.) Wild., *Mimosa catechu* L.f., *Acacia catechu* var. *catechuoides* (Roxb.) Prain, *Mimosa catechuoides* Roxb. Vern.: *Karungali, Katchukattai, Senkarungali* (Tam.).

Deciduous trees to 15 m tall; bark dark greyish-brown to dark brown, rough, narrow rectangular strips; branchlets brown, glabrous. Leaves bipinnate, alternate, pinnae 10-20 pairs, 3.6-8 cm long, opposite, with a gland in between the terminal 6 pairs on the lower side; leaflets 30-50, sessile, stipels absent, 2.5-8 x 0.5-1.5 mm, linear-oblong, base unequally truncate, apex round, obtuse, or mucronate, margin entire, pubescent; lateral nerves paired, 3-10 mm long, straight or hooked; rachis 8-19.5 cm, slender, pulvinate, grooved above. Flowers pale yellow, sessile, in long solitary or in groups of 2-4 axillary spikes; bracts cauducous. Sepals cupular-campanulate, 1-1.5 x 1.3-1.5 mm, teeth triangular or deltoid. Petals 2.5-3 mm long, lobes oblong, ovate to linear-lanceolate. Stamens many, 4.5-5 mm long. Ovary stipitate, 0.8-1.2 mm long, oblong-ellipsoid; style 4-5 mm long; stigma terminal. Pod straight, 5-10 x 1-1.6 cm, unlobed or sinuate along margins, beaked at apex, brown, narrowed at base into a stipe, dehiscent. seeds 3-10, orbicular or ovate, flattened. Fl. & Fr.: May – February.

Western Ghats: Deciduous forests of Gujarat, Maharashtra, Goa, Karnataka, Kerala and Tamil Nadu. India: Most of the states. World: Sri Lanka, Pakistan, Nepal, Bhutan, Myanmar, South China and Thailand.

IUCN: LC (IUCN, 2024).

Senegalia ferruginea (DC.) Pedley., Bot. J. Linn. Soc. 92: 250. 1986. Synonyms: *Acacia ferruginea* DC., *Mimosa ferruginea* Roxb. Vern.: *Seemai Velvael* (Tam.).

Treess to 20 m tall; bark dark brown, rough, deeply fissured; branchlets brownish or blackish. Leaves bipinnate, alternate, pinnae 4-6 pairs, opposite, even pinnate, 4.5-5.5 cm long, slender, with a gland in between terminal 2 pinnae pairs; leaflets 20-44, opposite, sessile, 5-8 x 2-3 mm, linear-oblong, base oblique, apex obtuse, margin entire, glabrous; midrib subcentral, lateral nerves obscure; stipular spines in pairs, nodal, 0.5-6 mm long, hooked or sometimes straight; rachis 5-8 cm, grooved above, pulvinate, glabrous. Flowers pale yellow, 4-4.5 cm long, in spikes arranged in loose panicles; bracts and bracteoles cauducous. Sepals campanulate, 1-1.5 x 1.5-1.8 mm, 5-lobed, puberulous. Petals 2.5-3 mm, twice longer than the calyx, lobes oblong, glabrous. Stamens many, 3-5 mm long, basally connate. Ovary stipitate, terete; style 2.5-3 mm long, filiform; stigma small, terminal. Pod dark brown, 7.5-17.5 x 1.6-1.8 cm, round and apiculate or shortly attenuate at apex, glabrous, transversely reticulate-veined. Seeds 2-6, ovate, flat. Fl. & Fr.: May – October.

Western Ghats: Foot hills of Gujarat, Maharashtra, Goa, Karnataka, Kerala and Tamil Nadu. India: Andhra Pradesh, Delhi, Gujarat, Madhya Pradesh, Odisha, Rajasthan, Uttar Pradesh and West Bengal. World: Sri Lanka and Myanmar.

IUCN: VU- A1c. (IUCN, 2024).

Senegalia hohenackeri (Craib) Ragup., Seigler, Ebinger & Maslin. Phytotaxa 162: 175. 2014. Synonym: *Acacia hohenackeri* Craib Vern.: *Karunkodi* (Tam.).

Scarmbling trees to 8 m tall. Leaves bipairy pinnate, pinna 8-12 pairs, 7 cm long, with a gland between upper 1-2 pairs, leaflets 35-45 pairs, pubescent beneath, 8 x 16 mm; petiole with gland near the base. Petals creamy yellow, twice length of sepals. Pod flat, coriaceous, stalked to 18 mm long. Fl. & Fr.: September

Western Ghats: Deciduous forests of Karnataka and Tamil Nadu (Endemic).

IUCN: NE (IUCN, 2024).

Senegalia modesta (Wall.) P.J.H.Hurter. D.J.Mabberley, Plant-book, ed. 3: 1021. 2008. Synonyms: *Acacia modesta* Wall., *Mimosa dumosa* Roxb., *Mimosa obovata* Roxb.

Small deciduous tree to 10 m tall; bark brownish or greenish grey, rough, prickles in pairs, below the petiole. Pinnae generally 2-3 pairs rarely 1, to 2.5 cm long; leaflets 3-5 pairs, 4-10 x 3-7 mm, broadly ovate or obovate, oblique, obtuse, glaucous, veins prominent; rachis 1.2-5 cm long. Inflorescence a pedunculate spike, to 7.5 cm long, peduncle 1.3-2.5 cm long; pedicel 1 mm long. Sepals 1.5 mm long, broadly campanulate, glabrous. Petals 2-2.5 mm long. Stamens indefinite, filaments 5 mm long. Pod straight, stipitate, stipe 5-6 x 8-10 mm, thin, glabrous, apex deltoid, mucronate, late dehiscing. Seeds 3-5. Fl. & Fr.: April – August.

Western Ghats: Deciduous forests above 700 m in Maharashtra and Tamil Nadu. World: Afghanistan and Pakistan.

IUCN: NE (IUCN, 2024).

Senegalia senegal (L.) Britton. N.L.Britton & P.Wilson, Sci. Surv. Porto Rico & Virgin Islands 6: 538. 1930. Synonyms: *Acacia Senegal* (L.) Willd., *Mimosa Senegal* L.

Small trees to 6 m tall; young branchlets pubescent, old branches glaucous-grey, on older stems the bark peels off in thin flakes of a darker colour. Pinnae 3-5 pairs, opposite, sometimes alternate, 1.2-2.5 cm long; leaflets 8-15 pairs, 2-5 x 1-1.5 mm, linear, obtuse, subsessile. prickles in threes at the base of the petiole, two lateral ones nearly straight or slightly curved upwards, the third recurved 5 mm long; rachis 2.5-5 cm long, with glands between the lowest. Inflorescence a pedunculate spike, peduncle 8-18 mm long, spike 5-10 cm long. Sepals 1.5-2.5 mm long, broadly campanulate, glabrous. Petals 4 mm long. Stamens indefinite, filaments 6-7 mm long. Pod straight, 5-7.5 x 1.7-2.5 cm, thin, shortly stipitate, tip with a slightly curved beak. Seeds 5-6, disc like, almost circular, ovate to linear-ovate, 6-9 x 5-8 mm, smooth, dark brown to greyish green in colour. Fl. & Fr.: August – April.

Western Ghats: Scrub forests of Gujarat and Maharashtra. India: Most of the states. World: Pakistan, Oman and Tropical Africa.

IUCN: NE (IUCN, 2024).

Senna Mill.

1a. Leaflets 2 pairs; pod cylindric .... **S. fruticosa**

1b. Leflets more than 10 paris; pod compressed 2

2a. Inflorescence terminal; leaflets upt to 15 pairs 3

2b. Inflorescence axillary and terminal; leaflets 20-30 pairs 4
3a. Leaflets pubescent; petals sulpur yellow ....**S. surattensis**
3b. Leaflets glabrous; petals orange yellow .... **S. montana**
4a. Calyx ferrugineous; pod more than 15 cm long, to 20 seeds .....**S. siamea**
4b. Calyx glabrous; pod less than 10 cm long, to 10 seeds ...... **S. timoriensis**

Senna fruticosa (Mill.) H.S. Irwin & Barneby., Mem. New York Bot. Gard. 35: 121. 1982. Synonyms: *Cassia fruticosa* Mill.

Small trees to 4 m tall; young parts greyish pubescent. Leaves alternate, leaflets 2 pairs, 4-16.5 x 2-7 cm, upper pair largest, obliquely ovate, acute-acuminate, very unequal-sided at the base, glabrous above, puberulous beneath, membranous; petioles to 3 cm long, pubescent; petiolules 3 mm long; stipules 3-4 mm long, linear, acute, pubescent, caducous. Flowers in short, slender, 4-5 cm long racemes, terminal, congested, corymbiform; bracts linear or lanceolate-subulate, 2-5 mm long, deciduous. Sepals to 1 cm long, oblong, membranous. Petals 2-3 x 2 cm, obovate or suborbicular, pubescent externally, yellow; claw 1-3 mm long. Fertile stamens 7, anthers of three lower filaments 5-6 mm long, puberulous, with 1-porose beak, thick, strongly curved; anthers of 4 filaments 4-6 mm long, slightly incurved; staminodes absent. Ovary curved, appressed pubescent. Pods subterete, 15-25 x 1-1.5 cm, pendent, long-beaked. Seeds 2-seriate, ovoid-oblong, compressed. Fl. & Fr.: September - December.

Western Ghats: Naturalized in semi-evergreen forests of Kerala. World: Native to Tropical America.

IUCN: LC (IUCN, 2024).

Senna montana (Heyne ex Roth) V. Singh., J. Econ. Taxon. Bot. 16: 600. 1992. Synonyms: *Cassia montana* Heyne ex Roth, *Senna glauca* Roxb. Vern.: *Malai kondrai, Malai avarai* (Tam.).

Small trees to 10 m tall; young shoots yellow tomentose. Leaves to 18 cm long; leaflets oblong, obtuse mucronate, base rounded, to 3 x 1.2 cm. Flowers 2.5 cm across, yellow, in axillary and terminal short corymbose panicles. Petals obovate. Stamens 10, upper 3 staminodes. Pod compressed, beaked. Seeds to 20 in a pod, ovoid, longitudinal. Fl. & Fr.: June – December.

Western Ghats: Clonized in deciduous forests of Maharashtra, Goa, Karnataka, Kerala and Tamil Nadu (Endemic). India: Andhra Pradesh.

IUCN: NE (IUCN, 2024).

Senna siamea (Lam.) H.S. Irwin & Barneby., Mem. New York Bot. Gard. 35: 98. 1982. *Cassia siamea* L., *Senna sumatrana* Roxb., *Cassia siamea* var. *puberula* Kurz Vern.: *Manjal kondrai* (Tam.).

Evergreen trees to 15 m tal; bark dark brown, fissured on old wood. Leaves pinnate compound, leaflets 7-10 pairs, ovate-oblong, 2.5-6 x 1.5-2 cm, base and apex obtuse. Flowers in panicles, about 3 cm across, yellow. Fruit is pod, 20-25 cm long, compressed, woody with thick sutures. Fl. & Fr.: August – June.

Westen Ghats: Planted in road sides and avenues of Maharashtra, Karnataka, Kerala, Tamil Nadu. India: Most of the states. World: Native to Malaysia, Laos, Cambodia, Vietnam.

IUCN: LC (IUCN, 2024).

Senna surattensis (Burm.f.) H.S. Irwin & Barneby., Mem. New York Bot. Gard. 35(1): 81. 1982. Synonyms: *Cassia surattensis* Burm.f., *Cassia glauca* var. *suffruticosa* (J.Koenig ex Roth) Baker, *Cassia surattensis* var. *suffruticosa* (J.Koenig ex Roth) Chatterjee, *Cassia spinigera* var. *suffruticosa* (J.Koenig ex Roth) Ali Vern.: *Kovaalai* (Tam.).

Small trees to 7 m tall; branchlets ribbed, thinly pubescent. Leaves alternate leaflets 6-10 pairs, 2-5 x 0.8-2 cm, elliptic or oblong to obovate-oblong, obtuse at apex, oblique at base, glabrous or pubescent beneath; secondary veins 6-9 pairs; petioles 2-4 cm long; stipules upto 1.5 cm long, linear-ensiform, inflexed, pubescent; rachis 8-20 cm long, ribbed. Racemes 5 to 20-flowered, in the axils of upper leaves, corymbose; peduncles 2.5-8 cm long; pedicels 1.5-2.5 cm long, thinly pubescent; bracts ovate-lanceolate, 2-7 x 1-2.7 mm, acute-acuminate, reflexed, with 2 subclavate glands. Sepals obtuse, pubescent outside in bud, outer 2 half the length of inner ones, inner ones 5-8 mm long. Petals 0.8-2.5 cm long and about as wide, yellow. Stamens 10, subequal, all fertile; anthers with narrow and deflexed top. Ovary stipitate, silky hairy; style 1.5-3 mm long, pubescent; stigma punctiform. Pod thin-valved, 4-10 x 0.8-1.8 cm, sepatate, dark brown. Seeds 10-13, brown, 5-7 x 2.5-4 mm, oblong, narrowed at hilum end, blackish, shiny. Fl. & Fr.: September – March.

Western Ghats: Cultivated avenue trees in Maharashtra, Goa, Karnataka, Kerala and Tamil Nadu. India: Most of the states. World: Bangladesh, Myanmar and Bhutan.

IUCN: LC (IUCN, 2024).

Senna timoriensis (DC.) H.S. Irwin & Barneby., Mem. New York Bot. Gard. 35: 98. 1982. Synonyms: *Cassia timoriensis* DC., *Cassia goensis* Dalzell, *Cassia eglandulosa* Dum.Cours. Vern.: *Chunnambu avaram* (Tam.).

Small trees to 10 m tall; branchlets pubescent, angled. Leaflets 4-5 x 1.5 cm, oblong-elliptic, apex mucronate, base slightly unequal; petiole 2.5 cm, stipules large cordate, acuminate. Racemes axillary, shorter than the leaves; peduncles 10 cm. Flowers yellow. Sepals 1.5 x 1 cm, ovate, hairy. Petals 2 x 1 cm, veined. Stamens 6-7 antheriferous, staminodes 4-3, various in length. Pod flat, stipitate, pubescent when young, transversely septate. Fl. & Fr.: October – May.

Western Ghats: Cultivated trees in estates in Maharashtra, Goa, Karnataka, Kerala and Tamil Nadu. India: Andhra Pradesh, Arunachal Pradesh, Assam, Manipur, Meghalaya, Mizoram, Nagaland, Tripura and West Bengal. World: Myanmar, Laos, Cambodia, Vietnam, Malesia and Australia.

IUCN: LC (IUCN, 2024).

Tamarindus L.

Tamarindus indica L., Sp. Pl.: 34. 1753. Synonyms: *Tamarindus officinalis* Hook.f., *Tamarindus occidentalis* Gaertn.

Trees to 20 m tall; bark brown to brownish-black, rough with vertical fissures; branchlets warty, tomentose. Leaves paripinnate, alternate, leaflets 20-34, opposite, sessile, 1.5-4 x 0.4-1.3 cm, oblong, apex obtuse, base unequal, margin entire, glabrous; lateral nerves 10-15 pairs, pinnate, slender, obscure, looped at the margin forming intramarginal nerve; intercostae reticulate, obscure; stipules lateral, minute, cauducous; rachis 8-13 cm long, glabrous, pulvinate; Flowers yellow, 1 cm across, in lax terminal racemes; bracts and bracteoles ovate-oblong, coloured, cauducous; pedicels upto 5 mm. Seplas 4, tube narrowly turbinate, lined by disc; lobes 4, subequal, oblong. Petals 3, outer one, 1 x 0.3 cm, rolled up, pink dotted, lateral 2, 1-1.5 x 0.7-1 cm, clawed, subequal, oblong-lanceolate, lower pair scaly. Stamens 9 monadelphous, only 3 fertile, others reduced to bristle, base pubescent; anthers versatile. Ovary half inferior, stipitate, adnate to the disc, ovules many; style attenuate, tomentose; stigma globose. Pod oblong, 10-15 x 1-2 cm, fruit wall crustaceous, indehiscent. Seeds 3-8 or more, obovoid-orbicular, compressed, brown. Fl. & Fr.: December – April.

Western Ghats: Common in foot hills and scrub forests of Maharashtra, Goa, Karnataka, Kerala and Tamil Nadu. World: Native to Tropical Africa.

IUCN: LC (IUCN, 2024).

Vachellia Wight & Arn.

1a. Inflorescence spicate; pod curved, broader than long................**V. horrida**
1b. Inflorescence heads; pod linear, compressed 2
2a. Head white coloured; peduncle shorter than 1 cm 3
2b. Head usually yellow; peducnlce more than 1 cm long 4
3a. Heads in terminal panicle; bark yellowish white ....**V. leucophloea**
3b. Head in axillary clusters; bark dark brown .... **V. panifrons**
4a. Stipular thorns shorter than 4 cm long; pinna 4-6 pairs 5
4b. Stipular thorns longer than 4 cm long; pinna 2-4 pairs 7
5a. Pod constricted moniliform, glabrous or velvety ....**V. niloctica**
5b. Pod straight not constricted, tomentose or pubescens 6
6a. Leaflets 10-20 pairs; head orange-yellow ....**V. bolei**
6b. Leaflets 8-12 pairs; head pale yellow .... **V. tomentosa**
7a. Leaflets 2-4 pairs; stipular throns to 4 cm long .... **V. jacquemontii**
8a. Leaflets 4-8 paris; stipular thorns to 5.5 cm long .... **V. eburnea**

Vachellia bolei (R.P.Subhedar) Ragup., Seigler, Ebinger & Maslin. Nuytsia 24: 21 (2014). Synonyms: *Acacia wightii* Bakerr ex Benth., *Acacia bolei* R.P.Subhedar, *Vachellia wightii* Ragup., Seigler, Ebinger & Maslin

Thorny trees to 8 m tall; bark smooth, greyish. Leave bipairy pinnate, pinna 4-6 pairs, to 5 cm long, leaflets small, 0.8-1.3 x 0.4-1 cm, pubescent, base obtuse, acuminate at apex; stipular thorns unequl, to 4 cm long, stout at base, whitish. Flowers in axillary peduncled globose head, orange yellow. Fl. & Fr.: September – February.

Western Ghats: Deciduous forests of Kerala and Tamil Nadu (Endemic).

IUCN: CR-D (Exticnt according to POWO) (IUCN, 2024).

Note: Sampathkumar *et al.*, (2019) mad a critical taxonomic assessment on *Vachellia bolei*. It was found that Wight had collected specimens from sea coasts of Tirunelveli and Travancore with the same field number (896), and two specimens were deposited at Royal Botanic Gardens, Kew (GBIF 2018; http://specimens.kew.org/herbarium/ K000791142, K000791143), besides the single specimen housed at CAL. It may be pointed out that though the original set of Wight's collections studied by Arnott were deposited at the herbarium of Royal Botanic Garden, Edinburgh, there is no specimen of *V. bolei* (or *A. wightii* Baker ex Bentham) at E (online Herbarium Catalogue at E) nor any specimen at MH too (Henry *et al.* 1978). The specimen collected by Wight in June, 1835 from Tirunelveli sea coast deposited at Kew (referred here Wight 896A) had been designated as the Lectotype (K000791142) by Subhedar (1985:440), whereas the duplicate (Wight 896B) of this specimen has been designated as the Isolectotype is at CAL (CAL0000012942). The third specimen of the taxon, not mentioned by Subhedar (1985: 440), and collected again by R Wight in June 1835 from Travancore (referred here as Wight 896C) have dissected floral parts supplemented with hand drawings in the sheet by JS Gamble was also housed at K (K000791143). It is important to note that all the three specimens have only flowers but no fruits. Recently Deshpande *et al.* (2018) has made the second-step lectotypification of the *V. bolei* wherein the K specimen of Wight 896C (K000791143 image!) collected from the erstwhile Travancore coast has been designated as the II-step lectotype, while the I-step lectotype of the specimen collected from the Tiruneveli sea coast at Kew (Wight 896A: K000791142) has been attributed to Subhedhar (1985: 440). But the specimen collected from Tirunelveli coast in June 1835 by Wight (Wight 896A: K000791142 image!) has been identified as the isolectotype which is not acceptable since the two sheets (K000791142 & K000791143) are not duplicates though they have the same field numbers (Wight 896 A & C). As the two specimens collected from Tirunelveli coast by Wight in June, 1835 (Wight 896 A & B, one at K and the other CAL respectively) in fact are duplicates (K000791142 & CAL0000012942). It should be noted that the II-Step lectotype designated was in fact collected from Travancore without mentioning any specific locality thereof and it is likely to have been collected from the West Coast, and not likely from the East Coast part of the present Kanniyakumari district (formerly with Tirunelveli district till November 01, 1956), since the region Travancore per se refers only to the West Coast of Southwestern India.

Vachellia eburnea (L.f.) P.J.H.Hurter & Mabb. D.J.Mabberley, Plant-book, ed. 3: 1021. 2008. *Acacia eburnea* (L.f.) Willd., *Acacia minutifolia* Ragup., A.Mahad. & Thoth., *Acacia pseudowightii* Thoth., *Vachellia pseudoeburnea* (J.R.Drumm. ex Dunn) Ragup., Seigler, Ebinger & Maslin Vern: *Chillodai, Udaivel* (Tam.).

Small trees to 8 m tall; bark rough dark grey. Leaves bipairipinnate, pinnae 2-8 pairs, 6-13 mm long; leaflets 5-8 pairs, 2.5-3 mm long, linear, acute to subobtuse, glabrous; stipular spines straight, to 5.5 cm long, tapering from a broad base, white. Flowers in axillary, pedunculate globose heads, bright yellow, odour disagreeable; peduncle to 2.5 cm long with involucre of connate bracts about the middle. Sepals campanulate, 1-1.5 mm long. Petals 2-3 mm long. Ovary stalked, glabrous. Pod stalked, 7.5-15 cm long, 6-9 mm broad, slightly curved, indented on both the sutures, glabrous. Seeds 6-10. l. & Fr.: October – April.

Western Ghats: Scrub forests and foot hills of Maharashtra, Karnataka and Tamil Nadu. World: Sri Lanka, Pakistan, Afghanistan and Oman.

IUCN: NE (IUCN, 2024).

Vachellia horrida (L.) Kyal. & Boatwr., Bot. J. Linn. Soc. 172: 513. 2013. Synonyms: *Acacia horrida* (L.f.) Willd., *Mimosa horrida* L. Vern.: *Anaimullu, Karodai, Odaimullu* (Tam.).

Small trees to 8 m tall; branchlets black or brownish, terete, glabrous. Leaves bipinnate: pinnae 2-5 pairs, 0.5-1.5 cm long, sparsely puberulous to glabrous; leaflets 6-15 pairs, subsessile, linear or oblong, 0.8-4 x 0.4-1 mm, subequal or rounded at base, obtuse to rounded or subacture at apex, glabrous; main vein prominent, nervules obscure; stipular spines in pairs, 0.5-6 cm long, ivory-white or often brown, the larger ones hollow and horn-like; rachis 1-3 cm long, slender, sparsely yellowish puberulous to glabrous; petioles 0.5-1 cm long. Inflorescences: axillary, spicate, solitary or 2-together, glabrous, 2-5.5 mm long; peduncles 0.5-1 cm long. Flowers: creamy white or yellowish, fragrant. Sepals campanulate, 1-2 x 1.5-1.8 mm, glabrous; teeth triangular or ovate, 0.5-0.8 mm long. Petals 2-3 mm long, glabrous; lobes 4-5, triangular, 1-1.5 x 0.6-1 mm. Stamens 3-4.5 mm long. Ovary subsessile, subglobose; 1.5 mm diam, style 3-3.5 mm long. Pods falcate-oblong, 2.5-4.5 x 1-2 cm, inflated, straight to curved, unlobed, rounded to obtuse at apex, glabrous, dark reddish-brown or blackish when dry, stipe 5 mm long. Seeds 1-4, ovate to suborbicular, flattened. Fl. & Fr.: June – February.

Western Ghats: Common in scrub forests of Maharashtra, Goa, Karnataka, Kerala and Tamil Nadu. India: Arunachal Prdesh, Andhra Pradesh, Delhi, Gujarat, Odisha and Rajasthan. World: Myanmar and Africa.

IUCN: NE (IUCN, 2024).

Vachellia jacquemontii (Benth.) Ali., Pakistan J. Bot. 46: 3. 2014. Synonyms: *Acacia jacquemontii* Benth.

Small trees to 5 m tall; young shoots slightly puberulous, twigs zigzag, bark greyish brown. Pinnae 1-4 pairs, 5-15 cm long; leaflets 5-10 pairs, sessile, 2.5-4 x 1-1.5 mm, oblong, obtuse, glabrous; stipular spines, straight, white, connate at the base, 1.5-5 cm long. Rachis 0.8-5 cm long, glabrous, usually with a gland between the upper pair of pinnae. Inflorescence pedunculate heads in axillary fascicles of 2-8, rarely shortly racemose or umbellate. Flowers yellow, fragnant, peduncle 1.2-2.5 cm long; bracts 2-3, about the middle of the peduncle. Sepals 1 mm long, campanulate. Petals 2.5 mm long. Pod oblong-elliptic, 5-7.5 cm long, 0.8-1.7 cm broad, glabrous, reticulately veined, 5-6 seeded. Fl. & Fr.: December – July.

Western Ghats: Scrub forests of Karnataka. World: India and Pakistan.

IUCN: LC (IUCN, 2024).

Vachellia leucophloea (Roxb.) Maslin, Seigler & Ebinger. Blumea 58: 42. 2013. Synonyms: *Acacia leucophloea* (Roxb.) Willd., *Mimosa leucophloea* Roxb. Vern.: *Velvaelam* (Tam.).

Trees to 20 m tall; bark, young branches velvety pubescent. Pinnae 3-15 pairs, to 3.8 cm long; leaflets 6-30 pairs, sessile, 2.5-8 mm long, 1-2 mm broad, linear-oblong, glabrous or nearly so, obtuse, base oblique; stipular spines in pairs, straight, to 3-25 mm; rachis 2.5-10 cm long, usually with glands between the lowest and uppermost 2-3 pairs of pinnae. Inflorescence pedunculate heads arranged in large terminal tomentose panicles, lower branches axillary; peduncle 7-10 mm long, tomentose, bracts whorled and fused basally, about or below the middle of the peduncle. Sepals campanulate, 1.2 mm long, villous. Petals 2.5 mm long, hairy outside. Pod sessile, 10-15 x 7-10 mm, brown velvety when young, curved, semi-dehiscent. Seeds 10-20, brown. Fl. & Fr.: July – February.

Western Ghats: Foot hills and scrub forests up to 600 m in Maharashtra, Karnataka, Kerala and Tamil Nadu. India: Almost all the states except North-east. World: Sri Lanka, Pakistan, Myanmar, Thailand, Laos, Cambodia, Vietnam, Malaysia, Indonesia and Sunda Islands.

IUCN: LC (IUCN, 2024).

Vachellia nilotica (Schumach.) Kyal. & Boatwr. Bot. J. Linn. Soc. 172: 515. 2013 ssp. **adstringens**. Synonyms: *Acacia nilotica* ssp. *adstringens* (Schumach. &Thonn.) Roberry, *Mimosa adstringens* (Schumach. &Thonn.) Baker.f., *Acacia nilotica* var. *adansonii* (Guill. & Perr.) Brenan, *Acacia adansonii* Guill. & Perr. Vern.: *Karuvaelam* (Tam.).

Trees to 18 m tall; bark on trunk rough, fissured, blackish, grey or brown; young branches almost glabrous to subtomentose. Leaves with 1-2 petiolar glands, pinnae 2-11 pairs; leaflets 7-25 pairs, 1.5-7 x 0.5-1.5 mm, glabrous to pubescent. Inflorescence axillary pedunculate heads, 6-15 mm in diam; stipules spinescent, to 8 cm long. Flowers bright yellow, involucel from near base to half way up the peduncle. Sepals 1-2 mm long, pubescent or subglabrous. Petals 2.5-3.5 mm long, glabrous to more or less pubescent outside. Pods very variable, indehiscent, straight or curved, glabrous to velvety, 22 x 1.3-2.2 cm. Seed blackish brown, smooth, 7-9 x 6-7 mm, subcircular, compressed. Fl. & Fr.: August – December.

Western Ghats: Foot hills areas of Maharashtra. World: India, Pakistan, West Asia, Africa.

IUCN: LC (IUCN, 2024).

Key to the subspecies

1a. Pods necklace-like, narrowly and regulary constricted between the seeds 2

1b. Pods not necklace-like, margins straight or crenate, not regularly constricted between the seeds 3

2a. Crown narrow and erect, cuppressus like ....ssp. **cupressiformis**

2b. Crown not narrow and erect ....ssp. **indica**

3a. Branchlets densely toments; pods densey or persistently tomentose, with dinstinctly oftern irregularly crenate margin ....ssp. **adstringens**

3b. Branchletes densely pubescent to tomentose; pods densely subtomentose, with straight or slightly crenate margin ....ssp. **subulata**

Vachellia nilotica ssp. **cupressiformis** (J.L.Stewart) Ali., Pakistan J. Bot. 46: 2. 2014. Synonyms: *Acacia nilotica* ssp. *cupressiformis* (J.L. Stewart) Ali &Faruqi, *Acacia arabica* var. *cupressiformis* J.L. Stewart, Vajr. & Kamble, *Acacia nilotica* ssp. *cupressiformis* (J.L.Stewart) Ali & Faruqi

Fl. & Fr.: August – February.

Western Ghats: Lower tracts of Maharashtra. World: India, Pakistan.

IUCN: NE (IUCN, 2024).

Vachellia nilotica ssp. **indica** (Benth.) Kyal. & Boatwr., Bot. J. Linn. Soc. 172: 515. 2013. *Acacia nilotica* ssp. *indica* (Benth.) Brenan, *Acacia arabica* var. *indica* Benth., *Acacia nilotica* var. *indica* (Benth.) A.F.Hill

Fl. & Fr.: June - January.

Western Ghats: Common in foot hills of Maharashtra, Karnataka, Kerala, Tamil Nadu. World: India, Sri Lanka, Nepal, Pakistan, Myanmar, South West Asia, Tropical Africa.

IUCN: NE (IUCN, 2024).

Vachellia nilotica ssp. **subalata** (Vatke) Kyal. & Boatwr., Bot. J. Linn. Soc. 172: 515. 2013. *Acacia nilotica* ssp. *subulata* (Vatke) Brenan, *Acacia arabica* var. *vediana* T.Cooke, *Acacia nilotica* ssp. *vediana* (Cooke) Vajr. & Kamble

Fl. & Fr.: July – December.

Western Ghats: Occasional in the low altitudes of Maharashtra. World: India, Pakistan, Tropical Africa.

IUCN: NE (IUCN, 2024).

Vachellia planifrons (Wight & Arn.) Ragup., Seigler, Ebinger & Maslin. Phytotaxa 162: 177. 2014. Synonyms: *Acacia planifrons* Wight & Arn., *Acacia roxburghii* Wight & Arn., *Mimosa planifrons* K.D.Koenig ex Wight & Arn., *Mimosa umbrifera* Banks ex Wight & Arn. Vern.: *Kodai vaelam, Udai maram, Kodaiseethai* (Tam.).

Trees to 7 m tall; bark greyish-brown, thick with horizontal markings; branchlets warty. Leaves bipinnate, alternate, fascicled on abbreviated branchlets; pinnae 3-5 pairs, 0.7-1 cm, opposite, even pinnate; leaflets 10-30, opposite, 1-3 x 0.5-1 mm, linear-oblong, base oblique, apex obtuse, margin entire, glabrous, nerves obscure; stipular spines in pairs, unequal, short ones recurved, brown, long one straight, divergent; rachis 8-12 mm, slender, pulvinate, glabrous, with a gland towards the base. Flowers white, heads axillary, 2-3 together, peduncles with involucel at the base. Sepals campanulate, 1-2 x 1.5-2.5 mm, glabrous. Petals funnel shaped, 3-4 mm long, lobes ovate to oblong. Stamens many, 3-4 mm long; filaments basally connate; Ovary 1.5 mm long, sessile, glabrous; style 2.5 mm long. Pod slightly fleshy, 5-10 x 0.6-0.7 cm, entire along margin, acute at apex, glabrous. Seeds 6-9. Fl. & Fr.: October - April.

Western Ghats: Common in umbrella thorn forests to 600 m in Maharashtra, Karnataka, Kerala and Tamil Nadu. India: Most of the states. World: Sri Lanka.

IUCN: NE (IUCN, 2024).

Vachellia tomentosa (Rottler) Maslin, Seigler & Ebinger. Blumea 58: 43. 2013. Synonyms: *Acacia tomentosa* Willd., *Mimosa tomentosa* Rottl., *Acacia chrysocoma* Miq., *Mimosa cavaty-tooma* Roxb. ex DC. Vern.: *Anaimullu* (Tam.).

Trees to 6 m tall; bark smooth, greyih, pubescent when young. Leaves bipairy pinnate, to 5 cm long, pinnae 10-12 pairs, 2.5 cm long, leaflets 15-20 pairs, oblong, 0.2 x 0.1 cm; stipulate thorns to 3 cm long, whtish; petiole gland below the lowest pair of pinnae. Head globose, yellow, 1.2 cm across; bracts near the middle of peduncle. Petals greenish white, two times longer than the sepals, pubescent. Pod flat, slightly falcate, stalked, 6-10-seeded. Fl. & Fr.: December - March.

Western Ghats: Occasional in the dry forests near the streams of Gujarat, Maharashtra, Tamil Nadu. India: Andhra Pradesh, Gujarat, Madhya Pradesh, Odisha, Rajasthan, Uttar Pradesh and West Bengal. World: India, Sri Lanka, Bangladesh, Myanmar, Thailand, Vietnam, Malesia.

IUCN: NE (IUCN, 2024).

Xylia Benth.

Xylia xylocarpa (Roxb.) W. Theob., Burmah, ed. 3, 2: 541. 1883. Synonyms: *Acacia xylocarpa* (Roxb.) Willd., *Inga xylocarpa* (Roxb.) DC., *Mimosa xylocarpa* Roxb. Vern.: *Iruvel, Iruli* (Tam.).

Deciduous trees to 20 m tall; bark 10-15 mm thick, reddish-brown, mottled with white, rough, exfoliations irregular. Leaves bipinnate, alternate; pinnae 2, opposite, even pinnate 8-30 cm, slender, glabrous; leaflets 4-18, opposite, with a gland between each pairs, 3.5 x 2.5-6 cm, oblong-acuminate, elliptic or elliptic-oblong, base obtuse or acute, apex acute or acuminate, margin entire, glabrous; lateral nerves 6-12 pairs, intercostae reticulate;' stipules small, free, lateral, deciduous; rachis 4-18 cm long, stout, glabrous, pulvinate. Flowers bisexual, sessile, dull yellow, in axillary globose heads; bracteole club shaped. Sepals in tubular, 4 mm long, 5-toothed. Petals 5, 1.5 cm long, linear, slightly connate at base. Stamens 10, free, exserted, filaments slender; anthers tipped with a stalked gland which is early deciduous. Ovary sessile, style filiform; stigma terminal. Pod woody, 10-15 x 5-6 cm, broadly falcate, compressed, dehiscent. Seeds 4-10, oblong, compressed, brown, shining. Fl. & Fr.: February - December.

Western Ghats: Semi-evergreen forests of Maharashtra, Goa, Karnataka, Kerala and Tamil Nadu. India: Andhra Pradesh, Bihar, Gujarat, Madhya Pradesh, Meghalaya, Odisha and West Bengal. World: Myanmar, Laos, Cambodia, Vietnam, Malesia and Singapore.

IUCN: LC (IUCN, 2024).

Cultivated:

Acacia auriculiformis A.Cunn. ex Benth.; Exotic: Native to Australia, New Guinea.

Acacia baileyana E Muell.; Exotic: Native to South East Australia.

Acacia balfourii Woodrow; Exotic: Native to Aftica.

Acacia burkei Benth.; Exotic: Native to Southern Africa.

Acacia concurrens Pedley; Exotic: Native to Australia.

Acacia confusa Merr.; Exotic: Native to Philippines.

Acacia cornigera (L.) Willd.; Exotic: Native to Mexico.

Acacia cultriformis A. Cunn. ex G. Don; Exotic: Native to Australia.

Acacia dealbata Link; Exotic: Native to Australia.

Acacia decora Rchb.; Exotic: Native to Australia.

Acacia decurrens (J. C. Wendl.) Willd.; Exotic: Native to Australia.

Acacia elata A. Cunn, ex Benth.; Exotic: Native to Australia.

Acacia farnesiana (L.) Willd.; Exotic: Native to America.

Acacia holosericea A. Cunn. ex G. Don.; Exotic: Native to Australia.

Acacia linifolia (Vent.) Willd.; Exotic: Native to Australia.

Acacia longifolia (Andrews) Willd.; Exotic: Native to Australia.

Acacia maidenii F. Muell.; Exotic: Native to Australia.

Acacia mangiurn Willd.; Exotic: Native to Maluku Islands, Australia.

Acacia mearnsii De Wild.; Exotic: Native to Australia.

Acacia melanoxylon R. Br.; Exotic: Native to Australia.

Acacia mellifera (Vahl) Benth.; Exotic: Native to Africa, West Asia.

Acacia myrtifolia (Sm.) Willd.; Exotic: Native to Australia.

Acacia neriifolia A. Cunn, ex Benth.; Exotic: Native to Australia.

Acacia nilotica ssp. **tomentosa** (Benth.) Brenan; Exotic: Native to Egypt.

Acacia omalophylla A. Cunn. ex Benth.; Exotic: Native to Australia.

Acacia paradoxa DC.; Exotic: Native to Australia.

Acacia podalyriifolia A. Cunn. ex G. Don; Exotic: Native to Australia.

Acacia pruinose A. Cunn. ex Benth.; Exotic: Native to Australia.

Acacia pulchella R. Br.; Exotic: Native to Australia.

Acacia pycnantha Benth.; Exotic: Native to Australia.

Acacia retinodes Schltdl.; Exotic: Native to Australia.

Acacia richii A. Gray; Exotic: Native to Pacific Islands.

Acacia robusta Burch.; Exotic: Native to Africa.

Acacia sphaerocephala Schltdl. & Cham,; Exotic: Native to Mexico

Acacia verticillata (L'Hér.) Willd.; Exotic: Native to SE. Australia.

Albizia lucidior (Steud.) C. Nielsen & H. Hara; Native: India, Nepal, Bhutan, Bangladesh, Myanmar, China, Singapore, Thailand, Laos, Cambodia, Vietnam.

Albizia niopoides (Spruce ex Benth.) Burkart; Exotic: Native to Central and South America, West Indies.

Amherstia nobilis Wall.; Exotic: Native to Myanmar

Bauhinia acuminata L.; Native: India, Bangladesh, Sri Lanka, Myanmar, China, Malaysia

Bauhinia monandra Kurz; Exotic: Native to Myanmar

Bauhinia rufescens Lam.; Exotic: Native to Tropical Africa

Brownea grandiceps Jacq.; Exotic: Native to Tropical America

Brya ebenus (L.) DC.; Exotic: Native to Cuba, Jamaica

Caesalpina sappan L.

Caesalpinia coriaria (Jacq.) Willd.; Exotic: Native to Tropical America

Caesalpinia pulcherrima (L.) Sw.; Exotic: Native to Tropical America

Caesalpinia punctata Willd.; Exotic: Native to Tropical America

Calliandra brevipes Benth.; Exotic: Native to South America.

Calliandra calothyrsus Meisn.; Exotic: Native to Central America.

Calliandra haematocephala Hassk.; Exotic: Native to Bolivia.

Calliandra inermis (L.) Druce; Exotic: Native to Mexico.

Calliandra surinamensis Benth.; Exotic: Native to Tropical America.

Cassia grandis L.f.; Exotic: Native to Tropical America

Cassia javanica ssp. **nodosa** (Buch.-Ham. ex Roxb.) K.Larsen & S.S.Larsen; Native to India, Bangladesh, Myanmar, Thailand, Malesia

Cassia renigera Wall. ex Benth.; Exotic: Native to Myanmar

Castanospermum austral A.Cunn. ex Mudie; Exotic: Native to Australia

Ceratonia siliqua L.; Exotic: Native to North Africa, Europe, Middle East.

Colvillea racemosa Bojer; Exotic: Native to Madagascar

Cynometra cauliflora L.; Exotic: Native to Malaysia

Delonix elata (L.) Gamble; Exotic: Native to Tropical Africa, West Asia

Delonix regia (Bojer ex Hook.) Raf.; Exotic: Native to Madagascar

Enterolobium cyclocarpum (Jacq.) Griseb.; Exotic: Native to Mexico, South America.

Erythrina crista-galli L.; Exotic: Native to South America

Falcataria moluccana (Miq.) Barneby & J. W. Grimes; Exotic: Native to Malesia.

Gliricidia sepium (Jacq.) Kunth; Exotic: Native to Tropical America

Haematoxylum campechianum L.; Exotic: Native to Tropical America

Hymenaea verrucose Gaertn.; Exotic: Native to Tropical East Africa, Madagascar, Mauritius, Seychelles Islands.

Indigofera zollingeriana Miq.; Exotic: Native South East Asia, Laos, Vietnam, Malesia

Laburnum anagyroides Medik.; Exotic: Native to Europe

Leucaena leucocephala (Lam.) de Wit; Exotic: Native to Mexico, Central America.

Lonchocarpus sericeus (Poir.) Kunth ex DC.; Exotic: Native to Tropical America, West Africa

Lysiloma latisiliquum Benth.; Exotic: Native to West Indies.

Millettia peguensis Ali.; Exotic: Native to Myanmar, Thailand, Cambodia, Vietnam

Myroxylon balsamum (L.) Harms; Exotic: Native to South and Central America

Paraserianthes lophantha (Willd.) I. C. Nielsen; Exotic: Native to Malesia, Australia. Flowering: August-September.

Parkia biglandulosa Wight & Arn.; Exotic: Native to Africa.

Parkia timoriana (DC.) Merr.; Native: India, Myanmar, Bangladesh, Malay Peninsula.

Peltophorum pterocarpum (DC.) Backer ex K.Heyne; Exotic: Native to Sri Lanka, Thailand, Vietnam, Malaysia, Indonesia

Pericopsis mooniana (Thwaites) Thwaites; Exotic: Native to Sri Lanka, Malesia, Indonesia, Australia

Pithecellobium unguis-cati (L.) Benth.; Native: Mexico to Venezuela, Florida to Caribbean

Prosopis glandulosa var. **glandulosa**

Prosopis glandulosa var. **torreyana**

Psoralea pinnata L.; Exotic: Native to South Africa

Pterocarpus dalbergioides Roxb. ex DC.; Native Andaman Islands

Pterocarpus indicus Willd.; Exotic: Native to South East Asia, China, Cambodia, Vietnam, Malesia

Pterocarpus santalinus L.f.; Endemic to Eastern Ghats, Cultivated in Western Ghats.

Schizolobium parahyba (Vell.) S.F.Blake; Exotic: Native to South America

Schotia brachypetala Sond.; Exotic: Native to South Africa

Senna multiglandulosa (Jacq.) H.S.Irwin & Barneby; Exotic: Native to Tropical America

Senna multijuga (Rich.) H.S.Irwin & Barneby; Exotic: Native to Tropical America

Sesbania cinerascens Welw. ex Baker; Exotic: Native to Tropical Africa

Sesbania grandiflora (L.) Poir.; Exotic: Native to Indonesia

Sesbania speciosa Taub. ex Engl.; Exotic: Native to East Africa, New Guinea

Sophora secundiflora (Ortega) Lag. ex DC.; Exotic: Native to Tropical America

GENTIANACEAE Juss.

Fagraea Thunb.

Fagraea ceilanica Thumb., Kongl. Vetensk. Acad. Nya Handl. 3: 132. 1782. Synonyms: *Fagraea obovata* Wall, *Fagraea malabarica* Wight, *Fagraea obovata* var. *gardneri* (Thwaites) C.B.Clarke

Small trees to 6 m tall; usually epiphytic on tree trunks; branchlets stout with prominent leaf-scars. Leaves obovate or oblanceolate, 10-18 x 5-8 cm, base cuneate, margin entire, apex obtuse, fleshy, shiny, glabrous; petiole to 2.5 cm long. Inflorescence terminal or axillary, few-flowered corymbose cymes. Flowers large, 6-8 cm across, creamy white. Sepals 5, campanulate, deeply 5-lobed; lobes 8 mm long, obovate, acute or obtuse. Petals 5, funnel-form; tube 4-5 cm long; lobes 5, broad, twisted to right. Stamens 5, adnate at base of corolla-tube. Ovary 2-celled; ovules many in each cell; stigma capitate. Berry ellipsoid, 2-4 x 1.5-3 cm, shiny. Seeds many. Fl. & Fr: June - Sepember.

Western Ghats: Evergreen forests of Maharashtra, Karantaka, Kerala and Tamil Nadu. World: Sri Lanka and Malaysia.

IUCN: NE (IUCN, 2024).

HAMAMELIDACEAE R.Br.

Exbucklandia R.Br.

Exbucklandia populnea (R.Br. ex Griff.) R.W.Br., J. Washington Acad. Sci. 36: 348. 1946. Synonyms: *Bucklandia populnea* R.Br. ex Griff., *Liquidambar tricuspis* Miq., *Bucklandia populifolia* Hook.f. & Thomson

Trees to 20 m tall; branchlets pubescent, bark rough, greyish. Leaves broadly ovate-rounded or sometimes palmately 3-lobed in young leaves, 10-17 x 9-13 cm, abaxially glabrous, adaxially drying dark green, shiny, base broadly cuneate, margin entire, palmately 5-7-veined; stipules elliptic or obovate, 2-3 x 1-2 cm; petiole 3–6 cm, longer in young leaves, glabrous. Inflorescences capitate, sometimes arranged in racemes; heads 8-12-flowered; peduncle 1-2 cm, pubescent. Petals usually absent. Stamens to 5 mm, filaments slender, glabrous. Ovary yellow-brown pubescent; styles 3-4 mm, erect. Capsules ellipsoid, 7-9 x 5-6 mm, smooth, dehiscing above middle. Seeds narrowly winged. Fl. & Fr.: April – August.

Western Ghats: Cultivated trees in Kodaikanal of Tamil Nadu. World: China, Nepal, Malaysia and Sumatra.

IUCN: LC (IUCN, 2024).

Uses: The tree is cultivated inhigh altitudes for its evergreen dense green foliage as ornamental and avenue tree.

HERNANDIACEAE Blume

Gyrocarpus Jacq.

Gyrocarpus americanus Jacq., Select. Stirp. Amer. Hist.: 282. 1763. Synonyms: *Gyrocarpus jacquinii* Gaertn. *Gyrocarpus asiaticus* Willd. Vern.: *Thanakku, Venthanakku* (Tam.).

Deciduous trees to 20 m tall; bark silvery grey or greyish-yellow, smooth, leafscars prominent, shining. Leaves simple, alternate, crowded at apex, broadly ovate, base truncate or subcordate, 10-18 x 9-15 cm, apex acute or acuminate, margin entire or some times 3-lobed, glabrous above and appressed pubescent below; 3-ribbed from base, palmate, lateral nerves 3-5 pairs, forked, intercostae scalariform, faint; petiole 10-15 cm long, slender, puberulus. Flowers unisexual, greenish-yellow, in subterminal panicled cymes to 4 cm, tomentose. Male flowers 5 mm across. Tepals 4-7, to 1.5 mm, unequal, ovate, concave, pubescent. Stamens 4-7, inserted at the bottom of the perianth, alternating with as many staminodes; filaments 2 mm; staminodes glandular, obtuse, flat. Female flowers 3 mm across. Tepals 4-7, basally adnate to the ovary, pubescent. Ovary inferior, 1-celled, ovule 1, pendulous; stigma inferior. Drupe ovoid, 1 x 0.8 cm, pubescent, wings 2, oblanceolate. Seed one. Fl. & Fr.: June – Sepember.

Western Ghats: Common in deciduous forests of Maharashtra, Karnataka, Kerala and Tamil Nadu. India: Andhra Pradesh, Odisha and Telangana. World: Malaysia, Africa, Pacific Islands and Tropical America.

IUCN: LC (IUCN, 2024).

ICACINACEAE Miers

Mappia Jacq.

Mappia nimmoniana (Grah.) Byng & Stull., Global Fl. 4: 145. 2018. Synonyms: *Nothapodytes nimmoniana* (J.Graham) Mabb., *Premna nimmoniana* J. Graham, *Nothapodytes foetida* (Wight) Sleumer, *Stemonurus foetidus* Wight, *Mappia foetida* (Wight) Miers, *Mappia wightiana* Miers, *Mappia tomentosa* Miers, *Mappia ovata* Miers, *Mappia oblonga* Miers Vern.: *Araali* (Tam.).

Trees to 15 m tall; branches conspicuously angular. Leaves alternate, elliptic-ovate or lanceolate-oblong, 10-20 x 5-12 cm, membranous to thinly leathery, abaxially glabrous or sparsely hispid or pubescent, adaxially sparsely tomentose; lateral veins 7 or 8 pairs, unequal; petiole 3-5 cm, adaxially grooved. Cymes or corymbs terminal; peduncle flattened or angled, tomentose. Sepals campanulate, 2.5-3 mm in diam, strigose or glabrous, 5-toothed. Petals lanceolate-oblong, 4.2-5 x 1.5-1.8 mm, outside appressed strigose, inside villous, apices incurved. Filaments 3-4 mm; anthers ovoid, 1-1.2 mm. Disc slightly lobed, outside glabrous, margin and inside hirsute. Ovary ovoid, 1-1.5 mm in diam, densely hirsute or pubescent; style 1.1-1.4 mm long, densely hirsute to glabrous; stigma 2-lobed. Drupe black, oblong-ovoid, 1-2 x 0.8-1 cm. Fl. & Fr.: September – November.

Western Ghats: Evergreen forests above 900 m in Maharashtra, Karnataka, Kerala and Tamil Nadu. India: West Bengal. World: Sri Lanka, China and Malesia.

IUCN: NE (IUCN, 2024), Vulnerable (Narasimhan and Irwin, 2021).

JUGLANDACEAE DC. ex Perleb

Juglans L.

Juglans regia L., Sp. Pl.: 997. 1753. Vern.: *Akrout* (Tam.).

Trees to 25 m tall. Leaves 25-30 cm; leaflets 5-9, entire on mature trees, sometimes obscurely serrulate on young plants; lateral leaflets subsessile or minutely petiolule, elliptic-ovate to long elliptic, 6-15 x 3-6 cm, abaxially glabrous except for tufts of hairs in vein axils, without glandular hairs, base oblique; petiole 5-7 cm long, rachis glabrescent, without glandular hairs;

terminal petiolule 2.5-6 cm long. Male spike 5-10 cm long. Stamens 6-30. Fruiting spike usually with 1-3 nuts. Nut subglobose, 4-6 cm, husk glabrous, irregularly dehiscent, wrinkled. Fl. & Fr.: June – October.

Western Ghats: Cultivated trees in Kerala and Tamil Nadu. India: Most of the States. World: Native to Europe, Temperate Asia, cultivated in tropics for its fruits.

IUCN: LC (IUCN, 2024).

Uses: The nut (walnut) is edible and economically viable for trade.

LAMIACEAE Martinov

1a. Leaves digitate; Pyrene 1 .... **Vitex**
1a. Leaves simple 2
2a. Corolla regular, stamens equal 3
2b. Corolla 2-lipped; stamens 4 didynamous 4
3a. Flowers 4-merous; drupe with 4 pyrenes .... **Callicarpa**
3a. Flowers 4-6-merous; drupe with 4 celled endocarps .... **Tectona**
4a. Drupe with 1 pyrene, normally 4-celled 5
4a. Druper with 4 pyrenes, 1-3 often suppressed .... **Clerodendron**
5a. Flowers small .... **Premna**
5b. Flowers large .... **Gmelina**

Callicarpa L.

1a. Leaves alternate, margin crenate-serrate; flowers greenish-white .... **C. arborea**
1b. Leaves opposite, margin entire or dentate; flowers pinkish-purple .... **C. tomentosa**

Callicarpa arborea Roxb., Fl. Ind. 1: 405. 1820. Synonyms: *Aganon umbellata* Raf., *Callicarpa magna* Schauer, *Callicarpa tomentosa* var. *magna* (Schauer) Bakh., *Callicarpa villosissima* Ridl.

Deciduous trees to 12 m tall; bark brownish, rough, exfoliations small, more or less triangular; branchlets pubescent. Leaves simple, alternate, clustered at the tips of branchlets, obovate, oblong-obovate, 15-30 x 5-15 cm, apex round or shortly acuminate, base cuneate or attenuate and decurrent on petiole, margin crenate-serrate, glabrous, membranous; lateral nerves 10-12 pairs, intercostae reticulate, prominent; petiole to 10 mm long, slender, pubescent. Flowers bisexual, greenish-white, 5-10 mm across, in terminal spikes; bracts 3, unequal. Seplas 4, campanulate, glabrous, adnate to ovary and not produced beyond the ovary; lobes 4, ovate. Petals 4, elliptic-oblong, revolute along margin, cauducous. Disc annular. Stamens numerous in several rows, slightly connate at base, inserted on the top of calyx; filaments subulate, purple, exserted, inner and outer rows without anthers. Ovary inferior, 4-5-celled, ovules many in each cell; style long, filiform; stigma capitate. Berry globose, 5-7.5 cm across, green, pericarp fibrous, crowned with sepals and style. Fl. & Fr.: April – September.

Western Ghats: Common in evergreen forests of Tamil Nadu. World: Sri Lanka, Nepal, Bangladesh, Myanmar, Cambodia, Vietnam, Thailand, Indonesia, Malaysia and New Guinea, Philippines.

IUCN: LC (IUCN, 2024).

Callicarpa tomentosa (L.) L., J.A.Murray, Syst. Veg., ed. 13.: 153. 1774. Synonyms: *Callicarpa arborea* Miq. ex C.B.Clarke, *Callicarpa farinosa* Roxb. ex C.B.Clarke, *Cornutia corymbosa* Lam., *Hedyotis arborescens* Noronha Vern.: *Pandakayamaram*, *Seembakutthu*, *Vettrilaipattai* (Tam.).

Trees to 10 m tall; bark brown, rough; branchlets fulvous tomentose. Leaves simple, opposite, ovate or elliptic, 13-30 x 8-20 cm, base truncate, acute or rounded, apex acuminate, margin entire or dentate, glabrous above except midrib, densely white stellate tomentose beneath, glaucous; lateral nerves 6-8 pairs, raised beneath, intercostae scalariform; petiole 25-60 mm long, tomentose. Flowers pinkish-purple, in axillary corymbose cymes, densely tomentose. Sepals campanulate, 2 mm long, lobes 5. Petals 4 mm long, 5 mm across, tube short, lobes 4, spreading. Stamens 4, exserted; filaments 6.5 mm, glandular; anthers sagitate. Ovary globose, superior, pubescent, 2 locular, ovules 2 in each cell; style slender; stigma capitate. Berry globose, 3 mm across, glabrous, black. Seeds 3-4, oblong, angular. Fl. & Fr.: December - June.

Western Ghats: Shola undergrowth above 1200 m in Maharashtra, Goa, Karnataka, Kerala and Tamil Nadu. India: Gujarat and Odisha. World: Sri Lanka, Bangladesh, Bhutan, Nepal, Myanmar, Indonesia, New Guinea, Malaysia, Philippines and Thailand.

IUCN: LC (IUCN, 2024).

Clerodendrum L.

1a. Leaves more than 5 cm long, densely villous beneath; panicels terminal, flowers white .... **C. infortunatum**

1b. Leaves less than 5 cm long, densely hirsute beneath; panciles axillary, flowers creamy white .... **C. phlomidis**

Clerodendrum infortunatum L., Sp. Pl.: 637. 1753. Synonyms: *Clerodendrum viscosum* Vent., *Volkameria infortunata* (L.) Roxb., *Clerodendrum cordatum* D.Don, *Clerodendrum viscosum* var. *helferi* Moldenke, *Clerodendrum viscosum* f. *rubrum* Moldenke Vern.: *Karukanni* (Tam.).

Small trees to 5 m tall; bark grey and corky; young shoots tawny villous. Leaves ovate or orbicular, 8-17 x 6-12 cm, base cordate, margin entire or denticulate, apex acuminate, villous beneath and tomentose above; basally 5-7 nerved; petiole to 8 cm long, villous. Panicle terminal, 14-20 cm long, villous; bracts foliaceous. Sepals 5, cupular, 0.8-1 cm long; lobes lanceolate, 3-5-ribbed, acuminate at apex, tomentose. Petals white, tubular; tube 1-1.5 cm long, villous without; lobes 6-10 mm long, oblong, hairy without. Stamens 4; filaments slender, purplish, 2-2.5 cm long; anthers oblong. Ovary globose; style slender; exserted, 2-2.5 cm long. Drupe globose, 6-8 mm across, bluish-black on ripening; fruiting calyx enlarged, pink. Seeds 2-4, globose, 2-3 mm across. Fl. & Fr.: April – May.

Western Ghats: Deciduous to evergreen forests of Maharashtra, Goa, Karnataka, Kerala and Tamil Nadu. India: Andaman & Nicobar Islands. World: Sri Lanka, Malaysia, Singapore and Thailand.

IUCN: LC (IUCN, 2024).

Clerodendrum phlomidis L.f., Suppl. Pl.: 292. 1782. Synonyms: *Clerodendrum multiflorum* (Burm.f.) Kuntze, *Volkameria multiflora* Burm.f. Vern.: *Takkari, Thazhuthaazhai, Vaadhamadakki* (Tam.).

Smll trees to 4 m tall; branchlets grey, pubescent. Leaves opposite, ovate to rhomboid-ovate, 1.5-5 x 1-3 cm, entire to sinuate-crenate, subacute-obtuse; petiole up to 2.5 cm long, densely hirsute. Cymes dichotomously branched, axillary, forming a rounded terminal panicle. Flowers creamy-white or pale yellowish, 1.5 cm across; pedicels 5-10 mm long, densely hirsute; bracts ovate lanceolate. Seplas campanulate, glabrous, pale or yellowish green, somewhat inflated, 5-lobed; lobes 4-5 mm long, ovate-triangulate, Petals 5, tubular, tube 2-2.5 cm long, much narrower than the calyx, pubescent externally; lobes 5, subequal, ovate-elliptic, 7-8 mm long, obtuse. Drupe obovoid, 8-12 mm long, black, usually 4-lobed, enclosed by the persistent calyx. Seeds oblong, white. Fl. & Fr.: October - March.

Western Ghats: Foot hills and scrub forests of Gujarat, Maharashta, Karnataka, Kerala and Tamil Nadu. India: Andhra Pradesh, Bihar, Delhi, Gujarat, Haryana, Himachal Pradesh, Jammu & Kashmir, Madhya Pradesh, Odisha, Punjab, Rajasthan, Uttar Pradesh and West Bengal. World: Sri Lanka, Pakistan, Myanmar, Bangladesh and Java.

IUCN: LC (IUCN, 2024).

Gmelina L.

1a. Leave ovate-obovate, 3-5-lobed, base cuneate; drupes yellow when ripe .... **G. asiatica**

1b. Leaves broadly ovate, not lobed, base cordate to subcordate; drupes black when ripe .... **G. arboera**

Gmelina arborea Roxb ex Sm., A.Rees, Cycl. 16: n.° 4. 1810. Synonyms: *Gmelina rheedei* Hook., *Gmelina arborea* var. *canescens* Haines, *Gmelina arborea* var. *glaucescens* C.B.Clarke, *Gmelina oblongifolia* Roxb., *Gmelina sinuata* Link Vern.: *Kumilamaram, Perumkumizha, Umithekku* (Tam.).

Trees to 15 m tall; bark grayish brown; branchlets 4-angled, lenticellate, densely yellow-brown tomentose. Leaves broadly ovate, 8-19 x 4.5-15 cm, thickly papery, base broadly cuneate to subcordate, apex acuminate; lateral nerves 3-5 pairs, abaxially prominent; petiole terete, 3.5-10 cm. Inflorescences terminal, narrow thyrses; peduncle 15-30 cm. Sepals 5, 3-5 mm long, with several black discoid gland patches; teeth 5, sharply triangular. Petals yellow, 3-4 cm, 2-lipped, sparsely glandular; lower lip 3-lobed, outside yellowish brown puberulent, inside glabrous; upper lip entire or slightly 2-cleft. Ovary glabrous, glandular. Stigma unequally 2-cleft. Drupes yellow when ripe and black when dry, ellipsoid to obovoid-ellipsoid. Fl. & Fr.: February - July.

Western Ghats: Deciduous forests of Maharashtra, Goa, Karnataka, Kerala and Tamil Nadu. India: Almost all the States. World: Sri Lanka, Nepal, Bangladesh, Pakistan, China, Bhutan, Myanmar, Vietnam, Cambodia, Thailand, Philippines, Malaysia and Tanzania.

IUCN: LC (IUCN, 2024).

Uses: Wood is used for pulp, particle board, plywood, matches, carpentary and pecking industries. It is also used for making musical instruments and partition boards.

Gmelina asiatica L., Sp. Pl.: 626. 1753. Synonyms: *Bignonia discolor* A.Rich., *Gmelina asiatica* f. *parvifolia* (Roxb.) Moldenke, *Premna parvifolia* Roth Vern.: *Kadambal, Kumil, Nilakkumil, Mulkkumil* (Tam.).

Small trees t0 6 m tall; baks smooth, greyish white; branchlets yellowish pubescens. Leaves ovate to obovate, 3-5 x 2-3 cm, papery, abaxially dark brown villous and glandular, adaxially glabrescent, base cuneate, margin entire or 3-5-lobed, apex acuminate; lateral nerves 3 or 4 pairs; petiole 0.5-4.5 cm long, pubescens. Inflorescences terminal, usually pendulous, few to many flowered; primary floral branches very short, 1-5-flowered; bracts leaflike, small to large. Flowers pendulous. Sepals campanulate, 3-6 2.5-4 mm, outside densely dark brown pubescent and with 2 to several disc-shaped glands, inside glabrous. Petals yellow, 2-5 cm, tube 2-lipped and 4-lobed, lower lip 3-lobed, upper lip entire, outside with scattered appressed hairs, inside with dense glandular hairs. Filaments with dense glandular hairs. Ovary 4-locular, glabrous; style slender, hardly exserted, apically curved; stigma unequally 2-cleft. Drupes yellow, ovoid to obovoid, glabrous. Fl. & Fr.: May – August.

Western Ghats: Scrub forests of Tamil Nadu, Kerala, Maharashtra and Karnataka. India: Almost all the States except North-east. World: Bangladesh, Cambodia, Laos, Myanmar, Nicobar Is., Sri Lanka, Thailand and Vietnam.

IUCN: LC (IUCN, 2024).

Premna L.

1a. Calyx four-lobed; style yellow with reddish brown stigma **P. rajendranii**

1b. Calyx five-lobed; style not yellow and stigma not reddish brown 2

2a. Inflorescence thyrse .... **P. wightiana**

2b. Inflorescence panicle or corymb 3

3a. Petiole up to 3.5 cm; ovary subglobose or obovoid 4

3b. Petiole up to 12 cm; ovary truncate .... **P. tomentosa**

4a. Leaves broadly ovate to orbicular; corymb terminal only; found in foothills and lower slopes ....**P. mollissima**

4b. Leaves elliptic-ovate to broadly oblong; corymb terminal or axillary; found in coastal area ....**P. serratifolia**

Premna mollissima Roth, Nov. Pl. Sp.: 286. 1821. Synonyms: *Premna latifolia* var. *mollissima* (Roth) C.B.Clarke, *Premna flavescens* Buch.-Ham. ex C.B.Clarke, *Premna flavescens* var. *glabrior* C.B.Clarke, *Premna latifolia* Roxb., *Premna latifolia* var. *mucronata* (Roxb.) C.B.Clarke, *Premna mucronata* Roxb., *Premna latifolia* var. *viburnoides* C.B.Clarke Vern.: *Erumai-munnai, Munnai, Perijitti, Pasu Munnai* (Tam.).

Small trees to 9 m tall; bark dark grayish, somewhat rough; young parts pubescent and young stem spinous. Leaves simple, opposite, lanceolate-ovate, elliptic-ovate to rhomboid, 3-9 x 1.5-6 cm, base attenuate-cuneate, margins entire, apex acute or shallow acuminate, densely pubescent both above and beneath; lateral veins 4-8 on either side of the midvein, ascending, arcuate; petiole slender, quadrangular, grooved above. Inflorescence axillary, terminal, divaricate or compact cymes, 4-7 x 4-6 cm, peduncles densely pubescent, bracts linear-lanceolate, 1-2 mm long. Flowers creamish white, yellowish green, pedicels 1-3 mm long, calyx campanulate, 5 toothed, 2 lipped, teeth obtuse, lower lip 3 toothed, upper lip 2 toothed, pubescent outside, about 2.5 x 1.5 mm. Petals infundibular, 4 lobed, 2 lipped, upper lip suborbicular, apex emarginate, lower lip 3 lobes, obovate, apex acute, about 3 x 2 mm. Stamens 4, didynamous, slightly exserted; filaments glabrous, filiform, about 2-3 mm long; anthers globose, 2 loculed, about 0.5 mm long. Ovary obovoid or subglobose, about 2 x 1.5 mm; style slender about the same length as the stamens; stigma bi-lobed. Drupe obovoid, about 6 x 5 mm, glabrous, green when young, black when ripe. Fl. & Fr.: April – August.

Western Ghats: Scrub to semi-evergreen forests of Maharashtra, Karnataka, Kerala and Tamil Nadu. India: Andhra Pradesh and Telangana. World: Sri Lanka, Pakistan, Bangladesh, China, Myanmar, Thailand, Laos, Cambodia, Vietnam, Java and Philippines.

IUCN: NE (IUCN, 2024).

Premna rajendranii K.M.P. Kumar, Sunilk., Sreeraj & V.T. Antony., Webbia 68: 127. 2013.

A small tree, 3–4 m high, stem glabrous. Leaves 4–6×2– 4 cm, ovate, cordate or cordato-ovate with acute apex, villous on both sides; petiole 1.2–1.8 cm long, hairy. The leaves are foetid when bruised. Panicles terminal in a compact corymb. Flowers white-cream, 5–6 mm long. Calyx cup-shaped, two-lipped and four-lobed, lobes obtuse, densely pubescent, 2–2.2 × 0.4 mm. Corolla creamy white, 4 mm wide, zygomorphic, two-lipped with four lobes, the upper lip is emarginate, 2 × 1.2 mm, obtuse; the lower one three-lobed, 2.1 × 1.4 mm, obtuse, the middle lobes. larger, somewhat exceeding the upper one. Corolla tube 2– 3 mm long, throat is densely villous. Stamens 4, didynamous, included within corolla, glabrous; filament pairs small 2 mm and large 2.6 mm long. Anthers 0.5 mm long. Style yellow with reddish brown stigma, globose, glabrous, bifid, 4–5 mm long. 'Ovary 0.7 × 1 mm; drupe globose, 4 mm wide, black when ripe. Fl. & Fr.: January – October.

Western Ghats: Kerala and Tamil Nadu (Endemic)

IUCN: NE (IUCN, 2024).

Premna serratifolia L., Mant. Pl. 2: 253. 1771. Synonyms: *Premna obtusifolia* f. *serratifolia* (L.) Moldenke, *Cornutia corymbosa* Burm.f., *Premna abbreviata* Miq., *Premna ceramensis* Miq., *Premna integrifolia* var. *angustior* C.B.Clarke, *Premna perrottetii* C.B.Clarke, *Premna punctulata* C.B.Clarke, *Premna spinosa* Roxb. Vern.: *Narumunnai* (Tam.).

Small treess to 8 m tall; branchlets with elliptic yellowish lenticels, densely pubescent when young. Leaves oblong to broadly ovate, 3-15 x 2.5-9.5 cm, subglabrous or pubescent only along veins, base broadly cuneate, rounded, or truncate, margin entire, slightly undulate or crenate, apex acute to rarely acuminate or obtuse; petiole 0.3-5 cm long, puberulent. Inflorescences 1.5-15 x 2.5-24 cm; peduncle 0.8-3 cm long; bracts linear to lanceolate, to 6 mm, puberulent. Seapsl cup-shaped, 1.5-3 mm long, 2-lipped, lower lip subentire to shortly 3-dentate, upper lip longer than lower lip and 2-dentate, outside puberulent and yellow glandular. Petals yellowish green, outside glandular, villous in throat, slightly 2-lipped; lower lip 3-lobed, lobes subequal or middle lobe slightly longer and broader; upper lip entire or emarginate. Ovary glabrous, apically glandular. Style 3.5-4 mm. Drupe globose, 2-4 mm in diam. Fl. & Fr.: Throughout the year.

Western Ghats: Deciduous forests of Maharashtra, Goa, Karnataka, Kerala and Tamil Nadu. India: Andhra Pradesh, Odisha and West Bengal. World: Sri Lanka, Myanmar, Bangladesh, China, Laos, Cambodia, Vietnam, Malaysia, Africa, Australia, New Guinea and Philippines.

IUCN: LC (IUCN, 2024).

Premna tomentosa Willd., Sp. Pl., ed. 4. 3: 314. 1800. Synonyms: *Gumira adenosticta* (Schauer) Kuntze, *Premna cumingiana* f. *dentata* Moldenke, *Premna latifolia* Thwaites, *Premna tomentosa* f. *jejuna* Moldenke Vern.: *Kozhukkattai thekku, Malaithekku, Peurangainaari, Podangunaari* (Tam.).

Trees to 15 m tall; bark greyish-brown, smooth; young branchlets with stellate tomentum. Leaves simple, opposite, broadly ovate, 6-25 x 5-15 cm, base obtuse or cordate, apex acuminate, margin entire, densely tawny yellow stellately tomentose, yellowish; lateral nerves 5-7 pairs, densely tomentose beneath, intercostae scalariform; petiole 20-50 mm long, slender, tomentose. Flowers yellow, small, in terminal and axillary corymbs, 7 cm long. Sepals 5, tubular, 5-ribbed, lobes 5, stellate pubescent. Petals 7.5 mm across, 4 mm long, tube cylindric, hairy, lobes 5, orbicular. Stamens 4, didynamous, 1.5 and 2 mm long; anthers rounded, 0.5 mm. Ovary truncate, superior, puberulus; style 3 mm; stigma unequally bifid. Drupe ovoid, 5-6 x 4 mm, black, hairy. Seeds 4. Fl. & Fr.: May - Septemebr.

Western Ghats: Common in scrub forest and deciduous forests of Karnataka, Kerala and Tamil Nadu. India: Andhra Pradesh and Odisha. World: Sri Lanka and Myanmar.

IUCN: LC (IUCN, 2024).

Premna wightiana Schauer, A.P.de Candolle, Prodr. 11: 635. 1847. Synonyms: *Gumira wightiana* (Schauer) Kuntze, *Gumira thyrsoidea* (Wight) Kuntze, *Premna thyrsoidea* Wight Vern.: *Pinja* (Tam.).

Trees to 8 m high; bark greyish-brown; branchlets pubescent. Leaves simple, opposite, lamina 6.5-10 x 3.5-5 cm, ovate or ovate-elliptic, base obtuse, apex acuminate, margin entire or serrate in the upper half, glabrous or slightly pubescent; lateral nerves 4-5, pubescent beneath, intercostae scalariform; petiole 10-30 mm long, pubescent, grooved above. Flowers yellowish-white, inflorescence terminal thyrsoid panicles. Sepals in tubular, obscurely 2 lipped, 2 mm long, upper lip 2, lower lip 3 lobed, acute, hairy. Petals 3.5 mm long, lobes 4, unequal, 2-lipped, hairy inside. Stamens 4, didynamous, inserted below the throat of corolla tube. Ovary globose, superior, hairy at apex, 4-celled, one ovule in each cell; style slender; stigma bifid. Drupe obovoid, 5 x 5 mm, tubercled. Seeds oblong. Fl. & Fr.: February – April.

Western Ghats: The mixed deciduous forests and scrubs above 500 m in Kerala and Tamil Nadu. India: Odisha. World: Sri Lanka.

IUCN: LC (IUCN, 2024).

Tectona L.f.

Tectona grandis L.f., Suppl. Pl.: 151. 1782. Synonyms: *Jatus grandis* (L.f.) Kuntze, *Tectona grandis* f. *abludens* Koord. & Valeton, *Tectona grandis* f. *tomentella* Moldenke, *Tectona theca* Lour. Vern.: *Thekku* (Tam.).

Trees to 40 m tall; branchlets gray to grayish brown, 4-angled, yellowish to grayish brown stellate tomentose. Leaves ovate-elliptic to ovate, 15-45 x 8-23 cm, abaxially densely grayish brown to yellowish brown minutely stellate tomentose, adaxially scabrous and puberulent along veins, base cuneate and de-current, margin entire, apex acuminate to obtuse; veins 7-12 pairs; petiole robust, 2-4 cm long, dawny toemntose. Panicles 25-40 x 30 cm. Flowers fragrant. Sepals cupular, tube 2.5-3 mm, with white stellate hairs. Petals white; tube 2.5-3 mm, outside puberulent glandular; lobes. 2 mm long, obtuse. Ovary strigose; style 3-4 mm. Fruit globose, 1.2-1.8 cm in diam., minutely tomentose. Fl. & Fr.: August - December.

Western Ghats: Typical deciduous forests above 700 m in Gujarat, Maharashtra, Goa, Karnataka, Kerala and Tamil Nadu. India: Most of the States. World: Sri Lanka, Myanmar and Malesia.

IUCN: EN- A2cd. (IUCN, 2024).

Uses: Teak wood is highest quality in south India for its economic importane and trade. It is used for making furniture, construction of boat and ships, indoor flooring, and counter tops.

Vitex L.

1a. Inflorescence terminal, with occasional additional branches in upper leaf axils 2

1b. Inflorescence axillary 5

2a. Small trees; leaves membranous, white-or grey pubescent beneath, entire or sometimes toothed 3

2b. Trees; leaves chartaceous or coriaceous, not grey-pubescebt beneath 4

3a. Leaves 1-3 foliolate, the leaflets sessile; panicles 3-15 cm .... **V. trifolia**

3b. Leaves 3-5 foliolate, the leaflet petiolules; panicles 10-25 cm .... **V. negundo**

4a. Petiole angular or winged with an auricled base; panicles with racemose branches of small grey-pubescent cymes, bracts small and caducous .... **V. altissima**

4b. Petiole not winged; panicles pyramidal, fulvous, with short racemose branches, bracts foliaceous, conspicuous and persistant .... **V. pinnata**

5a. Leaves submembranous, 3-foliate, corolla puberulous and glandular; drupr obovoid .... **V. peduncularis**

5b. Leaves coriaceous, 3-5-foliate, corolla long villous on palate; druple ellipsoid .... **V. leucoxylon**

Vitex altissima L.f., Suppl. Pl.: 294. 1782. Synonyms: *Vitex alata* Willd., *Vitex altissima* f. *alata* (Willd.) Moldenke, *Vitex latifolia* Wight ex Steud., *Vitex zeylanica* Turcz. Vern.: *Mayiladi, Mayilainochi* (Tam.).

Trees to 30 m tall; bark greyish-yellow, scaly; branchlets lenticellate, minutely tomentose. Leaves compound, trifoliate, opposite, estipulate; leaflets 3-5, sessile, 6-18 x 2-5 cm, lanceolate, elliptic-lanceolate, or oblanceolate, base cuneate or acute, apex acuminate, margin entire, glabrous above and pubescent or glabrescent along the nerves below; lateral nerves 10-16 pairs, intercostae reticulate; rachis 35-60 mm long, slender, pubescent, winged in sapling leaves. Flowers white, tinged with blue, in terminal panicles, 5-8 m long; bracts lanceolate, 3 mm long, cauducous. Sepals densely tomentose, shortly 5 lobed, ovate. Petals tubular, 5 mm long, 2 lipped, upper lip of 2 and lower lip of 3 lobes, obtuse, throat villous. Stamens 4, didynamous, exserted; filaments hairy at base. Ovary globose, 1 mm, superior, densely hairy, 2-4-celled, 4 ovuled; style filiform; stigmas bifid. Drupe globose, 5 x 5 mm, glabrous, supported by enlarged calyx. Seeds 4, obovate. Fl. & Fr.: March - October.

Western Ghats: Deciduous forests of Maharashtra, Goa, Karnataka, Kerala and Tamil Nadu. India: Andhra Pradesh, Odisha and Telangana. World: Sri Lanka, Bangladesh, Laos, Cambodia, Vietam and Malesia.

IUCN: LC (IUCN, 2024).

Uses: Wood is useful for timbers and used for making furniture, garden implements and home applayences.

Vitex leucoxylon L.f., Suppl. Pl.: 293. 1782. Synonym: *Vitex saligna* Roxb. Vern.: *Neernochi* (Tam.).

Deciduous trees to 20 m tall; bark yellowish-grey or greyish-brown, rough. Leaves digitately compound, opposite, decussate; leaflets 3-5, whorled, elliptic, elliptic-oblong or lanceolate, 2.5-16 x 1.5-4 cm, base acute or attenuate, apex acute or obtusely acuminate, margin entire, glabrous above, puberulent beneath; lateral nerves 5-14 pairs, intercostae reticulate, prominent; rachis 5.5-8.5 cm, slender, swollen at the base, glabrous. Flowers in axillary corymbose cymes, to 17 cm long; pedicels to 2 mm. Sepals cupular, sparsely appressed puberulent; tube 2 mm long, 5-toothed. Petals white, bilipped, purplish-pubescent in the throat; lobes 5. Stamens 4, didynamous, inserted below the throat of the corolla tube, slightly exserted; filaments thickened and villous towards the base; anthers purple. Ovary superior, globose; style filiform; stigma bifid. Drupe obovoid, 1.2-2 cm long, purple-black, fleshy. Fl. & Fr.: March - October.

Western Ghats: Semi-evergreen forests along the water course of Maharashtra, Goa, Karantaka, Kerala and Tamil Nadu. India: Andhra Pradesh, Madhya Pradesh, Odisha dand Uttar Pradesh. World: Sri Lanka, Myanmar, Pakistan, Java and Malaysia.

IUCN: LC (IUCN, 2024).

Vitex negundo L. Sp. Pl.: 638. 1753. Synonyms: *Agnus-castus negundo* (L.) Carriere, *Vitex agnus-castus* var. *negundo* (L.) Kuntze, *Vitex negundo* var. *incisa* (Lam.) C.B.Clarke, *Vitex spicata* Lour., *Vitex negundo* L. var. *purpurascens* Sivar. & Moldenke Vern.: *Notchi, Vennochi, Nirkundi, Nallanochi, Karunochi* (Tam.).

Shruby trees to 7 m tall; bark pale, purple pubescent all over, aromatic. Leaves 3-5-foliolate; leaflets 6-13 x 2-5 cm, narrowly oblong or elliptic to lanceolate, base acute, apex acuminate. Panicles terminal, 10-25 cm long. Sepals5, toothed, obconic, 3 mm long, teeth triangular. Petals deep purple to violet in colour, 7 mm across, hypocrateriform; tube 3-5 mm long, puberulent without, upper lipd 2-lobed, lower 3-lobed with the middle lobe larger, obovate, undulate-margined, other lobes shorter, subequal, obtuse. Stamens 4, filaments purple. Ovary 1 mm long; style purple; stigma 2-fid. Drupe globose, 3-5 mm across, purple or black. Fl. & Fr.: Throughout the year.

Western Ghats: common in foot hills of Gujarat, Maharashtra, Goa, Karnataka, Kerala and Tamil Nadu. India: Andhra Pradesh, Assam, Bihar, Gujarat, Jammu & Kashmir, Madhya Pradesh, Manipur, Meghalaya, Odisha, Punjab, Rajasthan, Sikkim, Tripura, Uttar Pradesh and West Bengal. World: Sri Lanka and Tropics.

IUCN: LC (IUCN, 2024).

Vitex peduncularis Wall. ex Schauer, A.P.de Candolle, Prodr. 11: 687. 1847. Synonyms: *Vitex alata* Roxb., *Vitex peduncularis* var. *roxburghiana* C.B.Clarke, *Vitex peduncularis* f. *roxburghiana* (C.B.Clarke) Moldenke

Trees to 18 m tall; bark greyish, smooth; branchelets slender, obtusely 4-angular or subterete, young parts pubescent, glabrous when mature. Leaves usually 3, rarely 5-foliolate; leaflets lanceolate, oblong-lanceolate or narrow-elliptic, acuminate or attenuate at base, entire somewhat undulate, acuminate at apex; middle leaflets 10-20 x 3-5 cm; petiolules 0.5-1.5 cm long; lateral leaflets 8.5-15 x 3-4 cm; petiolules 0.5-1 cm long; subglabrous or glabrous above, with dense minute rounded yellow resinous glands beneath; lateral nerves 18-22 pairs, distinct beneath; petioles flattened, slender, 5-10 cm long, pubescent. Panicles axillary, 20-28 cm long; peduncles flattened, slender, 6.5-14 cm long, densely pubescent; bracts linear, 1 mm, pubescent. Flowers numerous; pedicels 1-2 mm long. Sepals campanulate, minutely 5-toothed with minute rounded yellow glands outside. Petals infundibular, 2-lipped, 5-lobed, yellow or cream-coloured; upper lip 2-lobed; lobes ovate, 1 mm long; lower lip 3-lobed; midlobe suborbicular or ovate, entire, obtuse, 2 mm long; lateral lobes oblong, acute, 1.2 x 1 mm; tube narrow, 2 x 2 mm, grey-pubescent with rounded resinous glands outside. Stamens 4, didynamous; filaments slender, 2-3 mm long, included, sparsely hairy at base; anthers oblong. Ovary globose, 1 mm, glabrous; style stout, 3 mm long; stigma 2-lobed. Drupes globose, 0.7 cm, red; fruiting-calyx platelliform, glabrous. Fl. & Fr.: March – May.

Western Ghats: Semi-evergreen forests of Karnataka, Kerala and Tamil Nadu. World: India: Andhra Pradesh, Assam, Bihar, Madhya Pradesh, Meghalaya, Odisha, Tripura, Uttar Pradesh and West Bengal. Sri Lanka, Nepal, Bangladesh, Pakistan, Myanmar, Thailand, Laos, Cambodia, Vietnam and Malesia.

IUCN: LC (IUCN, 2024).

Vitex pinnata L., Sp. Pl.: 638. 1753. Synonyms: *Vitex puberula* Miq., *Vitex bracteata* Horsf. ex Miq., *Vitex heterophylla* Blume ex Miq., *Vitex pubescens* var. *pantjarensis* Hochr., *Wallrothia articulata* Roth Vern.: *Mayiladi* (Tam.).

Trees to 15 m tall; bark thick, white; young shoot grey pubescent; branchlets 4-angled, lenticellata. Leaves 3-foliolate, sometimes 5-foliolate, opposite, elliptic, lanceolate or oblanceolate, 7-16 x 1.8-4.9 cm, base acute, apex acuminate, margin entire or coarsely serrate, yellow gland dotted beneath, chartaceous; lateral nerves 8-22, intercostae reticulate, prominent; rachis 35-70 mm, slender, grooved above, flattened; petiolule 7-10 mm, subsessile. Flowers cream with yellow palate, in lax axillary panicles of dichasia to 25 cm. Sepals 4 mm long, yellow spotted without, 5 toothed. Petals 8 mm across, creamy, throat yellow, pubescent outside, 2-lipped, upper 2 and lower 3 lobed, upper lip 2 mm, acute, lower lip 4 mm, ovate, lateral lobes broadly oblong, subacute, puberulous and glandular. Stamens 4, didynamous, exserted; anthers 0.7 mm. Ovary 1 mm, superior; style 3 mm, filiform; stigma bifid. Drupe obovoid, 1.3 cm long, supported by enlarged calyx. Seeds 4. Fl. & Fr.: January - June.

Western Ghats: Deciduous forest and stream side of Karnataka, Kerala and Tamil Nadu. India: Andhra Pradesh, Assam, Meghalaya, Odisha, Tripura, Uttar Pradesh and West Bengal. World: Sri Lanka, Myanmar, Bangladesh, Laos, Cambodia, Vietnam, Celebes, Thailand, Malaysia and Guyana.

IUCN: LC (IUCN, 2024).

Vitex trifolia L., Sp. Pl.: 638. 1753. Synonyms: *Vitex agnus-castus* var. *trifolia* (L.) Kurz, *Vitex trifolia* var. *trifoliolata* Schauer, *Vitex variifolia* Salisb.

Small trees to 1.5-5 m tall; branchlets densely pubescent. Leaves 3-5-foliolate; leaflets sessile, oblong, lanceolate, or obovate, abaxially densely gray tomentose, adaxially green and glabrous or subglabrous, base cuneate, margin entire, apex obtuse; veins 8 pairs and slightly prominent on both surfaces; central or single leaflet 2.5-9 x 1.7-3 cm; petiole 1-3 cm long, pubescent. Panicles 3-15 cm; peduncle densely gray tomentose. Sepals 5, slightly 5-dentate, outside gray pubescent, inside glabrous. Petals purplish to bluish purple, 6-10 mm, outside scaly white, pubescent at filament bases and on inside of lower lobe. Stamens exserted. Ovary glabrous, with or without glands; style glabrous. Drupe black, subglobose, 5 mm in diam. Fl. & Fr.: August - February.

Western Ghats: Deciduous forests of Maharashtra, Goa, Karnataka, Kerala and Tamil Nadu. India: Andhra Pradesh, Assam, Manipur and West Bengal. World: Sri Lanka, Afghanistan, Iran, China, Japan, South East Asia, Madagascar and Australia.

IUCN: NE (IUCN, 2024).

Planted:

Clerodendrum glabrum var. minutiflorum (Baker) Fosberg; Exotic: Native to South Africa

Clerodendrum heterophyllum (Poir.) R.Br.; Exotic: Native to Mascarene Island

Vitex megapotamica (Spreng.) Moldenke; Exotic: Native to South America

Vitex parviflora A. Juss.; Exotic: Native to Thailand, Indonesia, Malaysia, Philippines, New Guinea.

LAURACEAE JUSS.

Key to the genera

1a. Stamens of 3rd row, when present, with extrorse anthers 2
1b. Stamens of all rows with introrse anthers, the lowers cells sometime lateral 8
2a. Anthers 2 celled 3
2b. Anthers 4 celled........................5
3a. Fruit surrounded by the enlarged perianth tube ....**Cryptocarya**
3b. Fruits not surrounded by the enlarded perianth tube........................4
4a. Fruiting perianth persistent........................ **Apollonias**
4b. Fruiting perianth deciduous ....**Beilschmiedia**
5a. Leaves opposite or subopposite rarely alternate ....**Cinnamomum**
5b. Leaves alternate........................6
6a. Fruiting perianth-tube enlarged, fruiting pedicels enlarged, thick usually warted ....**Alsedaphne**
6b. Fruiting perianth tube and pedicels hardly enlarged 7
7a. Fruiting perianth-lobes reflexed ....**Machilus**
7b. Fruiting perianth-lobes erect, cup-like ....**Phoebe**
8a. Fertile stamens 9 or 12 or more........................9
8b. Fertile stamens 6, involucral bracts early deciduous................. **Neolitsea**
9a. Involucral bracts deciduous in flower, leaves usually subverticillate ....**Actinodaphne**
9b. Involucral bracts persistant in flower, leaves alternate or sub-opposite .. .. **Litsea**

Actinidaphne Nees

1a. Inflorescences (either male or female or both) with an appreciable axis (racemiform or thyrsiform) 2
1b. Inflorescences (either male or female or both) without such axis (fasciculate or umbellate). 4
2a. Leaves verticillate as well as closely alternate to subopposite................... .... **A. wightiana**
2b. Leaves strictly verticillate 3

3a. Mature leaves glabrous beneath; fruits oblong-ellipsoid, 9–12 mm long, 7–8 mm in diameter .... **A. madraspatana**

3b. Mature leaves tomentellous beneath; fruits subglobose, 8–10 mm indiameter .... **A. bourdillonii**

4a. Leaves glabrous beneath 5

4b. Leaves variously pubescent to tomentose beneath 6

5a. Young shoots orange-brown sericeous to tomentose; lateral nerves ofleaves 10–16 pairs; persistent perianth subtending the fruits entire .... **A. elegans**

5b. Young shoots tawny tomentellous to brown puberulous; lateral nerves ofleaves 6–12 pairs; persistent perianth subtending the fruits undulate todivided .... **A. campanulata**

6a. Perianth subtending the fruits bearing persistent to semipersistent tepals .... **A. lanata**

6b. Perianth subtending the fruits entire, devoid of tepals 7

7a. Leaves strongly to weakly trinerved at base, in the latter case the firstpair of lateral nerves longer than the subsequent pairs .... **A. bourneae**

7b. Leaves penninerved 8

8a. Persistent perianth subtending the fruits discoid, 1–2 mm high, often reflexed or deflexed .... **A. angustifolia**

8b. Persistent perianth subtending the fruits concave to cupular, 3–5 mm high 9

9a. Leaves predominantly elliptic to oblong or ovate .... **A. tadulingamii**

9b. Leaves predominantly obovate, obovate-elliptic to obovate-oblong or suborbicular 10

10a. Lateral nerves of leaf 6-7 pairs; inflorescence lateral fascicles; staminodes 3 .... **A. lancolata**

10b. Lateral nerves of leaf 10-15 pairs; inflorescence latral umbels; staminodes 9 .... **A. lawsonii**

Actinodaphne angustifolia (Bl.) Nees. N. Wallich, Pl. Asiat. Rar. 3: 34. 1832. Synonyms: *Actinodaphne gullavara* M.R. Almeida , *Litsea angustifolia* Blume, *Lauras gullavara* Buch. -Ham. ex Nees, *Tetranthera verticillata* Wall. ex Meisn.

Trees to 15 m tall; bark brown, smooth; branchlets terete, rusty tomentose. Leaves simple, alternate, spiral, subverticilate, narrow-elliptic to oblanceolate,

12-27 x 4-6.5 cm, apex acuminate, base cuneate to acute, margin undulate, chartaceous, glaucous, fulvous tomentose when young, later glabrous except midrib beneath; secondary nerves 7-10 pairs, tertiary nerves nearly closely and horizontally percurrent, not prominent; petiole to 2.5 cm long, terete, rusty tomentose. Flowers unisexual, dioecious. Female flowers in umbels on short peduncle. Fl. & Fr.: October – June.

Western Ghats: Evergreen forests of Maharashtra, Goa, Karnataka, Kerala and Tamil Nadu. World: Bangladesh, Nepal and Myanmar.

IUCN: LC (IUCN, 2024).

Actinodaphne bourdillonii Gamble., Bull. Misc. Inform. Kew 1925: 129. 1925.

Trees to 10 m tall; bark greyish-brown; branchlets densely tomentose. Leaves simple, opposite or subopposite, estipulate, in whorls of 7-10, elliptic-lanceolate, oblong-lanceolate or oblanceolate, 11.5-26 x 3-8.3 cm, base acute or attenuate, apex acuminate, margin entire, glabrous above and softly grey or cinnamomeous-tomentose beneath, coriaceous; lateral nerves 7-9 pairs, intercostae scalariform, prominent; petiole 10-20 mm long, tomentose. Flowers yellowish-green, small. Male flowers: 8-10 together on peduncled umbels; pedicels to 6 mm long; bracts 12 x 8 mm orbicular. Tepals 6, oblong, subequal, Stamens 9, in 3 rows; filaments 2-glandular, densely hispid; anthers oblong, introrse. Female flowers: 5-10 together in peduncled umbels on lateral tubercles; pedicels to 1.5 cm long, bracts imbricate, cauducous; tepals 6, subequal; staminodes 9, in 3 rows, those of the row 3 with 2-glands. Ovary half inferior, ovoid, attenuate into the style; stigma dilated. Berry globose, 5 mm across, black, smooth. Fl. & Fr.: Throughout the year.

Western Ghats: Evergreen forests of Karnataka, Kerala and Tamil Nadu (Endemic).

IUCN: NT (Low risk/near threatened) (IUCN, 2024).

Actinodaphne bourneae Gamble., Bull. Misc. Inform. Kew 1925: 128. 1925.

Small trees to 10 m tall; branchlets and tender parts fulvous-tomentose. Leaves elliptic, rarely lanceolate or obovate-elliptic, 12 x 5 cm, subcoriaceous, more densely fulvous-tomentose below; nerves 5-7 pairs, base cuneate or acute, apex acute to acuminate; petiole 2 cm long. Flower cluster compact, of 3-6 flowers; pedicel 4 mm; bracts 5-8 x 4-6 mm, tawny-villous. Perianth yellowish, 5 mm wide; tube 4 mm, villous; lobes 6, ovate, 3 mm, densely villous outside. Staminate: stamens 9. Pistillate: staminodes 9. Berry globose, apically with pointed tip, basally covered the thick perianth cup, red when ripe. Fl. & Fr.: April – November.

Western Ghats: Evergreen forests of Kerala and Tamil Nadu (Endemic).

IUCN: VU- B1ab(iii)+2ab(iii). (IUCN, 2024).

Actinodaphne campanulata Hook.f., Fl. Brit. India 5: 148. 1886. var. **campanulata**

Trees to 15 m tall; shoots densely brown tomentose. Leaves simple, in whorls of 3-8 or opposite or subopposite, obovate or elliptic-obovate, 4-6 x 1.5-3 cm, base acute or cuneate, apex obtuse, margin entire, pubescent above, glabrous and glaucous beneath, coriaceous, reticulate; lateral nerves 5-8 pairs, intercostae scalariform, faint; petiole 5-13 mm long, stout, tomentose. Flowers small, unisexual, yellowish-green, in lateral and axillary umbellate fascicles; bracts scaly, 2 mm long, ovate. Perianth campanulate in female flowers, tube 0.1 cm long; lobes 6 in male flowers; ovate, silky brown hairy without. Stamens 9 in 3 rows; filaments villous, 2 glandular; anthers introrse; staminodes 9, in 3 rows, those of the third row linear with 2 glands. Ovary half inferior, 0.05 cm across, attenuate into the style; stigma dilated. Berry 0.4-0.6 cm across, with persistent perianth lobes. Fl. & Fr.: January – May.

Western Ghats: Evergreen forests above 1200 m in Kerala and Tamil Nadu (Endemic).

IUCN: VU- B1ab(iii)+2ab(iii). (IUCN, 2024).

Key to the varieties

1a. Leaves obovate or elliptic obovate, lateral nerves 5-8 pairs; filament glabrous .... var. **campanulata**

1b. Leaves elliptic-oblong or elliptic-lanceoalte, lateral nerves 8-11 pairs; filament villous .... var. **obtusa**

Actinodaphne campanulata Hook.f. **obtusa** Gamble

Fl. & Fr.: January – May.

Western Ghats: Evergreen forests of Kerala and Tamil Nadu (Endemic).

IUCN: NE (IUCN, 2024).

Actinodaphne elegans Thwaites, Enum. Pl. Zeyl. 256. 1861. Synonym: *Actinodaphne salicina* Meisn., *Iozoste elegans* (Thw.) Kuntze.

Trees to 15 m tall; bark orange-brown sericeous; branchlets greyish, brown to dark brown, terete. Leaves 3–7-verticillate, narrowly oblong-elliptic, oblong-lanceolate, or elliptic-lanaceolate to lanceolate, 5–16 x 1–3 cm, acute or cuneate at base, apicul-ate to acuminate at apex, stiffly chartaceous to thinly coriaceous, glabrous, brown above when dry, paler and sometimes glaucous beneath; penninerved; lateral nerves 10–16 pairs, slender, secondary nerves more or less prominent or sometimes faint above, obscure beneath; petioles to 14 mm long, 1.5–2 mm thick, glabrous. Male inflorescences fasciculate, few to many-flowered. Flowers pediceled 3 mm long; tepals broadly triangular-ovate, tawny-tomentellous outside. Stamens to 3 mm long. Female inflorescences umbellate, few-flowered. Flowers pedicels 4–6 mm long, 1 mm thick. Tepals broadly triangular-ovate, 3 x 2 mm, tomentellous outside; ovary subglobose, 1.5 mm in diam.; style 2 mm long. Berry subglobose, 10 mm in diam, in persistent perianth cup, 5-7 mm diam, fruiting pedicel to 10 mm long.

Western Ghats: Rare in evergreen forests and shola undergrowth of Kerala and Tamil Nadu. World: Native to Sri Lanka.

IUCN: NE (IUCN, 2024).

Actinodaphne lanata Miesn., Prodr. 15(1): 219. 1864.

Small trees to 6 m tall; bark brown, smooth; young shoots with densely rusty tomentose. Leaves verticillate of 5–7 per node, elliptic-oblong to narrowly lanceolate, 5–8 x1.4–2.5 cm, apex attenuate to acuminate, base acute to cuneate, margin entire, densely white-woolly when young, glabrous on both surfaceswhen mature, glaucous beneath; secondary veins 7–9 pairs, basal pair 7 mm away from leaf base, opposite and oblique; petioles slender, 1–2 cm long, slightly grooved above, rusty tomentose. Inflorescences fasciculate, ferrugineous tomentose, borne on twigs between whorls of leaves, 8-flowers per fascicle, sessile or short stalked; peduncles 2 mm long, fulvous tomentose. Staminate flowers not seen. Pistillate flowesr 7 mm long; pedicels 3–5 mm long, slightly angular, fulvous tomentose. Perianth lobes oblong-elliptic or orbicular, acute or rounded at apex, margin ciliate, fulvous tomentose outside, glabrous inside, gland-dotted, 3-veined, membranaceous, outer 2.0 x 1.5 mm long, concave, keeled at back; staminodes 9, nearly spathulate, linear-lanceolate, pilose at base, thirdrow with bi-glandular, glands 0.5 mm across, kidney-shaped. Ovary obovoid, 1.5 x 1.0 mm, glabrous; style stout, sparsely pilose, curved, 1 mm long, yellow; stigma peltate, white. Berry ellipsoid to obovoid, 1.5 mm long, apex with persistent stylar remnents, dryingblack, fulvous tomentose outside. Fl. & Fr.: Feb.-May

Western Ghats: Evergreen forests above 1100 m in Kerala and Tamil Nadu (Endemic).

IUCN: VU - B1ab(iii)+2ab(iii). (IUCN, 2024).

Actinodaphne lanceolata Dalzell & Gibson, Bombay Fl. 312. 1861. *Actinodaphne angustifolia* Hook.f. & Thomson ex Meisn., *Actinodaphne hookeri* Meisn., *Actinodaphne hookeri* var. *glabrata* Meisn., *Iozoste hookeri* (Meisn.) Kuntze

Trees to 15 m tall; bark rough, lenticellate; branchlets terete, ferrugineous tomentose when young and slightly sparse in maturity. Leaves clubbed at the ends of the branchlets, lanceolate or elliptic-lanceolate, 4-11 x 2-3.5 cm, base acute ro obtuse, apex acute, dense tomentose both sides, margin entire, ciliate; secondary nerves 6-7 pairs, prominent, densely pubescent below; petiole to 1.5 cm long, pubescent. Flowers in axillary clusters, 3-5 in each cluster, tomentose. Perianths 8-10 in two series, tomentose above, glabrous inside, elliptic-obovate, rounded apex. Stamens 6-8 in each flower, filament to 0.8 mm long, ciliate hairy; anthers linear-elliptic, lobes with 4-5-flaps; staminodes 3. Osvary globes, glabrous; style to 0.3 mm long, cylindric; stigma capitate. Fl. & Fr.: April – July.

Western Ghats: Evergreen forests of Maharashtra (Chakrabarty *et al.*, 2021) (Endemic).

IUCN: DD (IUCN, 2024).

Actinodaphne lawsonii Gamble., Bull. Misc. Inform. Kew 1925: 129. 1925.

Trees to 6 m tall. Leaves simple, opposite or subopposite, whorled, obovate or elliptic obovate, 12-18 x 4-7.5 cm, base cuneate to attenuate, apex obtuse or obtusely acute; margin entire, glabrous above except at nerves, densely rufous tomentose beneath; lateral nerves 8-10, rufous tomentose, intercostae scalariform, faint; petiole 15-23 mm long, pubescent, grooved above. Flowers unisexual, greenish, in 6-8-flowered clusters, rufous tomentose; pedicels 1 cm long. Perianth tube short; lobes 6, subequal. Stamens 9, in 3 rows; filaments 2-glandular; anthesr 4-celled, introrse; staminodes 9, in 3 rows, those of row 3 linear and 2-glandular. Ovary half inferior, ovoid; stigma dialated. Berry covered the enlarged perianth tube. Fl. & Fr.: February – June.

Western Ghats: Wet evergreen forest of Karnataka, Kerala and Tamil Nadu (Endemic).

IUCN: VU- B1ab(iii)+2ab(iii). (IUCN, 2024).

Actinodaphne maderaspatana Bedd. ex Hook.f., Fl. Brit. India 5: 149. 1886.

Trees to 20 m tall. Leaves 4-8 in a whorl, penninerved, up to 23 x 6 cm, oblong-lanceolate, acute at both ends, glabrous, glaucous beneath; lateral nerves 8-10 pairs, very oblique; petioles .8-1.2 cm long, glabrous. Flowers small, dioecious, in dense globose silkily tomentose clusters, females very shortly peduncled; bracts imbricate, caducous. Perianth-tube short; lobes 6, subequal. Stamens in male flowers 9, first row opposite the perianth-lobes; filaments 2-glandular; anthers all oblong, introrse, 4-celled; filaments slender. Staminodes in female flowers also 9, spathulate or linear, 2-glandular. Ovary in female flowers ovoid, attenuate into the style; stigma dilated; in male flowers 0 or a pistillode. Fruit a berry. Fl. & Fr.: July – March.

Western Ghats: Evergreen forests above 1000 m in Karnataka, Kerala and Tamil Nadu (Endemic).

IUCN: NE (IUCN, 2024).

Actinodaphne tadulingamii Gamble., Bull. Misc. Inform. Kew 1925: 130. 1925.

Trees to 7 m tall; bark grey mottled with green, lenticellate; branches rusty tomentose. Leaves simple, 3-5 in a whorl, opposite or subopposite, elliptic-lanceolate, 8-24 x 2.5-7.5 cm, base cuneate-attenuate, apex acuminate, margin entire, glabrate above and fulvous villous beneath, subcoriaceous; lateral nerves 10-15 pairs, intercostae scalariform, prominent; petiole 10-20 mm long, tomentose. Flowers unisexual, in lateral umbels or short subracemes; peduncle 4-5 mm long, pubescent; pedicels 0.5-0.6 cm long, pubescent; bracts orbicular, fulvous villous. Perianth tubular, campanulate and 5 mm across in males, cupular and 5-7 mm across in females; lobes 6. Stamens 9, in 3 rows, filaments 2-glandular; staminodes 9, in 3 rows, fulvous, first and third row linear and 2-glandular. Ovary half inferior, globose, sunk in perianth, 0.1-0.15 cm across, fulvous-villous; stigma dilated. Berry ovoid, 8-10 mm across, black, bsally cover cup shaped perianth; pedicel 1 cm long. Fl. & Fr.: August – December.

Western Ghats: Evergreen forests of Karnataka, Kerala and Tamil Nadu (Endemic).

IUCN: NT (Low risk/near threatened) (IUCN, 2024).

Actinodaphne wightiana (Kuntze) Noltie., Regnum Veg. 145: 495. 2005. Synonyms: *Actinodaphne malabarica* N. Balak., *Actinodaphne hirsute* Hook.f. Vern.: *Kambilivirinji* (Mal.)

Trees to 18 m tall; bark dark grey, smooth, lenticellate; young shoots densely fulvous tomentose. Leaves simple, subverticillate, opposite or subopposite, elliptic or elliptic-lanceolate, 7-25 x 2.5-6 cm, apex acuminate, base acute or attenuate, margin entire, glabrous above, villous beneath; lateral nerves 8-12 pairs, intercostae scalariform, obscure; petiole 7-20 mm long, stout, tomentose. Flowers unisexual, yellowish. Male flowers in densely tomentose racemes of 2.5 cm long; bracts 10 x 7 mm, orbicular, densely silky brown tomentose. Tepals 6, subequal, oblong, obtuse, silky brown tomentose. Stamens 9, in 3 rows; filaments 2-galndular, densely hispid; anthers 4-celled, oblong, introrse. Female flowers staminodes 9, in 3 rows, third row with 2-glands. Ovary half inferior, ovoid, attenuate into the style; stigma dilated. Berry globose, 7-10 mm across, seated on perianth cup, red. Fl. & Fr.: December – April.

Western Ghats: Evergreen forests of Karnataka, Kerala and Tamil Nadu (Endemic).

IUCN: LC (IUCN, 2024).

Alseodaphne Nees

1a. Fruit upto 2.5 cm diam, ellipsoid, not curved, dark red .... **A. semecarpifolia**

1b. Fruit upto 3 cm diam, obconical, curved, black .... **A. habrotricha**

Alseodaphne semecarpifolia Nees. N. Wallich, Pl. Asiat. Rar. 2: 72. 1831. Synonyms: *Alseodaphne semecarpifolia* Nees var. *angustifolia* Meisn, *Alseodaphne semecarpifolia* var. *malabarica* Robi & Udayan, *Alseodaphne semecarpifolia* var. *parvifolia* Wight ex Hook.f. Vern.: *Vandukadipattai, Yavarani* (Tam.)

Trees to 25 m tall; bark yellowish-brown, lenticellate, vertically furrowed; branchlets slender, smooth, glabrous, lenticellate. Leaves simple, alternate, crowded near the tip of branchlets, obovate, elliptic-obovate or oblong, 14-17 x 8-11 cm, base acute or cuneate, apex round, obtuse or emarginate or subacute, margin entire, coriaceous, glabrous, glaucous beneath, surface areolate; lateral nerves 6-10 pairs, intercostae reticulate, prominent; petiole 6-25 mm long, glabrous. Flowers yellowish-green, in slender lax axillary panicles longer than leaves; pedicels filiform. Perianth tube shallow; lobes 6, 3 + 3, subequal, ovate, concave, pubescent within, margin ciliate. Stamens 9 fertile, in 3 rows, introrse, with 2 large glands on the inner bases of the filaments, ovate, subsessile; anthers 4-celled, staminodes 3, 1.5 mm long, cordate, stipitate. Ovary half inferior, globose; style 1 mm; stigma peltate. Berry ellipsoid, 20-25 mm long, dark red. Seed oblong-globular. Fl. & Fr.: January – March.

Western Ghats: Semi-evergreen to evergreen forests of Maharashtra, Goa, Karnataka, Kerala and Tamil Nadu. World: Sri Lanka.

IUCN: LC (IUCN, 2024).

Alseodaphne habrotricha Kosterm., Candollea 28: 110. 1973. Synonym: *Alseodaphne semecarpifolia* Nees var. *rufa* Dalz. ex Hook.f.

Trees to 12 m tall; bark brown; branchlets ash-colored when mature, slender, scattered with sub-orbicular lenticels. Leaves alternate, crowded towards the apex of branches, obovate-oblong or elliptic-oblong, 7.5–14 x 3–5.8 cm, thickly coriaceous, glossy and smooth adaxially, white glaucous abaxially, midrib greenish-yellow, elevated abaxially, slightly impressed adaxially; lateral veins 5–8 pairs, very conspicuous abaxially, slightly elevated adaxially, arcuately connected at ends; petiole slightly thick, 5–20 mm long, greenish-yellow, concave above, glabrous. Inflorescence axillary cymose panicles or pseudo-terminal, clustered at apex of branchlets, 10–16 cm long, many flowered; peduncles 6–13 cm long, sparsely silky-puberulent along rachis, puberulous on nodes; bracts and bracteoles deciduous. Flowers small, greenish-yellow, 6–13 x 2.5 mm; pedicels 4–10 mm long, sparsely puberulous. Perianth lobes 6, orbicular-ovate, acute, thin, tri-nerved, punctate, membranaceous, margin hyaline and ciliate; outer ones 2 x 1.5 mm, inner ones 2 x 2 mm. Fertile stamens 9 in three whorls, 1.3 mm long in 1st whorl, 1.4 mm long; filaments complanate, thin, narrow, puberulous outside, shortly stalked or sessile orbicular-cordate glands, attached slightly above the base of filaments; anther lobes ovate, obtuse, 0.6 mm long, Staminodes minute, sagittate, 1 mm long, sparsely villous at base. Ovary ovate, 0.8 mm long, glabrous, attenuate into 0.5 mm long style; stigma peltate, 3-lobed. Berry ellipsoid, 1.8– 3 x 0.8–1.4 cm, fleshy, smooth, glossy, apex obtuse, black on maturity, fruiting pedicel fleshy, obconical, 8–12 mm long, 4 mm at apex, slightly curved, greenish. Seeds ellipsoid, 22 x 9 mm long, greenish-yellow, smooth. Fl. & Fr.: March – August.

Western Ghats: Evergreen forests of Maharashtra (Endemic).

IUCN: EN- B1ab(iii)+2ab(iii). (IUCN, 2024).

Apollonias Nees

Apollonias arnottii Nees, Syst. Laur.: 670. 1836. Synonym: *Apollonias arnottii* var. *beddomei* M. Gangop.

Trees to 12 m tall; young shoots glabrescent. Leaves simple, scattered near the end of branchlets, elliptic, elliptic-lanceolate or narrowly lanceolate, 5-16 x 1.4-4 cm, base attenuate or acute, apex acuminate or caudate-acuminate, margin entire, pubescent on both sides when young, glabrous on ageing; lateral nerves 5-13 pairs, intercostae reticulate, prominent; petiole 5-18 mm long, grooved above, glabrous. Flowers yellowish, small, 3-4 mm long, in axillary or subterminal racemose panicles of 2.5-5 cm long; peduncle and the pedicels slender; bracts and bracteoles minute. Perianth tube short; lobes 6, subequal, in 2 rows. Stamens 9 perfect, those of the first and second rows introrse, opposite the perianth lobes, third row opposite the first and extrorse with a pair of stalked glands at the base. Ovary half inferior, sessile, 1-celled; ovule 1, pendulous; style curved; stigma depressed capitate. Drupe ovoid, 10-12 mm long, pale green, seated on enlarged perianth lobes. Fl. & Fr.: November – May.

Western Ghats: Evergreen forests of Kerala and Tamil Nadu (Endemic).

IUCN: NE (IUCN, 2024).

Beilschmiedia Nees

1a. Leaf bud elongate, perulate; inflorescence glabrous; bracts absent or small, caducous .... **B. wightii**

1b. Leaf bud small, not perulate; inflorescence pubescent; bracts large, concave, pubescent 2

2a. Inflorescence supra-axillary racemes .... **B. tirunelvelica**

2b. Inflorescence axillary panicles 3

3a. Fruit ovoid; lateral veins 5-8 pairs .... **B. dalzellii**

3b. Fruits cylindrical; lateral veins 8-12 pairs .... **B. gemmiflora**

Beilschmiedia dalzellii (Meisn.) Kostrem., Reinwardtia 6: 282. 1962. Synonyms: *Beilschmiedia fastigiata* var. *dalzellii* Meisn.

Leaves simple, alternate, sometimes subopposite, clustered at twig ends. Leaves elliptic to ovate-lanceolate, 9.5-15 x 3.5-6 cm, apex acute to shortly acuminate with blunt tip, base acute to attenuate, margin entire, glabrous, smooth and shining above, coriaceous; secondary nerves 10 pairs, slender, gradually curved, tertiary nerves strongly reticulate; petioles 1-1.5 cm long. Fl. & Fr.: November – June.

Western Ghats: Rare in evergreen forests above 1000 m in Maharashtra and Karnataka (Endemic).

IUCN: VU- B2ab(i,ii,iii,iv). (IUCN, 2024).

Beilschmiedia gemmiflora (Blume) Kosterm., J. Sci. Res. (Jakarta) 1: 115. 1952. Syn.: *Beilschmiedia bourdillonii* Brandis.

Trees to 35 m tall; branchlets terete, angular when young; bark light brown. Leaves alternate, elliptic, 9–15 x 4–8 cm, base obtuse to cuneate, apex acute, margin slightly revolute, glabrous, not glaucous, midrib slightly impressed or almost immersed above; lateral veins 8–10 pairs, slightly raised above, raised below, tertiary veins not percurrent, slightly raised below; petioles 1.5–2 cm long, canaliculate. Inflorescences axillary, enveloped by an involucre when young, racemose to paniculate, 1–3 cm long, to 20 flowers; floral pedicels 1 mm long. Flowers greenish, 3 x 3 mm. Tepals 6, equal, oblong, outside pubescent, inside almost glabrous. Stamens 9, first and second whorls 1.3 mm long, third whorl 1.5 mm long, all filaments longer than anthers, pubescent. Stamens in third whorl with a pair of glands; anthers glabrous, obtuse at apex; staminodes 3, sagittate. Pistil almost glabrous; ovary ovoid, 0.5 mm long; style terete, 1.2 mm long. Infructescences 5–8 cm long; fruit pedicels 3 mm diam, not constricted at apex. Fruits ellipsoid, 3 x 1.7 cm, drying blackish brown, surface smooth. Fl. & Fr.: November – June.

Western Ghats: Rare in evergreen forests of Maharashtra, Karnataka and Tamil Nadu (Endemic).

IUCN: LC (IUCN, 2024).

Beilschmiedia tirunelvelica V. S Manickam, C. Murugan, G. J. Jothi & V. Sundaresan., Nord. J. Bot. 24(4): 407. 2005.

Tree 3-5 m tall; bark smooth, dark brown; branchlets flat, glabrous. Leaves opposite, simple, oblong-lanceolate, 7-12 x 3-5 cm, thinly coriaceous, glabrous, base cuneate, margin entire, apex acuminate; nerves 5-7 pairs, prominent on both sides, reticulation prominent; petiole 1-1.5 cm long, grooved above, glabrous. Racemes, supra-axillary, 2-5 cm long, glabrous; peduncles up to 1 cm long. Flowers pale green, 5-6 mm in diam; pedicels slender, 5-7 mm long, black when dry. Perianth tube 2-3 mm long, obconical; lobes 4, shallowly lobed, ovate, 2.5 mm long, deciduous, apiculate. Stamens 4 in 2 whorls; filaments 1.5 mm long; anthers 2-celled, introse; staminodes 2. Ovoid-cordate, hairy, 1.2 mm long. Ovary ovoid, 0.5 mm long, glabrous, 1-celled, free from perianth tube; ovule 1, pendulous; style 0.75 mmlong, slender, articulated, caducous; stigma simple, acute. Drupe oblong, 4 x 2 cm, glabrous, pink-purple. Fl. & Fr.: March – April.

Western Ghats: Rare in evergreen forests above 1300 m in Kanyakumari (Endemic).

IUCN: EN- B1ab(iii)+2ab(iii). (IUCN, 2024).

Beilschmiedia wightii (Nees) Benth. ex Hook.f., Fl. Brit. India 5: 124. 1886. Vern.: *Nagaramaram* (Tam.)

Trees to 25 m tall; bark surface brown, smooth, branchlets glabrous. Leaves simple, alternate and opposite, elliptic, elliptic-ovate, elliptic-oblong or elliptic-lanceolate, 7.5-15 x 2-6.2 cm, apex obtusely acuminate, base acute or alternate, margin entire, glabrous; lateral nerves 8-10 pairs, intercostae reticulate; petiole 1-2.5 cm, glabrous. Flowers greenish-yellow, in axillary and terminal, glabrous panicles, small, ebracteate; peduncle and pedicel slender. Perianth tube short, lobes 6, 3+3 subequal. Stamens 9 perfect, in 3 rows, 2 outer series with eglandular filaments and introrse anthers, third series with 2 glandular filaments with extrorse anthers; staminodes 3, glands 6, stipitate, as large as staminodes, between the second and third row of stamens. Ovary half inferior, ovoid, glabrous; style slender, stigma subcapitate or obtuse. Berry ovoid, 17-18 mm long, smooth, purple. Fl. & Fr.: January – February.

Western Ghats: Evergreen sholas of Karnataka, Kerala and Tamil Nadu (Endemic).

IUCN: NT (Low risk/near threatened) (IUCN, 2024).

Cinnamomum Schaeffer

1a. Flowers a few in axillary cymes....................2
1b. Flowers many in terminal or axillary panicles 13
2a. Leaves alternate, tip distinctly caudate ....**C. caudatum**
2b. Leaves opposite or subopposite, tip obtuse to acuminate 3
3a. Leaves ovate, tawny pubescent beneath 4
3b. Leaves linear laceolate or elliptic, glabrous beneath 7
4a. Perianth tube to 4 mm long, red in colour; ovary glabrous 5
4b. Perianth tube to 9 mm long, brownish pubescent; ovary pubescent .... **C. travancoricum**
5a. Petiole length more than 2 cm long; panicle glabrous 6
5b. Petiole length less than 1.5 cm long; panicle pubescent 9
6a. Petiole to 15 mm long, glabrous; anther filament glabrous; berry to 14 mm long ....**C. filipedicellatum**
6b. Petiole to 6 mm long, puberulous; anther filament pubescent; berry to 10 mm long ....**C. mohananii**
7a. Basal veins visible more than half of the leaves ....**C. chemungianum**

7b. Basal veins visible less than half of the leaves 8
8a. Leaves broadly ovate, silky pubescent beneath .... **C. ovalifolium**
8b. Leaves ovate-lanceolate, tomentose beneath .... **C. litsaeifolium**
9a. Peduncle to 4 cm long; ovary ellipsoid 10
9b. Peduncle to 10 cm long; ovary obovoid .... **C. agasthyamalayanum**
10a. Leaves linear-ovate, lateral pair of nerves reached above half; flowers yellowish 12
10b. Leaves elliptic-ovate, lateral pair of nerves reached to the top; flowers dark pink 11
11a. Leaves with strong camphor smell; anthers 4-locellate ... **C. nilagiricum**
11b. Leaves with less camphor smell; anthers not 4-locellate .... **C. dubium**
12a. Peduncle less than 10 cm long; lateral nerves prominent; apex of the fruit depressed ....**C. riparium**
12b. Peduncle more than 15 cm long; lateral nerves faint; apex of the fruits apiculate .... **C. heyneanum**
13a. Peducnle terete, glabrous or hairy 16
13b. Peduncle goorved, ribbed or angular, pubescent 14
14a. Leaves with 3-5-nerved; flowers bisexual; fruits to 1.5 cm diam............... ....**C. macrocarpum**
14b. Leaves 3-nerved; flowers unisexual; fruits to 0.8 cm diam 15
15a. Fruits cylindrical oblongoid; calyx cup with persistent lobes ....**C. malabatrum**
15b. Fruits subglobose-ellipsoid; calyx cup rounded margin ....**C. iners**
16a. Inflorescence yellowish tomentose; perianth lobes persistent in fruit 17
16b. Inflorescence simple pubescent; perianth lobes oblong in fruit, deciduous 20
17a. Leaves glabrous, broadly ovate; perianth densely dawny tomentose above ....**C. sulphuratum**
17b. Leaves brown tomentose, elliptic-ovate; perianth sparse yellow pubescent above...18
18a. Inflorescence uniformely dichotomous paniculate cymes; leaf base rhomboid-cuneate..**C. walaiwarense**
18b. Inflorescence irregularly paniculate or cymes; leaf base rounded-obtuse 19

19a. Leaves to 6 cm long, 3-5-ribbed; sepals densely woolly tomentose; stamens all 9-perfect .... **C. perrottetii**

19b. Leaves to 12 cm long, 3-ribbed; sepals glabrous or minutely pubescent; sataminode present .... **C. gamblei**

20a. Leaves 12-28 cm long, elliptic to oblong-lanceolate .... **C. malabaricum**

20b. Leaves less than 15 cm long, ovate to ovate lanceolate 21

21a. Peduncle glabrous or glabrescent; pedicel equal to or longer than perianth 22

21b. Peduncle glaucous (white); pedicle shorter than perianth .... **C. wightii**

22a. Peduncle less than 10 cm long; fruits ellipsoid with cupular calyx cup.... 24

22b. Peduncle more than 15 cm long; fruits ovoid-ellipsoid with lobulate calyx cup 23

23a. Leaves glacous beneath; flowers bisexual .... **C. nitidum**

23b. Leaves glabrous beneath; flowers unisexual .... **C. goaense**

24a. Leaves trinerved; fruiting calyx cup minute at the base of fruit................ .... **C. keralaense**

24b. Leaves 6-nerved; fruiting calyx enlarged at the base of fruit .... **C. verum**

Cinnamomum agasthyamalayanum Robi, Sujanapal & Udayan, Int. J. Advanced Res. 2(10): 1012. 2014.

Trees to 8 m tall; bark smooth or somewhat rough, dark brown with strong odor of camphor; branchlets slender, smooth, terete, glabrous. Leaves simple, opposite or sub-opposite, lanceolate to elliptic-lanceolate, 7-14 x 2-5 cm, base acute, apex attenuate or sub-acuminate, glabrous, glaucous below, smooth and glossy above, chartaceous to thinly coriaceous; midrib slightly raised above, prominent below, trinerved; lateral veins 2 pairs, opposite or sub-opposite, intercostal veins scalariform, prominent abaxially; petioles to 1 cm long, slender, shallowly grooved above, glabrous. Inflorescences sub-terminal, cymose-panicle, 5–10 cm long, slender, few flowered. Flowers greenish-yellow, sparsely minutely appressed sericeous; pedicels to 7 mm long, slender, sparsely sericeous. Tepals 6 in 2 whorls, equal, 3 x 1.5 mm, ovate or elliptic, acute or obtuse at apex, thick, minutely appressed sericeous outside, margin ciliate. stamens 9 in 3 whorls of 3 each, to 3 mm long; outer whorl 3, anthers ovate; filaments 1 mm long, minutely pilose; inner whorl 3, anthers ovate, latrorse; filaments with 2-glands attached near the basal portion; staminodes

3, to 1 mm long, semi-sagittate, sparsely pilose. Ovary obovoid, to 2 mm long, glabrous; style 1.5 mm long, glabrous; stigma small, peltate. Fl. & Fr.: January – May.

Western Ghats: Evergreen forests above 800 m in Kerala (Endemic).

IUCN: NE (IUCN, 2024).

Cinnamomum caudatum Nees, N. Wallich, Pl. Asiat. Rar. 2: 76. 1831. Synonym: *Neocinnamomum caudatum* (Nees) Merr.

Trees to 15 m tall; bark gray-black; branchlets terete, brown when dry, puberulent. Leaves alternate, ovate or ovate-oblong, 5-12 x 3-4.5 cm, glabrous on both surfaces, trinerved, midrib and basal lateral veins elevated abaxially, slightly impressed adaxially, base cuneate, broadly cuneate, or subrounded, apex acuminate, acumen obtuse; petiole 8-12 mm, subglabrous, slightly sulcate adaxially. Glomerules usually 5-6-flowered, on 0.5-1 mm stalk, remotely arranged into a panicle, panicle axillary or terminal, up to 10 cm, erect, unbranched or few branched; rachis of panicle rusty pubescent; bracts subulate, less than 1 mm, densely rusty puberulent; pedicels 2-6 mm long. Flowers small, yellow-green, 4-8 mm. Perianth lobes 6, subequal, triangular-ovate, 0.2 mm long, slightly thickened, densely rusty puberulent outside and inside. Fertile stamens 9, 1 mm long; filaments villous, nearly as long as anthers; anthers nearly square, with 2 upper extrorse cells and 2 lower lateral cells or with 4 cells arranged in an almost transverse series; staminodes small, nearly estipitate. Ovary ellipsoid-ovoid, less than 1 mm; style longer; stigma discoid. Fruit narrowly ellipsoid, 1.5-2 x 1 cm, red when ripe; perianth cup in fruit crateriform, 6-8 mm wide, with wilting persistent lobe at top; fruit stalk 0.5-1 cm, slightly dilated toward apex. Fl. & Fr.: August–December

Western Ghats: Rare in evergreen forests of Karnataka. World: North East India, Nepal, Myanmar and South China.

IUCN: NE (IUCN, 2024).

Cinnamomum chemungianum M. Mohanan & A.N. Henry., J. Bombay Nat. Hist. Soc. 88: 97. 1991.

Trees to 6 m tall. Leaves simple, opposite, ovate, base acute, obtuse or round, 3-7 x 2-4 cm, apex caudate-acuminate, margin entire, glabrous above, sparsely appressed hairy beneath; lateral nerves 3-ribbed from the base, prominent; petiole 10-20 mm long, slender, pubescent. Flowers unisexual, 11 mm long. Perianth tube 3 mm long; lobes 1.5 mm long, red, ovate, acute.

Stamens many in 3 rows, 1.5 mm long; anthers truncate, glandular, 4-loculed; filaments villous; staminodes 0.5 mm long, ovate, acute. Ovary inferior, 2 x 1.5 mm, globular, 1-ovuled; styles 1 mm long; stigma faintly 3-lobed. Berry ellipsoid, 1.4 x 1 cm. Fl. & Fr.: October – March.

Western Ghats: Evergreen forests of Kerala and Tamil Nadu (Endemic).

IUCN: CR-D (IUCN, 2024).

Cinnamomum dubium Nees. N. Wallich, Pl. Asiat. Rar. 2: 73. 1831. Synonyms: *Laurus dubia* (Nees) Wall. ex Nees, *Cinnamomum multiflorum* (Roxb.) Wight, *Cinnamomum villosum* Wight, *Laurus multiflora* Roxb. Vern.: *Kaattukaruvapattai, Val Kurundu* (Tam.).

Trees to 7 m tall; bark smooth; branchlets slender, sparcely grey tomentose. Leaves alternate, sub-opposite or opposite, ovate or ovate-lanceolate, 1.8-8.5 x 1.3-3.5 cm, acute to rounded at base, acute or gradually acuminate at apex, glabrous, glossy beneath; 2 basal nerves arising a little above the base and reaching nearer to the apex, prominent on both surfaces, secondary nerves faint; petiole to 0.7 cm long, glabrescent. Panicles axillary or subterminal, few flowered, to 5 cm long, main peduncle slender. Flowers dark pink; pedicels to 7 mm long, slender. Perianth tube 1.5 mm, pilose. Tepals 2.5 x 1.8 mm, ovate, subacute or obtuse at apex, minutely silky on both surfaces. Stamens 9, 1.5 mm long; filaments 0.7 mm long, pilose; anthers oval or sub rectangular, outer one with 4 introrse cells, 3 inner stamens and glands forming a loosly adhering ring around the ovary; staminodes 3, to 1.8 mm long, semi-sagitate. Ovary subglobose, to 1.2 mm long, glabrous; style equaling the ovary; stigma peltate. Fruit ellipsoid, 9-1.3 x 6-10 mm, smooth, black on ripening; fruiting pedicel obconical, to 12 mm long, tepals persistent. Seed 1, greyish-white. Fl. & Fr.: October – April.

Western Ghats: Evergreen forests of Kerala and Tamil Nadu. World: Sri Lanka.

IUCN: LC (IUCN, 2024).

Cinnamomum filipedicellatum Kostrem., Bull. Bot. Surv. India 25: 92. 1985. Synonym: *Cinnamomum gracile* Hook.f., non Miq. Vern.: *Kaattu karuvapattai* (Tam.).

Small trees to 5 m tall; branchlets slender, glaucous. Leaves elliptic-lanceolate to subovate, obtusely acumi-nate, base acute, to 12 x 6 cm, the 3 basal ribs reaching half length of lamina; pedicel to 2.5 cm long, terete, pubescent when young, glabrous at maturity. Panicles slender, glabrous. Flowers yellow with pink shades. Fl. & Fr.: December – February.

Western Ghats: Evergreen forests of Kerala and Tamil Nadu (Endemic).

IUCN: EN-D (IUCN, 2024).

Cinnamomum gamblei Geethakum., Deepu & Pandur., Phytotaxa 326: 252. 2017.

Trees to 14 m tall; bark brown; branchlets angular, covered with appressed hairs when young. Leaves opposite or subopposite, ovate to broadly elliptic to lanceolate, 5-16.7 x 2.2-8.5 cm, coriaceous, smooth, glossy dark green above, pale beneath, glabrous, base cuneate or obtuse, apex acute; midrib prominent below, lateral nerves arising 10-15 mm above the base, prominent; petioles stout, 1-1.8 cm long, flat above. Panicles terminal or from the axils of terminal leaves or extra-axillary, few flowered; peduncle thickened, 3-13.5 cm long; pedicels sparsely appressed pilose, 3-7 mm long. Flowers yellowish green, 7-12 x 3-3.5 mm, appressed pubescent. Tepals fleshy, 3+3 in 2 whorls, ovate to elliptic, 4.5 mm long, densely sericeous, apex acute, incurved. Stamens 9 in 3 whorls of 3 each, all 4-locellate; anthers oblong, fleshy, introrse; filaments 1-1.2 mm long, pilose, glands shortly stipitate, broadly ovate, 1 mm long; staminodes shorter than the stamens, 1.7 mm long, sub-cordate, stipes pilose. Ovary ellipsoid, 1.6 mm long, glabrous; style 1.6-2 mm long, thick; stigma peltate. Fruits ellipsoid, 2.5 cm long; cupule cup shaped, base conical, with remnants of tepal bases. Fl. & Fr.: March – October

Western Ghats: Kerala and Tamil Nadu (Endemic).

IUCN: NE (IUCN, 2024).

Cinnamomum goaense Kosterm., Bull. Bot. Surv. India 25: 94. 1985

Trees to 9 m tall; branchlets slender, angular, subappressed pilose. Leaves opposite or sub-opposite, sub-ovate or ovate-oblong or ovate-lanceolate, 7-26 x 5.2-7.3 cm, coriaceous, glabrous, paler beneath, acuteat base; midrib slender, prominent, lateral nerves arising to 5 mm above base; petioles slender, 1.6 cm long, slightly canaliculated. Panicles axillary, many-flowered, minutely sericeous; main peduncle slender, 8-18 cm long; branchlets 2–6, slender, 6 cm long. Flowers pale yellow, 5-8 x 5 mm, aggregated near apex of inflorescence branches. Tepals 6 in 2 whorls of 3 each, ovate or ovate-oblong, 2-3 x 2-2.5 mm, sericeous within, apex obtuse to acute. Stamens 9 in 3 whorls of 3 each, all 4-locellate; anthers ellipsoid, introrse; filaments 0.8 mm long, hirsute; glands shortly stipitate, flat, 0.6 mm long, attached to basal portion of the filaments; staminodes shorter than anthers, 1.3 mm long, sagittate, stipe spilose. Ovary ellipsoid, 1 mm long, glabrous, style 1mm long, cylindrical; stigma peltate. Fruits ellipsoid, 6 x 10 mm; cupule cup-shaped, base conical, with remnants of tepal bases. Fl. & Fr.:

Western Ghats: Evergreen forests of Maharashtra and Karnataka (Endemic).

IUCN: EN- B1ab(i,ii,iii)+2ab(i,ii,iii). (IUCN, 2024).

Cinnamomum heyneanum Nees. N. Wallich, Pl. Asiat. Rar. 2: 76. 1831. Synonyms: *Cinnamomum burmanni* f. *heyneanum* (Nees) H.W.Li, *Cinnamomum iners* var. *subvenosum* Meisn. Vern.: *Aatuvayana* (Mal.).

Trees to 10 m tall; branchlets slender, minutely grey tomentellous, wavy. Leaves opposite and sub-opposite, linear-oblong, 11-23 x 4 cm, attenuate-acuminate, base attenuate-acute, above glabrous, smooth, midrib prominulous, the sub-basal lateral nerves faint, below initially sparsely, smoothly reticulate, midrib prominent; lateral nerves reaching half of the lamina length, faint lateral nerves; secondary nerves faint, parallel, sub-horizontal, 2 mm apart; in the gradually attenuate leaf tip numerous more distinct lateral nerves. Petiole very slender, sparsely, minutely tomentose, 5-8 mm long, flattened above. Infructescences in the axils of terminal leaves with slender, glabrous, 5-8 cm long peduncle unbranched or with a single, apical, 1 cm long branch. Fruit ellipsoid, smooth, 7 x 12 mm; fruit pedicel obconical, 8-10 mm long, slender. Fl. & Fr.: November – March.

Western Ghats: Evergreen forests of Karnataka and Kerala (Endemic).

IUCN: EN- B1ab(i,ii,iii)+2ab(i,ii,iii); D. (IUCN, 2024).

Cinnamomum iners (Reinw. ex Nees & T. Nees) Blume., Bijdr. Fl. Ned. Ind.: 570. 1826. Synonyms: *Cinnamomum nitidum* var *iners* (Reinw. ex Nees & T. Nees) Miq., *Laurus iners* Reinw. ex Nees & T. Nees. Vern.: *Edana, Elavangam, Erikkolam, Ilavargam, Kachakam, Karoon, Karuva, Kunthakaimaram, Santhamaram, Thamala, Vayana* (Mal.).

Trees to 16 m tall; bark smooth, inner bark yellowish; branchlets stout or slender, terete, drying dark brown to black. Leaves opposite to subopposite, ovate, oblong-elliptic to lanceolate, 8-12 by 4-7 cm, base rounded to cuneate, apex acute, drying pale green, trinerved, coriaceous, shiny and sparsely hairy above and below, appressed and straight, without domatia; lateral veins raised on both sides, extending to the leaf tip; intercostal slender, subscalariform, faint, reticulate; petiole stout, subterete, shallowly grooved above, glabrescent, to 0.5 cm long. Inflorescences axillary or subterminal, slender, paniculate-cymose; rachis angular, 1-2 mm broad, densely to sparsely covered with short straight appressed hairs; bracts caducous. Flowers hairy, drying silky and greyish; pedicels slender, 3-5 mm long; hypanthium 1-3 mm high. Perianth lobes elliptic, 1.5-2.5 mm long, appressed pilose on both sides. Fertile stamen 1.5-2.5 mm long; anthers oblong ovate with truncate or obtuse tip; glands stalked on each side at the lower half of filaments; staminodes 1-1.5 mm long, apex hastate. Ovary subglobose, 1-1.5 mm long, stigma trilobed. Fruits ellipsoid or obovoid with pointed tip, 8 mm diam; cupule inconspicuous, very shallow, 1 mm high, 2 mm diameter, appressed hairy, perianth lobes persistent; fruiting pedicel slender, 3-4 mm long, minute-appressed hairy. Fl. & Fr.: March – May.

Western Ghats: Kerala, Karnataka and Tamil Nadu. World : Sri Lanka.

IUCN: NE (IUCN, 2024).

Cinnamomum keralaense Kostrem., Bull. Bot. Surv. India 25: 98. 1985.

Trees to 25 m tall; bark grey, thin, smooth with pustular lenticels, odourless and tasteless or with very faint smell; branchlets glabrous. Leaves simple, opposite and subopposite, lamina ovate, elliptic-ovate or elliptic-lanceolate, 6.5-12.5 x 3-6.5 cm, base acute or some times cuneately decurrent into the petiole, apex obtuse or obtusely acuminate, margin entire, glabrous, smooth and glossy above, coriaceous; 3-ribbed from a little above the base, intercostae scalariform, faint; petiole 15-25 mm long, glabrous. Flowers bisexual, 6 mm across, in axillary, 5-10 cm long slender, lax pubescent panicles; pedicels 4 mm long, slender, pubescent. Tepals 6, ovate, obtuse, densely tomentose. Stamens 9 perfect, those of first and second rows opposite the perianth lobes, introrse and eglandular, those of third row opposite the first row, bearing 2 large glands at the base; staminodes 3, cordate and stipitate. Berry, upto 8 x 20 mm, narrow ellipsoid; cupular, subconical, slightly fleshy. Fl. & Fr.: June.

Western Ghats: Evergreen forests of Kerala (Thekkadi and Munnar) and Tamil Nadu (Endemic).

IUCN: EN- C2a(i). (IUCN, 2024).

Cinnamomum litsaeifolium Thw., Enum. Pl. Zeyl.: 253. 1861. Synonyms: *Cinnamomum litsaeifolium* var. *willdenowii* Lukman.

Trees to 20 m tall; bark with shallow round cavities, light brown, paler inside, mildly fragrant; branchlets smooth, glossy. Leaves opposite, ovate-oblong to ovate-lanceolate, 4- 9 x 2-– 4 cm, rounded or subacute at base, blunt or subacute at apex, coriaceous, glabrous; midrib faint, two basal nerves arise a little above base; secondary nerves faint; petioles 1.2-1.8 cm long, canaliculate above, glabrous. Inflorescences panicle, terminal and axillary, 5.5-8 cm long, lax; peduncles 1.7-3.5 cm long, minutely sericeous, densely so towards apex. Flowers 1.2 cm long, densely silvery-sericeous, greenish white; pedicels 5 mm long, densely sericeous. Perianth tube 2.5-3 mm long, pilose; lobes subequal, ovate, 3.5 mm long, subacute at apex, thick, sericeous inside, densely silky outside, caducous. Fertile stamens 9 in 3 rows. Filaments pilose; anthers 1.5 mm long; outer 2 rows with 4 introrse cells; inner one broader with 2 lateral cells; basal glands stipitate; staminodes 3, sagittate, 2 mm long. Ovary ovoid, 2 mm long, glabrous; style as long as ovary; stigma peltate. Berry ellipsoid, 0.8-2 cm long, smooth, base surrounded by a shallow perianth tube. Seed 1, ellipsoid. Fl. & Fr.: March - October

Western Ghats: Evergreen forests of Kerala and Tamil Nadu. World: Sri Lanka.

IUCN: NE (IUCN, 2024)

Cinnamomum macrocarpum Hook.f., Fl. Brit. India 5: 133. 1886. Synonym: *Cinnamomum palghatensis* M. Gangop. Vern.: *Karuva, Karunthagiri, Illavapattai, Periyalavangapattai* (Tam.).

Trees to 15 m tall; bark greyish, smooth, with clove smell; branchlets thick, angular, glabrous or with a few, appressed hairs. Leaves simple, opposite and subopposite, ovate, elliptic-ovate or elliptic-lanceolate, 7-25 x 2-12 cm, base cuneate or decurrent into petiole, apex obtuse, margin entire, glabrous, densely reticulate beneath; 3-5-ribbed from base, of which 2 middle arise from a little above the base, intercostae scalariform, faint; petiole 10-28 mm long, glabrous. Flowers bisexual, upto 7 mm long, white, sparsely to more densely minutely sericeous, in axillary few flowered panicle upto 7 cm long; peduncle flattened, branchlets few, 1-2.5 cm long; pedicel upto 1 cm long, obconic, sparsely appressed pilose. Perianth tube 1-2 mm long, funnel shaped. Tepals 6, 3-4 mm long, ovate-oblong, acute, sericeous inside. Stamens 9 perfect, in 3 rows, 2-2.5 mm long; filaments pilose; anthers oblong, 4-celled, longer than the broad; staminodes fourth row opposite the second row, slightly shorter than the stamens with hastate-sagittate small apical part and long pilose stipe. Ovary half inferior, ellipsoid; style 1.5-2 mm long, thick; stigma minute, peltate. Berry ellipsoid, 2 x 2.5 cm, subtended by thick fleshy cup of 15 mm long. Fl. & Fr.: January – November.

Western Ghats: Evergreen forests of Maharashtra, Goa, Karnataka, Kerala and Tamil Nadu. World: Myanmar.

IUCN: VU- A2cd; D1. (IUCN, 2024).

Cinnamomum malabatrum (Burm.f.) J. Presl., F. Berchtold & J.S.Presl, P□ir. Rostlin Aneb. Rostl. 2: 46. 1825. Synonyms: *Laurus malabathrum* Burm.f., *Laurus malabatrum* Burm.f., *Cinnamomum nicolsonianum* Manilal & Shylaja Vern.: *Vayana* (Mal.).

Trees to 20 m tall; bark reddish-brown, smooth or slightly, longitudinally cracked. Leaves simple, opposite or subopposite, elliptic-oblong, oblong or oblong-lanceolate, 10-30 x 3-9 cm, apex acute or acuminate, base acute, margin entire, glabrous above, highly aromatic when bruised; 3-ribbed from at or a little above the base, intercostae scalariform, glabrescent, slender; petiole 10-20 mm long, glabrous. Flowers bisexual, in axillary and pseudo terminal paniculate cymes upto 25 cm long, with minutely subadpressed hairy; pedicels 3-4 mm long. Perianth tube shallow, 1 mm broad. Tepals 6, 3-3.5 mm long, fleshy, ovate, acute, with dense subadpressed pilose hairs. Stamens 9 perfect, in 3 rows, 2-2.5 mm long; filaments pilose, slender, slightly longer than the anthers; anthers oblong to subovate-oblong, 4-celled; staminodes of fourth row 1.5 mm long, hastate, cordate, opposite the second row. Ovary half inferior, 2 mm, ellipsoid; style thick, cylindrical; stigma small, peltate. Berry ellipsoid-terete, 8 x 10 mm, subtended by shallow fleshy cup, pilose. Fl. & Fr.: December – July.

Western Ghats: Evergreen forests of Karnataka, Kerala and Tamil Nadu (Endemic).

IUCN: LC (IUCN, 2024).

Cinnamomum mohananii Gangapr., S.P. Mathew & R. Jagad., Int. J. Advanced Res. 2(9): 611. 2014.

Small trees to 4 m tall; bark smooth; young stem, petioles and inflorescence minutely pubescent. Leaves opposite, sub-coriaceous or membranous, 2-9 x 1-4 cm, ovate-lanceolate to elliptic, acuminate at apex, obtuse-cuneate at base, glaucous or not beneath, glabrous; basal nerves 3-5, reticulation clear in the lower surface, faint on the upper surface; petioles to 6 mm long, puberulous. Cymes axillary or arising 2-6 mm above the immediate leaf axils, shorter than the leaves, 4-10-flowered; bracts 1-2 mm long, ovate-lanceolate, grey pubescent; pedicels up to 4 cm long, pubescent. Perianth lobes 6, outer 3, 2.5-3.5 mm long, elliptic, ovate-lanceolate, pubescent; inner perianth lobes 3, 1.5-1.75 mm ovate-obovate, glabrescent. Stamens in 3 rows of 3 each, outer 2 series with introrse and inner row with extrorse anthers havingglands at the base; staminodes 3, stipitate. Ovary sessile, glabrous; style simple; stigma capitate or obscurely 3-lobed. Berries 8-10 x 6-7 mm, oblong ovoid, seated on the enlarged perianth lobes, deep purple turning black on drying. Fl. & Fr.: April – August.

Western Ghats: Kerala

IUCN: NE (IUCN, 2024).

Note: Original protolague mentoned as *C. mohananii*, but later POWO record showed as *C. mohanensis*, and also some other authors published literature cited as *C. mohananianum*. To avoid the confusion, we have mentioned only the name which was published in original protolague.

Cinnamomum nilagiricum Geethakum., Pandur. & Deepu, Phytotaxa 224: 283. 2015.

Trees to12 m tall; bark brown; branchlets slender, terete, covered with apressed hairs when young. Leaves opposite or subopposite, rarely alternate, ovate to elliptic to lanceolate, 7.5-12.5 x 2.6-4.3 cm, coriaceous, smooth, glabrous, with strong smell of camphor, shortly attenuate at the base, acuminate to caudate at the apex; tertiary nerves parallel, 2–3 mm apart, finer venation reticulate, impressed adaxially, raised abaxially; petioles slender, to 1.0 cm long. Panicles terminal or from the axils of terminal leaves or extra axillary on new growth, many flowered, minutely sericeous; main peduncle to 7 cm long; pedicels 4–5 mm long, slender. Flowers greenish white, 9-12 mm long, 2.5-3 mm across, pubescent. Tepals 6 in 2 whorls, ovate, 3 mm long, minutely sericeous, sparsely within, densely towards the base and margin, apex obtuse to acute. Stamens 9 in 3 whorls of 3 each; whorls 1st and 2nd with anthers 1–1.2 mm long, 4-locellate, fleshy, introrse; filaments to 1 mm long, hirsute; 3rd whorl extrorse, to 2 mm long, anthers oblong, 4-locellate; glands subsessile, broadly ovate, 0.5 mm long, attached near the basal portion of the filaments; staminodes shorter than the anthers, to 1 mm long, cordate, stipes pilose. Ovary ellipsoid, 1 mm long, glabrous; style 1.2 mm long, rather thick; stigma peltate. Fruits ellipsoid, 8 mm long; cupule cup shaped, base conical, with remnants of tepal bases. Fl. & Fr.: November – March.

Western Ghats: Evergreen forests of Kerala and Tamil Nadu (Endemic).

IUCN: NE (IUCN, 2024).

Cinnamomum nitidum (Roxb.) Hook., Exot. Fl. 3: t. 176. 1825. Synonyms: *Laurus nitida* Roxb., *Cinnamomum brumanni* (Nees & T. Nees) Blume *'non* Blume'

Evergreen trees to 20 m tall; branches robust and angular, glabrescent. Leaves subopposed, ovate to elliptic, 8-15 x 3-6 cm, glabrescent above, glaucous beneath; 3- strong nerves from the base of the petiole, lateral nerves prominent, intercostal reticulate; petiole to 5 cm long, redish-brown, slender, grooved above, glabrous. Flowers small, bisexual, pubescent, grouped in axillary or terminal panicles, peduncle to 6-12 cm long. Berry ovoid, 10 x 7 mm. Fl. & Fr.: March –April.

Western Ghats: Evergreen forests of Maharashtra and Tamil Nadu. World: Myanmar, Malaysia, Maluku Islands.

IUCN: LC (IUCN, 2024).

Cinnamomum ovalifolium Wight, Icon. Pl. Ind. Orient. 1: t. 125. 1839.

Small trees to 5 m tall; branchlets subterete to 4-angular, greenish-yellow, silvery pubescent. Leaves opposite-decussate or sub-opposite, ovate to elliptic, 4-6 x 1-4 cm, obtuse or acute at apex, rounded at base, coriaceous, glabrous; secondary veins 2, impressed above, raised beneath, reaching near the tip of lamina, tertiary veins scalariform and finely reticulate, obscure above or slightly distinct below; petioles to 9 mm long, flat or slightly channeled above, stout, glabrescent. Inflorescences axillary and pseudo-terminal, simple panicle, 2-4 cm long, few-flowered; peduncle slender, glabrous or densely silvery-sericeous. Flowers few, 6 mm long, greenish-yellow, minutely sericeous to glabrous; pedicels to 2 mm long, slightly angular, densely, silvery-sericeous. Perianth tube to 1.5 mm long, funnel-shaped, densely silvery-sericeous; lobes 3-4 mm long, ovate, acute at apex, thick, slightly concave, pubescent inside, margin ciliate, gland-dotted, tri-nerved, membranous. Stamens 9 in 3 whorls, to 1-2 mm long; those of outer whorl introrse; anthers ovate, obtuse at apex, 4-locular, fleshy, longer than filaments; filaments thin, narrow, pilose, eglandular; middle whorl almost same as the outer; inner whorl extrorse; anthers 2-2.5 mm long, ovate-oblong, 4-locular, tip truncate; filaments with 2-glands attached at base; stipitate, ovate, pilose; staminodes 3, 1-1.5 mm long, sagittate, short-stipitate, acuminate at apex, sparsely pilose. Ovary ellipsoid, 3 mm long, glabrous; style 1 mm long, stout, slightly curved, glabrous; stigma lobed, peltate, creamy. Fl. & Fr.: May – August.

Western Ghats: Evergreen forests of Kerala. World: Sri Lanka.

IUCN: NE (IUCN, 2024).

Cinnamomum perrottetii Meissn., A.P.de Candolle, Prodr. 15(1): 22. 1864. Synonym: *Cinnamomum perrottetii* var. *nilgiriense* M.Gangop.

Trees to 8 m tall; bark smooth, brown, tasteless and odourless; branchlets stiff densely minutely tomentose. Leaves simple, opposite or subopposite, ovate or ovate-elliptic, 3.5-6 x 2.5-4 cm, apex obtuse or acute, base acute, subtruncate or round, margin entire, tomentose on both sides when young; 3-5-ribbed from near the base, prominent, intercostae scalariform, faint; petiole 10-15 mm long, grooved above, tomentose. Flowers bisexual, to 7 mm long, in axillary, few-flowered, minutely tomentose 2-10 cm long panicles; pedicels 2-3 m long, thick. Perianth tube 1.5-2 mm long, funnel shaped. Tepals 6, 5 mm long, oblong-ovate, fleshy, densely appressed pilose inside. Stamens 9 perfect, bearing 2 large glands attached to the middle of the filament; staminodes of the fourth row opposite the second row, 1.5 mm long, pilose. Ovary half inferior, ellipsoid, as long as style; stigma small, peltate. Berry ellipsoid, subtended by densely woolly tepals, base of cup fleshy, obconical. Fl. & Fr.: January.

Western Ghats: Evergreen forests of Kerala and Tamil Nadu (Endemic).

IUCN: EN- C2a(i). (IUCN, 2024).

Cinnamomum riparium Gamble., Bull. Misc. Inform. Kew 1925: 128. 1925. Vern.: *Aattuvayana* (Mal.).

Small trees to 7 m tall. Leaves opposite to subopposite, 7-11 x1.5-3.5 cm, acuminate at apex, broadly attenuate at base, 3-ribbed, 0.5 cm above from base. Panicles few-flowered, lax; peduncles 3-4 cm long; pedicels slender, 0.8-1 cm long. Flowers 0.3-0.35 cm across. Perianth lobes 6, 0.1 cm long. Stamens in 3 rows; first and second introrse, 3rd extrorse, glands attached to middle of filaments. Ovary ellipsoid; stigma peltate. Fl. & Fr.: December - March

Western Ghats: Evergreen forests along the water course of Karnataka, Kerala and Tamil Nadu (Endemic).

IUCN: EN-D (IUCN, 2024).

Cinnamomum sulphuratum Nees., N.Wallich, Pl. Asiat. Rar. 2: 74. 1831. Synonyms: *Cinnamomum sulphuratum* var. *wagneri* Lukman., *Laurus cassia* Wall. Vern.: *Kula, Kaatu karuva* (Tam.).

Trees to 8 m tall; bark smooth, odorless, reddish brown; branchlets slender, angular, densely, minute sub-appressed, pilose. Leaves simple, opposite or sub-opposite, elliptic, elliptic-ovate, elliptic-lanceolate or elliptic-oblong, 4-21 x

2-9 cm, base acute obtuse or shortly cuneate, apex acute or obtusely acuminate or acuminate, margin entire, glabrous, glaucous and yellowish pubescent beneath, coriaceous; 3-ribbed from at or a little above the base, intercostae scalariform, faint; petiole 8-15 mm long, yellow pilose when young. Flowers bisexual, in axillary or terminal lax, yellow fulvous pilose panicles. Perianth tube 1-1.5 mm long, densely yellow pilose on both sides. Tepals 6, 2-4 mm long, ovate, acute, yellow pilose on both sides. Stamens 9 perfect, those of first and second rows opposite the perianth lobes, bearing 2 large glands at the base; staminodes 3, narrowly sagitate, stipitate, pilose. Ovary half inferior, ellipsoid, as long as the style, stigma minute, peltate. Berry ovoid, 1 x 1.5 cm, cup-shaped, appressed-pilose, ovate, base of the cup fleshy, obconic. Fl. & Fr.: December – April.

Western Ghats: Evergreen forests of Karnataka, Kerala and Tamil Nadu (Endemic). World: Myanmar

IUCN: VU- A2cd. (IUCN, 2024).

Cinnamomum travancoricum Gamble., Bull. Misc. Inform. Kew 1925: 128. 1925.

Trees to 8 m tall; branchlets slender, angular, minutely appressed pilose. Leaves simple, opposite, elliptic, elliptic-oblong or elliptic ovate, 5-9 x 1.5-4 cm, base acute, apex acute or acuminate, margin entire, appressed pilose when young, glabrous above when mature; 3-ribbed from a little above the base, prominent beneath, intercostae scalariform, faint; petiole 5-15 mm long, slender, appressed pilose. Flowers bisexual, in axillary and pseudo terminal, 1-3 cm long, light brown, subappressed pubescent racemes; peduncle slender; pedicel 1-3 mm long. Tepals 6, 4 mm long, oblong, acute, densely subappressed pilose, subequal. Stamens 9 perfect, 2.5-3 mm long, bearing 2 large glands at the base; staminodes 3, cordate and stipitate; staminodes 3, 1-1.5 mm long, sagittate, stipe pilose. Ovary half inferior, sessile, glabrous, at the bottom of the perianth tube, narrowed into thick style, stigma peltate. Fl. & Fr.: November – June.

Western Ghats: Evergreen forests of Karnataka, Kerala and Tamil Nadu (Endemic).

IUCN: CR-D (IUCN, 2024).

Cinnamomum verum J. Presl., F. Berchtold & J.S. Presl, P□ir. Rostlin Aneb. Rostl. 2: 36. 1823. Synonyms: *Cinnamomum zeylanicum* Blume, *Cinnamomum*

alexii Kosterm., *Cinnamomum aromaticum* J.Graham, *Cinnamomum iners* Wight, *Laurus cassia* Burm.f., *Cinnamomum mathewianum* Remya *et al.*, Vern.: *Illavangapattai, Karuvapattai, Karipattamaram, Pattaimaram* (Tam.).

Trees to 20 m tall; bark brown, rough, cracks vertical; branchlets glabrous. Leaves simple, opposite or subopposite, ovate, elliptic ovate or elliptic-lanceolate, 9.5-14 x 3.5-5.5 cm, apex acute to acuminate, base acute, margin entire, glabrous, coriaceous; 3-ribbed from base, prominent, glabrous; lateral nerves 3-6 pairs, intercostae reticulate; petiole 8-20 mm, glabrous, slightly grooved above. Flowers bisexual, in terminal and axillary; pedicel 7 mm long, pale yellow, 5 mm long, 6 mm across. Perianth 8 mm, silky, tube campanulate, lobes 6, 3 mm long, oblong-lanceolate. Stamens 9 perfect, introrse and eglandular, bearing 2 large glands at the base; staminodes 3, cordate and stipitate. Ovary half inferior, sessile. Berry ellipsoid to oblong-ovoid, 1-2 cm long, dark purple, surrounded by the enlarged perianth. Fl. & Fr.: February – June.

Western Ghats: Evergreen forests of Maharashtra, Goa, Karnataka, Kerala and Tamil Nadu. World: Sri Lanka, Myanmar, Bangladesh and Malaysia.

IUCN: NE (IUCN, 2024).

Uses: Bark is used as condiments and tradable. Leaves used for spice and condiments.

Cinnamomum walaiwarense Kostrem., Bull. Bot. Surv. India 25: 119. 1983 publ. 1985. Vern.: *Ealavangam* (Mal.).

Trees to 6 m tall; bark very smooth, light brown; branchlets slender, the apical part densely subsericeous. Leaves opposite, chartaceous, oblong, 3-7 x 9 cm, acuminate, base acute, above glabrous, smooth; main nerves thin, below densely, minutely sericeous; secondary nerves faint, parallel, horizontal, 2-4 m apart; petiole 5-15 mm long, concave above. Inflorescence axillary, very slender, up to 10 cm long, with many fruits with long slender glabrous peduncle and few, minutely sericeous. Perianth cup deep, rather fleshy, conspicuously longitudinally ribbed, more than half of the basal part of the tepals, persistent, obconial. Fl. & Fr.: July – August.

Western Ghats: Evergreen forests above 1200 m in Kerala and Tamil Nadu (Endemic).

IUCN: CR- C2a(i). (IUCN, 2024).

Cinnamomum wightii Meisn., A.P.de Candolle, Prodr. 15(1): 11. 1864. Synonyms: *Cinnamomum zeylanicum* var. *wightii* (Meisn.) Bedd., *Cinnamomum ovalifolium* Gardner ex Meisn. Vern.: *Chembila, Kaattukaruvapattai* (Tam.).

Trees to 8 m tall; bark 5-10 mm, grey, smooth, pustular, without aromatic smell; branchlets stiff, smooth, glabrous. Leaves simple, opposite or spirally arranged, elliptic, ovate, elliptic-ovate or elliptic-oblong, 3.5-20 x 2.5-7.5 cm, apex obtusely acute or obtusely shortly acuminate, base cuneate, slightly decurrent, margin entire, glabrous, minutely pitted beneath, coriaceous; 3-5 or sometimes 7-ribbed from at or a little above the base, intercostae scalariform, very faint; petiole 10-20 mm long, flattened above. Flowers bisexual, 5-6 mm long, densely fulvous-sericeous, in axillary and terminal, short and compact but long peduncled panicle of 10-14 cm long; peduncle stout, adpressed fulvous pilose. Perianth tube 1 m long, funnel shaped. Tepals 6, 3-5 mm long, thick, ovate to oblong-ovate, acute. Stamens 9 perfect, in 3 rows, upto 2 mm long, introrse and eglandular; filaments pilose, anthers 4-celled, broadly ovate; staminodes 3, sagittate, pilose, stipitate. Ovary ellipsoid, half inferior; style as long with minute, peltate stigma. Berry ellipsoid, upto 1 x 1.5 cm, subtended by fleshy obconical pedicel, margin 1 cm across. Fl. & Fr.: March – May.

Western Ghats: Evergreen foersts of Karnataka, Kerala and Tamil Nadu (Endemic).

IUCN: EN- B1ab (i, ii,iii,iv) (IUCN, 2024).

Cryptocarya R.Br.

1a. Fruits globose–subglobose or occasionally obovoid 2

1b. Fruits oblongoid, ellipsoid-ovoid or occasionally cylindric 3

2a. Fruits conspicuously stipitate (stipe 5–8 mm long) .... **C. beddomei**

2b. Fruits not stipitate.. **C. wightiana**

3a. Venation on leaves forming a conspicuously elevated fine reticulum, particularly on upper leaf surface .... **C. membrnacea**

3b. Venation on leaves not forming an elevated reticulum.............................4

4a. Leaves not glaucous beneath; panicles terminal aggregate ..... **C. procera**

4b. Leaves glacous beneath; panciles both terminal aand axillary lax branched 5

5a. Fruits more than 3 cm long..6

5b. Fruits less than 3 cm long 7

6a. Fruits longitudinally striate or shallowly ribbed; pedicels of fruits slender .... **C. lawsonii**

6b. Fruit surface smooth, glabrous; pedicels of fruits stout .... **C. praetervisa**

7a. Mature branchlets and lower surface of leaves tomentose to villous 8

7b. Mature branchlets glabrous and lower surface of leaves sparsely pubescent to glabrous 9

8a. Lateral nerves of leaves 6-10 pairs; drupe cylindric, brown when ripe..... .... **C. anamalayana**

8b. Lateral nerves of leaves 10-13 pairs: drupe conical, blueish-purple when ripe .... **C. griffithiana**

9a. Leaf apex obtuse, emarginate or acute .... **C. stocksii**

9b. Leaf apex acuminate, acute-apiculate .... **C. muthuvariana**

Cryptocarya anamalayana Gamble., Bull. Misc. Inform. Kew 1925: 126. 1925.

Evergreen trees to 8 m tall; bark greyish-brown, smooth; young shoots fulvous-tomentose. Leaves simple, alternate, elliptic-oblong or elliptic-obovate, 8.5-20 x 3.5-9 cm, base obtuse, acute or cuneate, apex obtusely acuminate, margin entire, glabrous except nerves, shiny above, glaucous beneath; lateral nerves 6-10 pairs, nerves and nervules fulvous tomentose beneath; petiole 5-10 mm long, tomentose. Flowers bisexual, in short axillary fulvous-tomentose cymose panicles upto 3 cm long; pedicels scarcely any; bracts and bracteoles oblong, conspicuous, apparently persistent. Perianth 2-5 mm long, lobes 6, in 2 rows, subequal, tomentose. Stamens 9 perfect, in 3 rows, filaments and introrse anthers, fourth series replaced by 3 stipitate staminodes; anthers 2-celled; connective often produced. Ovary sessile, half inferior, enclosed in the tube; style short exserted. Drupe enclosed in accrescent perianth-tube. Fl. & Fr.: November – June.

Western Ghats: Evergreen forests of Kerala and Tamil Nadu (Endemic).

IUCN: EN- B1ab(iii)+2ab(iii). (IUCN, 2024).

Cryptocarya beddomei Gamble., Bull. Misc. Inform. Kew 1925: 127. 1925.

Evergreen trees to 8 m tall; bark pale reddish-brown, smooth; branchlets fulvous pubescent. Leaves simple, alternate, elliptic, ovate or oblong-ovate, 9.5-14 x 4-6.5 cm, apex obtuse or shortly obtusely acute, base round or truncate, margin entire, glabrous, reddish glaucous beneath; lateral nerves 5-9 pairs, intercostae scalariform, prominent; petiole upto 10 mm long, pubescent. Flowers bisexual, minute, in axillary or lateral fulvous pubescent panicles of

9 cm long; peduncle upto 3 cm long; pedicels short; bracts oblong, upto 3 mm long, deciduous; bracteoles minute, deciduous. Perianth tube narrowed at the top; lobes 6, subequal, in 2 rows, pubescent. Stamens 9 perfect, in 3 rows, with 2 glandular filaments and extrorse anthers with obtuse connective tip; filaments villous; anthers 2-celled, 3 stipitate staminodes in fourth series. Ovary sessile, half inferior, enclosed in the perianth tube; style short, exserted. Drupe 10 x 6 mm enclosed in accrescent perianth tube, puberulous. Fl. & Fr.: March – May.

Western Ghats: Evergreen forests of Karnataka, Kerala and Tamil Nadu (Endemic).

IUCN: LC (IUCN, 2024).

Cryptocarya griffithiana Wight., Icon. Pl. Ind. Orient. 5: t. 1830. 1852.

Trees to 25 m tall; bark rough, fissured longitudinally; branchlets terete, densely borown ferrugineous tomentose. Leaves alternate, simple, 5-8 x 2-4 cm, densely hairy below, dark green above with dense haris, margin entire, base obtuse, acuminate at apex; lateral nerves 10-13 pairs, impressed abive, prominent below, brown ferruginous; petiole to 3 cm long, slender, channeled above, strong hairy. Flowers on axillary and terminal panicles, small, 7 mm diam. Fruits blue-purple, conical, fleshy drupes, 2.5 mm long. Fl. & Fr.: November – February.

Western Ghats: Semi-evergreen forests of Kerala and Tamil Nadu. World: India, Myanmar, Malaysia.

IUCN: LC (IUCN, 2024).

Cryptocarya lawsonii Gamble., Bull. Misc. Inform. Kew 1925: 127. 1925. Synonyms: *Cryptocarya sheikelmudiyana* A.K.H.Bachan & P.K.Fasila

Lofty trees to 25 m tall; branchlets glabrous lenticellate, black. Leaves simple, alternate, elliptic, elliptic-oblong, oblong or oblong-lanceolate, 7-12.5 x 2.5-5 cm, base acute, apex obtuse, emarginate or somewhat acute, margin entire, glabrous, purple-glaucous beneath; lateral nerves 6-8 pairs, intercostae scalariform, faint; petiole 10-12 mm long, glabrous. Flowers bisexual, minute, in terminal and axillary panicles of 9 cm long; peduncle and pedicel slender, slightly fulvous-pubescent; bract and bracteoles small, deciduous. Perianth tube ultimately narrowed at the top; lobes 6 in 2 rows, subequal. Stamens 9 perfect, in 3 rows, with eglandular filaments and introrse anthers, fourth series replaced by 3 stipitate staminodes; connective often produced; anthers 2-celled. Ovary half inferior, sessile, enclosed in the tube; style short, exserted.

Drupe ovoid, 30-37 x 15-18 mm, black, longitudinally ribbed. Fl. & Fr.: April - June.

Western Ghats: Evergreen forests of Kerala and Tamil Nadu (Endemic).

IUCN: VU- C2a(i); EN (B1, B2, B2(a), C, C2 and D); CR-D (IUCN, 2024).

Cryptocarya membranacea Thw., Enum. Pl. Zeyl.: 254. 1861. Vern.: *Paalai* (Tam.).

Trees to 25 m tall; bark smooth, flaked when mature; branchlets tereted, densely tomentose. Leaves alternate, ovate-elliptic to ovate-laceolate, 6-10 x 3-4 cm, obtuse at base, caudate aucuninate at apex, glaucous above, dense white tomentum beneath; lateral nerves 12-14 pairs, prominent, redish below, secondary nerves finely reticulate; petiole to 3 cm long, slender, densely tomentose. Flower on terminal panicles, minute. Drupe ovoide, 5 mm diam, greenish, persistem stylar apex. Fl. & Fr.: May - June.

Western Ghats: Evergreen forests ofTamil Nadu. World: Native to Sri Lanka.

IUCN: EN- B1+2c. (IUCN, 2024).

Cryptocarya muthuvariana R. Jagad., Sureshkumar, Gangapr. & S.P. Mathew., Ann. Bot. Fennici 58: 379. 2021.

Evergreen trees to 15 m tall; bark yellowish brown; branchlets terete, glabrous, appressed with fulvous pubescence when young. Leaves simple, alternate, oblong to narrowly oblong, elliptic–oblong, rarely oblong-lanceolate, 8-18 x 2.5-4 cm, cuneate to acute at base, rarely oblique or asymmetrical, entire at margin, acute-apiculate, rarely acuminate at apex, glabrous above, fulvous puberulent on veins below, thick coriaceous; lateral nerves 7–8 pairs, usually subopposite, obscure above and conspicuous below, glabrous; tertiary nerves obscure to faint above, forming intercostate reticulate venation; petiole to 1.2 cm long, stout, canaliculate, glabrous, brown, minutely fulvous or rusty when young. Inflorescence an axillary or terminal paniculate cyme, to 16 cm long, fulvous pubescent; peduncle 2.5–8 cm long, terete, with appressed fulvous pubescence. Flowers bisexual, 4-4.5 x 2.4-2.6 mm, yellowish to pale green, fulvous tomentose; pedicels 0.4-0.6 mm long, fulvous pubescent. Perianth tube 1.8-2.4 x 1.3-1.5 mm, obconical, fleshy, densely fulvous tomentose. Tepals 6 in two whorls of 3 each, nearly equal. Outer tepals 1.8-2.1 x 1.3-1.5 mm, ovate, obtuse at apex, truncate at base, thick, both surfaces fulvous tomentose; inner tepals 1.6-1.8 x 1.2-1.35 mm, shape similar to outer tepal. Fertile stamens 9, in

3 whorls, first two whorls introrse and third whorl extrorse; anthers 2-celled, connective often produced into a mucro. Outer stamen 1.2-1.3 x 0.6-0.7 mm, introrse; anthers ovoid, glabrous; filament stout, fulvous tomentose. Middle stamens introrse; anther ovoid; filament fulvous tomentose; staminodes cordate, sessile, villous. Ovary to 1 mm long, elliptic–oblong, sessile, glabrous, enclosed in perianth tube; style stout; stigma simple, notched. Drupe fully enclosed in accrescent perianth tube, ellipsoid–oblongoid, slightly narrowed towards apex, smooth, glabrous, not lenticellate, fruiting pedicel or ultimate branches of infrutescences subtending fruits thickened; terete, thick, glabrous. Seed ellipsoid–oblongoid, attenuate at apex, wrinkled, dark brown, glabrous. Fl. & Fr.: December–May

Western Ghats: Kerala (Endemic).

IUCN: NE (IUCN, 2024).

Cryptocarya praetervisa M. gangop., Chakrab, & A.S. Chauhan., J. Econ. Taxon. Bot. 26: 472. 2002.

Evergreen trees to 18 m tall; bark roug, fissured; branchlets black, prominently lenticellate, rough, glabrous. Leaves simple, alternate, 5-10 x 2-6 cm, suborbicular, obovate-oblong or ovate, base round or subacute, apex round, truncate, emarginate or apiculate, margin entire, scattered pilose on midrib above, ferrugineous tomentose to pilose on major nerves beneath, glaucous beneath, coriaceous; lateral nerves 5-7 pairs, intercostae scalariform, prominent; petiole 6-10 mm long, stout, grooved above, ferrugenous tomentose. Flowers bisexual, 5-6 mm long, yellow, in axillary and terminal, ferrugineous-tomentose, congested, few flowered, stout, panicles, usually shorter than the leaves. Perianth tube narrowed at the top; lobes 6, subequal. Stamens 9, all perfect, 2 outer series with eglandular filaments and introrse anthers, third series with 2 glandular filaments and extrorse anthers; fourth series with stipitate staminodes; anthers 2-celled; connective often produced. Ovary sessile, enclosed in the tube; style short, exserted. Drupe cylindric-oblong, 3 x 1.1 cm, woody, smooth, with short obtuse beak, black, fruiting pedicel to 3.5 x 3 mm. Fl. & Fr.: December – March.

Western Ghats: Evergreen forests of Karnataka, Kerala and Tamil Nadu (Endemic).

IUCN: EN- B2ab(iii). (IUCN, 2024).

Cryptocarya procera Talbot., For. Fl. Bombay 2: 385. 1911.

Trees to 15 m tall; branchlets glabrous. Leaves obovate or ovate-lanceolate, obtuse at base, acute at apex, glabrous above, pale but not glaucous beneath; secondary nerves 7-10 pairs, slender, prominent, hairy beneath. Panicles spreading, grey pubescent. Perianth lobes ovate. Ovary glabrus. Fl. & Fr.: December – April.

Western Ghats: Evergreen forests of Karnataka (Endemic).

IUCN: DD (IUCN, 2024).

Cryptocarya stocksii Meisn., A.P.de Candolle, Prodr. 15(1): 71. 1864.

Evergreen trees to 10 m tall; bark greyish-brown, smooth; branchlets fulvous-tomentose and lenticellate. Leaves simple, alternate, elliptic-obovate, elliptic or ovate-oblong, 6-10.5 x 2.5-5.5 cm, base obtuse, apex rounded or slightly emarginate, margin entire, recurved, glabrous above, pubescent and glaucous beneath, coriaceous; lateral nerves 5-8 pairs, intercostae scalariform, prominent; petiole 7-15 mm long, stout, pubescent. Flowers bisexual, in short congested axillary or terminal fulvous tomentose panicles of 5-10 cm wide; pedicels to 0.5 cm. Perianth pubescent, tube ultimately narrowed at the top; lobes 6 4 mm, ovate, pubescent, subequal, in 2 rows. Stamens 9 perfect, in 3 rows, 2 outer series with eglandular filaments and introrse anthers, third series with 2 glandular filaments and extrorse anthers; fourth series replaced by 3 stipitate staminodes; filaments hairy; anthers 2-celled; connective often produced. Ovary globose, 2 mm across, glaucous, sessile, enclosed in the tube; style short, exserted. Berry ovoid, black. Fl. & Fr.: October - April.

Western Ghats: Evergreen forests of Karnataka, Kerala and Tamil Nadu (Endemic).

IUCN: VU- B1+2c. (IUCN, 2024).

Cryptocarya wightiana Thw., Enum. Pl. Zeyl.: 254. 1861. Synonyms: *Cryptocarya bourdullonii* Gamble, *Cryptocarya floribunda* Wight, *Cryptocarya wightiana* var. *parvifolia* Meisn. Vern.: *Paalai* (Tam.).

Evergreen trees to 18 m tall; bark reddish-brown, lenticellate; branchlets rusty-pubescent. Leaves simple, alternate, elliptic or elliptic-oblong, 6.5-20 x 2.5-10 cm, base subacute, obtuse, round or truncate, apex acute or acuminate, margin entire, glabrous except sometimes the midrib and nerves; lateral nerves 7-12 pairs, intercostae scalariform, faint; petiole 10-25 mm long, glabrous. Flowers bisexual, 3-4 mm across, yellowish, in spreading axillary brown tomentose panicles, upto 15 cm long; pedicels short; bracts and bracteoles caducous. Perianth cupular at base, ultimately narrowed at the top; lobes 6, subequal, in 2 rows, ovate, tomentose. Stamens 9 perfect, 2 outer series with eglandular filaments, third series with 2 glandular filaments; fourth series replaced by 3 stipitate staminodes; filaments pubescent; anthers 2-celled; connective often produced. Ovary sessile, half inferior, glabrous, enclosed in the tube; style short, exserted; stigma truncate. Drupe globose with an obtuse beak, 10-15 mm across, glabrous, black, enclosed in perianth-tube. Fl. & Fr.: March – September.

Western Ghats: Evergreen forests of Maharashtra, Goa, Karnataka, Kerala and Tamil Nadu. World: Sri Lanka and Myanmar.

IUCN: VU-A1c. (IUCN, 2024).

Litsea Lam.

1a. Umbels born on 0.5–1 cm long brachyblasts 2

1b. Umbels born on 1–10 cm long brachyblasts 17

2a. Umbels sessile or subsessile, clustered 3

2b. Umbels pedunculate, solitary or rarely clustered 7

3a. Leaves opposite, fulvous-pubescent below .... **L. travancorica**

3b. Leaves alternate, pubescent or glabrous below 4

4a. Secondary veins 12–18 pairs; lamina fulvous-pubescent, impressinerved; perianth cup sessile or short stalked.................................... **L. bourdillonii**

4b. Secondary veins 6–13; lamina glabrous or puberulent, not impressinerved; perianth cup stalked 5

5a. Lamina 6–11x1.3–2.0 cm, linear-lanceolate; flowers 4 per umbel; fruits 1×1 cm, ovoid .... **L. mysorensis**

5b. Lamina 6–18x3–6 cm, elliptic-lanceolate or elliptic-ovate, oblong; flowers 4–6 per umbel; fruits 1–1.5×0.8 cm, ellipsoid 6

6a. Bark pale yellow; leaves glossy above, chartaceous; secondary veins 6–8 pairs; 10–12 umbels in a cluster; flowers 5–6 per umbel .... **L. laevigata**

6b. Bark pale brown; leaves not glossy above, thickly coriaceous; secondary veins 5–10 pairs; 2–4 umbels in a cluster; flowers 4–5 per umbel .... **L. coriacea**

7a. Umbels axillary or laterally, solitary 8

7b. Umbels clustered or born on umbel-bearing short reduced branchlets 10

8a. Leaves glabrous; secondary veins 6–8 pairs, tertiary veins reticulate; flowers 5 per umbel; fruit stalk glabrous .... **L. ligustrina**

8b. Leaves tomentose; secondary veins 7–16 pairs, tertiary veins scalariform-reticulate; flowers 4–12 per umbel; fruit stalk pubescent 9

9a. Shrubs or small trees; leaves sub-verticillate, apex abruptly acute or obtuse; secondary veins 11–16 pairs; tepals 6–8; stamens 16 .... **L. ghatica**

9b. Trees; leaves alternate or spirally arranged, apex acute or obtusely acuminate; secondary veins 7–15 pairs; tepals 0; stamens 18–20 .... **L. deccanensis**

10a. Fruits globose or sub-globose; perianth tube enlarged and turbinate... 11

10b. Fruits ovoid or ellipsoid; perianth tube cup-shaped 12

11a. Branchlets glabrous; leaves alternate, glabrous, white glaucous below; flowers 1–3 in a cluster, creamy-white; fruits sub-globose ....**L. venulosa**

11b. Branchlets greyish-tomentose; leaves sub-verticillate, tomentose, not glaucous below; flowers 2–5 in a cluster, purple; fruits globose .... **L. quinqueflora**

12a. Branchlets, petiole, and midrib reddish-brown, glabrous; secondary veins 6–10 pairs, filiform..13

12b. Branchlets, petiole, and midrib greenish-yellow, tomentose; secondary veins 6–14 pairs, thick 14

13a. Shrubs to small tree 5m, lateral veins 4-7 pairs, ovary 2 mm long and ellipsoid, style slightly curved ....**L. gorayana**

13b. Small trees up to 10 m tall, lateral veins 6-10 pairs, ovary 1 mm long and globose, style straight ....**L. udayanii**

14a. Large trees; leaves whitish glaucous below, tertiary veins scalariform-reticulate and prominent; fruits ovoid or broadly ovoid, ellipsoid, 0.8–3 cm long; perianth cup lenticellate or smooth 15

14b. Shrubs to small trees; leaves bluish glaucous below, tertiary veins reticulate and faint; fruits ellipsoid, 1.3 cm long; perianth cup smooth.... .... **L. kakkachensis**

15a. Flowers creamy white; cupule pubescent or hairy 16

15b. Flowers yellow cupule glabrous.. **L. keralana**

16a. Stamens 9; berry 1cm; cupule slightly pubescent ....**L. monopetala**

16b. Stamens 12-16; berry 2 cm; cupule hairy throughout . **L. megamalayana**

17a. Flowers 9–15 per umbel; tepals 0–3; fruits globose; enlarged perianth tube shallow cup-shaped or nearly flattened .... **L. glutinosa**

17b. Flowers 4–6 per umbel; tepals 6–8; fruits depressed globose and ellipsoid; enlarged perianth tube deep cup-shaped 18

18a. Fruits 2x2.5 cm depressed globose, seated on a large cup-like enlarged perianth tube; fruit stalk short and angular .... **L. nigrescens**

18b. Fruits 1.5x1 cm, ovoid or ellipsoid, seated on a small cup-like enlarged perianth tube; fruit stalk long, terete or angular 19

19a. Leaves opposite, rigidly coriaceous, not glaucous below; tertiary veins reticulate and faint below; brachyblasts 2.5–6 cm long .... **L. oleoides**

19b. Leaves alternate, coriaceous, glaucous below; tertiary veins scalariform-reticulate and prominent below; brachyblasts 5–10 cm long 20

20a. Branchlets, leaves densely tomentose or pubescent; lamina 16–28x3.5–10.5 cm; pale glaucous beneath 21

20b. Branchlets, leaves glabrous or puberulent; lamina 8.5–18x3–7.5 cm, white glaucous beneath 22

21a. Secondary veins 10–16 pairs; inflorescence 6 cm long, involucral bracts 4–7; flowers 5–8 per umbel; fruits ovoid, sparsely silky pubescent ....**L. floribunda**

21b. Secondary veins 6–10 pairs; inflorescence 10 cm long, involucral bracts 5–8; flowers 5 per umbel; fruits ellipsoid, glabrous ....**L. wightiana**

22a. Leaves linear lanceolate, up to 8.5 cm long, densely puberulent below; brachyblasts up to 5 cm long ....**L. beddomei**

22b. Leaves oblanceolate, elliptic-oblong or elliptic-lanceolate, up to 18 cm long, glabrous below; brachyblasts up to 7 cm long 23

23a. Fruits 2–2.5x1 cm, ovoid to ellipsoid... .**L. stocksii**

23b. Fruits 1.5x0.8 cm, ellipsoid 24

24a. pubescent branchlets and petioles; fruiting pedicels angular .... **L. anamalayana**

24b. glabrous branchlets and petiole; fruiting pedicels terete......... **L. glabrata**

Litsea anamalayana Robi & Udayan., Edinburgh J. Bot. 78-373: 1. 2021.

Small trees to10 m tall; bark brown, lenticellate, inner bark pale yellow; branchletspubescent, subterete, lenticellate. Leaves simple, alternate, linear-oblong, lanceolate, 11–21 x 3–4 cm, base narrowly attenuate, apex acuminate, margin revolute, glabrous adaxially and puberulous abaxially, smooth above; lateral veins 11–14 pairs, intercostae scalariform-reticulate, prominent below; petiole 1.5–2 cm long, pubescent. Inflorescences umbel-bearing reduced branchlets, in axils of leaves or lateral, brachyblast 3–7 cm long, pubescent bearing 4–8 umbels; involucral bracts 6, 7 x 5 mm, suborbicular, ovate, concave, caducous. Flowers 7 in each umbel, creamy white, 6 mm long; pedicels 3 mm long. Tepals 6, 3 x 2 mm, ovate or elliptic, acute or obtuse at apex, equal, membranaceous, gland-dotted, glabrous on both sides, creamy white. Female flowers staminodes 12, linear, falcate, 2–3 mm long, glabrous 2-glandular; elliptic, 0.7 mm long glands. Ovary ellipsoid, 1.5 mm long, glabrous; style 3.5 mm long, straight and bend at apex, creamy white, glabrous; stigma lobed, papillose, creamy white. Berry ellipsoid, 1.5 cm long, obtuse at apex, glabrous, greenish, white specked, seated on the enlarged cup-shaped perianth tube. Fl. & Fr.: September – May

Western Ghats: Evergreen forest at 1300 m in Kerala and Tamil Nadu (Endemic).

IUCN: NE (IUCN, 2024).

Litsea beddomei Hook.f., Fl. Brit. India 5: 177. 1886. Synonym: *Malapoenna beddomei* (Hook.f.) Kuntze

Trees, to 5 m tall. Leaves simple, alternate, 15 x 3.5 cm, linear-oblong or lanceolate, obtuse, glabrous above, white-glaucous beneath when dry; main nerves 8-10 pairs rather close and prominent. Male flowers 0.7 cm long, pubescent; pedicel 0.8 cm long. Perianth tube hemispheric, 1 cm across. Berry oblong, 1.2 cm long, apiculate, seated on an enlarged perianth tube with a more or less thickened pedicel, shining black when dry. Seeds 2, 0.8 cm long. Fl. & Fr.: November – April

Western Ghats: Evergreen forests of Kerala and Tamil Nadu (Endemic).

IUCN: EN- B2ab(i,ii,iii). (IUCN, 2024).

Litsea bourdillonii Gamble, Bull. Misc. Inform. Kew 1925: 131. 1925.

Dioecious trees to 8 m tall; branchlets brownish-black tomentose. Leaves simple, alternate, obovate or obovate-oblong, 15-21 x 6-12 cm, base cuneate or acute, apex obtusely acute; margin entire, glabrous above except midrib, fulvous pubescent, punctate and glaucous beneath; lateral nerves 12-18 pairs, intercostae scalariform, prominent; petiole 10-17 mm long, dark-brown tomentose. Flowers unisexual, in umbellule which may be sessile or subsessile, in axillary or lateral clusters of 2-5. Flowers 4-6 in an umbellule; bracts 4, 5 mm long, orbicular, concave, densely white-tomentose. Perianth tube short, lobes 6, densely white-tomentose. Stamens 12, in 4 rows; filaments glabrous, those of rows 1 and 2 usually eglandular, those of rows 3 and row 4 with 2-glands; anthers 4-celled, introrse; staminodes in female flowers as the stamens of males but those of the outer rows clavate or linear, those of the inner rows subulate and 2-glandular. Ovary half inferior, ovoid; style thick; stigma dilated. Berry oblong, 10-12 mm long, seated on a cup-shaped perianth tube. Fl. & Fr.: December – May.

Western Ghats: Wet evergreen forests of Karnataka, Kerala and Tamil Nadu (Endemic).

IUCN: NT (Low risk/near threatened) (IUCN, 2024).

Litsea coriacea (Heyne ex Meisn.) Hook.f., Fl. Brit. India 5: 166. 1886. Synonyms: *Tetranthera coriacea* Heyne ex Meisn., *Tetranthera micrantha* Meisn., *Malapoenna coriacea* Kuntze Vern.: *Maravettithalai* (Tam.).

Dioecious trees to 12 m tall; bark pale brown, thinly scaly; branchlets puberulous, slender. Leaves simple, alternate, elliptic-lanceolate or elliptic-ovate, 8.5-17.5 x 2.5-7.5 cm, apex acute or acuminate, base acute or cuneate, margin entire, glabrous above, minutely puberulous beneath; lateral nerves 6-9 pairs, intercostae reticulate, obscure; petiole 5-10 mm long, pubescent. Flowers unisexual, white, in 4-flowered axillary, subsessile, crowded, umbels; bracts 4, silky. Perianth tube funnel shaped, finely silky, lobes 6. Stamens 12, in 4 rows, those of rows 1 and 2 usually eglandular, glands of those opposite to tepals very large; staminodes in female flowers as the stamens of males, but those of inner rows subulate and 2-glandular. Ovary half inferior, ovoid; style slender; stigma small, erect. Brry ovoid, 10-12 mm long, greenish-white, seated on a thickened perianth tube. Fl. & Fr.: Throughout the year.

Western Ghats: Evergreen forests of Goa, Kerala and Tamil Nadu (Endemic).

IUCN: NT (Low risk/near threatened) (IUCN, 2024).

Litsea deccanensis Gamble, Fl. Madras: 1235. 1925.

Dioecious trees to 10 m tall; bark somewhat rough, very dark, pustular; branchlets stout, densely pale brown-tomentellous. Leaves simple, alternate, crowded at the end of branchlets, elliptic, obovate or elliptic-oblong, 7.5-24 x 4-9 cm, base acute or round, apex acute or acuminate, margin entire, glabrous, finely tomentose beneath, chartaceous; lateral nerves 7-15 pairs, intercostae subscalariform, faint; petiole 9-40 mm long, stout, tomentose. Flowers unisexual, pale yellow, in axillary, pedunculate, solitary or in many flowered umbellate heads; peduncle 10-17 mm long, drooping, tomentose; involucral bracts 4, orbicular, concave, tomentose without; pedicel upto 3 mm long, pubescent. Perianth tube short, hairy outside, segments 0. Stamens 18-20 in male flowers; in 4 rows, filaments long, slender, densely villous, 1 and 2 row usually eglandular, rows 3 and 4 with 2 glands; anthers 4-celled, all introrse; staminodes in female flowers as the stamens of male but outer rows clavate or linear, inner rows subulate and 2-glandular. Ovary half inferior, enclosed in the perianth tube, glabrous; style 2-3 mm long; stigma dilated, irregularly lobed. Berry globose, 10-12.5 mm across, black, seated on flat, thickened cup. Fl. & Fr.: December – September.

Western Ghats: Evergreen forests of Maharashtra, Goa, Karnataka, Kerala and Tamil Nadu. World: Pakistan.

IUCN: NE (IUCN, 2024).

Litsea floribunda (Blume) Gamble, Fl. Madras: 1238. 1925. Synonyms: *Cylicodaphne floribunda* Blume, *Neolitsea floribunda* M.R. Almeida Vern.: *Pattuthalai* (Tam.).

Dioecious trees to 15 tall; bark light brown, smooth, brittle, lenticellate; branchlets brown to black pubescent. Leaves simple, alternate, elliptic, obovate-oblong, elliptic-oblong or ovate-oblong, 7.5-28 x 3.7-10.5 cm, apex acute, or acuminate, base acute, margin entire, glabrous above except midrib above and rusty tomentose beneath; lateral nerves 10-15 pairs, intercostae subscalariform, prominent; petiole 10-15 mm long, pubescent; Flowers unisexual, in axillary racemose umbellule; peduncle upto 1 mm long, silky pubescent; bracts 4, orbicular, silky tomentose. Flowers 6-8 in an umbellule. Perianth tube turbinate, silky tomentose, lobes 6. Stamens 12, in 4 rows; filaments hairy, rows 1 and 2 usually eglandular, rows 3 and 4 with 2-glands; anthers 4-celled, introrse; staminodes in female flowers as the stamens of males but those of the outer rows clavate or linear, those of the inner rows subulate and 2 glandular. Ovary half inferior, stigma capitate. Berry oblong, 12-18 mm long, seated on a turbinate perianth tube. Fl. & Fr.: August – March.

Western Ghats: Evergreen forests of Karnataka, Kerala and Tamil Nadu (Endemic).

IUCN: NT (Low risk/near threatened) (IUCN, 2024).

Litsea ghatica Saldanha, Fl. Karnataka 1: 67. 1985. Synonyms: *Litsea lakshmammaniana* Kesh.Murthy & Yogan.

Small trees to 10 m tall; young branchlets tomentose, elongating through terminal scales. Leaves subverticillate, obovate, 10-20 x 4-9 cm, apex abruptly acute or obtuse, tapering to subacute or rounded base, glabrous except nerves above, more or less softly tomentose beneath; nerves 10-14 pairs, prominent beneath, looped; petiole to 1 cm long. Flowers unisexual, 4-5, in solitary umbels; peduncle 5-8 mm long, cauliferous; bracts ovate, usually in 2 pairs, tomentose without. Perianth 6-lobed; tube short or absent. Stamens 16, biseriate; filaments hairy; anthers 4-celled; staminodes with stalked glands. Fl. & Fr.: August – December.

Western Ghats: Evergreen forests of Maharashtra, Karnataka and Kerala (Endemic).

IUCN: VU- B1ab(i,ii,iii)+2ab(i,ii,iii). (IUCN, 2024).

Litsea glabrata (Wall. ex Nees) Hook.f., Fl. Brit. India 5: 174. 1886. Synonyms: *Tetranthera glabrata* Wall. ex Nees., *Lepidadenia glabrata* Wight

Dioecious trees to 12 m tall; bark pale brown smooth; branchlets glabrous. Leaves simple, alternate, lanceolate, oblong, linear-oblong or elliptic-lanceolate, 9.5-18 x 3-5 cm, base acute or cuneate, apex acute, margin entire, glabrous, reddish-brown beneath when dry, coriaceous; lateral nerves 9-12 pairs, obscure, intercostae reticulate; petiole 15-30 mm long, grooved above, glabrous. Flowers unisexual, umbellules in axillary racemes of 2-6 cm long. Perianth tube short; lobes 6, ovate, hispid outside. Stamens 12, in 4 rows filaments glandular hairy; staminodes in females as the stamens in males, but those of inner rows subulate and 2-glandular. Berry ovoid, bright-red, glabrous; fruiting perianth obconic. Fl. & Fr.: September – May.

Western Ghats: Evergreen forests of Karnataka, Kerala and Tamil Nadu (Endemic).

IUCN: LC (IUCN, 2024).

Litsea glutinosa (Lour.) C.B. Robs., Philipp. J. Sci., C 6: 321. 1911. Synonyms: *Sebifera glutinosa* Lour., *Litsea sebifera* (Willd.) Pers., *Tomex sebifera* Willd., *Litsea chinensis* Lam., *Litsea tersa* (L.) M.R. Almieda, *Glabraria tersa* L., *Litsea glutinosa* var. *longifolia* (Haines) K.K.Khanna, *Litsea sebifera* var. *normalis* Haines, *Tetranthera fruticosa* Roxb. Vern.: *Elumburukki, Musaippeyetti, Urali* (Tam.).

Dioecious trees to 10 m tall; bark brown; branchlets rather slender, stiff, minutely tomentellous towards apex. Leaves simple, pilose when young, elliptic, elliptic-oblong or obovate, 6-23.5 x 3-10 cm, base acute, apex acute, obtuse or acuminate, margin entire, glabrous above, grey-pubescent beneath;

lateral nerves 7-12 pairs, intercostae reticulate. Flowers unisexual, 5-6 mm across, yellow, in many flowered umbels; peduncles upto 5 mm long, slender, densely pilose. Perianth short; tube silky, funnel shaped, segments 0. Stamens 9-12, all fertile, in 4 rows; filaments to 2 mm, very hairy; fourth series staminodes 9-12, to 2.5 mm, glandular, glands sessile, orbicular. Ovary half inferior, 1 mm long; style 1.5 mm, stigma dilated. Berry globose, 5-6 mm across, depressed, purple, on flat disc. Fl. & Fr.: January – September.

Western Ghats: Semi-evergreen to evergreen forests of Maharashtra, Karnataka, Kerala and Tamil Nadu. India: Andhra Pradesh and Odisha. World: Sri Lanka, Bangladesh, Nepal, China, Myanmar, Laos, Cambodia, Vietnam, Malesia and Australia.

IUCN: LC (IUCN, 2024).

Litsea gorayana Udayan & Robi., Ann. Bot. Fenn. 54: 287. 2017.

Small trees to 5 m tall; bark brown, smooth, lenticellate; branchlets glabrous, thick. Leaves alternate, elliptic, 4.5–6.5 x 2–3 cm, base acute, apex acuminate to acute, glabrous on both sides, silvery glaucous below; lateral veins 4–7 pairs, faint abaxially and prominent adaxially; intercostae faint; petioles 5–10 mm long, glabrous, slightly grooved above. Inflorescence an umbel, on 1 cm long brachyblasts in axils of leaves, puberulent; each brachyblast bearing 2–4 umbels in racemose manner; peduncle of umbels 5 mm long, slender, silky adpressed-pubescent; involucral bracts 6, 5 x 4 mm, orbicular, concave, adpressed-pubescent, caducous. Flowers white, 5 mm long; pedicels up to 4 mm long, angular; tepals 6, 1.5–2 x 1 mm, ovate, acute at apex, sub-equal, membranaceous, glabrous inside and pubescent outside, white. Female flowers staminodes 12, those of outer whorls linear, 2 mm long, those of inner whorls subulate, 1 mm long, pubescent, 2-glandular; glands ovate, 0.5 mm long, stipitate. Ovary 2 mm long, ellipsoid; style 1 mm long, slightly curved, stout, greenish-yellow, glabrous; stigma capitate, papillose, white. Berry ellipsoid, glabrous, glossy, seated on cup-shaped perianth tube. Fl. & Fr.: January – September.

Western Ghats: Evergreen forests of Kerala (Endemic).

IUCN: NE (IUCN, 2024).

Litsea kakkachensis R. Ganesan, Rheedea 21: 143. 2011.

Small trees to 3 m tall; bark smooth in young plants, covered with reddish brown, flaking papery cork. Leaves alternate, obovate to oblanceolate, 3.3–10.7

x 1.3–3.4 cm, acute or cuneate at base, entire to revolute at margins, acute to obtuse at apex, glabrous above, purplish brown on dry; lateral nerves 8–10 pairs, looping at margins, obscure above, prominent below; petioles 5–13 mm long. Male umbellules 1 to 3 per leaf axil; peduncles 12–23 mm long, glabrous; involucral bracts 4, cucullate, 4.5-4.8 mm, thick, 4 or 5-nerved at base. Flowers 6 per umbellule, 5.4 mm long; pedicels to 0.6 mm, densely white-sericeous. Tepals 6, linear-oblong to obovate, 3–3 x 7.1–1.2 mm, rounded at apex, densely gland-dotted, sericeous at margins, cream-coloured. Torus concave, densely white-sericeous; rudimentary ovary absent. Stamens 6, 2-whorled; outer stamens longer and slender than inner ones; outer filaments 6 mm long; anthers basifixed; inner stamens juxt-opposed outer stamens; filaments to 0.5 mm long, base of inner stamen with 2-glandular at base; anthers introrse, 4-celled, valvular. Female umbellules peduncles 4–4.8 mm long. Flowers 4 in each umbellule; pedicels 2.6 1.5 mm, laterally compressed, sericeous. Tepals 6, linear-oblong, to 0.8 mm long, entire or rarely distantly serrate at margins, acute at apex, cream-coloured. Hypanthium concave, densely sericeous inside; appendages 6, in 2-whorled; gland bean-shaped, 1.2 mm long. Pistil 3 mm long; ovary urnshaped, to 0.9 mm; style 1.8 mm long, terminating into a discoid, lobulate, echinate stigma. Berry ellipsoid, 13.8 mm diam, apiculate, red, seated on fleshy perianth tube, to 7 mm across, white gland-dotted. Seed 1, grey. Fl. & Fr.: April – September.

Western Ghats: Rare in evergreen forests of Tamil Nadu (Endemic).

IUCN: EN- B1ab(iii)+2ab(iii). (IUCN, 2024).

Litsea keralana Kostrem., Ceylon J. Sci., Biol. Sci. 12: 138. 1977. Synonym: *Litsea insignis* Gamble

Dioecious trees to 20 m tall; bark greyish-brown, lenticellate; branchlets ferruginous tomentose. Leaves simple, alternate, broadly, elliptic-oblong or obovate, 10-25 x 3.5-14 cm, base slightly acute, obtuse or truncate, apex obtuse, margin entire, glabrous above and roughly fulvous-pubescent beneath, coriaceous; lateral nerves 7-12 pairs, intercostae scalariform, prominent; petiole 10-25 mm long, stout, ferruginous. Flowers yellowish, in axillary, lateral or cauliflorus umbellules 10-20 together; peduncle 7-10 mm long; bracts orbicular, densely hairy. Flowers 6-8 in an umbellule, sessile. Perianth lobes 5, oblong, densely hairy. Stamens 12, in 4 rows; filaments hairy, first and second rows eglandular, 3 and 4 rows with 2-glands; anthers 4-celled, introrse; staminodes as the stamens of male but those of inner rows subulate and 2-glandular. Ovary enclosed irregularly lobed perianth cup, half inferior; style thick; stigma diliated. Berry ellipsoid, 2.5-3 x 1.5-2 cm, obtuse, reddish-brown, glabrous, seated on the enlarged obconic perianth tube. Fl. & Fr.: December – February.

Western Ghats: Evergreen forests above 1200 m in Kerala and Tamil Nadu (Endemic).

IUCN: VU- B1ab(i,ii,iii)+2ab(i,ii,iii). (IUCN, 2024).

Litsea laevigata (Nees) Gamble, Fl. Madras: 1236. 1925. Synonyms: *Tetranthera attenuata* var. *laevigata* Nees, *Tetranthera glauca* var. *laevigata* (Nees) Meisn.

Dioecious trees to 10 m tall; branchlets puberulus. Leaves simple, alternate, oblong-lanceolate or elliptic-lanceolate, 6.5-18 x 2.5-5 cm, base slightly obliquely acute or obtuse, apex acuminate or acute, margin entire, glabrous above and puberulent beneath; lateral nerves 6-8 pairs, intercostae reticulate, faint; petiole 8-15 mm long, puberulent. Flowers unisexual, yellowish, 10-12 in axillary or lateral subsessile umbellules; pedicelled; bracts orbicular, persistent. Perianth short, lobes 6, 2 mm long, densely hairy. Stamens 12, in 4 rows; filaments hairy, 1 and 2 rows eglandular, 3 and 4 rows with 2-glands; anthers 4-celled, introrse; staminodes in innermost rows subululate and 2-glandular. Ovary half inferior, enclosed in perianth tube; stigma irregularly lobed. Berry ellipsoid, 1.5 cm long, glabrous, yellow. Fl. & Fr.: October – June.

Western Ghats: Evergreen forests of Karnataka, Kerala and Tamil Nadu (Endemic).

IUCN: NT (Low risk/near threatened) (IUCN, 2024).

Litsea ligustrina (Nees) Fern.-Vill., F.M. Blanco, Fl. Filip., ed. 3. 4(13A): 180. 1880. Synonyms: *Actinodaphne quinqueflora* (Dennst.) M.R. Alemida & S.M. Almeida, *Litsea hookeriana* (Meisn.) Hook.f., *Litsea sebifera* var. *tomentosa* (Nees) Hook.f., *Litsea thwaitesii* N.P.Balakr., *Litsea tomentosa* (Nees) B.Heyne ex Hook.f., *Litsea undulata* Hook.f. Vern.: *Perumpanjatti* (Tam.).

Small trees to 10 m tall; branchlets densely tomentose. Leaves obovate, 20 x 9 cm, obtuse at apex, acute to the base; nerves 9-11 pairs, reticulate; petiole 2.5 cm long. Peduncle 1.5-2 cm long. Umbellules 10-20 mm across; bracts orbicular, densely tomentose outside. Flowers to 25 in an umbellule, pedicels 4 mm long. Perianth lobes 3-5 or absent, 2 mm long, densely hairy. Stamens 25-30, filaments bearded. Fl. & Fr.: October – March.

Western Ghats: Deciduous to evergreen forests of Kerala and Tamil Nadu. World: Bangladesh, Myanmar, Pakistan and Sri Lanka

IUCN: NT (Low risk/near threatened) (IUCN, 2024).

Litsea megamalayana Karupp., V.Ravich. & Bharath, Gard. Bull. Singapore 75(1): 170. 2023.

Small trees to 8 m tall; bark brown, branchlets brown, densely fulvous with sparsestellate hairs, lenticellate. Leaves simple, alternate, ovate-oblong, 3–10 x 3–4.5 cm, base attenuate, apex obtuse, margin entire, recurved, glabrous adaxially, hairy abaxially, also occasionally stellate hairy, coriaceous; lateral veins 5–8, prominent abaxially; petiole 1–1.5 cm long, greenish yellow, fulvous hairy, convex below, flat above. Inflorescence of many umbels in a short raceme, to 2.5 cm long. Flowers 4–6 in a cluster; peduncle 1–1.3 cm long, hairy; pedicel 5–7 mm long; bracts 4–5, broadly ovate, abaxial side densely silky hairy, adaxial side glabrous, margin entire, ciliate, 3.5 x 3 mm, apex rounded. Male flowers 5 or 6 in an umbel, 3–4 mm long, nearly 5 mm long on opening. Tepals 6–8, elliptic, 1.8 x 1.2 mm, apex acute orobtuse, entire, abaxial side silky hairy, adaxial side partially hairy. Stamens 12–16, appearing to be arranged in two whorls, base densely silky hairy, outer stamens 3–3.2 mm long, partially hairy, anthers 4-celled, 0.5 mm long; inner stamens 1.5–2 mmlong, with glands 0.7 mm long. Female flowers 4 in each umbel, 2–3mm long, nearly 4 mm long on opening. Tepals 6, 1.5 x 1 mm, elliptic, entire, abaxial side silky hairy, adaxial side glabrous, apex acute or obtuse, narrowed at base; staminodes 12, appearing to be arranged in two whorls, hairy abaxially, glabrous adaxially straight, white in colour. Ovary 0.8 mm long, hairy at base, style 1 mm long, stigma unevenly capitate, dark brown, glabrous. Berry ellipsoid, 2 cm long, obtuse at apex, greenish speckled white, seated on the perianth tube; fruiting pedicel to 6 mm long, hairy. Fl. & Fr.: September–November.

Western Ghats: Evergreen forests of Megamala hills in Tamil Nadu (Endemeic).

IUCN: NE (IUCN, 2024).

Litsea monopetala (Roxb.) Pers., Syn. Pl. 2: 4. 1806. Synonyms: *Tetranthera monopetala* Roxb., *Litsea polyantha* Juss., *Tetranthera quadriflora* Roxb., *Tetranthera verticillata* Buch.-Ham. ex Meisn. Vern.: *Maidalagadil, Musaippeyetti, Picinpatti* (Tam.).

Small trees 20 m tall; bark dark grey; branchlets stout, cylindric, dark brown, pubescent. Leaves alternate, broadly or narrowly oblong-ovate or elliptic-obovate, 8 - 20 x 3.4 - 9.5 cm, rounded or apiculate at apex, entire along margin, acute, rounded or obtuse at base, glabrous, beneath tomentose; secondary veins 8-10 pairs, tertiary veins mostly opposite, straight near margin; petiole 1 - 2.5 cm long, cylindric, glabrous. Inflorescence umbellules, 10 - 12 x 5 - 10 mm, cauliflorous, 5-8 flowered; peduncles 5-7 mm long, puberulous, brown; bracts 4-6, outer two or four 4-5 x 3.5 - 5 mm, concave, glabrous inside, densely tomentose outside. Flowers yellow green, 4-6 x 3 - 5 mm; pedicel 0.8 - 1 mm long, thick, puberulous. Perianth lobes 6, 1.5 - 2.5 mm long, oblong-elliptic, hairy at base, tube 1 mm long, funnel shaped, densely tomentose. Male flowers stamens 12, in 4 rows, outer 2 rows 3.5 - 4 mm long, exerted, inner 2 rows 2.5 - 3 mm long, glandular; anther 1 mm long, 4 celled, upper 2 cells introrse, lower 2 cells partly latrorse; glands 1 mm long, 2 each at base of inner rows of stamens, stalked, 2 lobed; pistillode 1 mm long, glabrous. Female flowers staminodes 12, in 4 rows, glands 1 mm long, 2 each at base ofinner rows of staminodes. Pistil 2-3.5 mm long, style 1 - 2.5 mm long, glabrous. Ovary 1 mm long, ovoid, glabrous. Berry globose-ellipsoid, 0.5 - 0.6 x 0.6 - 0.9 cm, green when young, black when ripe, seated on thickened perianth tube; fruiting pedicel 4-6 mm long, green, glabrous. Fl. & Fr.: June–May

Western Ghats: Evergreen forests of Maharashtra. India: Andhra Pradesh and Odisha. World: Bhutan, Myanmar, Nepal, Malay Peninsula and South West China.

IUCN: LC (IUCN, 20223).

Litsea mysorensis Gamble, Bull. Misc. Inform. Kew 1925: 130. 1925.

Dioecious trees to 10 m tall; branchlets slender, terete. Leaves simple, alternate, lanceolate, oblong, linear-lanceolate or elliptic-oblong, 5-12 x 1-2.5 cm, base acute, apex acute or acuminate, margin entire, glabrous; lateral nerves 6-8 pairs, intercostae reticulate, obscure; petiole upto 5 mm long, pubescent. Flowers unisexual, yellowish, in subsessile umbellules; bracts orbicular, 3 mm long, seriaceous out and glabrous within; 5 flowers in each umbellule. Perianth lobes 6, oblong, seriaceous out. Stamens 9 perfect, in 3 rows, 1 and 2 rows eglandular, 3 row glandular; filaments glabrous; staminodes 9, 2 glands at base. Ovary half inferior, ovate, glabrous; style thick; stigma capitate. Fruit a berry. Fl. & Fr.: November – February.

Western Ghats: Evergreen forests of Karnataka, Kerala and Tamil Nadu (Endemic).

IUCN: VU- B1ab(i,ii,iii)+2ab(i,ii,iii). (IUCN, 2024).

Litsea nigrescens Gamble, Fl. Madras: 1236. 1925.

Trees to 20 m tall; bark pale brown, smooth, slightly rugose; branchlets greyish, glabrous, stout. Leaves simple, alternate, oblong, elliptic-oblong or elliptic-lanceolate, 6-22 x 7-12 cm, base attenuate to cuneate, apex acuminate or acute, margins entire, glabrous, thinly coriaceous; midrib glabrous on both sides, slightly grooved above, prominent beneath; lateral veins 10–12 pairs, obscure beneath, intercostae reticulate obscure; petioles 10–30 mm long, grooved above, glabrous. Inflorescences umbels, on reduced branchlets with the appearance of a raceme of umbels, axillary; peduncles 6 mm long; involucral bracts 4, orbicular, 5 x 3 mm, concave, densely silky pubescent, coriaceous. Flowers unisexual, 3-6 in each umbel, white, 4 mm long, silky-pubescent. Tepals 6, lanceolate or oblong, acute, subequal, membranaceous, glabrous inside, silky-pubescent outside, creamywhite. Male flowers: Stamens 12, in 4 rows, unequal; filaments slender, villous, all with 2-glands at the base, 3 mm long in outer stamens, 2 mm long in inner stamens; anthers 1.5 mm long, 4-locular, the upper 2 smaller, lower 2 larger, introrse; pistillodes minute or rudimentary. Female flowers: Staminodes as many as stamens in male, outer ones linear, inner ones subulate, 2–3 mm long, 2- glandular at the base; ovary globose, 1 mm long; style 1 mm long, thick; stigma capitate, papillose. Berries depressed globose, 2–2.5 cm across, greenish, reddish to dark purple, white specked, smooth, glossy, seated on a perianth tube; perianth tube disc-shaped, persistent, 1 cm across, glabrous outside, margins entire; fruiting pedicels stout, angular, 7 mm long. Fl. & Fr.: December – March.

Western Ghats: Rare in evergreen forests of Kerala and Tamil Nadu (Endemic).

IUCN: VU- B2ab(i,ii,iii). (IUCN, 2024).

Litsea oleoides (Meisn.) Hook.f., Fl. Brit. India 5: 175. 1886. Synonyms: *Tetranthera oleoides* Meisn., *Malapoenna oleoides* Kuntze

Dioecious trees to 25 m tall; bark brown, smooth, slightly rugose. Leaves simple, alternate, oblong, elliptic-oblong or elliptic-lanceolate, 6.5-17 x 3-7.5 cm, base acute, apex acute or obtuse, margin entire, glabrous, reddish-brown beneath when dry; lateral nerves 8-12 pairs, intercostae obscure; petiole 10-30 mm long, grooved above, glabrous. Flowers unisexual, yellowish, in umbellules, in axillary racemes upto 6 cm long; peuncles to 6 mm long; bracts orbicular, pubescent. Perianth short; lobes 6, 2.5 x 3 mm, ovate, obtuse, pubescent. Stamens 9, in 3 rows, all filaments with sessile glands at the base, filaments hairy; staminodes 9, in 3-series, glandular at base. Ovary half inferior, papillose. Berry 2-2.5 cm across, subtended by the cupular perianth, red with white spots. Fl. & Fr.: December – June.

Western Ghats: Evergreen forests above 1000 m in Karnataka, Kerala and Tamil Nadu (Endemic).

IUCN: VU- B1ab(iii)+2ab(iii). (IUCN, 2024).

Litsea stocksii (Meisn.) Hook.f., Fl. Brit. India 5: 176. 1886 var. **stocksii.** Synonyms: *Cryptocarya neilgherrensis* Meisner., *Litsea vartakii* M.R.Almeida, *Cylicodaphne oblonga* var. *stocksii* Meisn., *Litsea josephi* S.M.Almeida Vern.: *Varikkeera* (Mal.).

Evergreen trees to 10 m tall; bark greyish-brown, smooth; branchlets fulvous-tomentose and lenticellate. Leaves simple, alternate, elliptic-obovate, elliptic or ovate-oblong, 6-10.5 x 2.5-5.5 cm, base acute, apex acuminate, acute, margin entire, recurved, glabrous above, pubescent and glaucous beneath; lateral nerves 5-10 pairs, intercostae scalariform, prominent; petiole 7-15 mm long, pubescent. Flowers bisexual, in short congested axillary or terminal fulvous tomentose panicles of 5-10 cm wide; pedicels to 0.5 cm. Perianth pubescent, tube ultimately narrowed at the top; lobes 6 4 mm, ovate, pubescent, subequal, in 2 rows. Stamens 9 perfect, in 3 rows, 2 outer rows with eglandular filaments and introrse anthers, third rows with 2 glandular filaments and extrorse anthers; fourth rows replaced by 3 stipitate staminodes; filaments hairy; anthers 2-celled; connective often produced. Ovary half inferior, globose, 2 mm across, glaucous, sessile, enclosed in the tube. Berry ovoid, black. Fl. & Fr.: October – April

Western Ghats: Evergreen forests of Maharashtra, Karnataka, Kerala and Tamil Nadu (Endemic).

IUCN: NT (Low risk/near threatened) (IUCN, 2024).

Key to the varieties

1a. Leaves elliptic-obovae, lateral nerves 7-9 pairs; berry obovoid, redish brown .... var. **stocksii**

1b. Leaves oblong elliptic or elliptic-lanceolate, lateral nerves 8-12 pairs; berry ovoid, red with whote spots .... var. **glabrescens**

Litsea stocksii var. **glabrescens** (Meisner) Hook. f. Synonym: *Cylicodaphne wightiana* Nees var. *glabrescens* Meisner

Fl. & Fr.: August – May

Western Ghats: Evergreen forests in Kerala and Tamil Nadu (Endemic).

IUCN: NE (IUCN, 2024).

Litsea travancorica Gamble, Bull. Misc. Inform. Kew 1925: 132. 1925.

Dioecious trees to 5 m tall; bark greyish-brown, smooth; branches some what pale, ferruginous tomentose. Leaves usually opposite or subopposite at the end of branchlets, alternate lower down, simple, oblong or obovate, 10.5-18 x 3-6 cm, base obtuse or acute, apex acuminate, margin entire, glabrous above, fulvous-pubescent especially on nerves beneath; lateral nerves 8-12 pairs, intercostae subscalariform, faint; petiole 5-10 mm long, ferruginous tomentose. Flowers unisexual, yellowish, in axillary crowded sessile umbellules; bracts orbicular, ferruginous. Perianth tube short, lobes 6, ovate, seriaceous. Stamens 12, in 4 rows; filaments 1 and 2 rows eglandular, 3 and 4 rows with 2-glands; anthers 4-celled; staminodes in female flowers as the stamens in male, but those of inner rows subulate and 2-glandular. Ovary half inferior, ovoid, glabrous; style stout; stigma capitate. Berry globose, 1 cm across, glabrous, seated on enlarged perianth tube. Fl. & Fr.: April – June.

Western Ghats: Evergreen forests of Karnataka, Kerala and Tamil Nadu (Endemic).

IUCN: VU- B2ab(i,ii,iii). (IUCN, 2024).

Litsea udayanii Robi, Phytotaxa 222(1): 45. 2015.

Small trees to 10 m tall; bark brown, smooth, lenticellate; branchlets adpressed sericeous or glabrous, purple colored. Leaves simple, alternate, elliptic-oblong or elliptic-lanceolate, 6–11 x 2–4 cm, base cuneate to rounded, apex obtuse, acute or attenuate, margin entire, glabrous on both sides, glaucous below; secondary veins 6–10 pairs, intercostae finely reticulate, prominent; petiole 5–10 mm long, glabrous, slightly grooved above, purple colored. Inflorescences on umbel-bearing reduced branchlets with the appearance of a raceme of umbels, in axils of leaves; peduncle of umbels 5 mm long, straight; involucral bracts 4, 3 x 3 mm, orbicular, densely adpressed hairy outside, glabrous and glossy inside. Flowers 4 in each umbel, 4 mm long; pedicels up to 2 mm long, silky-pubescent. Tepals 6, 1.3 x 0.8 mm, oblong, acute or incurved at apex, subequal, glabrous on both sides, white. Male flowers: not observed. Female flowers staminodes 12, outer rows linear; inner rows subulate, 1–1.3 mm long, glabrous, 2-glandular. Ovary 1 mm long, globose; style 1 mm long, pale-yellow, glabrous; stigma capitate, papillose, white. Berry ellipsoid, 1 cm long, glabrous, greenish, white specked, reddish on maturity, seated on the turbinate perianth tube. Fl. & Fr.: January – May.

Western Ghats: Evergreen forests of Kerala (Endemic).

IUCN: NE (IUCN, 2024).

Litsea venulosa (Meisn.) Hook.f., Fl. Brit. India 5: 161. 1886. Synonyms: *Tetranthera venulosa* Meisn., *Malapoenna venulosa* (Meisn.) Kuntze, *Litsea beei* N.Mohanan & E.S.S.Kumar, *Myristica glaucescens* Wall.

Small shrubby tree to 4 m tall; branchlets and leaves glabrous. Leaves oblong, obtusely caudate-acuminate, base obtuse, glaucous below, to 10 x 5 cm. Male umbellules solitary or in groups of 2 or 3, peduncles slender, to 1.5 cm long; female solitary. Perianth lobes 5 or 6, unequal, gland-dotted. Berry subglobose, 1 cm across, pedicels thickened towards apex, yellow. Fl. & Fr.: October – March.

Western Ghats: Evergreen forests of Kerala and Tamil Nadu (Endemic).

IUCN: NE (IUCN, 2024).

Litsea wightiana (Nees) Hook.f., Fl. Brit. India 5: 177. 1886 var. **wightiana**. Synonyms: *Cylicodaphne wightiana* Nees., *Cylicodaphne wightiana* var. *glabrescens* Meisn., *Litsea stocksii* (Meisn.) Hook.f. var. *glabrescens* (Meisn.) Hook.f.

Large evergreen trees to 25 m tall; bark light-brown, smooth; branchlets woody, cylindric, brown, young parts rusty-tomentose. Leaves alternate, elliptic to elliptic-oblong or oblong–lanceolate, 7 -15 cm x 2.5 - 6 cm, rarely obovate, acute or obtuse at apex, entire along margin, cuneate at base, glabrous except on veins, beneath rusty-tomentose, rarely purplish; secondary veins 6 - 10 pairs, reticulations strong; petioles stout, 1 - 1.5 cm long, densely puberulous. Umbellules 6-8 mm x 5.4 - 8 mm, axillary or terminal, arranged in stout racemes, 3 - 10 cm long, 4-6 flowered, brown-tomentose; peduncles 3.5 - 4 mm long, stout, silky-tomentose; bracts 4, in two rows, orbicular, thickly coriaceous, densely tomentose outside. Flowers 4-5 mm x 2.5 - 4 mm, creamish; pedicel 1.5-2 mm long, thick, green, densely tomentose. Perianth lobes 6, 2.5 - 3 mm long, elliptic, glabrous to silky-tomentose outside; perianth tube 1.5-2 mm long, turbinate. Male flowers stamens 12, in 4 rows, outer 2 rows 3.5 - 4,5 mm long, exerted; inner 2 rows 2.5 - 3 mm long, glandular; filaments slender, hairy at base; anthers 1-1.5 mm long, 4 celled, upper 2 cells smaller; pistillode 0.5 - 0.8 mm long, glabrous. Female flowers staminodes 12, in 4 rows, similar in male flowers. Pistil 2.5 - 3 mm long, style 2 - 2.5 mm long, glabrous, stigma discoid. Ovary 1 mm long, ovoid, glabrous. Berry sub-globose or ellipsoid, 8 - 12 x 5 - 8 mm, glabrous, seated on cup shaped perianth tube. Fl. & Fr.: August – November.

Western Ghats: Evergreen forests of Maharashtra, Karnataka, Kerala and Tamil Nadu (Endemic).

IUCN: NT (Low risk/near threatened) (IUCN, 2024).

Key to the varieties

1a. Branchlets glabrescent; leaves elliptic-lanceolate, tomentose beneath .... var **wightiana**

1b. Branchlets brown pubescent; leaves obovate-oblong, fulvous beneath .... var. **tomentosa**

Litsea wightiana (Nees) Hook.f. var. **tomentosa** (Meisn.) Gamble, Synonym: *Cylicodaphne wightiana* var. *tomentosa* Meisn.

Fl. & Fr.: November – April

Western Ghats: Evergreen forests of Kerala, Tamil Nadu (Endemic).

IUCN: NE (IUCN, 2024).

Neolitsea Merr.

1a. Young shoots glabrous; leaves finely and conspicuously areolat-reticulate .... **N. foliosa**

1b. Young shoots sericeous, tomentose or pubescent; leaves usually glaucous 2

2a. Leaves oblong-elliptic or suborbicular; tertiary nerves closely parallel; cupule subtending the fruits to 10 mm diam .... **N. fischeri**

2b. Leaves not as above; tertiary nerves neither closely parallel nor right angle to the midrib; cupule subtending the fruit to 8 mm diam 3

3a. Berry globose, 10-12 mm diam .... **N. cassia**

3b. Berry oblong-ellipsoid, 7-10 mm diam ....**N. zeylanica**

Neolitsea cassia (L.) Kosterm., J. Sci. Res. (Jakarta) 1: 85. 1952. Synonyms: *Laurus cassia* L., *Cinnamomum cassia* (L.) J.Presl Vern.: *Bill nisanghi, Masimara* (Kan.); *Keezhambalam, Pravaari, Vellakodala, Venkana* (Mal.); *Chirchira* (Mar.); *Molagachembakapalai, Pataapataa* (Tam.).

Deciduous trees to 18 tall; bark grey, rough, brittle. Leaves simple, alternate, opposite or in whorls, elliptic, elliptic-lanceolate or oblong-lanceolate, base acute or attenuate, apex acute, margin entire, glabrous above and glaucous beneath; main nerves 3-ribbed near the base, faint; lateral nerves 1-4 pairs, very faint, intercostae scalariform; petiole 10-15 mm long. Flowers unisexual, yellowish-green, densely seriaceous, in sessile umbels on extremely short branchlets at internodes; bracts 5 mm long, orbicular, seriaceous. Tepals 4, 2-3 mm long, narrowly ovate or oblong, acute. Stamens 6, in 3 rows; 2 outer rows eglandular, inner row biglandular opposite the first row; filaments 4-5 mm long, greenish-yellow, thread-like, glabrous; anthers 0.5 mm long, introrse. Ovary half inferior, ovoid; style long; stigma large peltate; staminodes 6, in 3 rows. Drupe globose, 10-12 mm across, seated on disc like crenate perianth cup. Fl. & Fr.: November – April.

Western Ghats: Evergreen forests of Kerala, Karnataka, Maharashtra and Tamil Nadu. India: Odisha. World: Sri Lanka, China, Thailand, Vietnam and Malaysia.

IUCN: NE (IUCN, 2024).

Key to the varieties

1a. Fruits globose to subglobose, 6-10 mm diam .... .var. **cassia**

1b. Fruits oblong, oblong-ellipsoid to ovoid-oblong, 12-13 mm diam............. .... .var. **rigescens**

Neolitsea cassia var. **rigescens** (Meisn.) Chakrab., Synonym: *Litsea zeylanica* var. *rigescens* Meisn.

Fl. & Fr.: February – March.

Western Ghats: Evergreen forests of Nilgiris in Tamil Nadu (Endemic).

IUCN: LC (IUCN, 2024).

Neolitsea fischeri Gamble., Bull. Misc. Inform. Kew 1925: 132. 1925.

Evergreen trees to 10 m tall; bark greyish-brown, smooth; young parts golden fulvous. Leaves simple, alternate, subopposite or in whorls, elliptic or elliptic obovate, 7-16 x 2-9 cm, base acute or cuneate, apex obtuse, margin entire, glaucous and white beneath; nerves 3-ribbed from base, lateral nerves 3-4, pinnate, intercostae scalariform, prominent; petiole 15-30 mm long, fulvous tomentose. Flowers unisexual, in sessile lateral umbellules; bracts 4, covered with scales. Perianth lobes 4, triangular, villous. Stamens usually 6, in 3 rows, 2 outer rows eglandular, 2 inner biglandular; sometimes 9 or 12 stamens in middle flowers; filaments villous; rudimentary pistil ovoid; staminodes 6, arranged as the stamens. Ovary half inferior, ovoid; style thick; stigma peltate. Drupe globose or obovoid, 10-15 x 10 mm, glabrous, seated on perianth tube. Fl. & Fr.: November – May.

Western Ghats: Very rare in evergreen forests of Kerala and Tamil Nadu (Endemic).

IUCN: VU- B1ab(iii)+2ab(iii). (IUCN, 2024).

Neolitsea foliosa (Nees) Gamble., First published in Fl. Madras: 1240. 1925. Synonyms: *Tetradenia foliosa* Nees, *Litsea foliosa* (Nees) Nees, *Tetranthera foliosa* Wall., *Litsea foliosa* var. *paniculata* Meisn.

Trees to 15 m tall; branchlets flattened towards apices, terete below. Leaves elliptic to oblong-elliptic to lanceolate-oblong or ovate-elliptic to ovate-oblong, 5-20 x 1.5-8 cm, acute, subacute to cuneate at base, usually decurrent into petiole at the extreme base, apiculate to acuminate at apex, thinly or occasionally stiffly coriaceous, usually glaucous; lateral nerves 3-5 pairs above the basal, nervules conspicuously reticulate-areolate above and beneath; petioles 10-25 mm long. Involucral bracts suborbicular, concave, 4-5 mm broad. Male umbels sessile or pulvinate, 2-5-together. Flowers pedicels 2-5 mm long. Sepals ovate to suborbicular 2-3 x 1-1.8 mm; filament 2-4 mm long, innermost with sessile glands; anthers oblong or squarish, 1 mm long. Female umbels 2-5-together. Flowers pedicellate, pedicels 2-6 mm long. Sepals as in male flowers; staminodes 2 mm long. Ovary subglobose, 1- 1.5 mm in diam; style 1.5-2 mm long; stigma conspicuous, peltate. Berry globose to subglobose, 7-12 mm in diam. fruiting pedicels 5–12 mm long, 1–2 mm thick, on fleshy perianth cup. Fl. & Fr.: February – April.

Western Ghats: Evergreen forests of Kerala and Tamil Nadu (Endemic). World: Bangladesh and Myanmar.

IUCN: LC (IUCN, 2024).

Key to the varieties

1a. Fruting perianth patelliform, entire .... var. **caesia**

1b. Fruiting perianth not patelliform 2

2a. Fruit oblong to oblong ellipsoid, 11-12 mm diam .... var. **foliosa**

2b. Fruit globose to subglobose, 7-11 mm diam .... var. **scrobiculata**

Neolitsea foliosa var. **caesia** (Meisn.) Gamble, Fl. Pres. Madras 2: 1240. 1925. Synonyms: *Camphorina camphora* (Thunb.) Farwell, *Camphorina cassia* (L.) Farwell.

Fl. & Fr.: February – April.

Western Ghats: Evergreen forests of Kerala and Tamil Nadu. World: Bangladesh and Myanamr.

IUCN: NE (IUCN, 2024).

Neolitsea foliosa var. **scrobiculata** (Meisn.) Chakrab., Phytotaxa 419: 213. 2019. Synonyms: *Litsea zeylanica* Nees & T. Nees, *Litsea scrobiculata* Meisn., *Neolitsea scrobiculata* (Meisn.) Gamble, *Litsea striolata* Blume, *Litsea foliosa* var. *impunctata* Meisn. Vern.: *Veganamaram* (Tam.).

Fl. & Fr.: September – March.

Western Ghats: Semi-evergreen to evergreen forests above 800 m in Kerala, Karnataka and Tamil Nadu. World: Sri Lanka, Bhutan, Bangladesh and Myanmar.

IUCN: NE (IUCN, 2024).

Neolitsea zeylanica (Nees & T. Nees) Merr., in Phillip. J. Sci. 1(Suppl.) 57. 1906. Synonyms: *Litsea zeylanica* Nees & T.Nees, *Tetradenia zeylanica* (Nees & T.Nees) Nees, *Laurus camphora* Thunb. ex Blume, *Laurus involucrata* Lam. *nom.illeg.*, *Laurus nitida* B.Heyne ex Wall., *Laurus verticillata* Royen ex Blume. *Litsea ceylanica* Nees & T.Nees *nom.illeg.*, *Litsea pseudoculitlawan* (Nees) Nees, *Malapoenna involucrata* Kuntze, *Malapoenna pseudoculitlawan* (Nees) Kuntze, *Neolitsea foliolosa* A.M.Cowan & Cowan, *Neolitsea involucrata* (Kuntze) Alston, *Tetradenia pseudoculitlawan* Nees, *Tetranthera foliosa* Wall. in Numer.

Trees up to 20 m tall; bark light brown, lenticellate; branchlets subterete, glabrous. Leaves simple, alternate, clustered at twig ends, elliptic to narrow elliptic, 7-11 x 3-3.5 cm, apex narrowly acute to acuminate, base attenuate, glaucous beneath, subcoriaceous; midrib raised above; trinerved just above the base with 4 pairs of secondary nerves, tertiary nerves closely reticulate; petiole 1.2-2 cm long, grooved above, glabrous. Flowers in axillary and lateral clusters. Tepals 4. Male flowers fertile, stamens 6; filaments villous, of third whorl each with 2 rounded small glands; rudimentary pistil small; style villous. Berry ellipsoid-oblong, fruiting perianth scarcely accescent and crenate. Seed 1. Fl. & Fr.: November – January.

Western Ghats: Evergreen forests of Kerala, Karnataka and Tamil Nadu. World: Sri Lanka, Nepal, China, Malay Peninsula.

IUCN: LC (IUCN, 2024).

Machilus Rumph. ex Nees

Machilus glaucescens (Nees) Wight, Icon. Pl. Ind. Orient. 5: t. 1825. 1852. Synonyms: *Laurus sylvestris* Buch.-Ham. ex Roxb., *Laurus villosa* Roxb., *Machilus macranthus* var. *brevifolius* (M. Gangop.) Kottaim., *Machilus villosus* (Roxb.) Hook.f., *Persea macrantha* var. *brevifolia* M. Gangop.

Tall tree 30 m tall, bark light brown, lenticellate, blaze pinkish. Branchlets terete, glabrous. Leaves simple, alternate, spiral, clustered at twig ends; petiole 1-4.5 cm long, glabrous; lamina 7-20 × 3-8 cm long, variable in shape, obovate or elliptic to elliptic-oblong, apex rounded or acuminate or acute, base acute to rounded, margin entire, coriaceous, glabrous, glaucous beneath, aromatic when squeeze; midrib canaliculate above; secondary_nerves 8-12 pairs. Flowers in terminal, spreading panicles. Flowers are bisexual, 1.0-1.2 cm across, pale yellow, in panicles from upper axils and terminal. Flower tube is very short, tepals 6, nearly equal, in 2 series, 4-5 mm, obovate, puberulous; persistent, spreading or reflexed in fruit. Berry is spherical, up to 2 cm across, black when ripe. Fl. & Fr.: May – August.

Western Ghats: Evergreen forests of Maharashtra. World: India and Sri Lanka.

IUCN: NE (IUCN, 2024).

Phoebe Nees

1a. Branchlets, leaves and infloresence glabrous; berry ellipsoid, glabrous ... . **P. lanceolata**

1b. Leaves elliptic or obovate, oblanceolate, acuminate, cuneate often narrowly at base 2

2a. Branchlets, leaves beneath and inflorescence minutely grey-puberulous .... **P. cathia**

2b. Branchlets, leaves beneath and inflorescence rusty-tomentose .... **P. wightii**

Phoebe lanceolata (Nees) Nees, Syst. Laur.: 109. 1836. Synonyms: *Ocotea lanceolata* Nees, *Persea lanceolata* (Nees) Oliv., *Laurus lanceolaria* Roxb., *Phoebe palghatensis* M.Gangop.

Evergreen trees, to 12 m high, bark greyish-brown, smooth, bark yellowish-white or grey, smooth with shallow fissures, exfoliating in thick flakes, fibrous. Leaves simple, alternate, estipulate, clustered at the tip of branchlets; lamina

8-25 x 2.5-4.5 cm, lanceolate, elliptic-obovate or elliptic-oblanceolate, base cuneate, apex long acuminate, margin entire, glabrous, coriaceous; lateral nerves 6-13 pairs, pinnate, prominent, intercostae scalariform, faint; petiole 5-20 mm long, slender, grooved above, glabrous. Flowers bisexual, 2-5 mm across, yellowish-green, in axillary, slender panicles. Perianth tube short; lobes 6, ovate, tomentose, unequal, accrescent. Stamens 9 perfect, in 3 rows, those of first and second row introrse, eglandular, those of third row extrorse with 2 large stipitate glands at base; staminodes of row 4 stipitate, cordate; anthers 4-celled. Ovary half inferior, sessile, globose, pubescent; style slender; stigma discoid. Fruit a drupe, 10-12 mm long, ovoid or ellipsoid, glabrous, black; supported by erect perianth lobes. Fl. & Fr.: December – July.

Western Ghats: Karnataka, Kerala, and Tamil Nadu.

World: India, Maynmar, Bhutan, Nepal, Bangladesh.

IUCN: LC (IUCN, 2024)

Phoebe cathia (D. Don) Kostrem. Nat. Hist. Bull. Siam Soc. 25: 44 (1974 publ. 1975). Synonyms: *Phoebe paniculata* (Nees) Nees, *Ocotea paniculata* Nees, *Cinnamomum cathia* D.Don, *Laurus cathia* Buch.-Ham. ex Meisn., *Laurus paniculata* Wall., *Laurus tomentosa* Buch.-Ham. ex Meisn., *Ocotea paniculata* Nees, *Ocotea paniculata* var. *minor* Nees, *Ocotea pubescens* Nees, *Persea cathia* Spreng., *Persea tomentosa* Spreng., *Phoebe pallida* subsp. *borii* M.Gangop., *Phoebe paniculata* var. *pubescens* (Nees) Meisn., *Phoebe prazeri* M.Gangop., *Phoebe pubescens* (Nees) Nees, *Phoebe wightii* Meisn. Vern.: *Karunthagiri* (Tam.).

Trees to 15 m tall; branchlets terete, lenticellate, flush grey pubescent. Leaves simple, alternate, spiral, oblanceolate, 9-21 x 4-9 cm, apex abruptly long acuminate, base cuneate, grey pubescent beneath at least when young; midrib slightly raised above; secondary nerves 7 pairs; tertiary nerves distantly percurrent; petiole 1-2 cm long, planoconvex in cross section, grey pubescent when young. Inflorescence axillary slender panicles, grey pubescent. Berry, ovoid, 1 cm across; seeds 1. Fl. & Fr.: December – January.

Western Ghats: Karnataka, Kerala, and Tamil Nadu. World: India, Bangladesh, and Myanmar.

IUCN: NE (IUCN, 2024).

Phoebe wightii Meissn., A.P.de Candolle, Prodr. 15(1): 38. 1864.

Evergreen trees, to 10 m high; bark brown, smooth; young parts rusty, pubescent. Leaves simple, alternate, estipulate, crowded at the tips of branchlets; petiole 7-15 mm long, stout, rusty pubescent when young; lamina 7.5-15 x 3-6 cm, elliptic, obovate or elliptic-obovate, base cuneate, apex acuminate, margin entire, glabrous above, rusty tomentose beneath, coriaceous; lateral nerves 6-7 pairs, pinnate, prominent, intercostae scalariform, faint. Flowers bisexual, yellowish-green, in axillary and lateral lax pubescent panicles; pedicel 5 mm; perianth tube short; lobes 6, 3-4.5 mm, subequal, ovate, tomentose; stamens 9 perfect, in 3 rows, those of first and second row introrseeglandular, those of third row extrorse with 2 large stipitate glands at their base; staminodes of the fourth row cordate, stipitate; anthers 4-celled; ovary half inferior, sessile, 1 mm, globose, pubescent, style 2 mm, stigma simple. Fruit a drupe, globose, 1 cm across; perianth lobes persistent, cupular, erect; epicarp fleshy; seed confirm to the fruit. Fl. & Fr.: November – April

Western Ghats: Karnataka, Kerala, Tamil Nadu (Endemic). World: India

IUCN: LC (IUCN, 2024).

Cultivated:

Beilschmiedia gammieana King ex Hook.f.; Native to India, Himalaya, Nepal, Bhutan

Cinnamomum bejolghota (Buch.-Ham.) Sweet; Native to India, Eastern Himalaya, Andaman Islands, Nepal, Tibet, Myanmar, China

Cinnamomum camphora (L.) J. Presl; Exotic: Native to Japan

Laurus nobilis L.; Exotic: Native to Mediterranean region.

Persea americana Mill.; Exotic: Native to Tropical America

LECYTHIDACEAE A. Rich.

1a. Fruit angular, fibrous, 1-seeded; stamens all perfect ....**Barringtonia**

1b. Fruit globose or ovoid, fleshy, many-seeded; stamens partly without anthers ....**Careya**

Barringtonia J.R. Forst. & G. Forst.

1a. Calyx-lobes valvate; fruit large, about 9 cm long, ovoid; flowers large, leaves large upto 35 cm long..**B. racemose**

1b. Calyx-lobes imbricate; fruit medium-sized, about 6 cm long, flowers small; Leaves small upto 15 cm long .... **B. acutangula**

Barringtonia acutangula (L.) Gaertn., Fruct. Sem. Pl. 2: 97. 1791. Synonyms: *Eugenia acutangula* L., *Barringtonia rubra* Baill. ex Laness., *Huttum acutangulum* (L.) Britten

Small trees to 20 m tall. Leaves elliptic or obovate to oblong, 6-15 x 2-6 cm, glabrous or slightly hairy underneath, subsessile, attenuate at the base, acute-acuminate. Inflorescence a densely flowered terminal raceme, pendulous, up to 50 cm long; bracts elliptic-lanceolate. Flowers 2-3 cm in diam; pedicel 3-7 mm long. Sepals 4, free at least above, semicircular, obtuse, 2-3.5 mm long, 1.5-3 mm broad. Petals 4 -5, elliptic, 6-12 mm long. Stamens in 3 whorls, innermost converted into staminodes. Ovary usually 2-loculed with 1-2 cm long style. Fruit ovoid, 2-6 cm long, 1-2.5 cm wide, quadrangular. Seeds ovoid, grooved, 1-4 x 0.5-1.5 cm. Fl. & Fr.: September – October.

Western Ghats: Semi-evergreen forests along the water course of Maharashtra, Goa, Karnataka, Kerala and Tamil Nadu. World: Sri Lanka, Tropical Asia and North Australia.

IUCN: LC (IUCN, 2024).

Barringtonia racemosa (L.) Spreng., Syst. Veg., ed. 16. 3: 127. 1826. Synonyms: *Eugenia racemosa* L., *Barringtonia ceylanica* (Miers) Gardner ex C.B.Clarke, *Stravadium racemosum* (L.) Sweet, *Barringtonia terrestris* (Miers) R.Knuth Vern.: *Samudhrachampa*, *Samutrakai* (Mal.); *Aratam* (Tam.).

Small trees to 20 m tall; bark smooth or fissured; branchelets pendulous, gray-brown. Leaves obovate-oblong, 20-35 x 6-14 cm, base cuneate, margin serrate-crenulate, apex acute or acumi-nate; petiole 2-15 mm, winged. Racemes usually terminal or in axils of fallen leaves, pendulous, many flowered; bracts triangular, 5-6 mm; bracteoles triangular, 1.5-2 mm; pedicel to 2.5 cm. Flowers 6-9 mm diam. Sepals cupular, 2-4-lobed, 10-12 cm at anthesis; tube 2.5 mm. Petals 4, green or tinged red or yellow, oblong, 1-1.3 x 0.5-0.8 cm. Stamens in 5 or 6 whorls, innermost sterile; tube 3.5-6 mm; filaments red, proximally white or pink, 3-3.5 cm; staminodes 10-13 mm. Ovary 2-4-loculed, 1.5-3.5 mm; ovules 2 or 3 per locule; style 4-6 cm. Fruit ovoid-cylindric, 4-angled, 5-9 x 3-4 cm. Seed ovoid, 2-4 cm. Fl. & Fr.: April – May.

Western Ghats: Semi-evergreen forests of Maharashtra, Goa, Karnataka, Kerala and Tamil Nadu. World: Sri Lanka, Malaysia and Polynesia.

IUCN: LC (IUCN, 2024).

Careya Roxb.

Careya arborea Roxb., Pl. Coromandel 3: 14. 1819. Synonyms: *Careya sphaerica* Roxb., *Barringtonia arborea* (Roxb.) F.Muell. Vern.: *Kumph* (Mar.); *Modama* (Tam.).

Deciduous trees to 12 m tall; bark brownish, rough, exfoliations small, more or less triangular; fibrous; branchlets pubescent. Leaves simple, alternate, clustered at the tips of branchlets, obovate, oblong-obovate, 15-30 x 5-15 cm, apex round or shortly acuminate, base cuneate or attenuate and decurrent on petiole, margin crenate-serrate, glabrous, membranous; lateral nerves 10-12 pairs, intercostae reticulate, prominent; petiole 10 mm long, slender, pubescent. Flowers bisexual, greenish-white, 5-10 mm across, in terminal spikes; peduncle woody; bracts 3, unequal. Sepals tubular, tube campanulate, glabrous, adnate to ovary and not produced beyond the ovary; lobes 4, ovate. Petals 4, elliptic-oblong, revolute along margin, cauducous. Disc annular. Stamens numerous in several rows, slightly connate at base, inserted on the top of calyx; filaments subulate, purple, exserted, inner and outer rows without anthers. Ovary inferior, 4-5-celled; ovules many in each cell; style long, filiform; stigma capitate. Berry globose, 5-7.5 cm across, green, crowned with sepals and style. Fl. & Fr.: March – June.

Western Ghats: Deciduous forests of Gujarat, Maharashtra, Goa, Karnataka, Kerala and Tamil Nadu. World: Sri Lanka, Nepal, Afghanistan, Myanmar and Malaysia.

IUCN: NE (IUCN, 2024).

Planted

Barringtonia asiatica (L.) Kurz; Native to India, Andaman Islands, Sri Lanka, Singapore, Malaya.

Couroupita guianensis Aubl.; Exotic: Native to South America

Gustavia angusta J.F.Gmel.; Exotic: Native to Tropical America

Gustavia gracillima Miers; Exotic: Native to South America

LOGANIACEAE R.BR. EX MART.

Strychnos L.

1a. Fruits 3–5 cm across, yellowish-red or orange, seeds 3–4........................... ....**S. nux-vomica**

1b. Fruits 1.5–2 cm across, black, seeds 1–2.................................**S. potatorum**

Strychnos nux-vomica L., Sp. Pl.: 189. 1753. Synonyms: *Strychnos vomica* St.-Lag., *Strychnos nux-vomica* var. *oligosperma* Dop Vern.: *Ettikai, Etti* (Tam.).

Trees to 20 m tall; bark thin, yellowish or blackish-grey, smooth or scurfy with lenticels. Leaves simple, opposite, 6-14 x 2-9 cm, elliptic, broadly elliptic or elliptic-ovate, apex acute or obtuse, base acute to attenuate, margin entire, glabrous; petiole to 13 mm, slender, glabrous; 3-5 nerved from base, intercostae reticulate. Flowers bisexual, greenish-white, in terminal sessile cymes, sessile, tomentose; bract 5 mm; bracteole 1.5 mm long. Sepals 4, lobes 4, 1 mm long, triangular, acute, pubescent. Petals greenish-yellow, tube 10 mm long, longer than lobes, slightly hairy near the base within, lobes 4, 4 x 1.5 mm, oblong, acute. Stamens 4, anthers sessile inserted at the throat of corolla tube. Ovary 2-celled, ovules many in each cell. Berry globose, 3-5 cm across, yellowish-red or orange, glabrous. Seeds 3 or 4, orbicular, flat, shiny, greenish-white. Fl. & Fr.: February – November.

Western Ghats: Common in deciduous to semi-evergreen forests of Maharashtra, Goa, Karnataka, Kerala and Tamil Nadu. India: Andhra Pradesh, Odisha, Gujarat, Rajasthan. World: Sri Lanka, Laos, Cambodia, Vietnam, Mayanmar, Thailand and Malesia.

IUCN: NE (IUCN, 2024).

Strychnos potatorum L.f., Suppl. Pl.: 148. 1782. Synonyms: *Strychnos heterodoxa* Gilg, *Strychnos tetankotta* Retz. Vern.: *Kataka* (Kan.); *Silamaram, Thettamkottai, Porithettham, Thettham* (Tam.).

Trees to 12 m tall; bark brownish-black, rough, corky with deep and narrow vertical cracks. Leaves simple, opposite, elliptic, ovate, elliptic-ovate, 5-15 x 3.5-7.5 cm, base obtuse or round, apex acute, margin entire, glabrous, shiny; nerves 3-5 from the base, lateral nerves from the midrib 4 pairs, intercostae reticulate, prominent; petiole 2-8 mm, glabrous. Flowers white, 0.85 cm long, in short axillary cymes, 6-8 mm long; pedicel 2 mm. Sepals 5, 5-lobed, small. Petals 5, 3.5 mm across, salver shaped, hairy within, tube cylindric, lobes 5, longer than tube. Stamens 5, inserted on the corolla tube; anthers 2-celled, introrse. Ovary 1 mm, superior, glabrous, ovules one in each cell; style glabrous; stigma obscurely 2 lobed. Berry globose, 16-18 mm across, black. Seeds 1-2, 10-12 mm across, round, compressed, yellowish. Fl. & Fr.: June – January.

Western Ghats: Deciduous forests above 600 m in Maharashtra, Karnataka, Kerala and Tamil Nadu. India: Andhra Pradesh, Odisha and Telangana. World: Sri Lanka and Myanmar.

IUCN: NE (IUCN, 2024).

LYTHRACEAE J.St.-Hil.

1a. Stamens all or in part inserted at rim of floral tube just below sepals; stamens usually many, seldom as few as 12 2

1b. Stamens all inserted at or near base of floral tube or somewhat above, but never at rim of floral tube; stamens usually 4-12, seldom numerous 3

2a. Flowers 1-3(-5), in terminal clusters; petals narrowly lanceolate, very small, or absent; fruit indehiscent, globose, leathery berries; seeds irregularly angular or falcate .... **Sonneratia**

2b. Flowers (3-)5 to many, in lax terminal corymbs; petals oblong to ovate; fruit loculicidally dehiscent capsules; seeds 2-tailed, filiform .... **Duabanga**

3a. Flowers solitary or paired; stamens 12; capsule circumscissile; seeds not winged .... **Pemphis**

3b. Flowers in panicles; stamens 6-100+, usually 18 or more; capsule loculicidally dehiscent; seeds unilaterally winged .... **Lagerstroemia**

Duabanga Buch.-Ham.

Duabanga grandiflora (Roxb. ex DC.) Walp., in Repert. Bot. Syst. 2: 114. 1843. Synonyms: *Lagerstroemia grandiflora* Roxb., *Leptospartion grandiflora* (Roxb. ex DC.) Griff., *Duabanga sonneratioides* Buch.-Ham., *Leptospartoion grandiflorum* Griff.

Large tree, up to 50 m tall; branches pendulous, glabrous throughout. Leaves rigid, distichous, lamina abaxially pale green, adaxially dark green, ovate to oblong-elliptic, broad, 11-15.5 x 4-7.5 cm, abaxially glaucous, margin entire, lateral veins 20-24 pairs, base cordate, apex shortly acuminate; petiole stout, 1.3 cm long. Corymbs 3-20-flowered, drooping at terminal branches. Flower 5-6 cm across, greenish whiter, pedicels 3-4 cm, stout. Floral tube broadly campanulate, 1.6-2.5 x 1.8-3 cm; sepals 5-6, green, thick fleshy, ovate, acuminate, enlarged in fruit. Petals 6, dull white, membranous, obovate, 2.5-3 x 1.5-2 cm, margin undulated. Stamens numerous, 50 or more, white, surpassing petals, anthers characteristically curved. Ovary subglobose, 2.5-3 cm across, style cylindrical, elongated up to 5 cm long, stigma capitate. Capsules subglobose, 3-4 x 4-4.5 cm, 6-9-valved. Seeds 4-6 mm long, brownish, smooth. Fl. & Fr.: December – May.

Western Ghats: Valparai hills and Anamalais of Tamil Nadu and Kerala. India: Meghalaya, Mizoram, Nagaland, Tripura and Assam. World: Burma, Malaysia, Nepal and Southern China.

IUCN: LC (IUCN, 2024).

Note: *Duabanga grandiflora* is observed along the road side and inside the tea plantations at Valparai estates. It may be planted during the British time for avenue and shade tree in tea plots but now extended their range in forests of Analamalais due to their overproduction of seeds. About 27 lofty trees were identified along the road side forests and plantations. They usually associated with some other trees like *Lagerstroemia parviflora, Bhesa indica, Cinchona officinalis, Mallotus tetracoccus, Syzygium caryophyllatum*, etc. It is reported by the author first for Western Ghats (Karuppusamy et al, 2021).

Lagerstroemia L.

1a. Clayx tube smooth without 2

1b. Calyx tube ribbed without .... **L. speciosa**

2a. Calyx-lobe with a ring; leaves greyish and glabrous or shortly pubescent beneath, rounded at base .... **L. parviflora**

2a. Calyx-tube without a ring; leaves glabrous or bluish-white beneath, narrowed at base .... **L. microcarpa**

Lagerstroemia speciosa (L.) Pers., Syn. Pl. 2: 72. 1806. *Lagerstroemia hirsuta* (Lam.) Willd., *Adambea hirsuta* Lam., *Lagerstroemia reginae* Roxb., *Lagerstroemia flos-reginae* Retz., *Munchansia speciosa* L.

Deciduous trees to 25 m tall; bark 4-6 mm thick, yellowish-brown, smooth. Leaves simple, opposite or subopposite, oblong, ovate-lanceolate, elliptic-lanceolate or elliptic-oblong, 7.5-25 x 3-10 cm, apex acute, base obtuse, round or acute, margin entire, glabrous; lateral nerves 11-19 pairs, parallel, prominent beneath, intercostae reticulate, prominent; stipules intrapetiolar, deciduous; petiole 6-12 mm long, glabrous; Flowers mauve or pink, 5-7 cm across, in terminal panicles; peduncle puberulous; pedicel 1 cm long. Calyx tube 1 x 1.3 cm, hemispherical, puberulous, ribbed without; lobes 6, triangular, acute. Petals 6, 4 x 2.5 cm, obovate, clawed, margins crispate. Stamens many, inserted near the base of the calyx tube; filaments exserted. Ovary half inferior, sessile, glabrous, 6-celled; ovules many; style 2 cm long, curved; stigma capitate. Capsule ovoid, 18-30 mm long, brown, woody, dehiscent. Seeds 10-12 mm long, winged, pale brown. Fl. & Fr.: August - October.

Western Ghats: Planted avenue trees in Maharashtra, Goa, Karnataka, Kerala and Tamil Nadu. World: Sri Lanka, Pakistan, Mayanmar, China, Malaysia and Indonesia.

IUCN: LC (IUCN, 2024).

Lagerstroemia microcarpa Wight, Icon. Pl. Ind. Orient. 1: t. 109. 1839. *Lagerstroemia lanceolata* Wall. ex Clarke, *Lagerstroemia thomsonii* Koehne, *Murtughas lanceolata* Kuntze

Deciduous trees to 30 m tall; bark greyish or greyish-white, smooth, peeling off in thin long and broad flakes; branches knotted. Leaves simple, opposite, distichous, elliptic, ovate, elliptic-lanceolate or ovate-lanceolate, 4.5-10 x 3.7-6.5 cm, apex acute or acuminate, base attenuate or acute, margin entire, glabrous, velvety pubescent beneath, chartaceous; lateral nerves 6-10 pairs, intercostae scalariform, faint; stipules 2, intrapetiolar, deciduous; petiole 6-15 mm long, glabrous; Flowers white, 5-6 mm across, in axillary and terminal panicles. Calyx smooth, hoary, without a ring; lobes 6, triangular, persistent, reflexed. Petals 6, 3 mm long, obovate, white, clawed. Stamens numerous, inserted at the base of calyx tube; filaments long, exserted. Ovary half inferior, sessile, glabrous, 4-6-celled; ovules many; style long, curved; stigma capitate. Capsule ovoid, 8-12 mm long, 4-6 valved, dehiscent. Seeds many, elongate, falcately winged, brownish. Fl. & Fr.: March - June.

Western Ghats: Deciduous to semi-evergreen forests of Maharashtra, Goa, Karnataka, Kerala and Tamil Nadu (Endemic).

IUCN: NE (IUCN, 2024).

Lagerstroemia parviflora Roxb., Pl. Coromandel 1: 48. 1796. Synonyms: *Murtughas parviflora* (Roxb.) Kuntze

Deciduous trees to 15 m tall; bark dark brown or greyish-black, vertically furrrowed, exfoliating in long, thin flakes. Leaves simple, opposite, elliptic, obovate-elliptic or oblong, 3.5-7.5 x 2-3 cm, base round or acute, apex acute, acuminate or obtuse-retuse, margin entire, glabrous or shortly pubescent beneath; lateral nerves 6-8 pairs, intercostae reticulate, prominent; stipules intrapetiolar, deciduous; petiole to 5 mm long, slender, glabrous. Flowers white, 2 mm across, fragrant, in lax axillary panicles. Sepals tubular, campanulate; tube 7 mm, with a ring inside, smooth with out; lobes 6, to 3 mm, appressed to the capsule in fruit. Petals 6, ovate-suborbicular, to 6 mm; claw 4 mm. Stamens many, inserted near the base of the calyx tube, exserted; filaments to 1 cm; anthers 0.5 mm. Ovary 2.5 mm, half inferior, sessile, glabrous, 4-6-celled; ovules many; style to 1 cm, curved; stigma capitate. Capsule ovoid, 3 x 1.5 cm, 3-4 valved, brown. Seeds many; wing terminal with a thick curved back. Fl. & Fr.: February – October.

Western Ghats: Deciduous forests of Maharashtra, Goa, Karnataka, Kerala and Tamil Nadu. India: Andhra Pradesh and Telangana. World: Nepal and Mayanmar.

IUCN: LC (IUCN, 2024).

Pemphis L.

Pemphis acidula J.R. Forst. & G. Forst., Char. Gen. Pl.: 68. 1776. Synonyms: *Lythrum pemphis* L.f., *Pemphis angustifolia* Roxb., *Pemphis setosa* Blanco

Small trees to 5 m tall. Leaves opposite, oblong-obovate, 1.5-3 x 0.7- cm, obtuse at apex, acute to cuneate at base, clothed in pale prostrate hairs on both the surfaces; lateral nerves 3-5 pairs, inconspicuous; petiole 2-3 mm long, grooved on the upper side. Flowers axillary, solitary; hypanthium 3-4 mm long, longitudinally ribbed, lobes 6, alternating with 6 small appendages. Petals 4-5 mm long, pink, obovate, wrinkled. Stamens 12, alternately long and short. Ovary 3-locular at the base and 1-locular towards the apex. Stigma globular. Fruits 5-6 x 4-5 mm, nearly globose, calyx lobes and style persistent. Seeds numerous, about 3 x 2 mm, angular-cuneate, margined. Fl. & Fr.: October – December.

Western Ghats: Evergreen forests of Kerala and Tamil Nadu. World: Sri Lanka and Old world Tropics.

IUCN: LC (IUCN, 2024).

Sonneratia L.f.

1a. Calyx lobes 4; petals 0 .... **S. apetala**

1b. Calyx lobes 6-8; petals 6–8 2

2a. Flowers white; calyx lobes 6-8; drupe 4 cm across .... **S. alba**

2b. Flowers pink; calyx lobes 6; drupe 7 cm across .... **S. caseolaris**

Sonneratia alba Sm., A. Rees, Cycl. 33: no. 2. 1816.

Evergreen trees to 7 m high; pneumatophores 75 x 0.8 cm, straight, stout, conical, outer thin layer flaky; bark brown, cracked; branchlets terete, glabrous. Leaves simple, opposite, broadly elliptic, ovate, obovate, oblong or suborbicular4.5-11 x 3-9 cm, base cuneate, apex obtuse, coriaceous, slightly fleshy, glabrous; lateral nerves 7-13 pairs, intercostae reticulate, obscure; petiole 5-10 mm long, glabrous, white or pink. Flowers white, 6.5 × 8 cm, in terminal axillary or leaf opposed clusters of 2, 3 or rarely solitary; pedicel 1-4 cm long terete, glabrous; bracteoles 2, green. Sepals green outside, white within, calyx cup 2 x 2.5 cm, angular, glabrous; lobes 6-8, 2 x 1 cm, oblong, acute, thick, greenish outside, reddish within. Petals linear, 6-8, 26 x 0.75 mm, white with reddish tinge towards apex, membranous, glabrous. Stamens numerous, free, inflexed in bud, inserted in several rows on the terminal raised rim of the calyx cup, white; anthers reniform, medifixed, bilobed. Ovary globose, 7 x 1.5 cm, glabrous, white; many celled; ovules many; style coiled in bud, to 6 cm long; stigma a capitate, pale green. Drupe 4 cm across, green, flattened above with persistent calyx. Fl. & Fr.: February - July.

Western Ghats: Low level tidal forests of Maharashtra, Goa, Karnataka, Kerala and Tamil Nadu. World: Sri Lanka, Myanmar, Malaysia, Japan, New Guinea, Micronesia, North Australia and East Africa.

IUCN: LC (IUCN, 2024).

Sonneratia apetala Banks., M. Symes, Account Embassy Kingd. Ava: 477. 1800. Synonyms: *Blatti apetala* (Banks) Kuntze, *Kambala pendula* Raf.

Trees to 20 m tall; pneumatophores to 1.5 m; ultimate branchlets pendulous. Leaves narrowly elliptic to lanceolate, gradually tapering toward apex, 5-13 x 1.5-4 cm, base attenuate, apex obtuse; petiole 5-10 mm long. Flowers 4-6-merous in a 3-7-flowered cyme. Floral tube 1.5-2.5 cm at maturity, smooth. Sepals green, slightly curving around base of fruit. Petals absent. Staminal filaments white. Stigma peltate, to 7 mm wide. Drupe 1-2 x 2-2.5 cm in diam. Seeds typically U-shaped or falcate. Fl. & Fr.: January - April.

Western Ghats: Typical tidal forest of Maharashtra and Karnataka. India: Andhra Pradesh and Odisha. World: Sri Lanka, Myanmar, Bangladesh and Malaysia.

IUCN: LC (IUCN, 2024).

Sonneratia caseolaris (L.) Engl., H.G.A. Engler & K.A.E.Prantl, Nat. Pflanzenfam., Nachtr. 1: 261. 1897. Synonyms: *Rhizophora caseolaris* L., *Sonneratia acida* L.f., *Sonneratia evenia* Blume, *Sonneratia ovalis* Korth.

Evergreen trees to 8 m tall; pneumatophores 1 m long and 6 cm diam, straight, corky, conical, brownish grey or orange-coloured; branchlets pendulousg, labrous. Leaves simple, opposite-decussate, broadly ovate or broadly elliptic, 4-11 x 3.5-6.5 cm, base cuneate, apex mucronate, emarginate or obtuse, glabrous, coriaceous, thick, slightly fleshy, green; lateral veins 10-15 pairs, intercostae reticulate, obscure; petiole 5-10 mm long, red, glabrous. Flowers pink, 6 x 5-7 cm, terminal, solitary; bracts rudimentary; pedicel 5-10 mm long, quadrangular, glabrous. Sepals green outside, white within, calyx cup 1.8 x 2 cm, smooth, glabrous, enclosing the basal portion of the pistil; lobes 6, 2 x 1 cm, oblong, elliptic, acute, thick, coriaceous, glabrous, persistent. Petals 6, 2 x 0.2 cm, free, purple to violet red, acuminate, apex wrinkled and coiled, cauducous. Stamens many, free, inflexed in bud, inserted in several rows on the terminal raised rim of the calyx cup, cauducous; filaments upto 4.2 cm long, reddish below, white above, thread like, anthers reniform, medifixed, bilobed. Ovary 1 cm across, globose, white, many celled, ovules many in each cell; style coiled in bud, to 6 cm, white, terete; stigma capitate. Drupe globose, 7 cm across, slightly flattened, calyx lobes horizontal, glabrous. Seeds many, embedded in the fleshy pulp of the placenta. Fl. & Fr.: February – August.

Western Ghats: Forests near the sea up to 400 m in Maharashtra, Goa, Karnataka and Kerala. India: Andaman & Nicobar Islands. World: Sri Lanka, Bangladesh, Myanmar, Thailand, Java, Solomon Islands, Malaysia, Sumatra, Philippines, New Guinea and North Australia.

IUCN: LC (IUCN, 2024).

Cultivated:

Lafoensia vandelliana DC. ex Cham. & Schltdl.; Exotic: Native to South America

Lagerstroemia floribunda Jack; Exotic: Native to Malaya, Thailand

Lagerstroemia indica L.; Exotic: Native to China

Lagerstroemia thorelii Gagnep.; Exotic: Native to Laos, Cambodia, Vietnam.

Hydnocarpus alpina

Hydnocarpus pendulus

Hydnocarpus pentandra

Saurauia napaulensis

Buchanania axillaris

Buchanania lanzan

Buchanania barberi

Buchanania lanzan

Holigarna arnottiana

Lannea coromandelica

Nothopegia heyneana

Nothopegia travancorica

Nothopegia vajravelui

Semecarpus anacardium

Spondias pinnata

Semecarpus anacardium

Alphonsea lutea *Goniothalamus wightii*

Meiogyne pannosa *Miliusa ammaii*

Miliusa nilagirica *Miliusa tomentosa*

Monoon fragrans *Orophea uniflora*

Polyalthia suberosa

Popowia beddomeana

Alstonia scholaris

Alstonia venenata

Holarrhena pubescens

Hunteria zeylanica

Tabernaemontana alternifolia

Wrightia tinctoria

Ilex denticulata *Ilex wightiana*

Heptapleurum agasthiyamalayanum *Heptapleurum bourdillonii*

Heptapleurum stellatum *Heptapleurum venulosum*

Polyscias acuminata *Caryota urens*

Cordia dichotoma

Cordia myxa

Commiphora caudata

Commiphora caudata var. pubescens

Calophyllum inophyllum

Agasthiyamalaia pauciflora

Celtis timorensis

Crateva adansonii

Cassine glauca

Euonymus kanyakumariensis

Euonymus laxiflorus

Glyptopetalum grandiflorum

Gymnosporia montana

Microtropis stocksii

Bhesa indica

Garcinia gummi-gutta

Garcinia mangostana

Garcinia rubroechinata

Garcinia travancorica

Terminalia anogeissiana

Terminalia arjuna

Terminalia paniculata

Alangium salviifolium

Alangium hexapetalum

Dichapetalum gelonioides

Dillenia pentagyna

Dillenia retusa

Diopyros candolleana

Diospyros chloroxylon

Diospyros ferrea *Diospyros foliosa*

Diospyros melanoxylon *Diospyros montana*

Elaeocarpus blascoi *Elaeocarpus gaussenii*

Elaeocarpus recurvatus *Elaeocarpus variabilis*

Rhododendron arboreum ssp. *nilagiricum*

Vaccinium symplocifolium

Rhododendron arboreum ssp. cinnamomeum

Erythroxylum moonii

Erythroxylum monogynum

Agrostistachys indica

Aleurites moluccanus

Blachia denudata

Blachia umbellata

Cleidion javanicum

Cleidion nitidum

Croton malabaricus

Epiprinus mallotiformis

Euphorbia santapaui **- Habit**

Euphorbia santapaui

Euphorbia vajravelui **var.** ***theniensis***

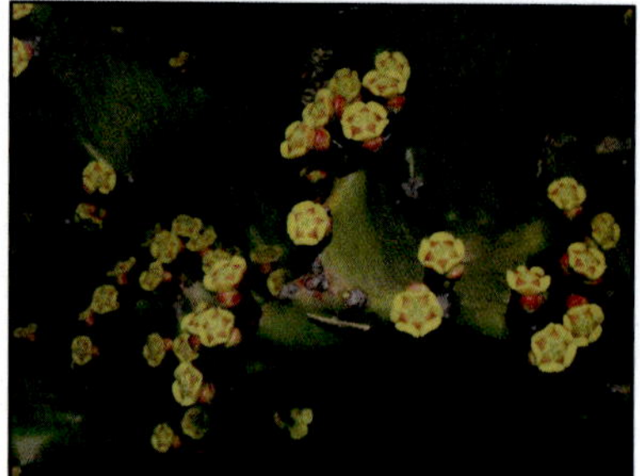

Euphorbia trigona

Euphorbia tortilis

Excoecaria oppositifolia var. *crenulata*

Falconeria insignis

Givotia moluccana

Glochidion ellipticum

Mallotus philippensis

Mallotus resinosus

Suregada lanceolata

Tritaxis beddomei

Acrocarpus fraxinifolius

Albizia amara

Albizia chinensis

Archidendron clypearia

Archidendron bigeminum

Bauhinia purpurea

Bauhinia variegata

Bauhinia racemosa

Butea monosperma

Dalbergia sissoo

Cassia fistula

Erythrina variegata

Humboldtia unijuga

Pterocarpus marsupium

Tamarindus indica

Fagraea ceilanica | *Mappia nimmoniana*

Callicarpa tomentosa | *Clerodendrum infortunatum*

Gmelina arborea | *Gmelina asiatica*

Vitex leucoxylon | *Vitex trifolia*

Actinodaphne bourneae

Actinodaphne campanulata

Actinodaphne malabarica

Alseodaphne semecarpifolia

Cinnamomum filipedicellatum

Cinnamomum iners

Cinnamomum sulphuratum

Cinnamomum wightii

Cryptocarya beddomei *Cryptocarya stocksii*

Litsea beddomei *Litsea bourdillonii*

Litsea deccanensis *Litsea floribunda*

Litsea kakkachensis *Litsea keralana*

Litsea ligustrina

Litsea beddomei

Litsea megamalayana

Neolitsea cassia

Neolitsea foliosa var. *scrobiculata*

Phoebe paniculata

REFERENCES

Ahmedullah, M.; Nayar, M. P. 1986. In Endemic Plants of the Indian Region, vol. 1; Botanical Survey of India: Calcutta.

Almeida, M. R. Flora of Maharashtra, Vol. 1, 2, 3,4 and 5; St. Xavier's College, Orient Press: Mumbai, 1996, 1998, 2001, 2003, 2009.

Ansari, M. Y. 1984. Asclepiadaceae—Genus Ceropegia. Fasc. Fl. India 16: 1–34.

Apte, G. S.; Bahulikar, R. A.; Kulkarni, R. S.; Lagu, M. D.; Kulkarni, B. G.; Suresh, H. S. 2006. Genetic Diversity Analysis in Gaultheria fragrantissima Wall. (Ericaceae) from the Two Biodiversity Hotspots in India using ISSR Markers. Curr. Sci. 91: 1634–1640.

Bahulikar, R. A.; Lagu, M. D.; Kulkarni, B. G.; Pandit, S. S.; Suresh, H. S.; Rao, M. K. V. 2004. Genetic Diversity Among Spatially Isolated Populations of Eurya nitida Korth. (Theaceae) Based on Inter-Simple Sequence Repeats. Curr. Sci. 86: 824–831.

Beddome, R. H. 1869–1874. In Flora Sylvatica for Southern India, vol. 2; Madras.

Bhushan K. S.; Kulkarni, A.; Vijayan, S.; Choudhary, R. K.; Datar, M. N. 2020. An assessment of the local endemism of flowering plants in the Northern Western Ghats and Konkan regions of India: checklist, habitat characteristics, distribution, and conservation. Phytotaxa 440 (1): 25–54.

Biju, S. D.; Bossuyt, F. 2003. New Frog Family from India Reveals an Ancient Biogeographical Link with the Seychelles. Nature 425: 711–714.

Biju, S. D.; Bossuyt, F.; Lannoo, M. M. 2009. Two New Philautus (Anura: Ranidae: Rhacophorinae) from Ponmudi Hill in the Western Ghats of India. J. Herptology 39: 349-353.

Blasco, F. 1970. Aspects of the Flora and Ecology of Savannas of South Indian Hills. J. Bombay Nat. Hist. Soc. 67: 522–534.

Blasco, F. 1970. Aspects of the Flora and Ecology of the Savannas of the South Indian Hills. J. Bombay Nat. Hist. Soc. 67: 522–534.

Blasco, F. 1971. Montagnes du sud del'nde: forests, savanes, ecologie. Inst, Fr. Pondicherry, Trv, Sec.Sci. Tech. Tome 10: 436.

Blatter, E.; McCann, C. 1935. The Bombay Grasses. Sci. Monogr. No.5. Imp. Counc. Agric. Res. India. 21: 324.

Bourdillon, T. F. 1908. In The Forest Trees of Travancore; The Travancore Govt. Press: Thiruvananthpuram.

Bourdillon, T. F. 1893. Report on the Forest of Travancore; Travancore Govt. Press: Trivandrum.

Bourgeon, G. 1989. In Reconnaissance Soil Map of Forest Area Western Karnataka and Goa; Institut Francais de Pondicherry: India.

Bourgeon, G. 1987. Les "sols rouges"oles Regions Semi-andes du sud de linde I & III. Proprietes of classification pedologique L. Agron. Trop. 42 (3): 153–170.

Champion, H. G. 1936. A Preliminary Survey of Forests Types of Indian and Burma. Indian Forest Rec. Bot. (n.s) Silva. 2 (1): 1–286.

Champion, H. G.; Seth, S. K. 1968. In A Revised Survey of the Forest Types of India; Government of India Press: Nasik, India.

Chandran, M. D. S.; Rao, G. R.; Gururaja, K. V.; Ramachandra, T. V. 2010. Ecology of the Swampy Relic Forests of Kathalekan from Central Western Ghats, India. Bioremediation, Biodiversity and Bioavailability. Global Sci. Books 4 (1): 54–68.

Chandrasekharan, C. 1962 a,b,c. Forest Types of Kerala State (1,2,3). Indian Forester 88: 660– 674, 731–747, 837–847.

Chatterjee, D. 1962. Floristic Pattern in Indian Vegetation. Proc. Summer School Botany, Darjeeling, 32–42.

Chatterjee, D. 1940. Studies of the Endemic Flora of India and Burma. J. Asiatic Soc. Bengal Sci. 5(1): 19–68.

Chatterjee, S.; Goswami, A.; Scotese; Christopher, R. 2013. The Longest Voyage: Tectonic, Magmatic, and Paleoclimatic Evolution of the Indian Plate During its Northward Flight from Gondwana to Asia. Gondwana Res. 23: 238–267.

Cooke, T. 1901-1908. Flora of Presidency of Bombay; Govt. of India.

Daniels, R. J. 2002. In National Biodiversity Strategy and Action Plan: Western Ghats Eco-region; Care Earth: Chennai.

Bossuyt, F.; Meegaskumbura, M.; Beenaerts, N.; Gower, D. J.; Pethiyagoda, R.; Roelants, K. 2004. Local Endemism within the Western Ghats-Sri Lanka Biodiversity Hotspot. Science 306: 479–481.

Dahanukar, N.; Raut, R.; Bhat, A. 2004. Distribution, Endemism and Threat Status of Freshwater Fishes in the Western Ghats of India. J. Biogeogr. 31: 123–136.

Dalzell, N. A.; Gibson, A. 1861. The Bombay Flora or Short Descriptions on All the Indigenous Plants Together with a Supplement of Introduced and Naturalized Species; Education Society's Press: Byculla.

Datar, M. N.; Watve, A. V. 2018. Vascular Plant Assemblage of Cliffs in Northern Western Ghats, India. J. Threat. Taxa 10: 11271–11284.

De Rosaryo, R. A. 1945. The Montane Grasslands of Ceylon. An Ecological Study with Reference to Afforestation Trop. Agriculture 101: 208–213.

Elizen, S. M.; Patil, D. N.; Al-Imam, O. A. O. 2013. Genesis of the Khanapur 'Red beds' Maharashtra, India. Intern. J. Eng. Sci. Res. Technol. 2 (6): 1422–1437.

Fraser, F. C. Odonata. 1936. In The Fauna of British India, Including Ceylon and Burma. Taylor and Francis: London.

Fyson, P. F. 1932. The Flora of South Indian Hill Stations, 2 vols.; Madras Govt. Press.

Fyson, P. F. 1915. The Flora of the Nilgiri and Pulney Hill Tops, 3 vols; Govt. Press: Madras.

Gaikwad, S.; Gore, R.; Garad, K.; Gaikwad, S. 2014. Endemic Floering Plants of Northern Western Ghats (Sahaydri ranges) of India: A Check List. Check List 10 (3): 461–472.

Gaikwad, S.; Gore, R.; Garad, K.; Gaikwad, S.; Mulani, R. 2005. Geophytes of Northern Western Ghats (Sahyadri Range) of India: A Check List. Check List 11(1). DOI: 10.15560/11.1.1543.

Gamble, J. S.; Fischer, C. E. C. 1915–1936. Flora of the Presidency of Madras, 3 vol.; Allard & Sons Ltd.: London.

Ganesan, V.; Thangaraj, P. S.; Sivaprasad, P. S.; Balakrishnan, S. 2019. Endemic Flora of Western Ghats—Anamalais, Vol. 1. Anamalai Tiger Reserve Foundation: Tamil Nadu.

Ganesh, T.; Ganesan, R.; Soubadra Devy, M.; Davidar, P.; Bawa, K. S. 1996. Assessment of Plant Biodiversity at a Mid-Elevation Evergreen Forest of Kalakad–Mundanthurai Tiger Reserve, Western Ghats, India. Curr. Sci. 71: 379–392.

Ghate, K. 2014. Management of Forests in Northern Western Ghats. J. Ecol. Soc. 26 & 27: 7–71.

Govindarajulu, E.; Swamy, B. G. L. 1958. Enumeration of Plants Collected in Mundanthurai and its Neighbourhood. J. Madras University 28B: 167–177.

Gower, D. J.; Dharne, M.; Bhatta, G.; Giri, V.; Vyas, R.; Govindappa, V. 2007. Remarkable Genetic Homogeneity in Unstriped, Long-Tailed Ichthyophis Along 1500 km of the Western Ghats, India. J. Zool. 272: 266–275.

Gupta, R. K. 1960. Ecological Notes on the Vegetation of Kodaikanal in South India. J. Indian Bot. Soc. 39: 601–607.

Gupta, S. C.; Chinnamani, S.; Rege, N. D. 1967. Ecological Relationship between High Altitude Grasslands in the Nilgiris. Indian For. 9: 164–168.

Vijayakumar, S. P.; Menezes, R. C.; Jayarajan, A.; Shanker, K. 2016. Glaciations, Gradients, and Geography: Multiple Drivers of Diversification of Bush Frogs in the Western Ghats Escarpment. Proc. Roy. Soc. B 283: 20161011. http://dx.doi.org/10.1098/rspb.2016.1011.

Hooker, J. D.; Thomson, T. 1855. In Flora Indica: Being a Systematic Account of the Plants of British India, Together with Observations on the Structure and Affinities of their Natural Orders and Genera; London, W. Pamplin.

Irwin, S. J.; Narasimhan, D. 2011. Endemic Genera of Angiosperms in India: A review. Rheedea 21 (1): 87–105.

Jacob, K.; Narayaswami, S. 1953. In The Structural and Drainage Patterns of the Western Ghats in the Vicinity of the Palghat Gap; Geological Survey of India: Calcutta.

Jeevith, S.; Ramachandran, V. S.; Ramsundar, S. 2014. Gymnosperms of Nilgiris District, Tamil Nadu. Res. Plant Biol. 4 (6): 10–16.

Jose, S.; Sreepathy, A.; Kumar, B. M.; Venugopal, V. K. 1994. Structural, Floristic and Edaphic Attributes of the Grassland-Shola Forests of Eravikulam in Peninsular India. Forest Ecol. Manage. 65: 279–291.

Joshi, J.; Karanth, K. P. 2011b. Coalescent Approach in Conjunction with Niche Modeling Reveals Cryptic Diversity Among Centipedes in the Western Ghats Biodiversity Hotspot of South India. PLoS ONE 7: e42225.

Joshi, J.; Karanth, K. P. 2011a. Cretaceous-Tertiary Diversification among Select Colopendrid Centipedes of South India. Mol. Phylogenet. Evol. 60: 287–294.

Joshi, J.; Karanth, P. 2013. Did Southern Western Ghats of Peninsular India Serve as Refugia for its Endemic Biota During the Cretaceous Volcanism? Ecol. Evolut. 3: 3275–3282.

Joshi, V. C.; Janarthanam, M. K. 2004. The Diversity of Life-Form Type, Habitat Preference and Phenology of the Endemics in the Goa Region of the Western Ghats, India. J. Biogeog. 31 (8): 1227–1237. https://doi.org/10.1111/j.1365–2699.2004.01067.x

Kammathy, R. V. 1983. Rare and Endemic Species of Indian Commelinaceae. In An Assessment of Threatened Plants of India; Jain, S. K., Rao, R. R., Eds.; Botanical Survey of India: Howrah, pp 213–221.

Karthikeyan, S. 1996. Northern Western Ghats and Northern West Coast. In Flora of India: Introductory Volume; Hajra, P. K., Sharma, B. D., Sanjappa, M., Sastry, A. R. K., Eds.; Botanical Survey of India: Calcutta, pp 375–390.

Karthikeyan, S.; Anandkumar, S. 1993. Flora of Yavatmal District, Maharashtra State; Botanical Survey of India, Fl. India ser. 3: Calcutta.

Karthikeyan, S.; Jain, S. K.; Nayar, M. P.; Sanjappa, M. 1989. Florae Indicae Enumeratio: Monocotyledonae; Botanical Survey of India: Calcutta.

Karthikeyan, S.; Nayar, M. P.; Sundara, R. A. 1981. Catalogue of Species Added to Cooke's Flora of the Presidency of Bombay During 1908–1978. Rec. Bot. Surv. India 21: 153–205.

Karunakaran, P. V.; Rawat, G. S.; Uniyal, V. K. 1998. In Ecology and Conservation of the Grasslands of Eravikulam National Park, Western Ghats; Wildlife Institute of India: Dehra Dun, India.

Keshava Murthy, K. R.; Yoganarasimhan, S. N. 1990. The Flora of Coorg (Kodagu), Karnataka, India; Vimsat Publishers: Bangalore.

Kothari, M. J.; Moorthy, S. 1993. Flora of Raigad District, Maharashtra State; Botanical Survey of India: Kolkatta.

Krishnamoorthy, K. 1960. Myristica Swamps in Evergreen Forests of Travancore. Tropical Moist Evergr. Symposium: FRI, Dehradun.

Krishnan, M. S. 1982. In Geology of India and Burma; CBS Publishers and Distributors: New Delhi.

Kulkarni, B. G. 1988. Flora of Sindhudurg; Botanical Survey of India: Calcutta.

Lakshminarasimhan, P.; Sharma, B. D. 1991. Flora of Nasik District. Fl. India ser. 3; Botanical Survey of India: Calcutta.

Legris, P. 2015. La vegetation de l'Inde. Inst. fr. Pondichéry, trav. sec.sci. tech. Tome 6, 1960. Marchese, C. Biodiversity Hotspots: A Shortcut for a more Complicated Concept. Glob. Ecol. Conserv. 3: 297–309.

Lekhak, M. M.; S. R. Yadav, S. R. 2012. Herbaceous Vegetation of Threatened High Altitude Lateritic Plateau Ecosystems of Western Ghats, Southwestern Maharashtra, India. Rheedea 22: 39–61.

Mahabale, T. S. 1987. Botany and Flora of Maharashtra. In Maharashtra State Gazetteer; Government of Maharashtra.

Malhotra, K.; Prasad, R. 1984. Some Important Aspects of Water Resources in Karnataka, with Particular Reference to Western Ghats. Ph.D. Thesis, The National Institute of Engineering, Mysuru, India.

Mani, M. S. 1974. In Ecology and Biogeography of India; Junk, Dr. W.; B.V. Publihsers: The Hague.

Manilal, K. S. 1995. Biodiversity of Silent Valley and Efforts for the Conservation of Tropical Rain Forests of India. In Taxonomy and Biodiversity; Pandey, A. K., Ed.; CBS Publishers & Distributors: New Delhi, pp 322–338.

Manilal, K. S. 1988. Flora of Silent Valley Tropical Rain Forest of India; Department of Science & Technology: Calicut.

Matthew, K. M. 1981-1984. The Flora of Tamil Nadu Carnatic, 3 vols.; The Rapinat Herbarium: Tiruchirapalli, India.

Matthew, K. M. 1999-2003. The Flora of the Palni Hills, South India, 3 Vols.; The Rapinat Herbarium, Tiruchirapalli, India.

Meher-Homji, V. M. 1965. Ecological Status of the Montane Grasslands of the South Indian Hills: A Phytogeographic Reassessment. Indian For. 91: 210–215.

Meher-Homji, V. M. 1967. Phytogeography of the South Indian Hill Stations. Bull. Torrey Bot. Club 94: 230–242.

Mehrotra, A.; Jain, S. K. 1982. Endemism in Indian Grasses—Tribe Andropogoneae. Bull. Bot. Surv. India 22: 51–58.

Menon, S.; Bawa, K. S. 1997. Application of Geographic Information Systems, Remote Sensing and a Landscape Ecology Approach to Biodiversity Conservation in Western Ghats. Curr. Sci. 73 (2): 134–145.

Mishra, D. K., Singh, N. P. 2001. Endemic and Threatened Flowering Plants of Maharashtra; Botanical Survey of India: Calcutta.

Mishra, D. K.; Singh, N. P. 2001. In Endemic and Threatened Flowering Plants of Maharashtra; Botanical Survey of India, Kolkata.

Mohanan, M.; Henry, A. N. 1994. Flora of Thiruvananthapuram District; BSI: Calcutta.

Mohanan, M.; Sivadasan, M. 2002. Flora of Agasthyamala; BSI: Calcutta.

Müller, R. D.; Yatheesh, V.; Shuhail, M. 2012. The Tectonic Stress Field Evolution of India since the Oligocene. Gondwana Res. 28: 612–624.

Myers, M.; Mittermeier, R. A.; Mittermeier, C. G.; da Fonseca, G. A. B.; Kent, J. 2000. Biodiversity Hotspots for Conservation Priorities. Nature 403: 853–858.

Myers, N. 1988-1990. Threatened Biotas: Hot Spots in Tropical Forests; Environmentalist 8: 1–20 and 10: 243 256.

Myers, N.; Mittermeier, R. A.; Mittermeier, C. G.; Fonesca, G. A. B.; Kents, J. 2000. Biodiversity Hotspots for Conservation Priorities. Nature 403: 853–858.

Naik, V. N. 1998. Flora of Marathwada of Maharashtra; Amurt Prakasahan: 2. Vols. Aurangabad.

Nair, K. K. N.; Jayakumar, R. 2008. Phytogeography, Endemism and Affinities of the Flora of New Amarambalam Reserve Forests in the Western Ghats of India. Indian J. Forestry 31: 85–94.

Nair, N. C., Daniel, P. 1986. Floristic Diversity of the Western Ghats and Its Conservation: A Review. Proc. Indian Acad. Sci. Suppl. 3: 127–163.

Nair, N. C., Henry, A. N. 1983. Flora of Tamil Nadu, India, Series 1: Analysis; Botanical Survey of India: Coimbatore.

Nair, S. C. 1991. The Southern Western Ghats: A Biodiversity Conservation Plan. INTACH: New Delhi, India.

Narasimhan, D.; Irwin, S. 2009. Endemic Angiosperms of Tamil Nadu. Envis News Lett. 5 (4): 26–34.

Nayar, M. P. 1977. Changing Patterns of the Indian Floras. Bull. Bot. Surv. India 19: 145–155.

Nayar, M. P. 1982. Endemic Flora of Peninsular India and Its Significance. Bull. Bot. Surv. India 22: 12–23.

Nayar, M. P. 1980. Endemism and Patterns of Distribution of Endemic Genera. J. Econ. Taxon. Bot. 1: 99–110.

Nayar, M. P. 1996. Hotspots of Endemic Plants of India, Nepal and Bhutan; Tropical Botanic Garden and Research Institute: Thiruvananthapuram.

Nayar, M. P.; Sastry A. R. K. 1987-1990. Red Data Book of Indian Plants, Vol. 1, 2, 3; BSI: Calcutta.

Nayar, T. S.; Beegam, A. R.; Sibi, M. 2014. In Flowering Plants of the Western Ghats, India, Vol. 1 & 2; Jawaharlal Nehru Tropical Botanic Garden and Research Institute: Palode, Thiruvananthapuram, Kerala, India.

Nayar, T. S.; Sibi, M.; Rasiya Begam, A.; Mohanan, M.; Rajkumar, G. 2008. Flowering Plants of Kerala. Status and Statistics. Rheedea 18(2): 95–106.

Noble, W. A. 1967. The Shifting Balance of Grasslands, Shola Forests, and Planted Trees in the Upper Nilgiris, Southern India. Indian For. 93: 691–693.

Oldeman, R. A. A. 1974. L'architecture de la Forêt Guyanaise. Mémoire ORSTOM No. 73.

Parthasarathy, N. A 1998. Phytogeographic Analysis of the Flora of Kalakkad Reserve Forest, Western Ghats. J. Indian Bot. Soc. 67: 342–346.

Pascal, J. P. 1982. Bioclomates of the Western Ghats at 1/250,000 (2 sheets). Inst. fr. Pondichéry, trav. sec. sci. tech. Hors série 17.

Pascal, J. P. 1991. Floristic Composition and Distribution of Evergreen Forests in the Western Ghats, India. Palaeobotanist 39: 110–126.

Pascal, J. P. 1982. Forest Map of South India—Sheet: Mercara-Mysore; Karnataka & Kerala Forest Departments and the French Institute: Pondicherry, India.

Pascal, J. P. 1988. In Wet Evergreen Forests of the Western Ghats of India: Ecology, Structure, Floristic Composition and Succession; French Institute of Pondicherry: India.

Pascal, J. P.; Ramesh, B. R.; De Franceschp, D. 2004. Wet Evergreen Forest Types of the Southern Western Ghats, India. Trop. Ecol. 45: 281–292.

Pascoe, E. H. 1962. In A Manual of the Geology of India and Burma, vol. III; Govt. of India Publication.

Patel, R. I. 1968. Forest Flora of Melghat; Bishen Singh Mahendra Pal Singh: Dehradun.

Patil, D. A. 2003. Flora of Dhule and Nandurbar Districts (Maharashtra), Bishen Singh Mahendra Pal Singh: Dehradun.

Patil, D. N. 1992. Laterites from the Semiarid Regions of Deccan Volcanic Province, India: A Reappraisal. Tropical Geomorphology Newsletter, 13: 8–10.

Potdar, G. G.; Salunkhe, C. B.; Yadav, S. R. 2012. Grasses of Maharashtra; Shivaji University: Kolhapur.

Prabhu Kumar, K. M.; Sreeraj, V.; Binu, T.; Manudev, K. M.; Rajendran, A. 2012. Validation and Documentation of Rare Endemic and Threatened (RET) Plants from Nilgiri, Kanuvai and Madukkarai Forests of Southern Western Ghats, India. J. Threat. Taxa 4 (15): 3436– 3442.

Putty, Y. M. R.; Madhusoodhanan, C. G. 2014. Water Resources and Hydrology of Western Ghats: Their Role and Significance in South India. NNRMS Bull. http://www.reserchgate.net/publication/268147276.

Puyravaud, J. P.; Davidar, P.; Pascal, J. P.; Ramesh, P. R. 2003. Analysis of Threatened Endemic Trees of the Western Ghats of India Sheds New Light on the Red Data Book on the Indian Plants. Biodivers. Conserv. 12: 2091–2106.

Radhakrishna, T.; Maluski, H.; Mitchell, J. G.; Joseph, M. 1999. 40Ar/39Ar and K/Ar Geochronology of the Dykes from the South Indian Granulite Terrain. Tectonophysics, 304: 109–129.

Rahangdale, S. R.; Rahangdale, S. S. 2018. Biodiversity of Durgawadi Plateaus; Department of Forests, Government of Maharashtra.

Rahangdale, S. S.; Rahangdale, S. R. 2017. Floristic Diversity of Bhimashankar Wildlife Sanctuary, Northern Western Ghats, Maharashtra, India. J. Threat. Taxa 9(8): 10493–10527. http://doi.org/10.11609/jott.3074.9.8.10493–10527

Rahangdale, S. S.; Rahangdale, S. R. 2014. Plant Species Composition on Two Rock Outcrops from the Northern Western Ghats, Maharashtra, India. J. Threat. Taxa 6(4), 5593–5612. http://doi.org/10.11609/JoTT.o3616.5593–612.

Rama Rao, M. 1914. In Flowering Plants of Travancore; Government Press: Trivandrum.

Ramachandran, V. S.; Nair, V. J. 1988. Flora of Cannanore District, Botanical Survey of India: Calcutta.

Ramaswamy S. N.; Radhakrishana Rao, M.; Govindappa, D. A. 2001. Flora of Shimoga District, Karnataka; Prasaranga, University of Mysore.

Ramesh, B. R. 2001. In Patterns of Richness and Endemism of Arborescent Species in the Evergreen Forests of the Western Ghats, India, Proceedings of International Conference on Tropical Ecosystems: Structure, Diversity and Human Welfare, Ganeshaiah, K. N., Shaanker, R. U., Bawa, K. S., Eds.; Oxford-IBH: New Delhi, pp. 539–544.

Ramesh, B. R.; Pascal, J. P. 1997. Atlas of the Endemics of the Western Ghats (India): Distribution of Tree Species in the Evergreen and Semi Evergreen Forests, Vol. 38; Institute Francais de Pondicherry, Publications du department d ecologie, p 403.

Ramkumar, M.; Menier, D.; Mathew, M.; Santosh, M.; Siddiqui, N. A. 2016. Early Cenozoic Rapid Flight Enigma of the Indian Subcontinent Resolved: Roles of Topographic Top Loading and Subcrustal Erosion. Geosci. Front. 8(1): 15–23. http:// dx.doi.org/10.1016/j.gsf.2016.05.004.

Rana, T. S.; Ranade, S. A. 2009. The Enigma of Monotypic Taxa and Their Taxonomic Implications. Curr. Sci. 96 (2): 219–229.

Ranganathan, C. R. 1938. Studies in the Ecology of the Shola Grassland Vegetation of the Nilgiri Plateau. Indian For. 64: 523–541.

Rao, C. K. 1972. Angiosperm Genera Endemic to Indian Floristic and Its Neighbouring Areas. Indian For. 98: 560–566.

Rao, R. R. 1973. Studies on Flowering Plants of Mysore District, 2 vols. Ph.D. Thesis submitted to Mysore University.

Rao, R. R.; Murugan, R.; Syamasundar, K. V.; Srinivasalu, B. 2006. Western Ghats - A Major Emporium of Wild Aromatic Plants: Diversity, Conservation and Bioprospection. In Concepts in Forestry Research; Todaria, N. P., Chamola, B. P., Chauhan, D. S., Eds.; International Book Distributors: Dehra Dun, pp 267–278.

Rao, R. R.; Razi, B. A. 1981. A Synoptic Flora of Mysore District; Today & Tomorrows Publishers: New Delhi.

Rao, R. S. 1985-1986. Flora of Goa, Diu, Daman, Dadra and Nagarhaveli; Botanical Survey of India: Calcutta.

Rao, R. S. M. 1914. Flowering Plants of Travancore; Government Press: Travancore.

Rau, M. A.; Narayana, B. M. 1985. A Review of the Tribe Vernonieae (Asteraceae) in South India. Bull. Bot. Surv. India 25: 19–25.

Raychuduri, S. P.; Aarwal, R. R.; Datta Biswas, N. R.; Gupta, S. P.; Thomas, P. K. 2963. In Soils of India; Indian Council of Agricultural Research: New Delhi.

Rice, C. G. 1988. Habitat, Population Dynamics and Conservation of the Nilgiri Tahr, Hemitragus hylocrius. Biol. Conserv. 44: 137–156.

Robin, V. V.; Sinha, A.; Ramakrishnan, U. 2010. Ancient Geographical Gaps and Paleo-Climate Shape the Phylogeography of an Endemic Bird in the Sky Islands of Southern India. PLoS ONE 5: e13321.

Rodgers, W. A.; Panwar, H. S. 1988. In Planning a Protected Area Network in India; Wildlife Institute of India: Dehra Dun.

Rodgers, W. A.; Panwar, H. S.; Mathur, V. B. 2002. In Wildlife Protected Area Network in India: A Review; Wildlife Institute of India: Dehradun.

Roelants, K.; Gower, D. J.; Wilkinson, M.; Loader, S. P.; Biju, S. D.; Guillaume, K. 2007. Global Patterns of Diversification in the History of Modern Amphibians. Proc. Natl. Acad. Sci. USA 104: 887–892.

Sahasrabudhe, Y. S.; Rajaguru, S. N. 1990. The Laterites of the Maharashtra State. Bull. Deccan College Research Institute 49: 357–370.

Saldanha, C. J. 1996. The Flora of Karnataka, Vol. 1 and 2; Oxford & IBH Publishing Co: New Delhi.

Saldanha, C. J., Nicolson, D. H. 1976. Flora of Hassan District, Karnataka, India. Amerind Publishers, New Delhi.

Saldanha, C. J.; Santapau, H. 1956. New Plants Records for Bombay—III. J. Bombay Nat. Hist. Soc. 53: 210–213.

Santapau, C. J. 1957. Flora of Purandhar; Oxford Book Co.: New Delhi.

Santapau, H. 1951. Acanthaceae of Bombay. Bot. Mem. Univ. Bombay 2: 1–104.

Santapau, H. 1953. The Flora of Khandala on the Western Ghats of India; Records of Botanical Survey of India, BSI: New Delhi.

Santhakumaran, L. N.; Singh, A.; Thomas, V. T. 1995. Description of a Sacred Grove in Goa (India), with Notes on the Unusual Aerial Roots Produced by its Vegetation. In Wood (OctobeDecember), pp. 24–28.

Sarvalingam, A.; Rajendran, A. 2016. Rare, Endangered and Threatened (RET) Climbers of Southern Western Ghats, India. Revista Chilena de Historia Natural 89: 9. DOI: 10.1186/s40693–016–0058–6.

Sasidharan, N. 2004. Flowering Plants of Kerala; Kerala Forest Research Institute: Peechi, Thrissur.

Sasidharan, N. 2006. Illustrated Manual on the Tree Flora of Kerala Supplemented with Computer Aided Identification; KFRI: Thrissur.

Sasidharan, N. 2009. Studies on the Flora of Shendrruny Wildlife Sanctuary with Emphasis on Endemic Species. KFRI Research Report: Peechi, Thrissur, 128, p 401.

Sasidharan, N.; Sivarajan, V. 1996. Flowering Plants of Thrissur Forest; Scientific Publishers: Jodhpur.

Sathish, B. N.; Viswanath, S.; Kushalappa, C. G.; Jagadish, M. R.; Ganeshaiah, K. N. 2013. Comparative Assessment of Floristic Structure, Diversity and Regeneration Status of Tropical Rain Forests of Western Ghats of Karnataka, India. J. Appl. Nat. Sci. 5(1): 157–164.

Savita, S. R.; Sanjaykumar, R. R. 2017. Floristic Diversity of Bhimashankar Wildlife Sanctuary, Northern Western Ghats, Maharashtra, India. J. Threat. Taxa 9(8): 10493–10527.

Senthilkumar, N.; Prakash, S.; Rajesh Kannan, C.; Arun Prasath, A.; Krishnakumar, N. 2014. Revisiting Forest Types of India (Champion and Seth, 1968): A Case Study on Myristica Swamp Forests in Kerala. Int. Nat. J. Adv. Res. 2(2): 492–501.

Sharma, B. D.; Singh, N. P.; Raghavan, R. S.; Deshpande, U. R. 1984. Flora of Karnataka: Analysis; Botanical Survey of India: Howrah.

Shetty, B. V.; Vivekanadan, K. 1971. Studies on the Vascular Floras of Anaimudi and Surrounding Regions; Kottayam District, Kerala. Bull. Bot. Surv. India 13: 16–42.

Singh, J. S. 2002. (Co-Ordinator) National Biodiversity Strategy and Action Plan (Natural Terrestrial Ecosystems)—a Report Submitted to Ministry of Environment and Forests: New Delhi.

Singh, J. S.; Chaturvedi, R. K. 2017. Diversity of Ecosystem Types in India: A Review. Proc. Indian Natn. Sci. Acad. 83(3): 569–594.

Singh, N. P., Lakshminarasimhan, S., Karthikeyan, S. Prasanna, P. V. 2001. Eds. Flora of Maharashtra State, Dicotyledons, Vol. 2; Botanical Survey of India: Calcatta.

Singh, N. P.; Karthikeyan, S. 2000. Flora of Maharashtra State (Dicotyledones) Vol. 1; Botanical Survey of India: Calcutta.

Singh, P. 2020. Floristic Diversity of India: An Overview; Biodiversity of the Himalaya and Jammu and Kashmir State, pp 41–79. DOI: 10.1007/978–981–32–9174–4_3.

Singh, P.; Karthigeyan, K.; Lakshminarasimhan, P.; Dash, S. S. 2015. Endemic Vascular Plants of India; Botanical Survey of India: Kolkata.

Singh, S. K.; Agarwala, D. K.; Jalal, J. S.; Dash, S. S.; Mao, A. A.; Singh, P. 2019. Orchids of India: A Pictorial Guide; Botanical Survey of India: Kolkata.

Sivarajan, V. V.; Mathew, P. 1997. In Flora of Nilambur (Western Ghats, Kerala); Dehradun.

Sreedharan, G.; Indulkar, M. 2018. New Distributional Record of the Northmost Myristica Swamp from the Western Ghats of Maharashtra. Curr. Sci. 115 (8): 1434–1436.

Sreekumar, P. V.; Nair, V. J. 1991. In Flora of Kerala – Grasses; Botanical Survey of India: Calcutta.

Subramanya, K. R. 1987. Evolution of the Western Ghats in India – A Simple Model. J. Geol. Soc. India 29: 446–449.

Subramanyam, K.; Nayar, M. P. 1974. Vegetation and Phytogeography of the Western Ghats; In Ecology and Biogeography in India, vol. 23; Mani, M. S., Ed The Hague, Netherlands, pp. 178–196.

Suresh, H. S.; Sukumar, R. 1999. Phytographical Affinities of Flora of Nilgiri Biosphere Reserve. Rheedea 9: 1–21.

Tetali, P.; Tetali, S.; Kulkarni, B. G.; Prasanna, P. V.; Lakshminarasimhan, P.; Lale, M.; Kumbhojkar, M. S.; Kulkarni, D. K.; Jagtap, A. P. 2000. In Endemic Plants of India (A Status Report of Maharashtra State); Naoroji Godrej Centre for Research: Shindewadi, Satara, India, pp. 87.

Thipperudrapa, S. 2009. Some Studies on the Intensity Pattern of Rainfall in Karnataka, with Particular Reference to Western Ghats. PhD Thesis, The National Institute of Engineering, Mysuru.

UNESCO. 2012. Decision Adopted by the World Heritage Committee at its 36th Session, [Online] http://whc.unesco.org/en/sessions/36COM.

Venu, P. 2006. Strobilanthes Blume (Acanthaceae) in Peninsular India; Botanical Survey of India: Kolkata.

Vidya, T. N. C.; Fernando, P.; Melnick, D. J.; Sukumar, R. Population Differentiation Within and Among Asian Elephant (Elephas maximus) Populations in Southern India.

Vinayaka, K. S.; Krishnamurthy, Y. L. 2016. Floristic Composition and Regeneration Analysis of Halika Ghat Region, Central Western Ghats, Karnataka. Trop. Plant Res. 3: 654–661.

Vishnu-Mittre; Gupta, H. P. 1968. A Living Fossil Plant Community in South Indian Hills. Cur. Sci. 37: 671–672.

Warrier, R. R.; Geetha, S.: Sivakumar, V.; Gurudev Singh, B.; Anandalakshmi, R. 2020. Threatened Tree Species of the Western Ghats: Status, Diversity and Conservation. In Conservation and Utilization of threatened medicinal plants; Rajasekharan, P. E., Wani, S. H., Eds.; Springer International Publishing: New York, pp 429–460.

Watve, A. 2008. Rocky Outcrops as Special Habitats in North Western Ghats, Maharashtra. In Special Habitats and Threatened Plants of India; Rawat, G. S., Ed.; ENVIS Bulletin: Wildlife and Protected Areas. Wildlife Institute of India: Dehradun, 11(1): 147–153.

Watve, A. 2013. Status Review of Rocky Plateaus in the northern Western Ghats and Konkan Region of Maharashtra, India with Recommendations for Conservation and Management. J. Threat. Taxa 5(5): 3935–3962; http://dx.doi.org/10.11609/JoTT.o3372.3935–62

WGEEP. 2011. Report on the Western Ghats Ecology Expert Panel; Submitted to the Ministry of Environment and Forests, Government of India: New Delhi.

Wight, R. 1840-1853. Icones Plantarum Indiae Orientalis; J. B. Pharoah: Madras.

Wight, R. 1840-1850. Illustrations to Indian Botany; Govt. Press: Madras.

Yadav, S. R.; Kamble, M. Y. 2008. Threatened Ceropegias of the Western Ghats and Strategies for Their Conservation. In Special Habitats and Threatened Plants of India; Rawat, G. S., Ed. ENVIS Bull. 11(1): 123–134.

Yadav, S. R.; Sardesai, M. M. 2002. Flora of Kolhapur District; Shivaji University: Kolhapur.

Yatheesh, V.; Dyment, J.; Bhattacharya, G. C.; Muller, R. D. 2013. Deciphering Detailed Plate Kinematics of the Indian Ocean and Developing a Unified Model for East Gondwana Land Reconstruction: An Indian Australian-French Initiative. DCS-DST News Lett. 23: 1–9.

Yoganarasimhan, S. N. 2000. Medicinal Plants of India, Vol.2, Tamil Nadu, R. R. I.: Bangalore.

Yoganarasimhan, S. N.; Subramanyam, K., Razi, B. A. 1981. Flora of Chickmagalur District, Karnataka; India, International Book Distributors: Dehra Dun.

INDEX TO FAMILIES, GENERA AND SPECIES

A

B

D

E

F

H

I

M

R

T

U

V